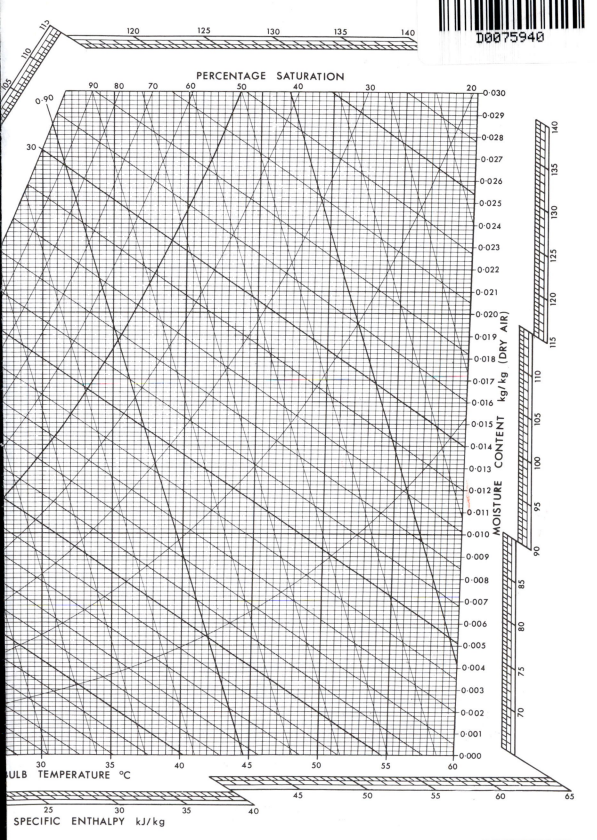

PERCENTAGE SATURATION

MOISTURE CONTENT kg/kg (DRY AIR)

BULB TEMPERATURE °C

SPECIFIC ENTHALPY kJ/kg

D0075940

Air Conditioning Engineering

Air Conditioning Engineering

Fifth Edition

W.P. Jones
MSc, CEng, FInstE, FCIBSE, MASHRAE

Oxford   Auckland   Boston   Johannesburg   Melbourne   New Delhi

Butterworth-Heinemann
Linacre House, Jordan Hill, Oxford OX2 8DP
225 Wildwood Avenue, Woburn, MA 01801-2041
A division of Reed Educational and Professional Publishing Ltd

 A member of the Reed Elsevier plc group

First published in Great Britain 1967
Second edition 1973
Third edition 1985
Fourth edition 1994
Reprinted 1996
Fifth edition 2001

British Library Cataloguing in Publication Data
Jones, W. P. (William Peter),
   Air conditioning engineering. – 5th ed.
   1. Air conditioning
   I. Title
   697.9'3

Library of Congress Cataloguing in Publication Data
Jones, W. P. (William Peter),
   Air conditioning engineering/W.P. Jones. – 5th ed.
    p. cm.
   Includes bibliographical references and index.
   ISBN 0 7506 5074 5
   1. Air conditioning. I. Title.

   TH7687.J618
   697.9'3–dc21   00–048640

ISBN 0 7506 5074 5

Typeset at Replika Press Pvt Ltd, Delhi 110 040, India
Printed and bound in Great Britain

Preface to the Fifth Edition

Although the fundamentals of the subject have not altered since the publication of the last edition there have been significant changes in the development and application of air conditioning. Among these are concerns about indoor air quality, revision of outside design data and the expression of cooling loads arising from solar radiation through glass by the CIBSE. The phasing-out of refrigerants that have been in use for many years (because of their greenhouse effect and the risks of ozone depletion) and the introduction of replacement refrigerants are far-reaching in their consequences and have been taken into account. The tables on the thermodynamic properties of refrigerant 22 have been deleted and new tables for refrigerants 134a and ammonia substituted. There have also been new developments in refrigeration compressors and other plant. Advances in automatic controls, culminating in the use of the Internet to permit integration of the control and operation of all building services worldwide, are very important. Revisions in expressing filtration efficiency, with an emphasis on particle size, have meant radical changes in the expression of the standards used in the UK, Europe and the USA. The above developments have led to changes in the content, notably in chapters 4 (on comfort), 5 (on outside design conditions), 7 (on heat gains), 9 (for the refrigerants used), 12 (automatic controls) and 17 (on filtration standards).

Two examples on heat gains in the southern hemisphere have been included.

As with former editions, the good practice advocated by the Chartered Institution of Building Services Engineers has been followed, together with the recommendations of the American Society of Heating, Refrigerating and Air Conditioning Engineers, where appropriate. It is believed that practising engineers as well as students will find this book of value.

W.P. Jones

Preface to the First Edition

Air conditioning (of which refrigeration is an inseparable part) has its origins in the fundamental work on thermodynamics which was done by Boyle, Carnot and others in the seventeenth and eighteenth centuries, but air conditioning as a science applied to practical engineering owes much to the ideas and work of Carrier, in the United States of America, at the beginning of this century. An important stepping stone in the path of progress which has led to modern methods of air conditioning was the development of the psychrometric chart, first by Carrier in 1906 and then by Mollier in 1923, and by others since.

The summer climate in North America has provided a stimulus in the evolution of air conditioning and refrigeration which has put that semi-continent in a leading position amongst the other countries in the world. Naturally enough, engineering enterprise in this direction has produced a considerable literature on air conditioning and allied subjects. The *Guide and Data Book* published by the American Society of Heating, Refrigeration and Air Conditioning has, through the years, been a foremost work of reference but, not least, the *Guide to Current Practice* of the Institution of Heating and Ventilation Engineers has become of increasing value, particularly of course in this country. Unfortunately, although there exists a wealth of technical literature in textbook form which is expressed in American terminology and is most useful for application to American conditions, there is an almost total absence of textbooks on air conditioning couched in terms of British practice. It is hoped that this book will make good the dificiency.

The text has been written with the object of appealing to a dual readership, comprising both the student studying for the associate membership examinations of the Institution of Heating and Ventilating Engineers and the practising engineer, with perhaps a 75 per cent emphasis being laid upon the needs of the former. To this end, the presentation follows the sequence which has been adopted by the author during the last few years in lecturing to students at the Polytechnic of the South Bank. In particular, wherever a new idea or technique is introduced, it is illustrated immediately by means of a worked example, when this is possible. It is intended that the text should cover those parts of the syllabus for the corporate membership examination that are relevant to air conditioning.

Inevitably some aspects of air conditioning have been omitted (the author particularly regrets the exclusion of a section on economics). Unfortunately, the need to keep the book within manageable bounds and the desire to avoid a really prohibitive price left no choice in the matter.

W.P. Jones

Acknowledgements

Originally this book was conceived as a joint work, in co-authorship with Mr. L.C. Bull. Unfortunately, owing to other commitments, he was compelled largely to forego his interest. However, Chapters 9 and 14 (on the fundamentals of vapour–compression and vapour–absorption refrigeration) are entirely his work. The author wishes to make this special acknowledgement to Mr. Bull for writing these chapters and also to thank him for his continued interest, advice and encouragement. Sadly, Mr. Bull is now deceased.

The helpful comment of Mr. E. Woodcock is also appreciated.

The author is also indebted to Mr. D.J. Newson for his contribution and comment.

The author is additionally grateful to the following for giving their kind permission to reproduce copyright material which appears in the text.

The Chartered Institution of Building Services Engineers for Figures 5.4 and 7.16, and for Tables 5.3, 5.4, 7.2, 7.7, 7.13, 7.14, 7.18, 16.1 and 16.2 from the CIBSE Guide.

H.M. Stationery Office for equation (4.1) from War Memorandum No. 17, Environmental Warmth and its Measurement, by T. Bedford.

Haden Young Ltd. for Tables 7.9 and 7.10.

The American Society of Heating, Refrigeration and Air Conditioning Engineers for Tables 7.5, 9.1, 9.2 and for Figure 12.12.

John Wiley & Sons Inc., New York, for Figure 13.8 from *Automatic Process Control* by D.P. Eckman.

McGraw-Hill Book Company for Table 7.12.

American Air Filter Ltd. (Snyder General) for Table 9.6.

Woods of Colchester Ltd. for Figure 15.23.

W.B. Gosney and O. Fabris for Tables 9.3 and 9.4.

Contents

1

The Need for Air Conditioning

1.1 The meaning of air conditioning

Full air conditioning implies the automatic control of an atmospheric environment either for the comfort of human beings or animals or for the proper performance of some industrial or scientific process. The adjective 'full' demands that the purity, movement, temperature and relative humidity of the air be controlled, within the limits imposed by the design specification. (It is possible that, for certain applications, the pressure of the air in the environment may also have to be controlled.) Air conditioning is often misused as a term and is loosely and wrongly adopted to describe a system of simple ventilation. It is really correct to talk of air conditioning only when a cooling and dehumidification function is intended, in addition to other aims. This means that air conditioning is always associated with refrigeration and it accounts for the comparatively high cost of air conditioning. Refrigeration plant is precision-built machinery and is the major item of cost in an air conditioning installation, thus the expense of air conditioning a building is some four times greater than that of only heating it. Full control over relative humidity is not always exercised, hence for this reason a good deal of partial air conditioning is carried out; it is still referred to as air conditioning because it does contain refrigeration plant and is therefore capable of cooling and dehumidifying.

The ability to counter sensible and latent heat gains is, then, the essential feature of an air conditioning system and, by common usage, the term 'air conditioning' means that refrigeration is involved.

1.2 Comfort conditioning

Human beings are born into a hostile environment, but the degree of hostility varies with the season of the year and with the geographical locality. This suggests that the arguments for air conditioning might be based solely on climatic considerations, but although these may be valid in tropical and subtropical areas, they are not for temperate climates with industrialised social structures and rising standards of living.

Briefly, air conditioning is necessary for the following reasons. Heat gains from sunlight, electric lighting and business machines, in particular, may cause unpleasantly high temperatures in rooms, unless windows are opened. If windows are opened, then even moderate wind speeds cause excessive draughts, becoming worse on the upper floors of tall buildings. Further, if windows are opened, noise and dirt enter and are objectionable, becoming worse on the lower floors of buildings, particularly in urban districts and industrial areas. In any case, the relief provided by natural airflow through open windows is only effective for a depth of about 6 metres inward from the glazing. It follows that the inner areas of deep buildings will not really benefit at all from opened windows. Coupled with the need for high intensity continuous electric lighting in these core areas, the lack of adequate ventilation means a good deal of discomfort for the occupants. Mechanical ventilation without refrigeration is only a partial solution. It is true that it provides a controlled and uniform means of air distribution, in place of the unsatisfactory results obtained with opened windows (the vagaries of wind and stack effect, again particularly with tall buildings, produce discontinuous natural ventilation), but tolerable internal temperatures will prevail only during winter months. For much of the spring and autumn, as well as the summer, the internal room temperature will be several degrees higher than that outside, and it will be necessary to open windows in order to augment the mechanical ventilation. See chapter 16.

The design specification for a comfort conditioning system is intended to be the framework for providing a comfortable environment for human beings throughout the year, in the presence of sensible heat gains in summer and sensible heat losses in winter. Dehumidification would be achieved in summer but the relative humidity in the conditioned space would be allowed to diminish as winter approached. There are two reasons why this is acceptable: first, human beings are comfortable within a fairly large range of humidities, from about 65 per cent to about 20 per cent and, secondly, if single glazing is used it will cause the inner surfaces of windows to stream with condensed moisture if it is attempted to maintain too high a humidity in winter.

The major market for air conditioning is to deal with office blocks in urban areas. Increasing land prices have led to the construction of deep-plan, high-rise buildings that had to be air conditioned and developers found that these could command an increase in rent that would more than pay for the capital depreciation and running cost of the air conditioning systems installed.

Thus, a system might be specified as capable of maintaining an internal condition of 22°C dry-bulb, with 50 per cent saturation, in the presence of an external summer state of 28°C dry-bulb, with 20°C wet-bulb, declining to an inside condition of 20°C dry-bulb, with an unspecified relative humidity, in the presence of an external state of –2°C saturated in winter.

The essential feature of comfort conditioning is that it aims to produce an environment which is comfortable to the majority of the occupants. The ultimate in comfort can never be achieved, but the use of individual automatic control for individual rooms helps considerably in satisfying most people and is essential.

1.3   Industrial conditioning

Here the picture is quite different. An industrial or scientific process may, perhaps, be performed properly only if it is carried out in an environment that has values of temperature and humidity lying within well defined limits. A departure from these limits may spoil the work being done. It follows that a choice of the inside design condition is not based on a statistical survey of the feelings of human beings but on a clearly defined statement of what is wanted.

Thus, a system might be specified to hold 21°C ± 0.5°C, with 50 per cent saturation $\pm 2\frac{1}{2}$ per cent, provided that the outside state lay between 29.5°C dry-bulb, with 21°C wet-bulb and – 4°C saturated.

2

Fundamental Properties of Air and Water Vapour Mixtures

2.1 The basis for rationalisation

Perhaps the most important thing for the student of psychrometry to appreciate from the outset is that the working fluid under study is a mixture of two different gaseous substances. One of these, dry air, is itself a mixture of gases, and the other, water vapour, is steam in the saturated or superheated condition. An understanding of this fact is important because in a simple analysis one applies the Ideal Gas Laws to each of these two substances separately, just as though one were not mixed with the other. The purpose of doing this is to establish equations which will express the physical properties of air and water vapour mixtures in a simple way. That is to say, the equations could be solved and the solutions used to compile tables of psychrometric data or to construct a psychrometric chart.

The justification for considering the air and the water vapour separately in this simplified treatment is provided by Dalton's *laws of partial pressure* and the starting point in the case of each physical property considered is its definition.

It must be acknowledged that the *ideal gas laws* are not strictly accurate, particularly at higher pressures. Although their use yields answers which have been adequately accurate in the past, they do not give a true picture of gas behaviour, since they ignore intermolecular forces. The most up-to-date psychrometric tables (CIBSE 1986) are based on a fuller treatment, discussed in section 2.19. However, the Ideal Gas Laws may still be used for establishing psychrometric data at non-standard barometric pressures, with sufficient accuracy for most practical purposes.

2.2 The composition of dry air

Dry air is a mixture of two main component gases together with traces of a number of other gases. It is reasonable to consider all these as one homogeneous substance but to deal separately with the water vapour present because the latter is condensable at everyday pressures and temperatures whereas the associated dry gases are not.

One method of distinguishing between gases and vapours is to regard vapours as capable of liquefaction by the application of pressure alone but to consider gases as incapable of being liquefied unless their temperatures are reduced to below certain critical values. Each gas has its own unique critical temperature, and it so happens that the critical temperatures of nitrogen and oxygen, the major constituents of dry air, are very much below the temperatures dealt with in air conditioning. On the other hand, the critical temperature of steam (374.2°C) is very much higher than these values and, consequently, the water vapour mixed with the dry air in the atmosphere may change its phase from gas to liquid if its pressure is increased, without any reduction in temperature. While this is occurring, the phase of the dry air will, of course, remain gaseous.

Figures 2.1(*a*) and 2.1(*b*) illustrate this. Pressure–volume diagrams are shown for dry air and for steam, separately. Point A in Figure 2.1(*a*) represents a state of dry air at 21°C. It can be seen that no amount of increase of pressure will cause the air to pass through the liquid phase, but if its temperature is reduced to −145°C, say, a value less than that of the critical isotherm, t_c (−140.2°C), then the air may be compelled to pass through the liquid phase by increasing its pressure alone, even though its temperature is kept constant.

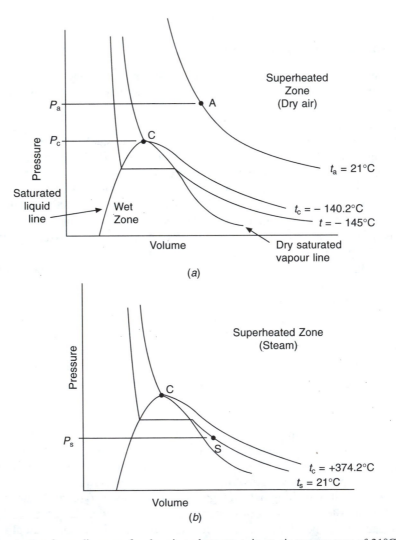

Fig. 2.1   Pressure–volume diagrams for dry air and steam. t_a is an air temperature of 21°C and t_s is a steam temperature of 21°C. t_c is the critical temperature in each case.

In the second diagram, Figure 2.1(*b*), a similar case for steam is shown. Here, point S represents water vapour at the same temperature, 21°C, as that considered for the dry air. It is evident that atmospheric dry air and steam, because they are intimately mixed, will have the same temperature. But it can be seen that the steam is superheated, that it is far below its critical temperature, and that an increase of pressure alone is sufficient for its liquefaction.

According to Threlkeld (1962), the dry air portion of the atmosphere may be thought of as being composed of true gases. These gases are mixed together as follows, to form the major part of the working fluid:

Gas	Proportion (%)	Molecular mass
Nitrogen	78.048	28.02
Oxygen	20.9476	32.00
Carbon dioxide	0.0314	44.00
Hydrogen	0.00005	2.02
Argon	0.9347	39.91

A later estimate by the *Scientific American* (1989) of the carbon dioxide content of the atmosphere is 0.035% with a projection to more than 0.040% by the year 2030. ASHRAE (1997) quote the percentage of argon and other minor components as about 0.9368%. From the above, one may compute a value for the mean molecular mass of dry air:

$$M = 28.02 \times 0.78084 + 32 \times 0.209476 + 44 \times 0.000314$$
$$+ 2.02 \times 0.0000005 + 39.91 \times 0.009347$$
$$= 28.969 \text{ kg kmol}^{-1}$$

Harrison (1965) gives the weighted average molecular mass of all the components as 28.9645 kg kmol^{-1} but, for most practical purposes, it may be taken as 28.97 kg kmol^{-1}.

As will be seen shortly, this is used in establishing the value of the particular gas constant for dry air, prior to making use of the General Gas Law. In a similar connection it is necessary to know the value of the particular gas constant for water vapour; it is therefore of use at this juncture to calculate the value of the mean molecular mass of steam.

Since steam is not a mixture of separate substances but a chemical compound in its own right, we do not use the proportioning technique adopted above. Instead, all that is needed is to add the masses of the constituent elements in a manner indicated by the chemical formula:

$$M = 2 \times 1.01 + 1 \times 16$$
$$= 18.02 \text{ kg kmol}^{-1}$$

More exactly, Threlkeld (1962) gives the molecular mass of water vapour as 18.015 28 kg kmol^{-1}.

2.3   Standards adopted

In general, those standards which have been used by the Chartered Institution of Building Services Engineers in their *Guide to Current Practice,* are used here.

The more important values are:

Density of Air	1.293 kg m^{-3} for dry air at 101 325 Pa and 0°C.
Density of Water	1000 kg m^{-3} at 4°C and 998.23 kg m^{-3} at 20°C.
Barometric Pressure	101 325 Pa (1013.25 mbar).

Standard Temperature and Pressure (STP) is the same as Normal Temperature and Pressure (NTP) and is 0°C and 101 325 Pa, and the specific force due to gravity is taken as 9.807 N kg^{-1} (9.806 65 N kg^{-1} according to ASHRAE (1997)). Meteorologists commonly express pressure in mbar, where 1 mbar = 100 Pa.

Both temperature and pressure fall with increasing altitude up to about 10 000 m. ASHRAE (1997) gives the following equation for the calculation of atmospheric pressure up to a height of 10 000 m:

$$p = 101.325 \, (1 - 2.255 \, 77 \times 10^{-5}Z)^{5.2559} \tag{2.1}$$

and

$$t = 15 - 0.0065Z \tag{2.2}$$

for the calculation of temperature up to a height of 11 000 m, where p is pressure in kPa, t is temperature in K and Z is altitude in m above sea level.

2.4 Boyle's law

This states that, for a true gas, the product of pressure and volume at constant temperature has a fixed value.

As an equation then, one can write

$$pV = \text{a constant} \tag{2.3}$$

where p is pressure in Pa and V is volume in m^3.

Graphically, this is a family of rectangular hyperbolas, each curve of which shows how the pressure and volume of a gas varies at a given temperature. Early experiment produced this concept of gas behaviour and subsequent theoretical study seems to verify it. This theoretical approach is expressed in the *kinetic theory of gases*, the basis of which is to regard a gas as consisting of an assembly of spherically shaped molecules. These are taken to be perfectly elastic and to be moving in a random fashion. There are several other restricting assumptions, the purpose of which is to simplify the treatment of the problem. By considering that the energy of the moving molecules is a measure of the energy content of the gas, and that the change of momentum suffered by a molecule upon collision with the wall of the vessel containing the gas is indication of the pressure of the gas, an equation identical with Boyle's law can be obtained.

However simple Boyle's law may be to use, the fact remains that it does not represent exactly the manner in which a real gas behaves. Consequently one speaks of gases which are assumed to obey Boyle's law as being ideal gases. There are several other simple laws, namely, Charles' law, Dalton's laws of partial pressures, Avogadro's law, Joule's law and Gay Lussac's law, which are not strictly true but which are in common use. All these are known as the *ideal gas laws*.

Several attempts have been made to deal with the difficulty of expressing exactly the behaviour of a gas. It now seems clear that it is impossible to show the way in which pressure–volume changes occur at constant temperature by means of a simple algebraic equation. The expression which, in preference to Boyle's law, is today regarded as giving the most exact answer is in the form of a convergent infinite series:

$$pV = A(1 + B/V + C/V^2 + D/V^3 + \ldots) \tag{2.4}$$

The constants A, B, C, D, etc., are termed the *virial coefficients* and they have different values at different temperatures.

It is sometimes more convenient to express the series in a slightly different way:

$$pV = A' + B'p + C'p^2 + D'p^3 + \dots \tag{2.5}$$

At very low pressures the second and all subsequent terms on the right-hand side of the equation become progressively smaller and, consequently, the expression tends to become the same as Boyle's law. Hence, one may use Boyle's law without sensible error, provided the pressures are sufficiently small.

The second virial coefficient, B, is the most important. It has been found that, for a given gas, B has a value which changes from a large negative number at very low temperatures, to a positive one at higher temperatures, passing through zero on the way. The temperature at which B equals zero is called the Boyle temperature and, at this temperature, the gas obeys Boyle's law up to quite high pressures. For nitrogen, the main constituent of the atmosphere, the Boyle temperature is about 50°C. It seems that at this temperature, the gas obeys Boyle's law to an accuracy of better than 0.1 per cent for pressures up to about 1.9 MPa. On the other hand, it seems that at 0°C, the departure from Boyle's law is 0.1 per cent for pressures up to 0.2 MPa.

We conclude that it is justifiable to use Boyle's law for the expression of the physical properties of the atmosphere which are of interest to air conditioning engineering, in many cases.

In a very general sort of way, Figure 2.2 shows what is meant by adopting Boyle's law for this purpose. It can be seen that the hyperbola of Boyle's law may have a shape similar to the curve for the true behaviour of the gas, provided the pressure is small. It also seems that if one considers a state sufficiently far into the superheated region, a similarity of curvature persists. However, it is to be expected that near to the dry saturated vapour curve, and also within the wet zone, behaviour is not ideal.

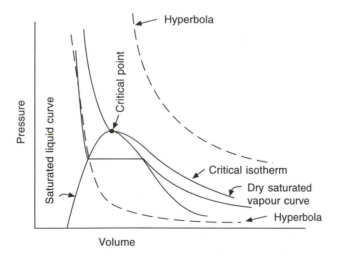

Fig. 2.2   Boyle's law and the true behaviour of a gas.

2.5   Charles' law

It is evident from Boyle's law that, for a given gas, the product pV could be used as an indication of its temperature and, in fact, this is the basis of a scale of temperature. It can

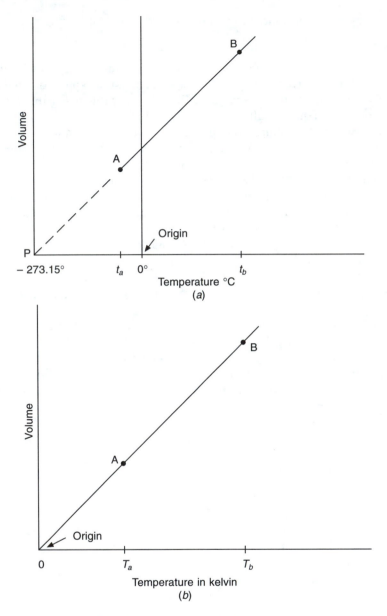

Fig. 2.3   Charles' law and absolute temperature.

be shown that for an ideal gas, at constant pressure, the volume is related to the temperature in a linear fashion. Experimental results support this, and reference to Figure 2.3 shows just how this could be so. Suppose that experimental results allow a straight line to be drawn between two points A and B, as a graph of volume against temperature. If the line is extended to cut the abscissa at a point P, having a temperature of −273.15°C, it is clear that shifting the origin of the co-ordinate system to the left by 273.15°C will give an equation for the straight line, of the form

$$V = aT,$$ (2.6)

where T is the temperature on the new scale and a is a constant representing the slope of the line.

Obviously

$$T = 273.15° + t \tag{2.7}$$

This graphical representation of Charles' law shows that a direct proportionality exists between the volume of a gas and its temperature, as expressed on the new abscissa scale. It also shows that a new scale of temperature may be used. This new scale is an absolute one, so termed since it is possible to argue that all molecular movement has ceased at its zero, hence the internal energy of the gas is zero and, hence also, its temperature is at an absolute zero. Absolute temperature is expressed in kelvin, denoted by K, and the symbol T is used instead of t, to distinguish it from relative temperature on the Celsius scale.

EXAMPLE 2.1

15 m^3 s^{-1} of air at a temperature of 27°C passes over a cooler coil which reduces its temperature to 13°C. The air is then handled by a fan, blown over a reheater, which increases its temperature to 18°C, and is finally supplied to a room.

Calculate the amount of air handled by the fan and the quantity supplied to the room.

Answer

According to Charles' law:

$$V = aT,$$

that is to say,

$$V_2 = V_1 \frac{T_2}{T_1}$$

Hence, the air quantity handled by the fan

$$= 15 \frac{(273 + 13)}{(273 + 27)}$$
$$= 14.3 \ m^3 \ s^{-1}$$

and the air quantity supplied to the room

$$= 15 \frac{(273 + 18)}{(273 + 27)}$$
$$= 14.55 \ m^3 \ s^{-1}$$

One further comment, it is clearly fallacious to suppose that the volume of a gas is directly proportional to its temperature right down to absolute zero; the gas liquefies before this temperature is attained.

2.6   The general gas law

It is possible to combine Boyle's and Charles' laws as one equation;

$$pV = mRT \tag{2.8}$$

where p = the pressure of the gas in Pa,
   V = the volume of the gas in m³,
   m = the mass of the gas in kg,
   R = a constant of proportionality,
   T = the absolute temperature of the gas in K.

Avogadro's hypothesis argues that equal volumes of all gases at the same temperature and pressure contain the same number of molecules. Accepting this and taking as the unit of mass the kilomole (kmol), a mass in kilograms numerically equal to the molecular mass of the gas, a value for the universal gas constant can be established:

$$pV_m = R_o T \qquad (2.9)$$

where V_m is the volume in m³ of 1 kmol and is the same for all gases having the same values of p and T. Using the values p = 101 325 Pa and T = 273.15 K, it has been experimentally determined that V_m equals 22.41383 m³ kmol⁻¹. Hence the universal gas constant is determined

$$R_o = \frac{pV_m}{T} = \frac{101\ 325 \times 22.41383}{273.15} = 8314.41 \text{ J kmol}^{-1}\text{K}^{-1}$$

Dividing both sides of equation (2.9) by the molecular mass, M, of any gas in question allows the determination of the particular gas constant, R, for the gas. This may then be used for a mass of 1 kg in equation (2.8) and we can write

$$pv = \frac{R_o T}{M} = RT \left(\text{whence } R = \frac{8314.41}{M} \right)$$

where v is the volume of 1 kg.
   If a larger mass, m kg, is used, the expression becomes equation (2.8)

$$pV = mRT \qquad (2.8)$$

where V is the volume of m kg and R has units of J kg⁻¹K⁻¹.

$$\text{For dry air, } R_a = \frac{8314.41}{28.97} = 287 \text{ J kg}^{-1}\text{K}^{-1}$$

$$\text{For steam, } R_s = \frac{8314.41}{18.02} = 461 \text{ J kg}^{-1}\text{K}^{-1}$$

A suitable transposition of the general gas law yields expressions for density, pressure and volume.

EXAMPLE 2.2

Calculate the density of a sample of dry air which is at a pressure of 101 325 Pa and at a temperature of 20°C.

Answer

$$\text{Density} = \frac{\text{mass of the gas}}{\text{volume of the gas}}$$

$$= \frac{m}{V}$$

$$= \frac{p}{R_a T}$$

$$= \frac{101\ 325}{287 \times (273 + 20)}$$

$$= 1.2049 \text{ kg m}^{-3}$$

We may compare this answer with that obtained by referring to the CIBSE tables of psychrometric data, which quote the volume of air at 20°C dry bulb and 0 per cent saturation as 0.8301 m^3 kg^{-1} of dry air.

The reciprocal of density is specific volume, hence the density of the air quoted by the tables is given by the reciprocal of 0.8301 m^3 kg^{-1} and is 1.2047 kg m^{-3}.

2.7   Dalton's law of partial pressure

This may be stated as follows:

If a mixture of gases occupies a given volume at a given temperature, the total pressure exerted by the mixture equals the sum of the pressures of the constituents, each being considered at the same volume and temperature.

It is possible to show that if Dalton's law holds, each component of the mixture obeys the general gas law. As a consequence, it is sometimes more convenient to re-express the law in two parts:

(i)   the pressure exerted by each gas in a mixture of gases is independent of the presence of the other gases, and
(ii)   the total pressure exerted by a mixture of gases equals the sum of the partial pressures.

Figure 2.4 illustrates this. It shows an air-tight container under three different conditions, from which it can be seen that the partial pressures exerted by the water vapour in (b) and (c) are equal, as are those exerted by the dry air in (a) and (c) and, that in (a), (b) and (c), the total pressure equals the sum of the partial pressures.

As in the two gas laws already considered, Dalton's law agrees with the results achieved by the kinetic theory of gases and, to some extent, finds substantiation in experiment.

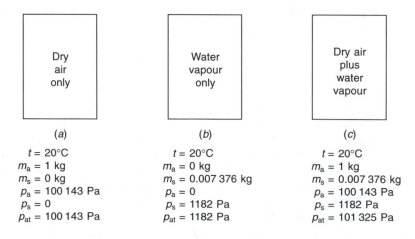

Fig. 2.4 Dalton's law of partial pressure referred to a mixture of dry air and water vapour.

It is now necessary to turn attention to the behaviour of water vapour at the state of saturation and to consider its partial pressure when it is in the superheated state and mixed with dry air.

2.8   Saturation vapour pressure

There are two requirements for the evaporation of liquid water to occur.

(i)   Thermal energy must be supplied to the water.
(ii) The vapour pressure of the liquid must be greater than that of the steam in the environment.

These statements need some explanation.

Molecules in the liquid state are comparatively close to one another. They are nearer to one another than are the molecules in a gas and are less strongly bound together than those in a solid. The three states of matter are further distinguished by the extent to which an individual molecule may move. At a given temperature, a gas consists of molecules which have high individual velocities and which are arranged in a random fashion. A liquid at the same temperature is composed of molecules, the freedom of movement of which is much less, owing to the restraining effect which neighbouring molecules have on one another, by virtue of their comparative proximity. An individual molecule, therefore, has less kinetic energy if it is in the liquid state than it does if in the gaseous state. Modern thought is that the arrangement of molecules in a liquid is not entirely random as in a gas, but that it is not as regular as it is in most, if not all, solids. However, this is by the way.

It is evident that if the individual molecular kinetic energies are greater in the gaseous state, then energy must be given to a liquid if it is to change to the gaseous phase. This explains the first stated requirement for evaporation.

As regards the second requirement, the situation is clarified if one considers the boundary between a vapour and its liquid. Only at this boundary can a transfer of molecules between the liquid and the gas occur. Molecules at the surface have a kinetic energy which has a value related to the temperature of the liquid. Molecules within the body of the gas also have a kinetic energy which is a function of the temperature of the gas. Those gaseous molecules near the surface of the liquid will, from time to time, tend to hit the surface of the liquid, some of them staying there. Molecules within the liquid and near to its surface will, from time to time, also tend to leave the liquid and enter the gas, some of them staying there.

It is found experimentally that, in due course, an equilibrium condition arises for which the gas and the parent liquid both have the same temperature and pressure. These are termed the saturation temperature and the saturation pressure. For this state of equilibrium the number of molecules leaving the liquid is the same as the number of molecules entering it from the gas, on average.

Such a state of thermal equilibrium is exemplified by a closed insulated container which has within it a sample of liquid water. After a sufficient period of time, the space above the liquid surface, initially a vacuum, contains steam at the same temperature as the remaining sample of liquid water. The steam under these conditions is said to be saturated.

Before this state of equilibrium was reached the liquid must have been losing molecules more quickly than it was receiving them. Another way of saying this is to state that the vapour pressure of the liquid exceeded that of the steam above it.

One point emerges from this example: since the loss of molecules from the liquid represents a loss of kinetic energy, and since the kinetic energy of the molecules in the

liquid is an indication of the temperature of the liquid, then the temperature of the liquid must fall during the period preceding its attainment of thermal equilibrium.

It has been found that water in an ambient gas which is not pure steam but a mixture of dry air and steam, behaves in a similar fashion, and that for most practical purposes the relationship between saturation temperature and saturation pressure is the same for liquid water in contact only with steam. One concludes from this a very important fact: saturation vapour pressure depends solely upon temperature.

If we take the results of experiment and plot saturation vapour pressure against saturation temperature, we obtain a curve which has the appearance of the line on the psychrometric chart for 100 per cent saturation. The data on which this particular line is based can be found in tables of psychrometric information. Referring, for instance, to those tables published by the Chartered Institution of Building Services Engineers, we can read the saturation vapour pressure at, say, 20°C by looking at the value of the vapour pressure at 100 per cent saturation and a dry-bulb temperature of 20°C. It is important to note that the term 'dry-bulb' has a meaning only when we are speaking of a mixture of a condensable vapour and a gas. In this particular context the mixture is of steam and dry air, but we could have a mixture of, say, alcohol and dry air, which would have its own set of properties of dry- and wet-bulb temperatures.

According to the *National Engineering Laboratory Steam Tables* (1964), the following equation may be used for the vapour pressure of steam over water up to 100°C:

$$\log p = 30.590\,51 - 8.2\log(t+273.16) + 0.002\,480\,4(t+273.16) - \frac{3142.31}{(t+273.16)}$$

$$(2.10)$$

where t is temperature in °C and p is pressure in kPa. Note that in equation (2.10) the saturation vapour pressure is expressed in terms of the triple point of water, $t+273.16$, not the absolute temperature, $t+273.15$. However, this is not the case in equation (2.11).

ASHRAE (1997) uses a very similar equation for the expression of saturation vapour pressure, based on equations calculated by Hyland and Wexler (1983). The results are in close agreement with the answers obtained by equation (2.10), as Jones (1994) shows.

Over ice, the equation to be used, from the National Bureau of Standards (1955), is:

$$\log p = 12.538\,099\,7 - 266\,391/(273.15 + t)$$ $$(2.11)$$

where p is pressure in Pa.

2.9   The vapour pressure of steam in moist air

It is worth pausing a moment to consider the validity of the ideal gas laws as they are applied to the mixture of gases which comprises moist air.

Kinetic theory, which supports the ideal gas laws, fails to take account of the fact that intermolecular forces of attraction exist. In a mixture such forces occur between both like molecules and unlike molecules. That is to say, between molecules of dry air, between molecules of steam and between molecules of steam and dry air. The virial coefficients mentioned in section 2.4 attempt to deal with the source of error resulting from attractive forces between like molecules. To offset the error accruing from the forces between unlike molecules, a further set of 'interaction coefficients' (sometimes termed 'cross-virial' coefficients) is adopted.

An explanation of the modern basis of psychrometry, taking these forces into account, is given in section 2.19. For the moment, and for most practical purposes, we can take it that the saturation vapour pressure in humid air depends on temperature alone; that is, it is uninfluenced by barometric pressure.

EXAMPLE 2.3

Determine the saturation vapour pressure of moist air (*a*), at 15°C and a barometric pressure of 101 325 Pa and (*b*) at 15°C and a barometric pressure of 95 000 Pa.

Answer

(*a*) From CIBSE tables of psychrometric data, at 15°C dry-bulb and 100 per cent saturation, the saturation vapour pressure is 1704 Pa.

(*b*) From the same source exactly, we determine that the saturation vapour pressure is 1704 Pa at 15°C dry-bulb and 100 per cent relative humidity. We can, of course, use the CIBSE tables of psychrometric data for determining this saturation vapour pressure, even though the question speaks of 95 000 Pa, since saturation vapour pressure does not depend on barometric pressure. On the other hand, it should be noted that at all relative humidities less than 100 per cent the vapour pressures quoted in these tables are valid only for the total or barometric pressure for which the tables are published, namely, 101 325 Pa.

To illustrate the distinction between saturated vapour pressure and superheated vapour pressure, consider a sample of liquid water within a closed vessel. On the application of heat evaporation occurs, and for every temperature through which the liquid passes there is an equilibrium pressure, as has already been discussed. Figure 2.5(*a*) shows a curve A, B, B′ representing the relationship between saturation vapour pressure and absolute temperature. If heat is applied to the vessel beyond the instant when the last of the liquid water turns to saturated steam, the change of state of the steam can no longer be represented by the curve. The point B represents the state of the contents of the vessel at the instant when the last of the liquid has just evaporated. The vessel contains dry saturated steam but, unlike the case so far, no liquid is present. By our earlier assumptions then, the contents of the vessel approximate an ideal gas and, therefore, may be taken to obey Charles' law for any further heating at constant volume. Equation (2.6) states this law, and further changes of state of the steam in the closed vessel may be represented by a straight line. This is shown in Figure 2.5(*a*) by the line BC.

The changes can also be shown on another sort of diagram, Figure 2.5(*b*), where pressure and volume are used as co-ordinates. The total volume of the liquid and vapour has remained constant throughout the application of all the heat, hence changes on the $p - V$ diagram must occur along a line of constant volume, for this example. At condition A the vessel contains saturated liquid and saturated vapour. Accordingly, on the $p - V$ diagram state A must lie within the wet zone. On the other hand, at point B the contents of the vessel are saturated steam only, hence B lies on the saturated vapour line. It can be seen that the change of state into the superheated zone at C is an extension of the vertical line AB as far as C.

It is seen later (sections 2.10 and 2.19) that, for a mixture of steam and dry air, there is a relationship between the mass of steam in the mixture and the vapour pressure it exerts. Since the thermodynamic properties of saturated steam and dry air are well established, it is possible, according to Goff (1949), to express the vapour pressure of the steam present in a mixture on a proportional basis, related to the mass of steam present in the mixture. Thus, for 1 kg of dry air only, the vapour pressure is zero but, for a mixture of saturated steam and 1 kg of dry air, the vapour pressure is given by equation (2.10) or (2.11). For a lesser amount of steam mixed with 1 kg of dry air, the vapour pressure exerted by the steam present would be between zero and the saturated pressure, in proportion to the mass of water vapour in the mixture. This leads to the concept of percentage saturation and is dealt with in section 2.11.

It is to be noted that, when steam is mixed with dry air but the steam is not saturated, it

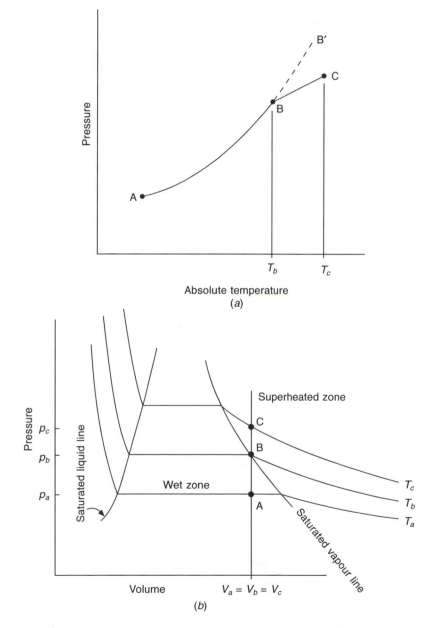

Fig. 2.5 (a) Saturation vapour pressure and temperature. Charles' law applies for the superheated vapour from *B* to *C*. (b) Pressure–volume diagram showing evaporation and superheating.

is in the superheated state. For example, such steam would be represented by the point C in Figure 2.5(*b*) rather than by the point *B*.

2.10   Moisture content and humidity ratio

Moisture content is defined as the mass of water vapour in kilograms which is associated with one kilogram of dry air in an air–water vapour mixture. It is sometimes called specific humidity or humidity ratio.

Starting with the definition, we can write—

moisture content = mass of water vapour per unit mass of dry air
$$= m_s/m_a$$

By using Dalton's law we can now apply the general gas law to each of the two constituents of moist air, just as though the other did not exist:

$$pV = mRT \text{ in general}$$

hence

$$p_sV_s = m_sR_sT_s \text{ for the water vapour}$$

and

$$p_aV_a = m_aR_aT_a \text{ for the dry air}$$

The general gas law may be rearranged so that mass is expressed in terms of the other variables:

$$m = \frac{pV}{RT}$$

By transposition in the equations referring to water vapour and dry air, we can obtain an expression for moisture content based on its definition:

$$\text{moisture content} = \frac{p_sV_sR_aT_a}{R_sT_sp_aV_a}$$

$$= \frac{R_ap_s}{R_sp_a}$$

since the water vapour and the dry air have the same temperature and volume.

The ratio of R_a to R_s is termed the relative density of water vapour with respect to dry air and, as already seen, it depends on the ratio of the molecular mass of water vapour to that of dry air:

$$\frac{R_a}{R_s} = \frac{M_s}{M_a} = \frac{18.02}{28.97} = 0.622$$

Hence we may write moisture content as $0.622 \, p_s/p_a$ and hence

$$g = 0.622 \frac{p_s}{(p_{at} - p_s)} \tag{2.12}$$

Since the vapour pressure of superheated steam mixed with dry air is proportional to the mass of the steam present, the above equation can be re-arranged to express such a vapour pressure:

$$p_s = \frac{p_{at}\, g}{(0.622 + g)} \tag{2.13}$$

EXAMPLE 2.4

Calculate the moisture content of 1 kg of dry air at 20°C mixed with saturated steam for barometric pressures of (*a*) 101.325 kPa and (*b*) 95 kPa.

Answer

(*a*) By equation (2.10), or from steam tables, or from CIBSE psychrometric tables, the saturation vapour pressure, p_{ss}, is 2.337 kPa. Hence, using equation (2.12)

$$g = 0.622 \,\frac{2.337}{(101.325 - 2.337)} = 0.014\ 68 \text{ kg per kg dry air}$$

CIBSE psychrometric tables give a value of 0.014 75 kg per kg dry air. The discrepancy is because the tables use a slightly different, more correct, equation that takes account of intermolecular forces (see section 2.19).

(*b*) Similarly, because the saturation vapour pressure is independent of barometric pressure p_{ss} is still 2.337 kPa, hence

$$g = 0.622 \,\frac{2.337}{(95 - 2.337)} = 0.01569 \text{ kg per kg dry air}$$

It is often convenient to express moisture content as g/kg dry air. The two above answers would then be 14.68 g/kg dry air and 15.69 g/kg dry air.

Note that although the saturation vapour pressure is independent of barometric pressure the moisture content of air at 100 per cent saturation is not. Figure 2.6 shows that the curve

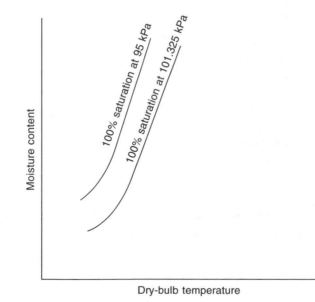

Dry-bulb temperature

Fig. 2.6   The effect of barometric pressure on percentage saturation.

for air at 100 per cent saturation corresponds to higher moisture contents as the barometric pressure decreases.

2.11   Percentage saturation

This is not the same as relative humidity but is sometimes confused with it. However, for saturated air and for dry air the two are identical and within the range of states used for comfort conditioning they are virtually indistinguishable.

Percentage saturation is defined as the ratio of the moisture content of moist air *at a given temperature*, t, to the moisture content of saturated air *at the same temperature t*.

It is also known as the degree of saturation.

Applying the general gas law to the superheated steam present in moist air which is not saturated, we may write

$$g = \frac{m_s}{m_a} = \frac{p_s V_s}{R_s T_s} \cdot \frac{R_a T_a}{p_a V_a}$$

$$= \frac{p_s}{p_a} \cdot \frac{R_a}{R_s}$$

since the steam and dry air occupy the same volume and are at the same temperature, being in intimate contact with one another.

We may then write

$$g = \frac{p_s}{(p_{at} - p_s)} \cdot \frac{R_a}{R_s}$$

for the unsaturated moist air.

Similarly, for saturated air the moisture content is given by

$$g_{ss} = \frac{p_{ss}}{(p_{at} - p_{ss})} \cdot \frac{R_a}{R_{ss}}$$

From the definition of percentage saturation we can write

$$\mu = \frac{g \text{ at a given temperature}}{g_{ss} \text{ at the same temperature}} \times 100 \tag{2.14}$$

$$= \frac{p_s}{p_{ss}} \cdot \frac{(p_{at} - p_{ss})}{(p_{at} - p_s)} \cdot 100 \tag{2.15}$$

From what has been said earlier it is clear that R_{ss} is the same as R_s and hence the ratio R_{ss}/R_s is absent from equation (2.15).

EXAMPLE 2.5

Calculate the percentage saturation of air at 20°C dry-bulb and a moisture content of 0.007 34 kg per kg dry air for (*a*) a barometric pressure of 101.325 kPa and (*b*) a barometric pressure of 95 kPa.

Answer

From Example 2.4 the moisture content of saturated air at 20°C is 0.014 68 kg per kg dry air at 101.325 kPa and 0.015 69 kg per kg dry air at 95 kPa. Hence, by equation (2.14)

(a) $\mu = \dfrac{0.00734}{0.01468} \times 100 = 50\%$ at 101.325 kPa

(b) $\mu = \dfrac{0.00734}{0.01569} \times 100 = 46.8\%$ at 95 kPa

2.12 Relative humidity

This is defined as the ratio of the partial pressure of the water vapour in moist air *at a given temperature, t,* to the partial pressure of saturated water vapour in air *at the same temperature, t.*
   Namely

$$\phi = \frac{p_s \text{ at a given temperature}}{p_{ss} \text{ at the same temperature}} \times 100 \qquad (2.16)$$

This is illustrated in Figure 2.7 which shows the pressure–volume changes for steam alone.

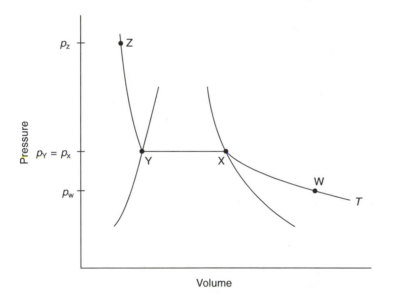

Fig. 2.7   Relative humidity is the ratio p_a/p_b for a constant absolute temperature T, shown on a pressure–volume diagram for steam.

That is to say, accepting Dalton's law, the water vapour content of moist air is considered separately from the dry air content. The line WXYZ is an isotherm for a value of absolute temperature denoted by T. If moist air at a relative humidity of less than 100 per cent contains steam with a partial pressure of p_w and a temperature T, it is represented by the point W in the superheated zone. Saturated steam at the same temperature and having a partial pressure p_x is represented by the point X. The relative humidity, by equation (2.16), is given by

$$\phi = \frac{p_w}{p_x} \times 100$$

EXAMPLE 2.6

Calculate the relative humidity of moist air at a dry-bulb temperature of 20°C and a moisture content of 0.007 34 kg per kg dry air for a barometric pressure of (*a*) 101.325 kPa and (*b*) 95 kPa.

Answer

(*a*) From equation (2.13)

$$p_s = \frac{101.325 \times 0.00734}{(0.622 + 0.00734)} = 1.1818 \text{ kPa}$$

From equation (2.10), or from steam tables, or from CIBSE psychrometric tables at 20°C saturated,

$$p_{ss} = 2.337 \text{ kPa}$$

Then, by equation (2.16)

$$\phi = \frac{1.1818}{2.337} \times 100 = 50.57\%$$

(*b*) Similarly

$$p_s = \frac{95 \times 0.00734}{(0.622 + 0.00734)} \; 1.1080 \text{ kPa}$$

p_{ss} is unchanged because it is independent of barometric pressure and depends only on temperature, which remains at 20°C. Hence

$$\phi = \frac{1.1080}{2.337} \times 100 = 47.41\%$$

Compare with the answers to Example 2.5.

2.13   Dew point

This is defined as the temperature of saturated air which has the same vapour pressure as the moist air under consideration.

It is not possible to express this definition in the form of a simple equation by means of the ideal gas laws. This is evident if we refer to equation (2.10). It is more convenient to refer to tabulated values when saturated vapour pressure is required. However, provided such tables are available, we can determine the dew point of air at a given psychrometric state and barometric pressure.

EXAMPLE 2.7

Calculate the dew point of moist air at a dry-bulb temperature of 20°C and a moisture content of 0.007 34 kg per kg dry air at a barometric pressure of 95 kPa.

Answer

In Example 2.6 the partial pressure of the superheated steam mixed with 1 kg of dry air at a barometric pressure of 95 kPa was calculated by equation (2.13) as 1.1080 kPa. At its dew point temperature, the steam present in the moist air will be saturated and will have this value as its saturated vapour pressure. Reference to steam tables or to the CIBSE psychrometric tables shows, by interpolation, that this corresponds to a saturated temperature of 8.49°C. This is the dew point temperature.

2.14 Specific volume

This is the volume in cubic metres of one kilogram of dry air mixed with g kilograms of water vapour. In the mixture each constituent occupies the same volume and is at the same temperature, but each exerts its own partial pressure. By Dalton's law the sum of these partial pressures is the total (barometric) pressure of the mixture. See Figure 2.4.

The general gas law, in the form of equation (2.8), may be transposed to express the specific volume:

$$V = \frac{mRT}{p} \tag{2.17}$$

This equation could be used to refer to the dry air, or to the water vapour, independently if Dalton's law is accepted. In doing so, the appropriate values for the mass, particular gas constant and partial pressure of the constituent considered must be used.

EXAMPLE 2.8

Calculate the specific volume of air at a dry-bulb temperature of 20°C and a moisture content of 0.007 34 kg per kg dry air at a barometric pressure of 95 kPa.

Answer

Considering the water vapour alone and adopting the subscript 's' to indicate this we have

$$V_s = \frac{m_s R_s T_s}{p_s}$$

From the answer to Example 2.6 we know that the partial pressure of the water vapour in the mixture is 1.1080 kPa and from section 2.6 we know that R_s is 461 J kg^{-1}K^{-1}, hence

$$V_s = \frac{0.007\,34 \times 461 \times (273 + 20)}{1108.0} = 0.895 \text{ m}^3$$

Alternatively the dry air could be considered and, adopting the subscript 'a' to denote this, we have:

$$V_a = \frac{m_a R_a T_a}{p_a}$$

We know from section 2.6 that the particular gas constant for dry air is 287 J kg^{-1}K^{-1} and hence

$$V_a = \frac{1 \times 287 \times (273 + 20)}{(95\,000 - 1108.0)} = 0.896 \text{ m}^3$$

A more exact agreement would be obtained if the influences of intermolecular forces were taken into account. This is discussed in section 2.19.

The thermodynamic properties of dry air and steam are well established. Hence the general principle followed by ASHRAE (1997), Goff (1949) and CIBSE (1986) in the expression of the volume of moist air is to add to the volume of 1 kg of dry air (v_a), a proportion of the difference between the volume of saturated air (v_s), and that of dry air (v_a). This gives rise to the following equation:

$$v = v_a + \mu(v_s - v_a)/100 \tag{2.18}$$

where μ is the percentage saturation.

EXAMPLE 2.9

Calculate the specific volume of moist air at a dry-bulb temperature of 20°C, 50 per cent saturation and a barometric pressure of 101.325 kPa.

Answer

From CIBSE psychrometric tables (or, less accurately, from a chart) the specific volume of dry air, v_a, is 0.8301 m^3 per kg dry air and the specific volume of saturated air, v_s, is 0.8497 m^3 per kg dry air. Then, by equation (2.18), the specific volume of moist air at 50 per cent saturation is

$$v = 0.8301 + 50(0.8497 - 0.8301)/100 = 0.8399 \text{ m}^3 \text{ per kg dry air.}$$

CIBSE psychrometric tables quote the same value.

2.15   Enthalpy: thermodynamic background

The *first law of thermodynamics* may be considered as a statement of the principle of the conservation of energy. A consequence of this is that a concept termed 'internal energy' must be introduced, if the behaviour of a gas is to be explained with reasonable exactness during processes of heat transfer. Internal energy is the energy stored in the molecular and atomic structure of the gas and it may be thought of as being a function of two independent variables, the pressure and the temperature of the gas.

We can consider heat being supplied to a gas in either one of two ways: at constant volume or at constant pressure. Since the work done by a gas or on a gas, during a process of expansion or compression, is expressed by the equation: work done = $\int p \, dV$, it follows that if heat is supplied to a gas at constant volume, no work will be done by the gas on its environment. Consequently the heat supplied to the gas serves only to increase its internal energy, U. If a heat exchange occurs at constant pressure, as well as a change in internal energy taking place, work may be done.

This leads to a definition of enthalpy, H:

$$H = U + pV \tag{2.19}$$

The equation is strictly true for a pure gas of mass m, pressure p, and volume V. However, it may be applied without appreciable error to the mixtures of gases associated with air conditioning.

It is desirable that the expression 'heat content' should not be used because of the way in which enthalpy is defined by equation (2.19). This, and the other common synonym for enthalpy, 'total heat', suggest that only the internal energy of the gas is being taken into account. As a result of this, both terms are a little misleading and, in consequence, throughout the rest of this book the term enthalpy will be used.

2.16 Enthalpy in practice

It is not possible to give an absolute value to enthalpy since no assessment is possible of the absolute value of the internal energy of a gas. The expression, mentioned earlier, of internal energy as a function of pressure and temperature, is a simplification. Fortunately, air conditioning involves only a calculation of changes in enthalpy. It follows that such changes may be readily determined if a datum level of enthalpy is adopted for its expression. Thus, we are really always dealing in relative enthalpy, although we may not refer to it as such.

The enthalpy, h, used in psychrometry is the specific enthalpy of moist air, expressed in kJ kg^{-1} dry air, defined by the equation:

$$h = h_a + gh_g \tag{2.20}$$

where h_a is the enthalpy of dry air, h_g is the enthalpy of water vapour, both expressed in kJ kg^{-1}, and g is the moisture content in kg per kg dry air.

The value of temperature chosen for the zero of enthalpy is 0°C for both dry air and liquid water. The relationship between the enthalpy of dry air and its temperature is not quite linear and values taken from NBS Circular 564 (1955), for the standard atmospheric pressure of 101.325 kPa and suitably modified for the chosen zero, form the basis of the CIBSE tables of the properties of humid air. An approximate equation for the enthalpy of dry air over the range 0°C to 60°C is, however

$$h_a = 1.007t - 0.026 \tag{2.21}$$

and for lower temperatures, down to −10°C, the approximate equation is

$$h_a = 1.005t \tag{2.22}$$

Values of h_g for the enthalpy of vapour over water have been taken from NEL steam tables (1964), slightly increased to take account of the influence of barometric pressure and modified to fit the zero datum. The enthalpy of vapour over ice, however, is based on Goff (1949).

For purposes of approximate calculation, without recourse to the CIBSE psychrometric tables, we may assume that, in the range 0°C to 60°C, the vapour is generated from liquid water at 0°C and that the specific heat of superheated steam is a constant. The following equation can then be used for the enthalpy of water vapour:

$$h_g = 2501 + 1.84t \tag{2.23}$$

Equations (2.21) and (2.23) can now be combined, as typified by equation (2.20), to give an approximate expression for the enthalpy of humid air at a barometric pressure of 101.325 kPa:

$$h = (1.007t - 0.026) + g(2501 + 1.84t) \tag{2.24}$$

EXAMPLE 2.10

Calculate the approximate enthalpy of moist air at a dry-bulb temperature of 20°C, 50 per cent saturation and a barometric pressure of 101.325 kPa. Use CIBSE psychrometric tables or a psychrometric chart to establish the moisture content.

Answer

From tables (or less accurately from a chart),

$g = 0.007\ 376$ kg per kg dry air

Using equation (2.24)

$$h = (1.007 \times 20 - 0.026) + 0.007\ 376 \times (2501 + 1.84 \times 20)$$
$$= 38.83 \text{ kJ per kg dry air}$$

CIBSE quote a value of 38.84 kJ per kg dry air.

For the range of temperatures from –10°C to 0°C equation (2.23) is also approximately correct for the enthalpy of water vapour over ice. Using equations (2.22) and (2.23) the combined approximate equation for the enthalpy of humid air over ice becomes

$$h = 1.005t + g(2501 + 1.84t) \tag{2.25}$$

EXAMPLE 2.11

Calculate the approximate enthalpy of moist air at a dry-bulb temperature of –10°C, 50 per cent saturation and a barometric pressure of 101.325 kPa. Use CIBSE psychrometric tables or a psychrometric chart to establish the moisture content.

Answer

From tables (or less accurately from a chart)

$g = 0.000\ 804$ kg per kg dry air

Using equation (2.25)

$$h = 1.005 \times (-10) + 0.000\ 804(2501 + 1.84 \times (-10))$$
$$= -8.054 \text{ kJ per kg dry air}$$

CIBSE tables quote –8.060 kJ per kg dry air.

As in the case of specific volume, the general principle followed by ASHRAE (1997), Goff (1949) and CIBSE (1986) for determining the enthalpy of moist air is to add to the enthalpy of dry air, h_a, a proportion of the difference between the enthalpy of saturated air, h_s, and the enthalpy of dry air, h_a. This is expressed by the following equation:

$$h = h_a + \mu(h_s - h_a)/100 \tag{2.26}$$

where μ is the percentage saturation.

EXAMPLE 2.12

Calculate the enthalpy of moist air at a dry-bulb temperature of 20°C, 50 per cent saturation and a barometric pressure of 101.325 kPa using equation (2.26).

Answer

From CIBSE psychrometric tables (or, less accurately, from a psychrometric chart), the enthalpy of dry air, h_a, is 20.11 kJ per kg dry air and the enthalpy of saturated air, h_s, is 57.55 kJ per kg dry air. Hence, by equation (2.26):

$$h = 20.11 + 50(57.55 - 20.11)/100 = 38.83 \text{ kJ per kg dry air}$$

CIBSE tables quote 38.84 kJ/kg dry air. The difference is due to rounding off.

EXAMPLE 2.13

Calculate the enthalpy of moist air at a dry-bulb temperature of 60°C, 50 per cent saturation and a barometric pressure of 101.325 kPa. Use CIBSE psychrometric tables to establish the moisture content.

Answer

From tables, g = 0.076 66 kg per kg dry air
Using equation (2.24)

$$h = (1.007 \times 60 - 0.026) + 0.076\ 66(2501 + 1.84 \times 60)$$
$$= 260.6 \text{ kJ per kg dry air}$$

CIBSE tables quote a value of 260.4 kJ per kg dry air.

2.17 Wet-bulb temperature

A distinction must be drawn between measured wet-bulb temperature and the temperature of adiabatic saturation, otherwise sometimes known as the thermodynamic wet-bulb temperature. The wet-bulb temperature is a value indicated on an ordinary thermometer, the bulb of which has been wrapped round with a wick, moistened in water. The initial temperature of the water used to wet the wick is of comparatively minor significance, but the cleanliness of the wick and the radiant heat exchange with surrounding surfaces are both important factors that influence the temperature indicated by a wet-bulb thermometer. On the other hand, the temperature of adiabatic saturation is that obtained purely from an equation representing an adiabatic heat exchange. It is somewhat unfortunate, in air and water-vapour mixtures, that the two are almost numerically identical at normal temperatures and pressures, which is not so in mixtures of other gases and vapours.

Wet-bulb temperature is not a property only of mixtures of dry air and water vapour. Any mixture of a non-condensable gas and a condensable vapour will have a wet-bulb temperature. It will also have a temperature of adiabatic saturation. Consider a droplet of water suspended in an environment of most air. Suppose that the temperature of the droplet is t_w and that its corresponding vapour pressure is p_w. The ambient moist air has a temperature t and a vapour pressure of p_s.

Provided that p_w exceeds p_s, evaporation will take place and, to effect this, heat will flow from the environment into the droplet by convection and radiation. If the initial value of t_w is greater than that of t, then, initially, some heat will flow from the drop itself to assist in the evaporation. Assuming that the original temperature of the water is less than the wet-bulb temperature of the ambient air, some of the heat gain to the drop from its surroundings will serve to raise the temperature of the drop itself, as well as providing for the evaporation. In due course, a state of equilibrium will be reached in which the sensible heat gain to the water exactly equals the latent heat loss from it, and the water itself will have taken up a steady temperature, t', which is termed the wet-bulb temperature of the moist air surrounding the droplet.

The condition can be expressed by means of an equation:

$$(h_c + h_r)A(t - t') = \alpha A h_{fg}(p'_{ss} - p_s) \tag{2.27}$$

where

h_c = the coefficient of heat transfer through the gas film around the drop, by convection

h_r = the coefficient of heat transfer to the droplet by radiation from the surrounding surfaces,

A = the surface area of the droplet,

h_{fg} = the latent heat of evaporation of water at the equilibrium temperature attained,

α = the coefficient of diffusion for the molecules of steam as they travel from the parent body of liquid through the film of vapour and non-condensable gas surrounding it.

The heat transfer coefficients $(h_c + h_r)$ are commonly written, in other contexts, simply as h.

Equation (2.27) can be re-arranged to give an expression for the vapour pressure of moist air in terms of measured values of the dry-bulb and wet-bulb temperatures:

$$p_s = p'_{ss} - \frac{(h_c + h_r)}{(\alpha h_{fg})} (t - t') \qquad (2.28)$$

The term $(h_c + h_r)/(\alpha h_{fg})$ is a function of barometric pressure and temperature.

Equation (2.28) usually appears in the following form, which is termed the psychrometric equation, due to Apjohn (1838):

$$p_s = p'_{ss} - p_{at}A(t - t') \qquad (2.29)$$

where

p_s = vapour pressure

p'_{ss} = saturated vapour pressure at a temperature t'

p_{at} = atmospheric (barometric) pressure

t = dry-bulb temperature, in °C

t' = wet-bulb temperature, in °C

A = a constant having values as follows:

	wet-bulb $\geq 0°C$	wet-bulb $< 0°C$
Screen	7.99×10^{-4} °C^{-1}	7.20×10^{-4} °C^{-1}
Sling or aspirated	6.66×10^{-4} °C^{-1}	5.94×10^{-4} °C^{-1}

Any units of pressure may be used in equation (2.29), provided they are consistent.

EXAMPLE 2.14

Calculate the vapour pressure of moist air at a barometric pressure of 101.325 kPa if the measured dry-bulb temperature is 20°C and the measured sling wet-bulb is 15°C.

Answer

The saturated vapour pressure at the wet-bulb temperature is obtained from CIBSE tables for air at 15°C and 100 per cent saturation and is 1.704 kPa. Alternatively, the same value can be read from steam tables at a saturated temperature of 15°C. Then, using equation (2.29):

$$p_s = 1.704 - 101.325 \times 6.6 \times 10^{-4}(20 - 15)$$
$$= 1.370 \text{ kPa}$$

CIBSE psychrometric tables quote 1.369 kPa for air at 20°C dry-bulb and 15°C sling wet-bulb.

The radiation component of the sensible heat gain to the droplet is really a complicating factor since it depends on the absolute temperatures of the surrounding surfaces and their emissivities, and also because the transfer of heat by radiation is independent of the amount of water vapour mixed with the air, in the case considered. The radiation is also independent of the velocity of airflow over the surface of the droplet. This fact can be taken advantage of to minimise the intrusive effect of h_r. If the air velocity is made sufficiently large it has been found experimentally that $(h_c - h_r)/h_c$ can be made to approach unity. In fact, in an ambient air temperature of about 10°C its value decreases from about 1.04 to about 1.02 as the air velocity is increased from 4 m s^{-1} to 32 m s^{-1}.

At this juncture it is worth observing that increasing the air velocity does not (contrary to what might be expected) lower the equilibrium temperature t'. As more mass of air flows over the droplet each second there is an increase in the transfer of sensible heat to the water, but this is offset by an increase in the latent heat loss from the droplet. Thus t' will be unchanged. What will change, of course, is the time taken for the droplet to attain the equilibrium state. The evaporation rate is increased for an increase in air velocity and so the water more rapidly assumes the wet-bulb temperature of the surrounding air.

It is generally considered that an air velocity of 4.5 m s^{-1} is sufficient to give a stable wet-bulb reading.

In general,

$$g = \frac{M_s}{M_a} \cdot \frac{p_s}{p_a}$$

see equation (2.12) and so

$$(g'_{ss} - g) = \frac{M_s}{M_a} \cdot \frac{(p'_{ss} - p_s)}{p_a}$$

it being assumed that p_a, the partial pressure of the non-condensable dry air surrounding the droplet, is not much changed by variations in p_s, for a given barometric pressure. The term g'_{ss} is the moisture content of saturated air at t'. Equation (2.28) can be re-arranged as

$$\frac{(g'_{ss} - g)}{(t - t')} = \frac{M_s}{M_a} \cdot \frac{(h_c + h_r)}{p_a h_{fg} \alpha}$$

Multiplying above and below by c, the specific heat of humid air, the equation reaches the desired form

$$\frac{(g'_{ss} - g)}{(t - t')} = (Le) \frac{c}{h_{fg}} \qquad (2.30)$$

where (Le) is a dimensionless quantity termed the Lewis number and equal to

$$\frac{M_s}{M_a} \cdot \frac{(h_c + h_r)}{p_a \alpha c}$$

In moist air at a dry-bulb temperature of about 21°C and a barometric pressure of 101.325 kPa, the value of

$$\frac{M_s}{M_a} \cdot \frac{h_c}{p_a \alpha c}$$

is 0.945 for airflow velocities from 3.8 to 10.1 m s^{-1} and the value of

$$\frac{(h_c + h_r)}{h_c}$$

is 1.058 at 5.1 m s^{-1}. So for these sorts of values of velocity, temperature and pressure the Lewis number is unity.

As will be seen in the next section, this coincidence, that the Lewis number should equal unity for moist air at a normal condition of temperature and pressure, leads to the similarity between wet-bulb temperature and the temperature of adiabatic saturation.

2.18   Temperature of adiabatic saturation

An adiabatic process is one in which no heat is supplied to or rejected from the gas during the change of state it undergoes.

Consider the flow of a mixture of air and water vapour through a perfectly insulated chamber. The chamber contains a large wetted surface, and water is evaporated from the surface to the stream of moist air flowing over it, provided the moist air is not already saturated. Feed-water is supplied to the chamber to make good that evaporated. Figure 2.8 shows such a situation.

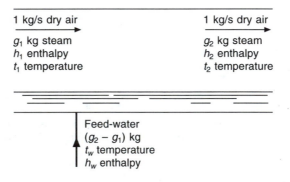

Fig. 2.8   Airflow through a perfectly insulated humidification chamber.

A heat balance may be established:

$$c_a(t_1 - t_2) + c_s g_1(t_1 - t_2) = (g_2 - g_1)[(t_2 - t_w) + h_{fg}]$$

This equation expresses the physical changes that have taken place:

(i) $g_2 - g_1$ kg of feed-water are sensibly heated from a temperature of t_w to a temperature t_2.

(ii) $g_2 - g_1$ kg of water are evaporated from the wetted surface within the chamber, at a temperature t_2. The latent heat required for this is h_{fg} kJ kg^{-1} at temperature t_2.

(iii) 1 kg of dry air is cooled from a temperature t_1 to a temperature t_2.

(iv) g_1 kg of steam are cooled from t_1 to t_2.

The left-hand side of the equation may be simplified by using a combined specific heat, c, for moist air:

$$c(t_1 - t_2) = (g_2 - g_1)\{(t_2 - t_w) + h_{fg}\}$$

In due course, if the chamber is infinitely long, the moist air will become saturated and will have a moisture content g_{ss} and a temperature t_{ss}.

The equation then becomes

$$c(t_1 - t_{ss}) = (g_{ss} - g_1)\{(t_{ss} - t_w) + h_{fg}\}$$

If the feed-water is supplied to the chamber at a temperature t_{ss}, then a further simplification results:

$$c(t_1 - t_{ss}) = (g_{ss} - g_1)h_{fg}$$

From this we can write an equation representing an adiabatic saturation process:

$$\frac{(g_{ss} - g_1)}{(t_1 - t_{ss})} = \frac{c}{h_{fg}} \tag{2.31}$$

This equation should be compared with equation (2.30). Both give a value for the slope of a line in a moisture content–temperature, co-ordinate system, which is shown in Figure 2.9. It can also be seen that if the air is saturated by passing it through an adiabatic saturation chamber, its wet-bulb temperature will fall, if as illustrated, the value of the Lewis number is greater than one. At the saturated condition, the temperature of adiabatic saturation, t^*, the wet-bulb temperature, t_2', and the dry-bulb temperature, t_2, are all equal and may be denoted by t_{ss}.

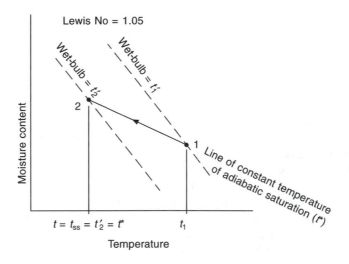

Fig. 2.9 A humidification process along a line of constant adiabatic saturation temperature has a different slope from that of a process along a line of constant wet-bulb temperature.

An illustration of a Lewis number being greater than one is that of moist air, initially at 21°C dry-bulb, 15°C wet-bulb (sling) and 101.325 kPa barometric pressure which is flowing over a wet-bulb thermometer or through an adiabatic saturation chamber at a velocity of 1 m s^{-1}. The lewis number is 1.05. If the velocity exceeded about 1 m s^{-1}, then the Lewis number would become less than unity and the wet-bulb temperature would increase as the moist air was humidified by a process of adiabatic saturation.

Outside the field of air conditioning, if toluene is evaporated into dry air by a process of adiabatic saturation, it can be observed that the wet-bulb temperature is very much more than the temperature of adiabatic saturation. This is because, at the temperatures and pressures considered here, the Lewis number equals about 1.8, instead of unity.

2.19　Non-ideal behaviour

The application of the ideal gas laws to mixtures of gases and vapours assumes the validity of Dalton's Law. To quote Goff (1949): 'By present-day standards, Dalton's Law must be regarded as an inaccurate conjecture based upon an unwarranted faith in the ultimate simplicity of nature.' The ideal laws and the kinetic theory of gases ignore the effects of inter-molecular forces. The more up-to-date approach of statistical mechanics, however, gives us a theory of thermodynamics which takes account of forces between like molecules, in terms of virial coefficients, and between unlike molecules in terms of interaction coefficients.

The methods of statistical mechanics, in conjunction with the best experimental results, yield accurate evaluations of the properties of dry air and water vapour, published in NBS Circular 564 (1955) and in the *NEL Steam Tables* (1964), respectively.

The association of dry air and water vapour in a mixture is expressed by the moisture content on a mass basis and so percentage saturation is particularly relevant, since it defines the fraction of saturated water vapour present in a mixture, for a particular temperature. It is typified by the equation:

$$g = \frac{\mu g_{ss}}{100} \tag{2.32}$$

and the moisture content at saturation, g_{ss}, is defined more precisely by the CIBSE (1986): than is possible with the ideal gas laws

$$g_{ss} = \frac{0.621\,97 f_s p_{ss}}{p_{at} - f_s p_{ss}} \tag{2.33}$$

The numerical constant in equation (2.33) is the ratio of the molecular mass of water vapour and dry air with respective values of 18.016 and 28.966 kg mol^{-1}. The molecular masses used by ASHRAE (1997) are slightly different and yield a value of 0.621 98, to replace the value used in equation (2.33).

The symbol f_s is a dimensionless enhancement factor and, according to Hyland and Wexler (1983), is dependent on: Henry's constant (relating to the solubility of air in water), saturation vapour pressure, temperature, barometric pressure, and other factors. The related equations are complicated. The value of f_s is approximately 1.004 at a barometric pressure of 101.325 kPa and a temperature of 0°C.

The data in Table 2.1 are based on Goff (1949) and form the basis of the calculations leading to the CIBSE psychrometric tables (1986) and chart according to Jones (1970). Other research by Hyland and Wexler (1983) yields slightly different values for the enhancement factor, namely, 1.0039 at a barometric pressure of 100 kPa over the temperature range from 0°C to 90°C. For all practical purposes f_s can be taken as 1.004.

At sufficiently low pressures the behaviour of a mixture of gases can be accurately described by

$$pv = RT - p \sum_{ik} x_i x_k A_{ik}(T) - p^2 \sum_{ijk} x_i x_j x_k a_{ijk}(T) \tag{2.34}$$

The term $A_{ik}(T)$ is a function of temperature and represents the second virial coefficient if $i = k$, but an interaction coefficient if $i \neq k$, the molecules being considered two at a time.

Table 2.1   Enhancement factor, f_s

kPa	0°C	10°C	20°C	30°C	40°C	50°C	60°C
101.325	1.0044	1.0044	1.0044	1.0047	1.0051	1.0055	1.0059
100	1.0044	1.0043	1.0044	1.0047	1.0051	1.0055	1.0059
97.5	1.0043	1.0043	1.0043	1.0046	1.0050	1.0054	1.0058
95	1.0042	1.0042	1.0043	1.0045	1.0049	1.0053	1.0057
92.5	1.0041	1.0041	1.0042	1.0045	1.0049	1.0053	1.0056
90	1.0040	1.0040	1.0041	1.0044	1.0048	1.0052	1.0056
87.5	1.0039	1.0039	1.0040	1.0043	1.0047	1.0051	1.0055
85	1.0038	1.0038	1.0040	1.0042	1.0046	1.0050	1.0054
82.5	1.0037	1.0037	1.0039	1.0042	1.0045	1.0049	1.0053
80	1.0036	1.0036	1.0038	1.0041	1.0045	1.0049	1.0052

The function $A_{ijk}(T)$ then refers to the molecules considered three at a time. The terms in x are the mole fractions of the constituents, the mole fraction being defined by

$$x_a = \frac{n_a}{n_a + n_w} \tag{2.35}$$

where n_a is the number of moles of constituent a and n_w is the number of moles of constituent w.

Also, since pressure is proportional to the mass of a gas (see equation (2.8)) it is possible to express the mole fraction in terms of the partial pressures of dry air (p_a) and water vapour (p_w):

$$x_a = \frac{p_a}{(p_a + p_w)} \tag{2.36}$$

If x_a is the mole fraction of constituent a then $(1 - x_a)$ is the mole fraction of constituent w and equation (2.34) can be rephrased as

$$pv = RT - [x_a^2 A_{aa} + x_a(1 - x_a)2A_{aw} + (1 - x_a)^2 A_{ww}]p - [(1 - x_a)^3 A_{www}]p^2 \tag{2.37}$$

where the subscripts 'a' and 'w' refer to dry air and water vapour, respectively. The third virial coefficient, A_{www}, is a function of the reciprocal of absolute temperature and is insignificant for temperatures below 60°C. Other third order terms are ignored.

If equation (2.37) is to be used to determine the specific volume of dry air then p_a and R_a replace p and R. It then becomes

$$v = \frac{R_a T}{p_a} - [x_a^2 A_{aa} + x_a(1 - x_a)2A_{aw} + (1 - x_a)^2 A_{ww}]$$

$$= \frac{82.0567 \times 101.325}{28.966} \cdot \frac{T}{(p_{at} - p_s)} - [x_a^2 A_{aa} + x_a(1 - x_a)2A_{aw} + (1 - x_a)^2 A_{ww}] \tag{2.38}$$

For water vapour mixed with dry air

$$x_w = \frac{n_w}{n_a + n_w} \qquad\qquad \text{by equation (2.35)}$$

but

$$g = \frac{n_w M_w}{n_a M_a} = \frac{0.621\,97 n_w}{n_a}$$

therefore

$$x_w = \frac{g}{0.621\,97 + g} \qquad (2.39)$$

Table 2.2 Virial and interaction coefficients for moist air

t (°C)	A_{aa}	A_{aa}	A_{ww}	A_{aw}
0	4.540×10^{-4}	4.56×10^{-4}	6.318×10^{-2}	1.45×10^{-3}
10	3.728×10^{-4}	3.76×10^{-4}	5.213×10^{-2}	1.36×10^{-3}
20	2.983×10^{-4}	3.04×10^{-4}	4.350×10^{-2}	1.27×10^{-3}
30	2.296×10^{-4}	2.37×10^{-4}	3.708×10^{-2}	0.19×10^{-3}
40	1.663×10^{-4}	1.76×10^{-4}	3.190×10^{-2}	0.12×10^{-3}

All the coefficients are in $m^3\ kg^{-1}$

The coefficient A_{aa} in the first column is due to Hyland and Wexler (1983) but the coefficient A_{aa} in the second column, and the coefficients A_{ww} and A_{aw}, are due to Goff (1949).

The enthalpy of dry air increases very slightly as the pressure falls at constant temperature but, for most practical purposes, it can be regarded as a constant value at any given temperature from one standard atmosphere down to 80 kPa. It follows that the most significant effect of a change in barometric pressure on the enthalpy of moist air lies in the alteration of the moisture content. A revised expression for the enthalpy of moist air is

$$h = h_a + \left[\frac{0.621\,97 f_s p_{ss}}{p_{at} - f_s p_{ss}}\right] h_g \qquad (2.40)$$

EXAMPLE 2.15

Calculate the enthalpy of moist air at 20°C and 82.5 kPa (*a*) when saturated, and (*b*) at 50 per cent saturation.

Answer

(*a*) From CIBSE tables, $h_a = 20.11$ kJ kg^{-1} and $p_{ss} = 2.337$ kPa at 20°C. From *NEL Steam Tables* (1964) or from the *CIBSE Guide* (1986) by interpolation $h_g = 2537.54$ kJ kg^{-1} at 20°C saturated. From Table 2.1 $f_s = 1.0039$ at 20°C and 82.5 kPa. If it is assumed that the liquid from which the steam was evaporated into the atmosphere was under a pressure of 82.5 kPa then an addition of $0.08 \times 82.5/101.325 = 0.07$ kJ kg^{-1} should be made to the NEL value for h_g giving a new figure of 2537.61 kJ kg^{-1}. Then

$$h = 20.11 + \frac{(0.62197 \times 1.0039 \times 2.337 \times 2537.61)}{(82.5 - 1.0039 \times 2.337)}$$

$$= 20.11 + 46.20$$

$$= 66.31 \text{ kJ } kg^{-1}$$

The *CIBSE Guide* (1986) gives a table of additive values to be applied to enthalpies read from the psychrometric tables for 101.325 kPa at any value of the temperature of adiabatic saturation. From this table the additive value at 20°C adiabatic saturation temperature and 82.5 kPa is 8.77 kJ kg^{-1}. The CIBSE psychrometric tables for 101.325 kPa and 20°C saturated quote 57.55 kJ kg^{-1}. Hence the value to be compared with our result is 66.32 kJ kg^{-1}.

(*b*) By equation (2.33)

$$g_{ss} = \frac{0.621\,97 \times 1.0039 \times 2.337}{(82.5 - 1.0039 \times 2.337)}$$
$$= 0.018\,205 \text{ kJ kg}^{-1}$$

By equations (2.20) and (2.32):

$$h = 20.11 + 0.5 \times 0.018\,205 \times 2537.61$$
$$= 43.21 \text{ kJ kg}^{-1}$$

2.20   The triple point

Figure 2.10 shows the relationship between the three phases of water; solid, liquid

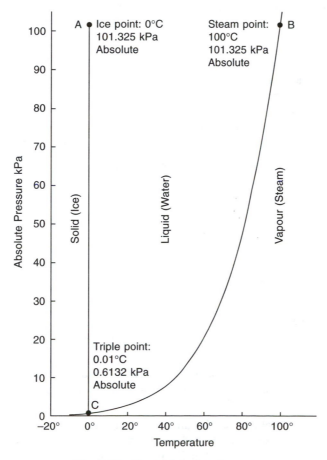

Fig. 2.10   The triple point of water.

and vapour, as changes of pressure and temperature occur. The triple point at 0.01°C (273.16 K) and 613.2 Pa is defined by international agreement as the temperature at which the solid, liquid and vapour forms of water can exist together in saturated equilibrium. As a matter of interest, a consequence of this definition for the Celsius scale of temperature is that the ice point and the steam point are determined experimentally and are not precisely 0°C (273.15 K) and 100°C (373.15 K), respectively.

We see in the figure that it is not possible for liquid water to exist at a pressure below that of the triple point. We also see that, because the solid–liquid line is very nearly vertical, liquid water temperatures less than 0°C are only possible at very high pressures.

Exercises

1. (*a*) Air at a condition of 30°C dry-bulb, 17°C wet-bulb and a barometric pressure of 105 kPa enters a piece of equipment where it undergoes a process of adiabatic saturation, the air leaving with a moisture content of 5 g kg^{-1} higher than entering. Calculate, using the data below, and the ideal gas laws,

(i)   the moisture content of the air entering the equipment,
(ii)   the dry-bulb temperature and enthalpy of the air leaving the equipment.

(*b*) Using the data below, calculate the moisture content of air at 17°C dry-bulb and 40 per cent saturation where the barometric pressure is 95 kPa.

Sat. vapour pressure at 17°C	= 1.936 kPa
Specific heat of dry air	= 1.015 kJ kg^{-1} K^{-1}
Specific heat of water vapour	= 1.890 kJ kg^{-1} K^{-1}
Latent heat of evaporation at 17°C	= 2459 kJ kg^{-1}
Latent heat of evaporation at 0°C	= 2500 kJ kg^{-1}
Constant for psychrometric equation for wet-bulb temperatures above 0°C	= 6.66 × 10^{-4}

Note: Psychrometric tables are not to be used for this question.

Answers

(*a*) (i) 6.14 g kg^{-1}, (ii) 18°C, 46.5 kJ kg^{-1}.
(*b*) 5.18 g kg^{-1}.

2. (*a*) Define the following psychrometric terms

(i)   Vapour pressure
(ii)   Relative humidity
(iii) Humid volume
(iv)   Dew point

(*b*) Name two distinct types of instrument which are used to determine the relative humidity of atmospheric air and briefly explain the principle in each case.

3. (*a*) Calculate the enthalpy of moist air at 28°C dry-bulb, a vapour pressure of 1.926 kPa and a barometric pressure of 101.325 kPa, using the ideal gas laws as necessary.
(*b*) Explain how the value would alter if the barometric pressure only were reduced.

Answer

58.93 kJ kg^{-1}.

4. (*a*) Calculate the dew point of air at 28°C dry-bulb, 21°C wet-bulb and 87.7 kPa barometric pressure. Use CIBSE tables where necessary to determine saturation vapour pressure.
  (*b*) Calculate its enthalpy.

Answers

(*a*) 18.11°C, (*b*) 66.69 kJ per kg dry air.

5. (*a*) Distinguish between saturation vapour pressure and the vapour pressure of moist air at an unsaturated state.
  (*b*) Illustrate relative humidity by means of a $p - V$ diagram for steam.
  (*c*) What happens if air at an unsaturated state is cooled to a temperature below its dew point?
  (*d*) Calculate the relative humidity and percentage saturation of air at 101.325 kPa, 21°C dry-bulb and 14.5°C wet-bulb (sling). You may use the CIBSE tables of psychrometric data only to determine saturation vapour pressures.

Answers

(*a*) 48.7%, 48.1%.

Notation

Symbol	Description	Unit
A	surface area of a water droplet	m^2
$A_{ik}(T)$	second virial coefficient, molecules considered two at a time	m^3 kg^{-1}
$A_{ijk}(T)$	third virial coefficient, molecules considered three at a time	m^3 kg^{-1}
A_{aa}	second virial coefficient of dry air	m^3 kg^{-1}
A_{ww}	second virial coefficient of water vapour	m^3 kg^{-1}
A_{www}	third virial coefficient of water vapour	m^3 kg^{-1}
A_{aw}	interaction coefficient	m^3 kg^{-1}
A, B, C, D	constants	
A', B', etc.	constants	
a	constant	
c	specific heat of moist air	J kg^{-1} K^{-1}
c_a	specific heat of dry air	J kg^{-1} K^{-1}
c_s	specific heat of water vapour (steam)	J kg^{-1} K^{-1}
f_s	dimensionless coefficient	
g	moisture content	kg per kg dry air
g_{ss}	moisture content of saturated air	kg per kg dry air
g'_{ss}	moisture content of saturated air at a temperature t'	kg per kg dry air
H	enthalpy	kJ
h	specific enthalpy of moist air	kJ per kg dry air

h_a	enthalpy of 1 kg of dry air	kJ kg^{-1}
h_g	latent heat plus sensible heat in the water vapour associated with 1 kg of dry air	kJ kg^{-1}
h_s	enthalpy of 1 kg of water vapour at a temperature t, the temperature of the dry bulb	kJ kg^{-1}
h_c	coefficient of heat transfer through the gas film surrounding a water droplet, by convection	W m^{-2} K^{-1}
h_r	coefficient of heat transfer to a water droplet by radiation from the surrounding surfaces	W m^{-2} K^{-1}
h	$= h_c + h_r$	W m^{-2} K^{-1}
h_{fg}	latent heat of evaporation	kJ kg^{-1}
(Le)	Lewis number	
M	molecular mass	kg kmol^{-1}
M_a	molecular mass of dry air	kg kmol^{-1}
M_s	molecular mass of water vapour (steam)	kg kmol^{-1}
m	mass of gas	kg
m_a	mass of dry air	kg
m_s	mass of water vapour (steam)	kg
n_a	number of molecules of constituent a	
n_w	number of molecules of constituent w	
p	gas pressure	N m^{-2}, Pa or kPa
p_a	pressure of dry air	N m^{-2}, Pa or kPa
p_{at}	atmospheric pressure (barometric pressure)	N m^{-2}, Pa or kPa
p_s	pressure of water vapour (steam)	N m^{-2}, Pa or kPa
p_{ss}	saturation vapour pressure	N m^{-2}, Pa or kPa
p'_{ss}	saturation vapour pressure at the wet-bulb temperature	N m^{-2}, Pa or kPa
R_o	universal gas constant	J kmol^{-1} K^{-1}
R	particular gas constant	J kg^{-1} K^{-1}
R_a	particular gas constant of dry air	J kg^{-1} K^{-1}
R_s	particular gas constant of water vapour (steam)	J kg^{-1} K^{-1}
R_{ss}	particular gas constant of dry saturated steam	J kg^{-1} K^{-1}
T	absolute temperature	K
t	temperature	°C
t_c	critical temperature	°C
t_d	dew-point temperature	°C
t_{ss}	saturation temperature	°C
t'	wet-bulb temperature	°C
t^*	temperature of adiabatic saturation	°C
U	internal energy	kJ
v	specific volume	m^3 kg^{-1}
V	gas volume	m^3
V_a	volume of dry air	m^3
V_m	volume of 1 kmol	m^3
V_s	volume of steam	m^3
Z	altitude above sea level	m
x	mole fraction	
α	coefficient of diffusion	kg N^{-1}

| μ | percentage saturation | % |
| ϕ | relative humidity | % |

References

Apjohn, J. (1838): *J. Trans. Roy. Irish Acad.* **18**, part 1, 1.

ASHRAE (1997): *Handbook*, Fundamentals, SI Edition, 6.1.

CIBSE Guide (1986): *C1* Properties of humid air, and *C2* Properties of water and steam.

Goff, A.W. (1949): Standardisation of thermodynamic properties of moist air, *Trans. ASHVE*, **55**, 125.

Harrison, L.P. (1965): Fundamental concepts and definitions relating to humidity. In *Science and Industry* 3:289. A Wexler and W.H. Wildhack, eds, Reinhold Publishing Corp., New York.

Hyland, R.W. and Wexler, A. (1983): Formulations for the thermodynamic properties of the saturated phases of H_2O from 173.15 K to 473.15 K, *ASHRAE Trans.* **89**(2A), 500–519.

Jones, W.P. (1970): The psychrometric chart in SI units, *JIHVE*, **38**, 93.

Jones, W.P. (1994): A review of CIBSE psychrometry, *Building Serv. Eng. Res. Technol.* **15**, 4, 189–198.

National Engineering Laboratory Steam Tables (1964): HMSO.

NBS Circular 564 (1955): Tables of thermal properties of air, November.

Scientific American (1989), **261**, number 3, Sept., 32.

Threlkeld, J.J. (1962): *Thermal Environmental Engineering*, Prentice-Hall, New York, p. 175

Bibliography

1. J.A. Goff and S. Gratch, Thermodynamic properties of moist air, *Trans. ASHVE*, 1945, **51**, 125.
2. W.H. McAdams, *Heat Transmission*, McGraw-Hill, 1942.
3. D.G. Kern, *Process Heat Transfer*, McGraw-Hill, 1950.
4. W.H. Walker, W.K. Lewis, W.H. McAdams and E.R. Gilliland, *Principles of Chemical Engineering*, McGraw-Hill, 1937.
5. W. Goodman, *Air Conditioning Analysis with Psychrometric Charts and Tables*. The Macmillan Company, New York, 1943.
6. J.A. Goff and S. Gratch, Low-pressure properties of water from –160 to 212F. *Trans. ASHVE*, 1946, **52**, 95–122.
7. W.P. Jones and B.G. Lawrence, *New psychrometric data for air*, National College for Heating, Ventilating, Refrigeration and Fan Engineering, Technical Memorandum No. 11.
8. D.B. Spalding and E.H. Cole, *Engineering Thermodynamics,* Edward Arnold (Publishers) Ltd, 1961.
9. R.G. Wyllie and T. Lalas, The WMO Psychrometer, Tech. Paper, *CSIRO*, Division of Applied Physics, Australia, 1981.
10. NACA, Standard atmosphere—tables and data for altitudes to 65,000 feet, *NACA Report 1235*: 66, Washington DC, 1955.
11. L.C. Bull, Design and use of the new IHVE psychrometric chart, *JIHVE*, October 1964, **32**, 268.

3

The Psychrometry of Air Conditioning Processes

3.1 The psychrometric chart

This provides a picture of the way in which the state of moist air alters as an air conditioning process takes place or a physical change occurs. Familiarity with the psychrometric chart is essential for a proper understanding of air conditioning.

Any point on the chart is termed a *state point*, the location of which, at a given barometric pressure, is fixed by any two of the psychrometric properties discussed in chapter 2. It is customary and convenient to design charts at a constant barometric pressure because barometric pressure does not alter greatly over much of the inhabited surface of the earth. When the barometric pressure is significantly different from the standard adopted for the chart or psychrometric tables to hand, then the required properties can be calculated using the equations derived earlier.

The British standard is that adopted by the Chartered Institution of Building Services Engineers for their Tables of Psychrometric Data and for their psychrometric chart. It is 101.325 kPa. The American standard is also 101.325 kPa and this value is used by the American Society of Heating, Refrigeration and Air Conditioning Engineers. It is also, incidentally, the value adopted by the Meteorological Office in Great Britain.

It is worth noting though, that there are a few differences of expression in the British and American charts. The most important of these is, of course, in the datum used for the enthalpy of dry air. Two other minor points of difference are that the American chart expresses the temperature of adiabatic saturation and specific humidity, whereas the British chart uses wet-bulb temperature (sling) and moisture content (which is the same as specific humidity).

The psychrometric chart published by the CIBSE uses two fundamental properties, mass and energy, in the form of moisture content and enthalpy, as co-ordinates. As a result, mixture states lie on the straight line which joins the state points of the two constituents. Lines of constant dry-bulb temperature are virtually straight but divergent, only the isotherm for 30°C being vertical. The reason for this is that to preserve the usual appearance of a psychrometric chart, in spite of choosing the two fundamental properties as co-ordinates, the co-ordinate axes are oblique, not rectangular. Hence, lines of constant enthalpy are both straight and parallel, as are lines of constant moisture content. Since both these properties are taken as linear, the lines of constant enthalpy are equally spaced as are, also, the lines of constant moisture content. This is not true of the lines of constant humid volume and constant wet-bulb temperature, which are slightly curved and divergent. Since their curvature is only slight in the region of practical use on the chart, they can be regarded as straight without significant error resulting. In the sketches of psychrometric charts used throughout this text to illustrate changes of state, only lines of percentage saturation are shown curved. All others are shown straight, and dry-bulb isotherms are shown as vertical, for convenience.

The chart also has a protractor which allows the value of the ratio of the sensible heat gain to the total heat gain to be plotted on the chart. This ratio is an indication of the slope of the room ratio line (see section 6.4) and is of value in determining the correct supply state of the air that must be delivered to a conditioned space. The zero value for the ratio is parallel to the isotherm for 30°C because the enthalpy of the added vapour depends on the temperature at which evaporation takes place, it being assumed that most of the latent heat gain to the air in a conditioned room is by evaporation from the skin of the occupants and that their skin surface temperature is about 30°C.

Figure 3.1 shows a state point on a psychrometric chart.

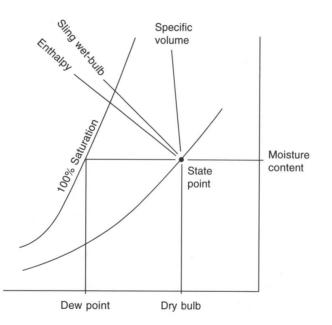

Fig. 3.1 The variables shown on the psychrometric chart.

3.2 Mixtures

Figure 3.2 shows what happens when two airstreams meet and mix adiabatically. Moist air at state 1 mixes with moist air at state 2, forming a mixture at state 3. The principle of the conservation of mass allows two mass balance equations to be written:

$m_{a1} + m_{a2} = m_{a3}$ for the dry air and

$g_1 m_{a1} + g_2 m_{a2} = g_3 m_{a3}$ for the associated water vapour

Hence

$$(g_1 - g_3)m_{a1} = (g_3 - g_2)m_{a2}$$

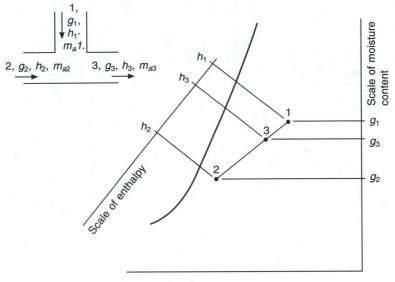

Fig. 3.2   The adiabatic mixing of two airstreams.

Therefore

$$\frac{g_1 - g_3}{g_3 - g_2} = \frac{m_{a2}}{m_{a1}}$$

Similarly, making use of the principle of the conservation of energy,

$$\frac{h_1 - h_3}{h_3 - h_2} = \frac{m_{a2}}{m_{a1}}$$

From this it follows that the three state points must lie on a straight line in a mass–energy co-ordinate system. The psychrometric chart published by the CIBSE is such a system—with oblique co-ordinates. For this chart then, a principle can be stated for the expression of mixture states. When two airstreams mix adiabatically, the mixture state lies on the straight line which joins the state points of the constituents, and the position of the mixture state point is such that the line is divided inversely as the ratio of the masses of dry air in the constituent airstreams.

EXAMPLE 3.1

Moist air at a state of 60°C dry-bulb, 32.1°C wet-bulb (sling) and 101.325 kPa barometric pressure mixes adiabatically with moist air at 5°C dry-bulb, 0.5°C wet-bulb (sling) and 101.325 kPa barometric pressure. If the masses of dry air are 3 kg and 2 kg, respectively, calculate the moisture content, enthalpy and dry-bulb temperature of the mixture.

Answer

From CIBSE tables of psychrometric data,

$$g_1 = 18.400 \text{ g per kg dry air}$$

$g_2 = 2.061$ g per kg dry air

$h_1 = 108.40$ kJ per kg dry air

$h_2 = 10.20$ kJ per kg dry air

The principle of the conservation of mass demands that

$$g_1 m_{a1} + g_2 m_{a2} = g_3 m_{a3}$$
$$= g_3(m_{a1} + m_{a2})$$

hence

$$g_3 = \frac{g_1 m_{a1} + g_2 m_{a2}}{m_{a1} + m_{a2}}$$

$$= \frac{18.4 \times 3 + 2.061 \times 2}{3 + 2}$$

$$= \frac{59.322}{5}$$

$$= 11.864 \text{ g per kg dry air}$$

Similarly, by the principle of the conservation of energy,

$$h_3 = \frac{h_1 m_{a1} + h_2 m_{a2}}{m_{a1} + m_{a2}}$$

$$= \frac{108.40 \times 3 + 10.20 \times 2}{3 + 2}$$

$$= 69.12 \text{ kJ per kg dry air}$$

To determine the dry-bulb temperature, the following practical equation must be used

$$h = (1.007t - 0.026) + g(2501 + 1.84t) \tag{2.24}$$

Substituting the values calculated for moisture content and enthalpy, this equation can be solved for temperature:

$$h = 69.12 = (1.007t - 0.026) + 0.011\,86(2501 + 1.84t)$$

$$t = \frac{39.48}{1.029} = 38.4°C$$

On the other hand, if the temperature were calculated by proportion, according to the masses of the dry air in the two mixing airstreams, a slightly different answer results:

$$t = \frac{3 \times 60 + 2 \times 5}{5}$$

$$= 38°C$$

This is clearly the wrong answer, both numerically and by the method of its calculation. However, the error is small, considering that the values chosen for the two mixing states spanned almost the full range of the psychrometric chart. The error is even less for states likely to be encountered in everyday practice and is well within the acceptable limits of accuracy. To illustrate this, consider an example involving the mixture of two airstreams at states more representative of common practice.

EXAMPLE 3.2

A stream of moist air at a state of 21°C dry-bulb and 14.5°C wet-bulb (sling) mixes with another stream of moist air at a state of 28°C dry-bulb and 20.2°C wet-bulb (sling), the respective masses of the associated dry air being 3 kg and 1 kg. With the aid of CIBSE tables of psychrometric data calculate the dry-bulb temperature of the mixture (*a*) using the principles of conservation of energy and of mass and, (*b*), using a direct proportionality between temperature and mass.

Answer

The reader is left to go through the steps in the arithmetic, as an exercise, using the procedure adopted in example 3.1.

The answers are (*a*) 22.76°C, dry-bulb, (*b*) 22.75°C dry-bulb.

The conclusion to be drawn is that the method used to obtain answer (*b*) is accurate enough for most practical purposes.

3.3   Sensible heating and cooling

Sensible heat transfer occurs when moist air flows across a heater battery or over the coils of a sensible cooler. In the heater, the temperature of the medium used to provide the heat is not critical. The sole requirement for heat transfer is that the temperature shall exceed the final air temperature. In sensible cooling there is a further restriction: the lowest water temperature must not be so low that moisture starts to condense on the cooler coils. If such condensation does occur, through a poor choice of chilled water temperature, then the process will no longer be one of sensible cooling since dehumidification will also be taking place. This complication will not be discussed further here but is dealt with in sections 3.4 and 10.6.

Figure 3.3 shows the changes of state which occur, sketched upon a psychrometric chart. The essence of both processes is that the change of state must occur along a line of constant moisture content. The variations in the physical properties of the moist air, for the two cases, are summarised below:

	Sensible heating	*Sensible cooling*
Dry-bulb	increases	decreases
Enthalpy	increases	decreases
Humid volume	increases	decreases
Wet-bulb	increases	decreases
Percentage saturation	decreases	increases
Moisture content	constant	constant
Dew point	constant	constant
Vapour pressure	constant	constant

EXAMPLE 3.3

Calculate the load on a battery which heats 1.5 m^3 s^{-1} of moist air, initially at a state of 21°C dry-bulb, 15°C wet-bulb (sling) and 101.325 kPa barometric pressure, by 20 degrees. If low temperature hot water at 85°C flow and 75°C return is used to achieve this, calculate the flow rate necessary, in kilograms of water per second.

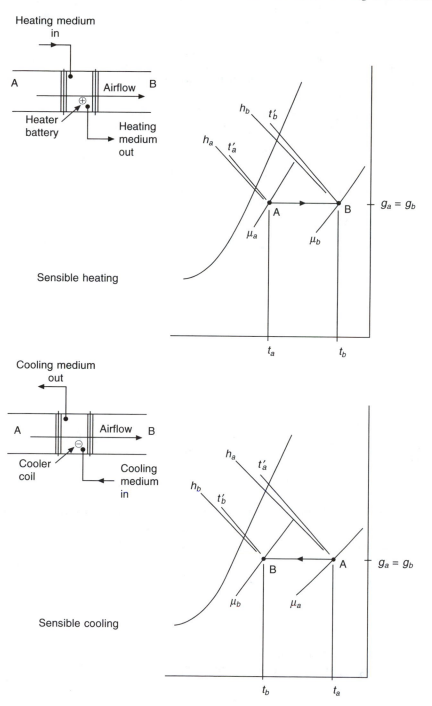

Fig. 3.3   Psychrometry for sensible heating and cooling.

Answer

$$\text{Heating load} = \begin{pmatrix} \text{mass flow of moist air expressed} \\ \text{in kg s}^{-1} \text{ of associated dry air} \end{pmatrix} \times$$

$$\begin{pmatrix} \text{increase in enthalpy of moist air expressed} \\ \text{in kJ per kg of associated dry air} \end{pmatrix}$$

From CIBSE tables of psychrometric data (or, less accurately, from the CIBSE psychrometric chart), the initial enthalpy is found to be 41.88 kJ kg^{-1}, the moisture content to be 8.171 g kg^{-1} and the humid volume to be 0.8439 m^3 per kg of dry air. Since the air is being heated by 20 degrees, reference must now be made to tables in order to determine the enthalpy at the same moisture content as the initial state but at a dry-bulb temperature of 41°C. By interpolation, the enthalpy of the moist air leaving the heater battery is found to be 62.31 kJ per kg of dry air.

$$\text{Heating load} = \left(\frac{1.5}{0.8439} \right) \times (62.31 - 41.88) = 36.3 \text{ kW}$$

$$\text{Flow rate of LTHW} = \frac{36.3}{(85° - 75°) \times 4.2} = 0.864 \text{ kg s}^{-1}$$

where 4.2 kJ/kg K is the specific heat capacity of water.

EXAMPLE 3.4

Calculate the load on a cooler coil which cools the moist air mentioned in example 3.3 by 5 degrees. What is the flow rate of chilled water necessary to effect this cooling if flow and return temperatures of 10°C and 15°C are satisfactory?

Answer

The initial enthalpy and humid volume are the same as in the first example. The final dry-bulb temperature of the moist air is 16°C but its moisture content is still 8.171 g per kg of dry air. At this state, its enthalpy is found from tables to be 36.77 kJ per kg of dry air.

$$\text{Cooling load} = \left(\frac{1.5}{0.8439} \right) \times (41.88 - 36.77) = 9.1 \text{ kW}$$

$$\text{Chilled water flow rate} = \frac{9.1}{5 \times 4.2} = 0.433 \text{ kg s}^{-1}$$

It is to be noted that, for sensible cooling, the selection of cooler coils and the choice of chilled water flow temperature require some care. See section 10.6.

3.4   Dehumidification

There are four principal methods whereby moist air can be dehumidified:

(i)   cooling to a temperature below the dew point,
(ii)   adsorption,

(iii)   absorption,
(iv)   compression followed by cooling.

The first method forms the subject matter of this section.

Cooling to a temperature below the dew point is done by passing the moist air over a cooler coil or through an air washer provided with chilled water.

Figure 3.4 shows on a sketch of a psychrometric chart what happens when moist air is cooled and dehumidified in this fashion. Since dehumidification is the aim, some of the spray water or some part of the cooler coil must be at a temperature less than the dew point of the air entering the equipment. In the figure, t_d is the dew point of the moist air 'on' the coil or washer. The temperature t_c, corresponding to the point C on the saturation curve, is termed the *apparatus dew point*. This term is in use for both coils and washers but, in the case of cooler coils alone, t_c is also sometimes termed the *mean coil surface temperature*. The justification for this latter terminology is offered in chapter 10.

For purposes of carrying out air conditioning calculations, it is sufficient to know the

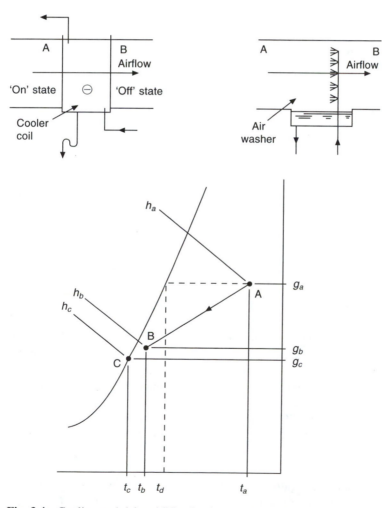

Fig. 3.4   Cooling and dehumidification by a cooler coil or an air washer.

state A of the moist air entering the coil, the state B of the air leaving the coil, and the mass flow of the associated dry air. What happens to the state of the air as it passes between points A and B is seldom of more than academic interest. Consequently, it is quite usual to show the change of state between the 'on' and the 'off' conditions as occurring along a straight line. In fact, as will be seen in chapter 10, the change of state follows a curved path, the curvature of which is a consequence of the heat transfer characteristics of the process, not of the construction of the psychrometric chart.

It can be seen from Figure 3.4 that the moisture content of the air is reduced, as also is its enthalpy and dry-bulb temperature. The percentage saturation, of course, increases. It might be thought that the increase of humidity would be such that the 'off' state, represented by the point B, would lie on the saturation curve. This is not so for the very good reason that no air washer or cooler coil is a hundred per cent efficient. It is unusual to speak of the efficiency of a cooler coil. Instead, the alternative terms, *contact factor* and *by-pass factor*, are used. They are complementary values and contact factor, sometimes denoted by β, is defined as

$$\beta = \frac{g_a - g_b}{g_a - g_c}$$

$$= \frac{h_a - h_b}{h_a - h_c} \tag{3.1}$$

Similarly, by-pass factor is defined as

$$(1 - \beta) = \frac{g_b - g_c}{g_a - g_c}$$

$$= \frac{h_b - h_c}{h_a - h_c} \tag{3.2}$$

It is sufficient, for all practical purposes, to assume that both these expressions can be rewritten in terms of dry-bulb temperature, namely that

$$\beta = \frac{t_a - t_b}{t_a - t_c} \tag{3.3}$$

and that

$$(1 - \beta) = \frac{t_b - t_c}{t_a - t_c} \tag{3.4}$$

It is much less true to assume that they can also be written, without sensible error, in terms of wet-bulb temperature, since the scale of wet-bulb values on the psychrometric chart is not at all linear. The assumption is, however, sometimes made for convenience, provided the values involved are not very far apart and that some inaccuracy can be tolerated in the answer.

Typical values of β are 0.82 to 0.92 for practical coil selection in the UK. In hot, humid climates more heat transfer surface is necessary and higher contact factors are common.

EXAMPLE 3.5

1.5 m³ s⁻¹ of moist air at a state of 28°C dry-bulb, 20.6°C wet-bulb (sling) and 101.325 kPa flows across a cooler coil and leaves the coil at 12.5°C dry-bulb and 8.336 g per kg of dry air.

Determine (*a*) the apparatus dew point, (*b*) the contact factor and (*c*) the cooling load.

Answer

Figure 3.5 shows the psychrometric changes involved and the values immediately known from the data in the question. From CIBSE tables (or from a psychrometric chart) it is established that $h_a = 59.06$ kJ kg^{-1} and that $h_b = 33.61$ kJ kg^{-1}. The tables also give a value for the humid volume at the entry state to the coil, of 0.8693 m^3 kg^{-1}.

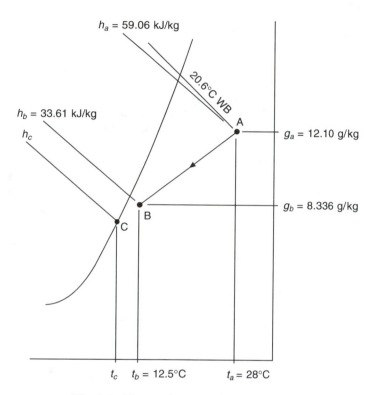

Fig. 3.5 The psychrometry for Example 3.5.

(*a*) Mark state points A and B on a psychrometric chart. Join them by a straight line and extend the line to cut the saturation curve. Observation of the point of intersection shows that this is at a temperature of 10.25°C. It is not easy to decide this value with any exactness unless a large psychrometric chart is used. However, it is sufficiently accurate for present purposes (and for most practical purposes) to take the value read from an ordinary chart. Hence the apparatus dew point is 10.25°C.

(*b*) Either from the chart or from tables, it can be established that the enthalpy at the apparatus dew point is 29.94 kJ kg^{-1}. Using the definition of contact factor,

$$\beta = \frac{59.06 - 33.61}{59.06 - 29.94} = 0.874$$

One might also determine this from the temperatures:

$$= \frac{28° - 12.5°}{28° - 10.25°} = 0.873$$

Clearly, in view of the error in reading the apparatus dew point from the chart, the value of β obtained from the temperatures is quite accurate enough in this example and, in fact, in most other cases.

(c) Cooling load = mass flow rate × decrease of enthalpy

$$= \frac{1.5}{0.8693} \times (59.06 - 33.61) = 43.9 \text{ kW}$$

3.5   Humidification

This, as its name implies, means that the moisture content of the air is increased. This may be accomplished by either water or steam but the present section is devoted to the use of water only, section 3.7 being used for steam injection into moving airstreams.

There are three methods of using water as a humidifying agent: the passage of moist air through a spray chamber containing a very large number of small water droplets; its passage over a large wetted surface; or the direct injection of water drops of aerosol size into the room being conditioned. (A variant of this last technique is to inject aerosol-sized droplets into an airstream moving through a duct.) Whichever method is used the psychrometric considerations are similar.

It is customary to speak of the humidifying efficiency or the effectiveness of an air washer (although neither term is universally accepted) rather than a contact or by-pass factor. There are several definitions, some based on the extent to which the dry-bulb temperature of the entering moist airstream approaches its initial wet-bulb value, and others based on the change of state undergone by the air. In view of the fact that the psychrometric chart currently in use by the Chartered Institution of Building Services Engineers is constructed with mass (moisture content) and energy (enthalpy) as oblique, linear co-ordinates, the most suitable definition to use with the chart is that couched in terms of these fundamentals. There is the further advantage that such a definition of effectiveness, E, is identical with the definition of contact factor, β, used for cooler coils.

Although humidifying efficiency is often expressed in terms of a process of adiabatic saturation, this is really a special case, and is so regarded here.

Figure 3.6(a) shows an illustration of the change of state experienced by an airstream as it passes through a spray chamber.

The effectiveness of the spray chamber is then defined by

$$E = \frac{h_b - h_a}{h_c - h_a} \tag{3.5}$$

and humidifying efficiency is defined by

$$\eta = 100E \tag{3.6}$$

It is evident that, because moisture content is the other linear co-ordinate of the psychrometric chart and the points A, B and C lie on the same straight line, C being obtained by the extension of the line joining AB to cut the saturation curve, then it is possible to put forward an alternative and equally valid definition of effectiveness, expressed in terms of moisture content:

$$E = \frac{g_b - g_a}{g_c - g_a} \tag{3.7}$$

and

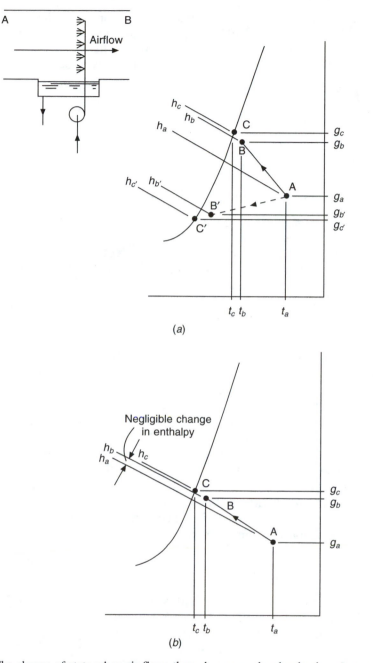

Fig. 3.6 (a) The change of state when air flows through a spray chamber having spray water at a controlled temperature. (b) Adiabatic saturation when the spray water is entirely recirculated and neither cooled nor heated.

$\eta = 100E$, as before

Figure 3.6(*a*) shows that h_b is greater than h_a. This implies that there is a heat input to the spray water being circulated through the spray chamber. Although it is not recommended it could be easily accomplished by means of, say, a calorifier in the return pipe from the washer tank. If the spray water had been chilled, instead of heated, then a change of state might have been as shown dotted on Figure 3.6(*a*), from the point A to the point B′. Under these circumstances, the effectiveness would have been expressed by

$$E = \frac{h_a - h_{b'}}{h_a - h_{c'}}$$

or

$$E = \frac{g_a - g_{b'}}{g_a - g_{c'}}$$

Consider the special case of adiabatic saturation: for this to occur it is necessary that

(i)   the spray water is totally recirculated, no heat exchanger being present in the pipelines or in the washer tank;
(ii)   the spray chamber, tank and pipelines are perfectly lagged; and
(iii)   the feed water supplied to the system to make good the losses due to evaporation is at the temperature of adiabatic saturation (see section 2.18).

Under these conditions it may be assumed that the change of state follows a line of constant wet-bulb temperature (since the Lewis number for air–water vapour mixtures is unity), the system having been given sufficient time to settle down to steady-state operation. There must be a change of enthalpy during the process because feed water is being supplied at the wet-bulb temperature (virtually) and this will not, as a general rule, equal the datum temperature for the enthalpy of water, 0°C. Strictly speaking then, it is incorrect to speak of the process as being an adiabatic one. The use of the term stems from the phrase 'temperature of adiabatic saturation' and so its use is condoned. One thing is fairly certain though: at the temperatures normally encountered the change of enthalpy during the process is neligible, and so effectiveness must be expressed in terms of change of moisture content. Figure 3.6(*b*) shows a case of adiabatic saturation, in which it can be seen that there is no significant alteration in enthalpy although there is a clear change in moisture content. It can also be seen that a fall in temperature accompanies the rise in moisture content. This temperature change provides an approximate definition of effectiveness or efficiency which is most useful, and, for the majority of practical applications, sufficiently accurate. It is

$$E \simeq \frac{t_a - t_b}{t_a - t_c} \tag{3.8}$$

The way in which the psychrometric chart is constructed precludes the possibility of this being an accurate expression; lines of constant dry-bulb temperature are not parallel and equally spaced, such properties being exclusive to enthalpy and moisture content.

EXAMPLE 3.6

1.5 m³ s⁻¹ of moist air at a state of 15°C dry-bulb, 10°C wet-bulb (sling) and 101.325 kPa barometric pressure, enters the spray chamber of an air washer. The humidifying efficiency of the washer is 90 per cent, all the spray water is recirculated, the spray chamber and the tank are perfectly lagged, and mains water at 10°C is supplied to make good the losses due to evaporation.

Calculate (*a*) the state of the air leaving the washer, (*b*) the rate of flow of make-up water from the mains.

Answer

(*a*) This is illustrated in Figure 3.7.

Using the definitions of humidifying efficiency quoted in equations (3.6) and (3.7), and referring to tables of psychrometric data for the properties of moist air at states A and C, one can write

$$\frac{90}{100} = \frac{g_b - 5.558}{7.659 - 5.558}$$

Hence,

$$g_b = 7.449 \text{ g per kg of dry air}$$

The state of the moist air leaving the air washer is 10°C wet-bulb (sling), 7.449 g per kg and 101.325 kPa barometric pressure. The use of equation (3.8) shows that the approximate dry-bulb temperature at exit from the washer is 10.5°C.

(*b*) The amount of water supplied to the washer must equal the amount evaporated. From tables (or less accurately from a psychrometric chart), the humid volume at state A is 0.8232 m³ kg⁻¹. Each kilogram of dry air passing through the spray chamber has its associated water vapour augmented by an amount equal to $g_b - g_a$, that is, by 1.891 g.

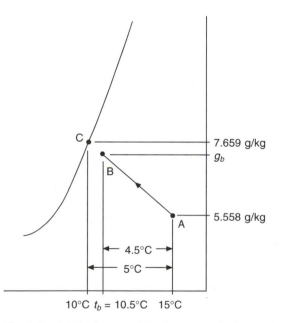

Fig. 3.7   Adiabatic saturation along a wet-bulb line.

Thus, the rate of make-up is

$$= \frac{1.5 \times 0.001\,891 \times 3600}{0.8232}$$

$$= 12.40 \text{ kg of water per hour}$$

It is to be noted that this is not equal to the pump duty; the amount of water the pump must circulate depends on the humidifying efficiency and, to achieve a reasonable value for this, the spray nozzles used to atomise the recirculated water must break up the water into very small drops, so that there will exist a good opportunity for an intimate and effective contact between the airstream and the water. A big pressure drop across the nozzles results from the atomisation, if this is to be adequate. Spray water pumps used with this type of washer have to develop pressures of the order of 2 bar, as a result.

3.6   Water injection

The simplest case to consider, and the one that provides the most insight into the change of state of the airstream subjected to humidification by the injection of water, is where all the injected water is evaporated. Figure 3.8 shows what happens when total evaporation occurs.

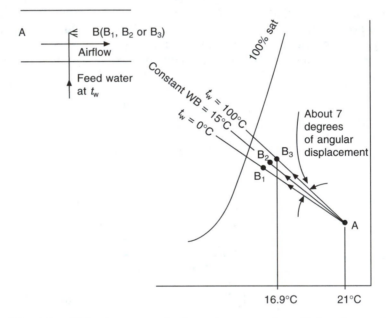

Fig. 3.8   A humidification process by the entire evaporation of injected spray water.

Air enters a spray chamber at state A and leaves it at state B, all the injected water being evaporated, none falling to the bottom of the chamber to run to waste or to be recirculated. The feed-water temperature is t_w. It is important to realise that since total evaporation has occurred, state B must lie nearer to the saturation curve, but just how much nearer will depend on the amount of water injected.

Two equations, a heat balance and a mass balance, provide the answer required.

Striking a heat balance, we can write

$$h_a + h_w = h_b$$

and, in a similar way, the mass balance may be written as

$$m_a + m_w = m_b$$

Knowing the amount of feed water evaporated, the mass balance gives the necessary information about the moisture content of the airstream leaving the area of the water injection. Applying the mass balance to the water vapour only (since the associated kilogram of dry air may be ignored),

$$g_b = g_a + m_w$$

where m_w is the amount of feed water evaporated in kg per kg of dry air flowing through the spray chamber.

Applying the heat balance:

$$h_a + h_w = h_b$$
$$= (1.007t_b - 0.026) + g_b(2501 + 1.84t_b) \qquad (2.24)$$

One thing is immediately apparent: if feed water is injected into the airstream at a temperature of 0°C there will be no alteration in the enthalpy of the airstream, since 0°C is the temperature datum of zero enthalpy for the water associated with 1 kg of dry air. Under these circumstances the change of state between A and B will follow a line of constant enthalpy.

One further conclusion may be drawn: if the feed is at a temperature equal to the wet-bulb of the airstream, the change of state will be along a line of constant wet-bulb temperature. This is a consequence of the fact that the Lewis number of air–water vapour mixtures at normally encountered temperatures and pressures, is virtually unity—as was discussed in sections 2.17 and 2.18.

To see what happens at other water temperatures, consider water at 100°C injected into a moving moist airstream and totally evaporated.

EXAMPLE 3.7

Moist air at a state of 21°C dry-bulb, 15°C wet-bulb (sling) and 101.325 kPa barometric pressure enters a spray chamber. If, for each kilogram of dry air passing through the chamber, 0.002 kg of water at 100°C is injected and totally evaporated, calculate the moisture content, enthalpy and dry-bulb temperature of the moist air leaving the chamber.

Answer

From CIBSE tables of psychrometric data,

$$h_a = 41.88 \text{ kJ per kg dry air}$$
$$g_a = 0.008\ 171 \text{ kg per kg dry air}$$

Since the feed water has a temperature of 100°C, its enthalpy is 419.06 kJ per kg of water injected, from CIBSE tables of properties of water at saturation.

Use of the equation for mass balance yields the moisture content of the moist air leaving the spray chamber:

$$g_b = 0.008\ 171 + 0.002$$
$$= 0.010\ 171 \text{ kg per kg dry air}$$

Use of the energy balance equation gives the enthalpy of the air leaving the chamber and hence, also, its dry-bulb temperature by equation (2.24)

$$h_b = 41.88 + 0.002 \times 418.06$$
$$= 42.716 \text{ kJ per kg dry air}$$
$$= (1.007t_b - 0.026) + 0.010\ 171(2501 + 1.84t_b)$$

thus,

$$42.716 = 1.007t_b - 0.026 + 25.44 + 0.0187t_b$$
$$t_b = 16.9°C$$

Reference back to Figure 3.8 shows a summary of what happens with different feed-water temperatures: change of state from A to B_1 is along a line of constant enthalpy and is for a feed-water temperature of 0°C; change of state from A to B_2 occurs when the water is injected at the wet-bulb temperature of the entry air and takes place along a line of constant wet-bulb temperature; and change of state from A to B_3 (the subject of example 3.7) is for water injected at 100°C. For all cases except that of the change A to B_1, an increase in enthalpy occurs which is a direct consequence of the enthalpy (and, hence, the temperature) of the injected water, provided this is all evaporated. In general, the condition line AB will lie somewhere in between the limiting lines, $AB_1(t_w = 0°C)$ and $AB_3(t_w = 100°C)$. The important thing to notice is that the angular displacement between these two condition lines is only about 7 degrees and that one of the intermediate lines is a line of constant wet-bulb temperature. It follows that for all practical purposes the change of state for a process of so-called adiabatic saturation may be assumed to follow a line of wet-bulb temperature. It is worth noting that although the process line on the psychrometric chart still lies within the 7° sector mentioned, the warmer the water the faster the evaporation rate, since temperature influences vapour pressure (see equations (4.2) and (4.3)).

Ultrasonic acoustic humidifiers are also used to atomise water. A piezo-electric crystal submerged in water converts a high frequency electronic signal into oscillations which drive particles of water from the surface into the airstream. The psychrometry is similar to that shown in Figure 3.8. As with all humidifiers that inject droplets into an airstream demineralised water must be used. Otherwise the salts of hardness and other pollutants remain in the humidified airstream as airborne solids after the droplets have evaporated and are delivered to the conditioned space.

The air washer is not used today for humidifying airstreams supplied to occupied spaces. It is not hygienic (see Pickering and Jones (1986)). It is also expensive in capital and maintenance costs, occupies a lot of space and is thermally inefficient when compared with a cooler coil. These objections are more than enough to offset the advantages it offers of giving thermal inertia to the system (with improved stability for dew-point control of the airstream leaving the washer) and simple 'free' cooling in winter.

3.7   Steam injection

As in water injection, steam injection may be dealt with by a consideration of a mass and energy balance. If m_s kg of dry saturated steam are injected into a moving airsteam of mass flow 1 kg of dry air per second, then we may write

$$g_b = g_a + m_s$$

and

$$h_b = h_a + h_s$$

If the initial state of the moist airstream and the condition of the steam is known, then the final state of the air may be determined, provided none of the steam is condensed. The change of state takes place almost along a line of constant dry-bulb temperature between limits defined by the smallest and largest enthalpies of the injected steam, provided the steam is in a dry, saturated condition. If the steam is superheated then, of course, the dry-bulb temperature of the airstream may increase by any amount, depending on the degree of superheat. The two limiting cases for dry saturated steam are easily considered. The lowest possible enthalpy is for dry saturated steam at 100°C. It is not possible to use steam at a lower temperature than this since the steam must be at a higher pressure than atmospheric if it is to issue from the nozzles. (Note, however, that steam could be generated at a lower temperature from a bath of warm water, which was not actually boiling.)

The other extreme is provided by the steam which has maximum enthalpy; the value of this is 2803 kJ kg^{-1} of steam and it exists at a pressure of about 30 bar and a temperature of about 234°C.

What angular displacement occurs between these two limits, and how are they related to a line of constant dry-bulb temperature? These questions are answered by means of two numerical examples.

EXAMPLE 3.8

Dry saturated steam at 100°C is injected at a rate of 0.01 kg s^{-1} into a moist airstream moving at a rate of 1 kg of dry air per second and initially at a state of 28°C dry-bulb, 11.9°C wet-bulb (sling) and 101.325 kPa barometric pressure. Calculate the leaving state of the moist airstream.

Answer

From psychrometric tables, $h_a = 33.11$ kJ per kg dry air,

$g_a = 0.001\ 937$ kg per kg

From NEL steam tables, $h_s = 2675.8$ kJ per kg steam.

$g_b = 0.001\ 937 + 0.01$
$\quad = 0.011\ 937$ kg per kg dry air, by the mass balance
$h_b = 33.11 + 0.01 \times 2675.8$
$\quad = 59.87$ kJ per kg dry air

Hence,

$59.87 = (1.007t_b - 0.026) + 0.011937(2501 + 1.84t_b)$
$\quad t_b = 29.2°C$

EXAMPLE 3.9

Dry saturated steam with maximum enthalpy is injected at a rate of 0.01 kg s^{-1} into a moist airstream moving at a rate of 1 kg of dry air per second and initially at a state of 28°C dry-bulb, 11.9°C wet-bulb (sling) and 101.325 kPa barometric pressure. Calculate the leaving state of the moist airstream.

Answer

The psychrometric properties of state A are as for the last example. The moisture content at state B is also as in example 3.8.

From NEL steam tables, h_s = 2803 kJ per kg of steam, at 30 bar and 234°C saturated. Consequently,

$$h_b = 33.11 + 0.01 \times 2803$$
$$= 61.14 \text{ kJ kg}^{-1}$$
$$g_b = 0.001\ 937 + 0.01$$
$$= 0.011\ 937 \text{ kg kg}^{-1} \text{ dry air}$$

Hence, as before:

$$61.14 = (1.007t_b - 0.026) + 0.011\ 937(2501 + 1.84t_b)$$
$$t_b = 30.4°C$$

Figure 3.9 illustrates what occurs. It can be seen that for the range of states considered, the change in dry-bulb value is not very great. In fact, the angular displacement between the two condition lines for the last two examples of steam injection is only about 3 or 4 degrees. We can conclude that, although there is an increase in temperature, it is within the accuracy usually required in practical air conditioning to assume that the change of state following steam injection is up a line of constant dry-bulb temperature.

In the injection of superheated steam, every case should be treated on its merits, as the equation for the dry-bulb temperature resulting from a steam injection process shows:

$$t_b = \frac{h_b + 0.026 - 2501g}{1.007 + 1.84g} \tag{3.9}$$

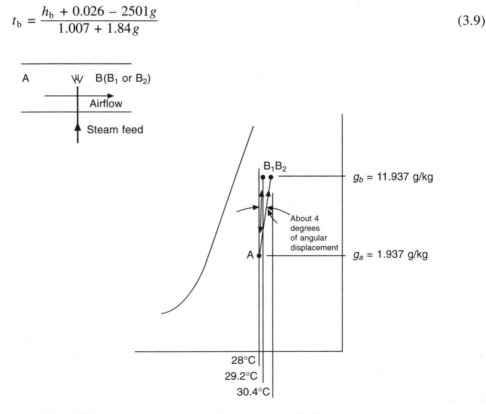

Fig. 3.9 A humidification process by the entire acceptance of injected dry saturated steam. See Examples 3.8 and 3.9.

The final enthalpy is what counts, but it is worth noting that the above expression holds for both steam and water injection.

3.8 Cooling and dehumidification with reheat

As was seen in section 3.4, when a cooler coil is used for dehumidification, the temperature of the moist air is reduced, but it is quite likely that under these circumstances this reduced temperature is too low. Although (as will be seen later in chapter 6), we usually arrange that under conditions of maximum loads, both latent and sensible, the state of the air leaving the cooler coil is satisfactory. This is not so for partial load operation. The reason for this is that latent and sensible loads are usually independent of each other. Consequently, it is sometimes necessary to arrange for the air that has been dehumidified and cooled by the cooler coil to be reheated to a temperature consistent with the sensible cooling load; the smaller the sensible cooling load, the higher the temperature to which the air must be reheated.

Figure 3.10(*a*) shows, in diagrammatic form, the sort of plant required. Moist air at a state A passes over the finned tubes of a cooler coil through which chilled water is flowing. The amount of dehumidification carried out is controlled by a dew point thermostat, C1, positioned after the coil. This thermostat regulates the amount of chilled water flowing through the coil by means of the three-way mixing valve R1. Air leaves the coil at state B, with a moisture content suitable for the proper removal of the latent heat gains occurring in the room being conditioned. The moisture content has been reduced from g_a to g_b and the cooler coil has a mean surface temperature of t_c, Figure 3.10(*b*) illustrating the psychrometric processes involved.

If the sensible gains then require a temperature of t_d, greater than t_b, the air is passed over the tubes of a heater battery, through which some heating medium such as low temperature hot water may be flowing. The flow rate of this water is regulated by means of a two-port modulating valve, R2, controlled from a thermostat C2 positioned in the room actually being air-conditioned. The air is delivered to this room at a state D. with the correct temperature and moisture content.

The cooling load is proportional to the difference of enthalpy between h_a and h_b, and the load on the heater battery is proportional to h_d minus h_b. It follows from this that part of the cooling load is being wasted by the reheat. This is unavoidable in the simple system illustrated, and is a consequence of the need to dehumidify first, and heat afterwards. It should be observed, however, that in general it is undesirable for such a situation to exist during maximum load conditions. Reheat is usually only permitted to waste cooling capacity under partial load conditions, that is, the design should be such that state B can adequately deal with both maximum sensible and maximum latent loads. These points are illustrated in the following example.

EXAMPLE 3.10

Moist air at 28°C dry-bulb, 20.6°C wet-bulb (sling) and 101.325 kPa barometric pressure flows over a cooler coil and leaves it at a state of 10°C dry-bulb and 7.046 g per kg of dry air.

(*a*) If the air is required to offset a sensible heat gain of 2.35 kW and a latent heat gain of 0.31 kW in a space to be air-conditioned, calculate the mass of dry air which must be supplied to the room in order to maintain a dry-bulb temperature of 21°C therein.

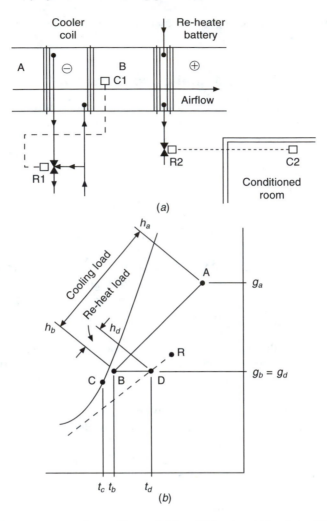

(a)

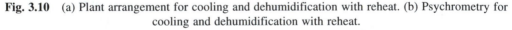

(b)

Fig. 3.10   (a) Plant arrangement for cooling and dehumidification with reheat. (b) Psychrometry for cooling and dehumidification with reheat.

(*b*)   What will be the relative humidity in the room?

(*c*)   If the sensible heat gain diminishes by 1.175 kW but the latent heat gain remains unchanged, at what temperature and moisture content must the air be supplied to the room?

Answer

(*a*) If m_a kg per s of dry air are supplied at a temperature of 10°C, they must have a sensible cooling capacity equal to the sensible heat gain, if 21°C is to be maintained in the room.
   Thus assuming the specific heat capacity of air is 1.012 kJ kg^{-1} K^{-1}

$$m_a \times c_a \times (21° - 10°) = 2.35 \text{ kW}$$

then,

$$m_a = \frac{2.35}{1.012 \times 11°} = 0.211 \text{ kg s}^{-1}$$

(*b*) 0.211 kg s^{-1} of dry air with an associated moisture content of 7.046 g per kg of dry air must take up the moisture evaporated by the liberation of 0.31 kW of latent heat. Assuming a latent heat of evaporation of, say, 2454 kJ per kg of water (at about 21°C), then the latent heat gain corresponds to the evaporation of

$$\frac{0.31 \text{ kJ s}^{-1}}{2454 \text{ kJ kg}^{-1}} = 0.000\ 126\ 3 \text{ kg of water per second}$$

The moisture associated with the delivery to the room of 0.211 kg s^{-1} of dry air will increase by this amount. The moisture picked up by each kg of dry air supplied to the room will be

$$\frac{0.1263}{0.211} = 0.599 \text{ g kg}^{-1}$$

Thus, the moisture content in the room will be equal to 7.046 + 0.599, that is, 7.645 g kg^{-1}. The relative humidity at this moisture content and 21°C dry-bulb is found from tables of psychrometric data to be 49.3 per cent. If the relevant data are taken from a psychrometric chart, instead of from psychrometric tables, then similar but slightly different results are obtained, of adequate practical accuracy in most cases. The percentage saturation at 21°C dry-bulb is then just under 49 per cent.

(*c*) If 0.211 kg s^{-1} of dry air is required to absorb only 1.175 kW of sensible heat then, if 21°C is still to be maintained, the air must be supplied at a higher temperature:

$$\text{temperature rise} = \frac{1.175}{1.012 \times 0.211} = 5.5 \text{ K}$$

Thus, the temperature of the supply air is 15.5°C.

Since the latent heat gains are unrelated, the air supplied to the room must have the same ability to offset these gains as before; that is, the moisture content of the air supplied must still be 7.046 g per kg of dry air.

EXAMPLE 3.11

(*a*) For sensible and latent heat gains of 2.35 and 0.31 kW, respectively, calculate the load on the cooler coil in example 3.10.

(*b*) Calculate the cooling load and the reheater load for the case of 1.175 kW sensible heat gains and unchanged latent gains.

Answer

(*a*) The load on the cooler coil equals the product of the mass flow of moist air over the coil and the enthalpy drop suffered by the air. Thus, as an equation:

$$\text{cooling load} = m_a \times (h_a - h_b)$$

The notation adopted is the same as that used in Figure 3.10. In terms of units, the equation can be written as

$$\text{kW} = \frac{\text{kg of dry air}}{\text{s}} \times \frac{\text{kJ}}{\text{kg of dry air}}$$

Using enthalpy values obtained from tables for the states A and B, the equation becomes

$$\text{cooling load} = 0.211 \times (59.06 - 27.81)$$
$$= 0.211 \times 31.25$$
$$= 6.59 \text{ kW}$$

(*b*) Since it is stipulated that the latent heat gains are unchanged, air must be supplied to the conditioned room at the same state of moisture content as before. That is, the cooler coil must exercise its full dehumidifying function, the state of the air leaving the coil being the same as in example 3.11(*a*), above. Hence the cooling load is still 6.59 kW. To deal with the diminished sensible heat gains it is necessary to supply the air to the room at a temperature of 15.5°C, as was seen in example 3.10(*c*). The air must, therefore, have its temperature raised by a reheater battery from 10°C to 15.5°C. Since the moisture content is 7.046 g kg^{-1} at both these temperatures, we can find the corresponding enthalpies at states B and D directly from a psychrometric chart or by interpolation (in the case of state D) from psychrometric tables.

Hence, we can form an equation for the load on the reheater battery:

$$\text{reheater load} = 0.211 \times (33.41 - 27.81) = 1.18 \text{ kW}$$

If the relative humidity in the room is unimportant there is no need to supply air to it at a controlled moisture content. This fact can be taken advantage of, and the running costs of the refrigeration plant which supplies chilled water to the cooler coil minimised. The method is to arrange for the room thermostat (C1 in Figure 3.11(*a*)) to exercise its control in sequence over the cooler valve and the reheater valve (R1a and R1b). As the sensible heat gain in the conditioned space reduces to zero, the three-way mixing valve, R1a, gradually opens its by-pass port, reducing the flow of chilled water through the cooler coil and, hence, reducing its cooling capacity. If a sensible heat loss follows on the heels of the sensible heat gain, the need arises for the air to be supplied to the room at a temperature exceeding that maintained there, so that the loss may be offset. The plant shown in Figure 3.11(*a*) deals with this by starting to open the throttling valve, R1b, on the heater battery.

The consequences of this are shown in Figure 3.11(*b*), as far as a partial cooling load is concerned. Because its sensible cooling capacity has been diminished by reducing the flow of chilled water through its finned tubes, the cooler coil is able to produce an 'off' state B′, at the higher temperature required by the lowered sensible heat gain, without having to waste any of its capacity in reheat. There is a penalty to pay for this economy of control: because of the reduction in the flow rate of chilled water, the characteristics of the coil performance change somewhat and the mean coil surface temperature is higher ($t_{c'}$ is greater than t_c), there is less dehumidification as well as less sensible cooling, and, as a result, the moisture content of the air supplied to the room is greater. In consequence, for the same latent heat gain in the room a higher relative humidity will be maintained, the room state becoming R′ instead of R.

EXAMPLE 3.12

If the plant in example 3.10 is arranged for sequence control over the cooler coil and heater battery, and if under the design conditions mentioned in that example the plant is able to maintain 21°C dry-bulb and 49.3 per cent relative humidity in the conditioned space, calculate the relative humidity that will be maintained there, under sequence control, if the sensible heat gains diminish to 1.175 kW, the latent gains remaining unaltered.

It is given that the mean coil surface temperature under the condition of partial load is 14°C. (See section 10.7 on the partial load performance of cooler coils.)

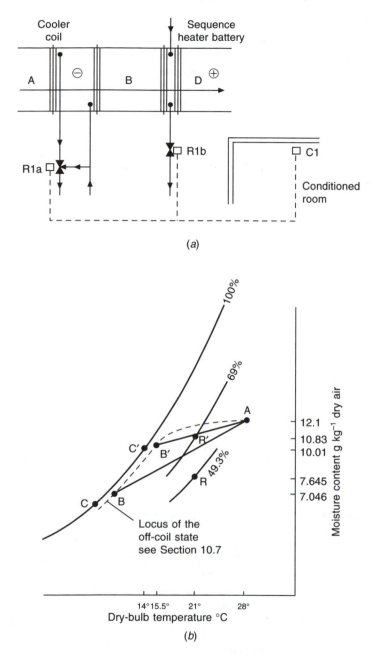

Fig. 3.11   (a) Plant arrangement for heating and cooling in sequence. (b) Psychrometry for heating and cooling in sequence, for Example 3.12.

Answer

As was mentioned in section 3.4, the mean coil surface temperature (identified by the state point C), lies on the straight line joining the cooler coil 'on' and 'off' states (points A and B) where it cuts the saturation curve. Figure 3.11(*b*) illustrates this, and also shows that the moisture content of state B′ is greater than that of B. It is possible to calculate the moisture content of the supply air by assuming that the dry-bulb scale on the psychrometric chart is linear. By proportion, we can then assess a reasonably accurate value for the moisture content of the air leaving the cooler coil under the partial load condition stipulated in this example:

$$\frac{g_a - g_{b'}}{g_a - g_{c'}} = \frac{t_a - t_{b'}}{t_a - t_{c'}}$$

From tables, or from a psychrometric chart, $g_a = 12.10$ g kg^{-1} and $g_{c'} = 10.01$ g kg^{-1}. Also, it was calculated in example 3.10(b) that the supply temperature must be 15.5°C dry-bulb for the partial sensible load condition.

Hence we can write

$$g_{b'} = g_a - (g_a - g_{c'}) \times \frac{(t_a - t_{b'})}{(t_a - t_{c'})}$$

$$= 12.10 - (12.10 - 10.01) \times \frac{(28 - 15.5)}{(28 - 14)} = 10.23 \text{ g kg}^{-1}$$

Since the moisture pick-up has been previously calculated as 0.599 g kg^{-1}, the moisture content in the conditioned room will be 10.83 g kg^{-1}; at 21°C dry-bulb the relative humidity will be about 69 per cent.

Sometimes, to deal with the possibility of high relative humidities at partial load conditions, a high limit humidistat (not shown in Figure 3.11(a)) is located in the treated room, with a set point of, say, 60% ± 5%. Upon rise in humidity to above the upper limit, the humidistat breaks into the temperature control sequence and regulates the cooler coil capacity through Rla (Figure 3.11(a)) in order to dehumidify the air supplied to the room, while allowing temperature control to be retained by C1 through R1b. Thus, for a short time, until the room humidity falls to the lower limit of the humidistat setting, the system operates with conventional reheat. After the humidity has been reduced, the system reverts to the control of the cooler coil and heater battery in sequence.

3.9   Pre-heat and humidification with reheat

Air conditioning plants which handle fresh air only may be faced in winter with the task of increasing both the moisture content and the temperature of the air they supply to the conditioned space. Humidification is needed because the outside air in winter has a low moisture content, and if this air were to be introduced directly to the room there would be a correspondingly low moisture content there as well. The low moisture content may not be intrinsically objectionable, but when the air is heated to a higher temperature its relative humidity may become very low. For example, outside air in winter might be at −1°C, saturated (see chapter 5). The moisture content at this state is only 3.484 g per kg of dry air. If this is heated to 20°C dry-bulb, and if there is a moisture pick-up in the room of 0.6 g per kg of dry air, due to latent heat gains, then the relative humidity in the room will be as low as 28 per cent. This value may sometimes be seen as too low for comfort. The plant must increase the temperature of the air, either to the value of the room temperature if there is background heating to offset fabric losses, or to a value in excees of this if it is intended that the air delivered should deal with fabric losses.

The processes whereby the moisture content of air may be increased were discussed in section 3.5. The actual method chosen depends on the application but, for comfort air conditioning or for any application where people are present in the conditioned space, the use of an air washer or any method involving the recirculation of spray water, or the exposure of a wetted surface area to the airstream, is not recommended. This is because of the risk to health caused by the presence of micro-organisms in the water. The use of dry steam injection is much preferred.

As an exercise in psychrometry, it is worth considering the case where an air washer is used. If 100 per cent fresh air is handled in cold weather it is pre-heated, passed through an air washer where it undergoes adiabatic saturation, and reheated to the temperature at which it must be supplied to the room. Pre-heating and adiabatic saturation permit the relative humidity in the room to be controlled, and reheating allows the temperature therein to be properly regulated, in winter.

Figure 3.12(*a*) shows, in a diagrammatic form, a typical plant. Opening the modulating valve R1 in the return pipeline from the pre-heater increases the heating output of the battery and provides the necessary extra energy for the evaporation of more water in the washer, if an increase in the moisture content of the supply air is required. Similarly, opening the control value R2, associated with the reheater, allows air at a higher temperature to be delivered to the room being conditioned. C1 and C2 are a room humidistat and a room thermostat, respectively.

EXAMPLE 3.13

The plant shown in Figure 3.12(*a*) operates as illustrated by the psychrometric changes in Figure 3.12(*b*).

Air is pre-heated from −5.0°C dry-bulb and 86 per cent saturation to 23°C dry-bulb. It is then passed through an air washer having a humidifying efficiency of 85 per cent and using recirculated spray water. Calculate the following:

(*a*)   The relative humidity of the air leaving the washer
(*b*)   The cold water make-up to the washer in litres s^{-1}, given that the airflow rate leaving the washer is 2.5 m^3 s^{-1}
(*c*)   The duty of the pre-heater battery in kW
(*d*)   The temperature of the air supplied to the conditioned space if the sensible heat losses from it are 24 kW and 20°C dry-bulb is maintained there
(*e*)   The duty of the reheater battery
(*f*)   The relative humidity maintained in the room if the latent heat gains therein are 5 kW

Answer

(*a*) At state O the moisture content is found from tables to be 2.137 g kg^{-1}. Consequently, at 23°C dry-bulb and 2.137 g kg^{-1} the wet-bulb (sling) value is found from tables to be 10°C. This state is represented by the point A in the diagram. Assuming that the process of adiabatic saturation occurs up a wet-bulb line, we can establish from tables that the moisture content at state C, the apparatus dew point, is 7.659 g kg^{-1}. If the further assumption is made that the dry-bulb scale is linear, we can evaluate the dry-bulb temperature at state B, leaving the washer, by proportion:

$$t_b = 23 - 0.85(23 - 10)$$
$$= 23 - 11.05$$
$$= 12°C \text{ dry-bulb}$$

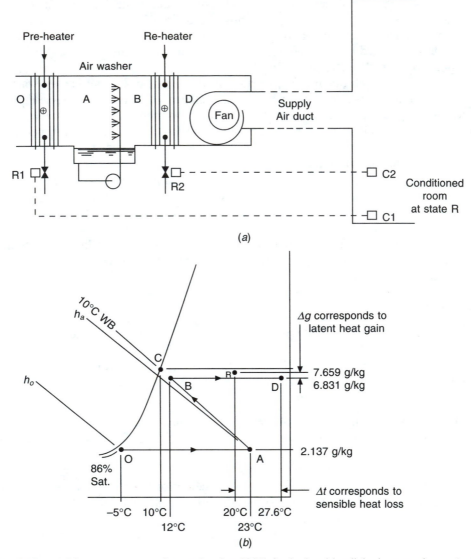

Fig. 3.12   (a) Plant arrangement for pre-heating 100% fresh air with adiabatic saturation and reheat. (b) The psychrometry for Example 3.13.

We know that the dry-bulb temperature at the point C is 10°C because the air is saturated at C and it has a wet-bulb temperature of 10°C.

The above answer for the dry-bulb at B is approximate (because the dry-bulb scale is not linear) but accurate enough for all practical purposes. On the other hand, we may calculate the moisture content at B with more exactness, since the definition of humidifying efficiency is in terms of moisture content changes and this scale is linear on the psychrometric chart.

$$g_b = 7.659 - (1 - 0.85) \times (7.659 - 2.137)$$
$$= 7.659 - 0.828$$
$$= 6.831 \text{ g kg}^{-1}$$

We can now refer to tables or to a chart and determine that the relative humidity at 12°C dry-bulb and 6.831 g kg^{-1} is 78 per cent. Percentage saturation is virtually the same as relative humidity. More accurately, we might perhaps determine the humidity at a state of 10°C wet-bulb (sling) and 6.831 g kg^{-1}. The yield in accuracy is of doubtful value and the method is rather tedious when using tables.

(*b*) The cold water make-up depends on the mass flow of dry air. It may be determined that the humid volume of the air leaving the washer at state B is 0.8162 m^3 kg^{-1} of dry air. The cold water fed from the mains to the washer serves to make good the losses due to evaporation within the washer. Since the evaporation rate is $(6.831 - 2.137)$ g kg^{-1}, we may calculate the make-up rate:

$$\text{make-up rate} = \frac{(6.831 - 2.137) \times 2.5}{1000 \times 0.8162}$$
$$= 0.0144 \text{ kg s}^{-1} \text{ or litres s}^{-1}$$

(*c*) The pre-heater must increase the enthalpy of the air passing over it from h_o to h_a. Reference to tables establishes that $h_o = 0.298$ kJ kg^{-1} and $h_a = 28.57$ kJ kg^{-1}. Then,

$$\text{pre-heater duty} = \frac{2.5 \times (28.57 - 0.298)}{0.8162}$$
$$= 3.063 \text{ kg s}^{-1} \times 28.27 \text{ kJ s}^{-1}$$
$$= 86.59 \text{ kW}$$

(*d*) 3.063 kg s^{-1} of air diffuses throughout the room and its temperature falls from t_d to 20°C as it offsets the heat loss. Assuming the specific heat capacity of water vapour is 1.89 kJ kg^{-1} K^{-1} and that of dry air 1.012 kJ kg^{-1} K^{-1}, a heat balance equation can be written:

$$\begin{aligned}
\text{Sensible heat loss} = \ &\text{(mass flow rate of dry air} \times \\
&\text{specific heat of dry air} + \text{associated} \\
&\text{moisture} \times \text{specific heat of water vapour)} \\
&\times \text{(supply air temperature} - \text{room temperature)} \\
24 = \ &3.063 \times (1.012 + 0.006\,831 \times 1.89) \times (t_d - 20) \\
= \ &3.063 \times 1.025 \times (t_d - 20)
\end{aligned}$$

whence $t_d = 27.6°C$

The term involving the specific heat of dry air plus the moisture content times the specific heat of water vapour, is termed the specific heat of humid air or the humid specific heat. In this case its value is 1.025 kJ kg^{-1} K^{-1}.

(*e*) The reheater battery thus has to heat the humid air from the state at which it leaves the washer to the state at which it enters the room, namely from 12°C dry-bulb and 6.831 g kg^{-1} to 27.6° dry-bulb and 6.831 g kg^{-1}.

$$\text{Reheater duty} = 3.063 \times 1.025 \times (27.6 - 12)$$
$$= 49 \text{ kW}$$

Alternatively, we can determine the enthalpies of the states on and off the heater, by interpolating from tables or reading directly from a psychrometric chart:

$$\text{reheater duty} = 3.063 \times (45.2 - 29.3)$$
$$= 48.7 \text{ kW}$$

(f) A mass balance must be struck to determine the rise in moisture content in the room as a consequence of the evaporation corresponding to the liberation of the latent heat gains:

latent heat gain in kW = (kg of dry air per hour delivered to the room)
× (the moisture pick-up in kg of water per kg of dry air)
× (the latent heat of evaporation of water in
kJ per kg of water)

$$5.0 = 3.063 \times (g_r - 0.006\ 831) \times 2454$$

whence $g_r = 0.007\ 496$ kg per kg dry air

From tables or from a chart it may be found that at a state of 20°C dry-bulb and 7.496 g per kg dry air, the relative humidity is about 51 per cent and, for use in Example 3.14(b), $h_r = 39.14$ kJ kg^{-1}.

3.10   Mixing and adiabatic saturation with reheat

Figure 3.13(a) shows a plant arrangement including an air washer which, although undesirable for comfort air conditioning, is retained here to illustrate the psychrometry involved. Air at state R is extracted from a conditioned room and partly recirculated, the remainder being discharged to atmosphere. The portion of the extracted air returned to the air conditioning plant mixes with air at state O, drawn from outside, and forms a mixture state M. The air then passes through an air washer, the spray water of which is only recirculated and adiabatic saturation occurs, the state of the air changing from M to W (see Figure 3.13(b)) along a line of constant wet-bulb temperature (see sections 3.5 and 3.6). An extension of the line MW cuts the saturation curve at a point A, the apparatus dew point. To deal with a particular latent heat gain in the conditioned room it is necessary to supply the air to the room at a moisture content g_s, it being arranged that the difference of moisture content $g_r - g_s$, in conjunction with the mass of air delivered to the room, will offset the latent gain. In other words, the air supplied must be dry enough to absorb the moisture liberated in the room.

It is evident that the moisture content of the air leaving the washer must have a value g_w, equal to the required value, g_s. This is amenable to calculation by making use of the definition of the effectiveness of an air washer, in terms of g_a, g_w and g_m (see section 3.5).

EXAMPLE 3.14

If the room mentioned in example 3.13 is conditioned by means of a plant using a mixture of recirculated and fresh air, of the type illustrated in Figure 3.13(a), calculate:

(*a*)   the percentage of the air supplied to the room by mass which is recirculated, and
(*b*)   the humidifying efficiency of the air washer.

Answer

(*a*) Since the wet-bulb scale is not linear, it is not accurate enough to calculate the mixing proportions on this basis. Instead, one must make use of changes of enthalpy or moisture content. Bearing in mind that lines of constant enthalpy are not parallel to lines of constant wet-bulb temperature, some slight inaccuracy is still present if the assumption is made that the change of state accompanying a process of adiabatic saturation is along a line of constant enthalpy. However, this is unavoidable, and so such an assumption is made.

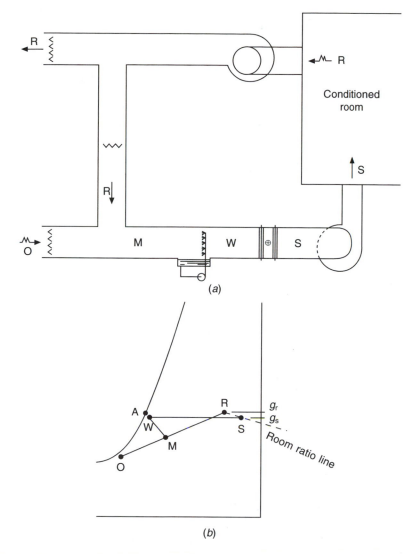

Fig. 3.13   (a) Plant arrangement to permit variable mixing proportions of fresh and recirculated air.
(b) Psychrometry for example 3.14.

Referring to Figure 3.13(*b*) it can be seen that

$$h_a \simeq h_w \simeq h_m$$

From tables and Figure 3.12(*b*) it is established that h_w (at 12°C dry-bulb and 6.831 g kg^{-1}) is 29.30 kJ kg^{-1}.

From the principles set out in section 3.2, governing the change of state associated with a mixing process, it is clear that the percentage of recirculated air, by mass,

$$= \frac{h_m - h_o}{h_r - h_o} \times 100$$

$$= \frac{29.30 - 0.298}{39.14 - 0.298} \times 100$$

$$= 75 \text{ per cent}$$

Thus, 75 per cent of the air supplied to the room, if recirculated and mixed with 25 per cent of air from outside, will have an enthalpy of 29.30 kJ kg^{-1} and a wet-bulb of 10°C (sling). If adiabatic saturation is then to produce a state of 12°C dry-bulb and 6.831 g kg^{-1}, the humidifying efficiency of the washer used can no longer be the value used for example 3.13, namely, 85 per cent. An entirely different washer must be used for the above calculations to be valid and this must have an effectiveness which may be calculated as follows:

(*b*) Since efficiency is expressed in terms of moisture content, it is necessary to determine the value of g_m, the values of g_a and g_w being already known.

$$g_m = 0.75 g_r + 0.25 g_o$$
$$= 0.75 \times 7.497 + 0.25 \times 2.137$$
$$= 6.157 \text{ g kg}^{-1}$$

$$\text{Humidifying efficiency} = \frac{g_w - g_m}{g_a - g_m} \times 100$$

$$= \frac{6.831 - 6.157}{7.659 - 6.157} \times 100$$

$$= 45 \text{ per cent}$$

In practical terms, this is a low efficiency.

If the washer used in this example had an efficiency of 85 per cent, as in example 3.13, then the calculations would not have been so easy. The line AWM would have had to have been at a lower wet-bulb value in order to fulfil two requirements:

(i) $g_w = 6.831$ g kg^{-1}

(ii) $\dfrac{g_w - g_m}{g_a - g_m} \times 100 = 85$ per cent

For this to be the case, the dry-bulb temperature of W must obviously be less than 10°C. The easiest way to achieve a practical solution is by drawing a succession of lines representing processes of adiabatic saturation on a psychrometric chart and calculating several values of efficiency until one of acceptable accuracy is obtained.

3.11   The use of dry steam for humidification

It is only if the air can accept the additional moisture, that humidification may be achieved by the injection of spray water or dry steam. If the state of the air into which it is proposed to inject moisture is saturated, or close to saturation, some or all of the moisture added will not be accepted and will be deposited downstream in the air handling plant or the duct system. It is therefore essential to ensure that the airstream is sufficiently heated prior to moisture injection.

EXAMPLE 3.15

A room is to be maintained at a state of 20°C dry-bulb and 50 per cent saturation by a plant handling 0.5 m^3 s^{-1} of outside air at a state of –2°C saturated. The airstream is heated to a temperature warm enough to offset a heat loss of 2.5 kW and dry steam is then injected to maintain the humidity required in the room. Calculate the supply air temperature and the heating and humidification loads. See Figure 3.14.

Answer

From psychrometric tables or a psychrometric chart, air at the outside state has a moisture content of 3.205 g kg^{-1}, an enthalpy of 5.992 kJ kg^{-1} and a specific volume of 0.7716 m^3 kg^{-1}. Assuming specific heats of 1.012 and 1.89 kJ kg^{-1} K^{-1}, respectively, for dry air and water vapour, the humid specific heat of the fresh air handled is

$$1 \times 1.012 + 0.003\,205 \times 1.89 = 1.018 \text{ kJ kg}^{-1} \text{ K}^{-1}$$

The heat loss is offset by the supply of air at a temperature t_s, warmer than the room temperature of 20°C. Hence

$$2.5 = (0.5/0.7716) \times 1.018 \times (t_s - 20)$$

whence $t_s = 23.8$°C.

From tables or a chart the enthalpy at state B, leaving the heater battery, is 32.10 kJ kg^{-1}. See Figure 3.14. Hence the heater battery load is

$$(0.5/0.7716) \times (32.10 - 5.992) = 16.92 \text{ kW}$$

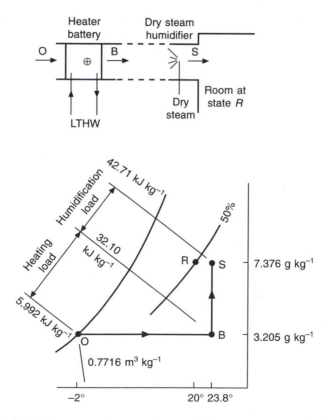

Fig. 3.14 The use of dry steam for humidification. See Example 3.15.

Alternatively

$$(0.5/0.7716) \times 1.018 \times (23.8 + 2) = 17.02 \text{ kW}$$

The first method is preferred because enthalpy values are based on well-established thermodynamic properties of dry air and water vapour whereas, on the other hand, the values of the specific heats used for the second method may not be exactly correct.

Assuming the change of state resulting from the injection of dry saturated steam is up a dry-bulb line, the state of the air supplied to the room is 23.8°C dry-bulb and 7.376 g kg^{-1}, at which the enthalpy is determined as 42.71 kJ kg^{-1} from tables or a chart. Hence the humidification load is

$$(0.5/0.7716) \times (42.71 - 32.10) = 6.88 \text{ kW}$$

3.12 Supersaturation

When hot, humid air passes over a cooler coil with a sufficiently low mean coil surface temperature it is possible for the state of the air leaving the coil to be momentarily supersaturated, as point B in Figure 3.15 shows. Such a state is unstable and the presence of condensation nuclei causes an immediate reversion to a stable, saturated state (point E in Figure 3.15). No heat is supplied to or rejected from the system and no work is done. Hence the enthalpy at a notional state D equals that at B. However, moisture is condensed out of the airstream and lost, representing a liquid enthalpy drop, above a datum of 0°C. Hence the enthalpy of the saturated air at state E must be less than that at D, by this amount. Thus we have

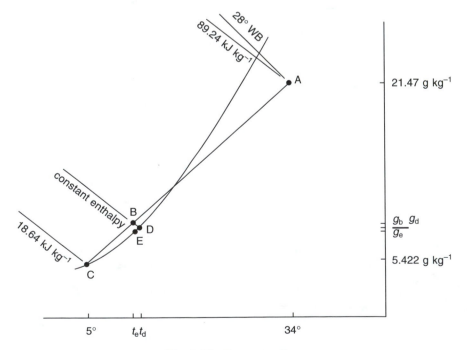

Fig. 3.15 Supersaturation.

$$h_b = h_d = h_e + (g_b - g_e) \times c \times (t_e - 0)$$

whence

$$h_e = h_b - (g_b - g_e)ct_e \tag{3.10}$$

where c is the specific heat capacity of liquid water.

The reduction in moisture content from g_b to g_e appears as a fog of liquid droplets, conveyed by the airstream leaving the cooler coil. Operational conditions for this exist in suitably warm and humid climates, as in the Gulf or on the West African coast, where room air conditioning units may sometimes be seen emitting plumes of mist.

EXAMPLE 3.16

Air at 34°C dry-bulb, 28°C wet-bulb (sling) enters a cooler coil having a mean coil surface temperature of 5°C and a contact factor of 0.75. Determine the unstable, supersaturated state, the moisture condensed as a visible mist and the stable, saturated state of the air leaving the coil.

Answer

Plotting states A and C on a psychrometric chart shows that, with a contact factor of 0.75, state B is supersaturated. Hence we may calculate:

$$h_b = 0.25h_a + 0.75h_c = 0.25 \times 89.24 + 0.75 \times 18.64$$
$$= 36.29 \text{ kJ kg}^{-1}$$
$$g_b = 0.25 \times 21.47 + 0.75 \times 5.422 = 9.434 \text{ g kg}^{-1}$$
$$t_b = 0.25 \times 34 + 0.75 \times 5 = 12.25°C$$

At the notional state B, $h_d = h_b = 36.29$ kJ kg^{-1}. Interpolating in psychrometric tables at 100 per cent saturation yields $t_d = 12.830$°C and $g_d = 9.262$ g kg^{-1}. As a first approximation we may assume $g_e = g_d$ and $t_e = t_d$. Then using equation (3.10) and taking the specific heat capacity of water to be 4.168 kJ kg^{-1} K^{-1} we have

$$h_e = 36.29 - (0.009\ 434 - 0.009\ 262) \times 4.168 \times (12.83 - 0)$$
$$= 36.28 \text{ kJ kg}^{-1}$$

Interpolating again at saturation in psychrometric tables yields $t_e = 12.826$°C and $g_e = 9.260$ g kg^{-1}. Hence, the moisture condensed as a mist is

$$g_b - g_e = 9.434 - 9.260 = 0.174 \text{ g kg}^{-1}$$

as a reasonable approximation.

3.13   Dehumidification by sorption methods

To achieve very low dew points it is necessary to dehumidify by a process of absorption or adsorption, the former being a chemical change and the latter a physical one. In either case, the process of dehumidification is termed desiccation. Refrigeration plant is often still needed.

Dew points as low as −70°C can be achieved by sorption methods.

(a) Absorption

Most substances take up moisture from the atmosphere to some extent but some materials have the ability to absorb very large amounts of moisture from the ambient air. An example of this is lithium chloride, which can remove from the atmosphere up to 100 times its own dry weight, under suitable circumstances. The amount of moisture absorbed from air by an aqueous solution of lithium chloride depends on the strength of the concentration, the temperature of the solution, the vapour pressure of the ambient air and the extent to which the air comes in contact with the absorbent solution. The stronger the solution of lithium chloride in water, the more is its ability to absorb moisture. Conversely, moisture can be driven off from the solution and the concentration of lithium chloride increased, by raising its temperature until the solution vapour pressure exceeds that of the ambient air. If the strong solution is then cooled it may be used again to absorb moisture from the air. This provides the opportunity for a regeneration cycle.

Two methods are in common use to effect desiccation by absorption:

(i) Figure 3.16 illustrates the first of these methods. A strong solution of lithium chloride is pumped from a sump and is sprayed over the outside of the tubes of a cooler coil, fed with chilled water. Moisture is absorbed from the airstream and a weak solution drains back to the sump. The concentration in the sump is kept at an adequately high value by pumping a proportion of it (10 per cent to 20 per cent) to a re-activation unit, where it is sprayed over the outside of the tubes of a heater coil, fed with steam at a gauge pressure of from 0.2 to 1.7 bar, or an equivalent heating medium.

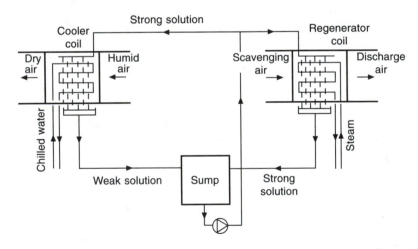

Fig. 3.16   Absorption process of drying using aqueous lithium chloride solution sprayed over a cooler coil.

The moisture content of the air supplied to the treated space is controlled by regulating the flow rate of chilled water through the cooling coil.

The heat removed by the cooling coil comprises the latent heat of condensation, plus the heat of solution, plus the sensible heat removed from the air, plus the heat added to the solution by the regeneration process. A sensible cooler coil may be needed after the desiccating unit.

(ii) The second method of desiccation by absorption uses a continuously rotating drum containing a material that has been impregnated with lithium chloride. Figure 3.17 illustrates this. A sector of the drum is used for continuous re-activation by passing high temperature air through it. This air is usually heated by means of steam or electricity. LTHW is inadequate.

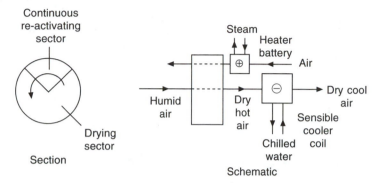

Fig. 3.17   Continuously re-activating, rotating absorption dehumidifier.

Control over the moisture content of the air supplied to the treated space is effected by regulating the energy input to the re-activation heater battery.

The heat removed by the desiccating sector of the drum consists of the latent heat of condensation plus the sensible heat conducted from the re-activation sector. A sensible cooler coil is usually required, after the rotary absorption drum, in the supply air to the treated space.

(b) Adsorption

This involves no chemical changes. The ability of solid surfaces to retain gaseous molecules in contact, is exploited by providing materials that have been manufactured specifically to have an enormously large internal surface area: one gram of some substances can have an internal surface area of more than 4000 m^2. Adsorbent materials are re-activated by heating them to a high temperature in order to drive off the moisture from the internal surfaces.

Typical substances are silica gel and activated alumina for drying, and activated carbon for the removal of other gases. Adsorbent materials can be tailor-made to take up specific gases, selectively.

EXAMPLE 3.17

1 m^3 s^{-1} of air at 30°C dry-bulb, 20°C wet-bulb (sling) is to be cooled and dehumidified to a state of 10°C dry-bulb and 4.70 g kg^{-1} by means of a continuously rotating, absorption dehumidifier. Making use of Figure 3.18 determine the sensible cooling load after the dehumidifier.

Answer

From psychrometric tables or a chart, the entering moisture content is found to be 10.38 g kg^{-1}, and the specific volume to be 0.8728 m^3 kg^{-1}. This value of moisture content is identified on the abscissa of Figure 3.18 and read upwards to the first curve, for an entering

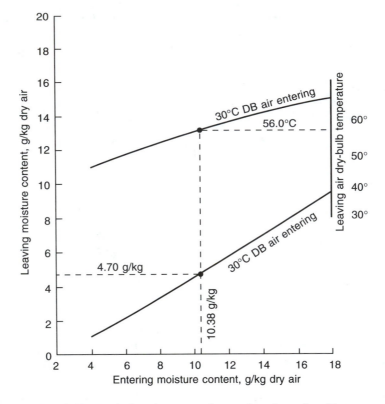

Fig. 3.18   Typical performance of a rotating absorption drier.

air dry-bulb temperature of 30°C. Reading across to the scale on the left-hand side gives a leaving moisture content of 4.7 g kg^{-1} dry air. (If a different leaving moisture content were wanted then reference would have to be made to another curve, not shown in the figure, for different operating conditions. Possibly a drier of different size would have to be considered.) Reference to the upper curve, for the same entering moisture content, gives a leaving dry-bulb temperature of 56°C.

Figure 3.19 illustrates the related psychrometry, the process being shown by the full lines; air enters the drier at state O and leaves it at state B. A sensible cooler coil is then used to cool the air from state B at 56° to state C at 10°. From tables or a chart the enthalpy at state B is determined as 68.60 kJ kg^{-1}. Similarly, at state C, the enthalpy is found to be 21.67 kJ kg^{-1}.

Hence the sensible cooling duty is

$$\frac{1}{0.8728} \times (68.60 - 21.67) = 53.77 \text{ kW}$$

An alternative approach would be first to cool and dehumidify the air by conventional means and then to pass it through the rotary drier. Sensible cooling is still required and the overall energy removal is the same because the enthalpies on and off the plant are unchanged but the economics might be different. This alternative process is shown by the broken lines in Figure 3.19: air is first cooled and dehumidified from O to W, then dried from W to E and finally sensibly cooled from E to C.

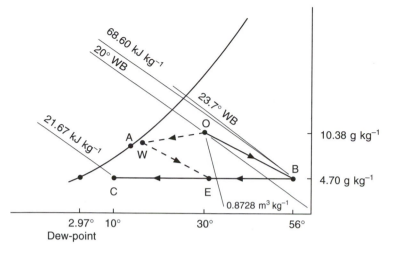

Fig. 3.19   Psychrometry for Example 3.17.

If an air washer is fitted in the airstream extracted from the conditioned room the air may be adiabatically cooled to a lower dry-bulb temperature. The air can then be passed through a rotary sensible heat exchanger in order to reduce the temperature of the hot, fresh airstream leaving the absorption dehumidifier without the use of mechanical refrigeration (see Figures 3.20(*a*) and (*b*)). Further mechanical refrigeration is certainly needed but it will be less than if the adiabatic cooling and sensible heat exchange described had not been adopted.

The air extracted from the conditioned space is hot after passing through the rotary sensible cooler and may be discharged to waste. However, the thermal efficiency of the process may be improved if it is fed through the heater battery used to regenerate the absorption dehumidifier.

Figure 3.20(*a*) shows the plant arrangement for this. Supply and extract fans are not shown in the figure, as a simplification. When such necessary fans are included, the air handled by them suffers a temperature rise of 1 K per kPa of fan total pressure if the driving motor is outside the airstream, or a rise of 1.2 K per kPa of fan total pressure if the driving motor is within the airstream. See section 6.5.

EXAMPLE 3.18

See Figures 3.20(*a*) and (*b*). It is assumed that the room treated by the plant considered in Example 3.17 has sensible and latent heat gains that give a state in the room (the point R in Figure 3.20(*b*)) of 20.5°C dry-bulb, 34 per cent saturation, 5.177 g kg^{-1} dry air and 12.0°C wet-bulb (sling) when 1 m^3 s^{-1} air is supplied at 10°C dry-bulb and 4.70 g kg^{-1} (the point S in Figure 3.20(*b*)). Air is extracted from the room, passed through an air washer having a humidifying efficiency of 90 per cent and undergoes adiabatic humidification from state R to state W. The air then flows through one side of a rotary sensible heat exchanger having an efficiency of 80 per cent. Air from the absorption dehumidifier, at 56°C dry-bulb and 4.70 g kg^{-1} dry air (state B), flows through the other side of the rotary sensible heat exchanger, leaving at state C. The air then passes through a sensible cooler coil and is supplied to the room at 10°C dry-bulb, 4.70 g kg^{-1} dry air (state S). The air

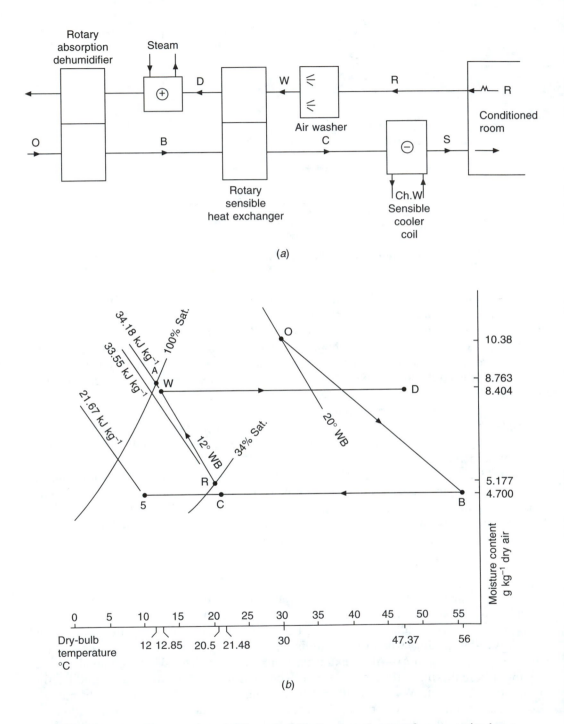

Fig. 3.20 (a) Plant arrangement for Example 3.18. Supply and extract fans are omitted as a simplification. (b) Psychometry for Example 3.18. Air temperature rises due to supply and extract fans are omitted as a simplification.

extracted from the room and leaving the washer at state W flows through the extract side of the rotary sensible heat exchanger, emerging at state D, with a high temperature. This air is fed through the steam heater battery used to regenerate the rotary absorption dehumidifier to improve the overall thermal efficiency of the operation.

Determine:

(a)   The state of the air leaving the washer (W).
(b)   The state of the dehumidified air leaving the supply side of the rotary sensible heat exchanger (C).
(c)   The state of the humidified extract air leaving the extract side of the rotary sensible cooler (D).
(d)   The load on the necessary sensible cooler coil.

Answer

(*a*) The adiabatic change of state through the air washer is along a wet-bulb line from 20.5°C dry-bulb, 12°C wet-bulb, and 5.177 g kg^{-1} (state R). From tables or a chart the moisture content at 12°C saturated is 8.763 g kg^{-1} dry air and its enthalpy is 34.18 kJ kg^{-1} dry air. Since the humidifying efficiency is 90 per cent the state leaving the washer (W) is determined as:

$$t = 20.5 - 0.9(20.5 - 12) = 12.85°C \text{ dry-bulb}$$
$$g = 5.177 + 0.9(8.763 - 5.177) = 8.404 \text{ g kg}^{-1} \text{ dry air}$$
$$h = 34.18 \text{ kJ kg}^{-1} \text{ dry air}$$

(because the points A and W are so close together their enthalpies are virtually the same).

(*b*) The dry-bulb temperature on the extract side of the rotary sensible heat exchanger receiving adiabatically cooled air from the washer at state W is 12.85°C. On the supply side of the rotary sensible heat exchanger air enters from the rotary absorption dehumidifier at 56°C dry-bulb and 4.70 g kg^{-1} dry air (state B). Since the rotary sensible heat exchanger is 80 per cent efficient the dehumidified air is cooled to:

$$56 - 0.80(56 - 12.85) = 21.48°C$$

but its moisture content remains at 4.70 g kg^{-1} dry air, giving state C. From tables or a chart the corresponding enthalpy is 33.55 kJ kg^{-1} dry air.

(*c*) On the opposite side of the rotary sensible heat exchanger the air has been warmed and its temperature is calculated as $12.85 + 0.8(56 - 12.85) = 47.37°C$ dry-bulb (state D). The moisture content remains at 8.404 g kg^{-1} dry air and the air is discharged into the steam heater battery to assist in the regeneration of the rotary absorption dehumidifier.

(*d*) Since 1 m^3 s^{-1} of air is supplied to the conditioned room at 10°C dry-bulb, 4.70 g kg^{-1} the corresponding enthalpy (from tables or a chart) is 21.67 kJ kg^{-1} dry air. The residual refrigeration load is:

$$[1/0.8728] \times (33.55 - 21.67) = 13.61 \text{ kW}$$

This is about 25 per cent of the sensible cooling load found in the answer to Example 3.17.

Exercises

1.   Dry saturated steam at 1.2 bar and 104.8°C is injected at a rate of 0.01 kg s^{-1} into an airstream having a mass flow rate of dry air of 1 kg s^{-1} and an initial state of 28°C dry-bulb,

11.9°C wet-bulb (sling). Calculate the leaving state of the airstream if all the injected steam is accepted by it. Use equation (2.24) as necessary.

Answers

29.3°C, 11.937 g kg^{-1}.

2.   (*a*) Define the term 'humidifying efficiency', as applied to an air washer. Explain why this is only approximately true if expressed in terms of dry-bulb temperature.
    (*b*) Write down an equation that defines an adiabatic saturation process in an air washer, identifying all symbols used.
    (*c*) 2 m^3 s^{-1} of air at 23.5°C dry-bulb, 11°C wet-bulb (sling) and 2.94 g kg^{-1} dry air enters an air washer that has a humidifying efficiency of 85 per cent. If the airstream undergoes a process of adiabatic saturation use your definition from part (*b*) to determine the dry-bulb temperature and moisture content of the air leaving the washer. It is given that the specific heat capacity of dry air is 1.012 kJ kg^{-1} K^{-1}, the latent heat of evaporation of water is 2450 kJ kg^{-1} and the specific heat of water vapour is 1.89 kJ kg^{-1} K^{-1}. Do not use psychrometric tables or a chart.

Answers

12.9°C, 7.36 g kg^{-1}.

3.   (*a*) An operating theatre is maintained at an inside temperature of 26°C dry-bulb when the outside air is at 45°C dry-bulb, 32°C wet-bulb (sling) and the sensible and latent heat gains are 9 kW and 3 kW, respectively. Determine the cooling load if 100 per cent fresh air is handled, the air temperature leaving the cooler coil being 14°C and the apparatus dew point 12°C. Assume a rise of 1.5°C across the supply fan (which is located after the cooler coil) and a further rise of 2°C because of heat gains to the supply duct.
    (*b*) Determine the percentage saturation maintained in the theatre under the conditions of part (*a*).

Answers

72.67 kW, 50.9 per cent.

Notation

Symbol	Description	Unit
c_a	specific heat of dry air	J kg^{-1} K^{-1} kJ kg^{-1} K^{-1}
E	effectiveness of an air washer or spray chamber	
g	moisture content	kg per kg dry air or g per kg dry air
h	enthalpy of moist air	kJ per kg dry air
h_s	enthalpy of steam	kJ kg^{-1}
h_w	enthalpy of feed-water	kJ kg^{-1}
m_a	mass or mass flow of dry air	kg or kg s^{-1}
m_s	mass flow of steam injected	kg s^{-1}

m_w	mass flow of feed water evaporated	kg s^{-1}
t	dry-bulb temperature	°C
t'	wet-bulb temperature	°C
t_d	dew point temperature	
	(or the dry-bulb temperature at a state D)	°C
t_w	feed-water temperature	°C
β	contact factor	
η	humidifying efficiency	
μ	percentage saturation	%

Reference

Pickering, C.A.C. and Jones, W.P. (1986): Health and hygienic humidification, Technical Note TN 13/86, BSRIA.

Bibliography

1. W. Goodman, *Air Conditioning Analysis with Psychrometric Charts and Tables*. The Macmillan Company, New York, 1943.
2. *ASHRAE Handbook Fundamentals* 1997 SI Edition.
3. *ASHRAE Handbook Equipment* 1996, SI Edition.
4. P.L. Martin and D.M. Curtis, *M-C Psychrometric charts for a range of barometric pressures*. Troup Publications Ltd.

4

Comfort and Inside Design Conditions

4.1 Metabolism and comfort

The human body differs from the insentient machine in tolerating only a narrow band of temperature (36°C to 38°C) and in having an appreciation of comfort. Defining the latter is not easy because of its subjective nature: in the long run a person's comfort can only be assessed by posing the question, 'Are you comfortable?' Even then the individual may have difficulty in phrasing a satisfactory answer. Establishing a numerical scale of comfort, interpreted from simple physical measurements, has not proved straightforward although Fanger (1972) has proposed a comfort equation (see section 4.7) that allows the determination of thermal comfort in terms of activity, clothing, dry-bulb temperature, the mean radiant temperature of the surrounding surfaces, relative humidity and air velocity.

The body eats and digests food which, through the intake of oxygen by breathing, is converted into energy for useful work with the liberation of heat and the emission of waste products. The chemical changes occurring in the liver and other parts of the body, and muscular contraction, release thermal energy which is transported throughout the body by the circulation of the blood so that it can be lost to the outside world and the temperature of the deep tissues kept at comparatively constant value (about 37.2°C) if good health is to be preserved. The efficiency of the body—defined as $W/(M - 44)$, where W is the rate of working, M the metabolic rate of energy production for the particular activity and 44 W m^{-2} is the basal metabolic rate per unit area of bodily surface—is not very great, varying from 0 per cent when at rest to a maximum of about 20 per cent when walking up a one in four gradient at 10 km h^{-1}. So most of the energy produced by the bodily metabolism must be dissipated as heat to the environment.

In good health, the thermo-regulatory system of the body exercises automatic control over the deep tissue temperature by establishing a correct thermal balance between the body and its surroundings. Apart from the matter of good health the comfort of an individual also depends on this balance and it might therefore be interpreted as the ease with which such a thermal equilibrium is achieved. ASHRAE Comfort Standard 55–74 goes further and defines comfort as '... that state of mind which expresses satisfaction with the thermal environment ...' but points out that most research work regards comfort as a subjective sensation that is expressed by an individual, when questioned, as neither slightly warm nor slightly cool.

4.2   Bodily mechanisms of heat transfer and thermostatic control

Heat is exchanged between the body and its environment by four modes: evaporation (E), radiation (R), convection (C) and, to an insignificant extent because of the small area of contact usually involved, conduction. There is also a small loss by the rejection of excreta but this is generally ignored. A heat balance equation can be written if we include an additional element, S, representing the positive or negative storage of heat in the body that would cause the deep tissue temperature to rise or fall:

$$M - W = E + R + C + S \tag{4.1}$$

For a normally clothed, healthy human being in a comfortable environment and engaged in a non-strenuous activity, S is zero and the thermo-regulatory system of the body is able to modify the losses by radiation and convection to maintain a stable, satisfactory temperature. Evaporative losses occur in three ways: by the exhalation of saturated water vapour from the lungs, by a continual normal process of insensible perspiration, and by an emergency mechanism of sweating. Insensible perspiration results from body fluids oozing through the skin under osmotic pressure according to Whitehouse *et al.* (1932) and forming microscopic droplets on the surface which, because of their small size, evaporate virtually instantaneously, not being felt or seen and hence termed insensible. Sweating is entirely different and in comfortable conditions should not occur: if body temperature tends to rise the thermo-regulatory system increases the evaporative loss by operating sweat glands selectively and flooding strategic surfaces. In extreme cases the body is entirely covered with sweat that must evaporate on the skin to give a cooling effect—if it rolls off or is absorbed by clothing its cooling influence will be nullified or much reduced.

Evaporative loss is a function of the difference in vapour pressure between the water on the skin and that of the ambient air. It also depends on the relative velocity of airflow over the wet surface. The two following equations are sometimes useful to solve numerical problems although equation (4.2), for parallel airflow over a lake surface (or the like) is considered to give an underestimate. Equation (4.3) describes the case of transverse airflow, as across a wet-bulb thermometer.

$$\text{Evaporative loss (W m}^{-2}) = (0.0885 + 0.0779v)(p_w - p_s) \tag{4.2}$$
$$\text{Evaporative loss (W m}^{-2}) = (0.018\,73 + 0.1614v)(p_w - p_s) \tag{4.3}$$

where the relative air velocity, v, is in m s^{-1} and p_w and p_s, the vapour pressures of the water and the ambient air respectively, are in Pa.

Losses occur by radiation if the skin temperature exceeds that of the surrounding surfaces and by convection if it is greater than the ambient dry-bulb. The average temperature of the surrounding surfaces is termed the mean radiant temperature, T_{rm}, and is defined as: the surface temperature of that sphere which, if it surrounded the point in question, would radiate to it the same quantity of heat as the room surfaces around the point actually do. The mean radiant temperature thus varies from place to place throughout the room. Bodily surface temperature is influenced by the type of clothing worn, its extent, the activity of the individual, the performance of the thermo-regulatory system and the rate of heat loss to the environment. Signals are sent by nerve impulses from the brain in response to changes in blood temperature to regulate the flow of heat from the warmer, deep tissues of the body to the cooler, surface tissues. This regulation is done two ways: by changing the rate of sweat production and by dilating or constricting the blood vessels (the vascular system) beneath the skin. Vaso-dilation increases the flow of blood to the surface and so the flow of heat from the deeper tissues. Conversely, vaso-constriction reduces the flow of blood, the skin temperature and the bodily heat loss. A skin temperature exceeding 45°C or less than 18°C triggers a response of pain according to ASHRAE (1997) and subjective sensations for mean skin temperatures of sedentary workers are: 33.3°C comfortable, 31°C uncomfortably cold, 30°C shivering cold and 29°C extremely cold. For a more strenuous activity such temperatures might be reported as comfortable. A skin temperature of 20°C on the hand may be considered uncomfortably cold, one of 15°C extremely cold and 5°C painful. Higher ambient air temperatures than the skin temperatures mentioned can be borne because of the insulating effect of the air surrounding the surfaces of the body and some tolerances quoted given by ASHRAE (1989) are: 50 minutes at 82°C, 33 minutes at 93°C and 24 minutes at 115°C, for lightly clad persons in surroundings with dew points less than 30°C. Tolerance decreases rapidly as the dew point approaches 36°C. A rise in body temperature of a few degrees, because of ill-health or inadequate heat loss, is serious with possibly fatal results at above 46°C when the thermo-regulation control centre in the brain may be irreversibly damaged: sweating can cease and vaso-constriction become uncontrolled with the onset of heat production by shivering. On the other hand, a fall in the deep temperature of the body to below 35°C can also cause a loss of control by the thermo-regulation system and although recovery from a temperature as low as 18°C has occurred, according to ASHRAE (1989), 28°C is taken as the lower survival limit. The normal response of the body to a fall in temperature after vaso-constriction is the generation of heat by muscular tension and, subsequently, by involuntary work or shivering.

According to Gagge *et al.* (1938) experimental evidence shows that unclothed and lightly clad people can be comfortable at operative temperatures (see section 4.8) of 30°C and 27°C, respectively. These temperatures are the mid-points of narrow ranges (29°C to 31°C and 25°C to 29°C) within which there is no change of evaporative loss and no body cooling or heating, the deep tissue temperature being maintained at a constant value without physiological effort. Above and below these ranges are zones of vaso-motor regulation against cold and heat wherein the control system of the body can keep the deep tissue temperature constant (albeit with some change in skin temperature) by vaso-constriction and dilation, respectively. Below the cold zone muscular tension etc. takes over until eventually deep temperature can only be maintained at a satisfactory level by putting on more clothing. At the upper end of the range of operative temperatures the zone of control over heat is much smaller and vaso-dilation only suffices until the skin temperature approaches to within 1°C of the deep body temperature, after which the sweat glands must work if the thermal balance between the person and the environment is to be maintained.

It follows from the foregoing that skin temperature and evaporative cooling from the skin will have a significant influence on comfort. Belding and Hatch (1955) defined a heat stress index as the ratio of the total evaporation loss in bodily thermal equilibrium to the maximum loss by evaporation if the skin were entirely wetted by regulatory sweating, is sometimes used. The ASHRAE scale of effective temperature (see section 4.8) makes use of the concept of heat stress.

4.3 Metabolic rates

Mackean (1977) expresses metabolism as 'All the chemical changes going on in the cells of an organism.' In the context of air conditioning the interest lies mainly in the quantities of sensible and latent heat dissipated by the body to its environment as a consequence of these changes and in accordance with equation (4.1). People vary in shape, surface area and mass, thus research workers currently prefer to express heat emission from the body in terms of the met unit, equal to 58.2 W m^{-2}. The related body surface area is difficult to measure but a favoured equation for its expression, developed by Du Bois and Du Bois (1916), defines it as the Du Bois surface area, A_D, where

$$A_D = 0.202m^{0.425}h^{0.725} \tag{4.4}$$

in which m is the body mass in kg and h its height in m. Using equation (4.4) a man of 70 kg (11 stone) with a height of 1.8 m (5′ 11″) has a surface area of 1.9 m^2. One met unit is often taken as corresponding to 100 W, approximately, representing the emission from a resting adult.

The metabolic rate depends on the activity and varies from about 84 W for a person of 1.8 m^2 surface area, as a basal rate, to a maximum of about 1200 W for a normal, healthy young man. Although trained athletes can work at maxima as high as 2000 W, an ordinary person can only maintain some 50 per cent of his maximum output continuously for any length of time. The maximum possible decays with age to roughly 700 W at the age of 70 years. Women have maximum levels at approximately 70 per cent of these values and children less still, because of their smaller surface areas, but for estimating the emission of heat from a mixed group of people the proportions of the normal male output taken by ASHRAE (1989) for women and children are usually 85 per cent and 75 per cent, respectively. Conservative values, as used for heat gain calculations, are quoted in Table 7.16, but more exact figures for different activities are given in Table 4.1.

Table 4.1 Typical metabolic rates for various activities, according to ASHRAE (1997)

Activity	Total heat production in watts
Sleeping	72
Sitting quietly	108
Standing (relaxed)	126
Walking on a level surface	
at 6.4 km/h (4 mph)	396
Reading (seated)	99
Writing	108
Typing	117
Dancing	82–256

The total heat productions are for continuous activity by an adult having a Du Bois surface area of 1.8 m^2. The figures are in good agreement with those published in the *CIBSE Guide* (1999), A1, Environmental criteria for design.

4.4 Clothing

The loss of heat from the body and the feeling of individual comfort in a given environment is much affected by the clothing worn and in a room with a mixed population of men and women wearing different garb, comfort for everyone may be almost impossible to achieve. There also appears to be a seasonal pattern in the clothing worn, according to Berglund

(1980), which prevails even if the temperature of the working environment is virtually constant throughout the year, lighter garments of less thermal insulation value being worn in the summer.

The unit used to describe the thermal insulating quality of the clothing worn is the clo with a physical value of $0.155 \text{ m}^2 \text{ K W}^{-1}$. Table 4.2 lists some thermal resistances proposed by Berglund (1980) for individual items, to be combined by equation (4.5), according to McCullough and Jones (1984).

$$I_{clo} = 0.835 \, \Sigma I_{clui} + 0.161 \qquad\qquad (4.5)$$

where I_{clui} is the effective thermal insulation of garment i, and I_{clo} is the thermal insulation of the total clothing ensemble.

Table 4.2   Thermal resistances for some items of clothing according to Berglund (1980)

Men	clo	Women	clo
Sleeveless singlet	0.06	Bra and pants	0.05
T-shirt	0.09	Half slip	0.13
Underpants	0.05	Full slip	0.19
Shirt, light-weight, short sleeves	0.14	Blouse, light-weight	0.20 (a)
Shirt, light-weight, long sleeves	0.22	Blouse, heavy-weight	0.29 (a)
Waistcoat, light-weight	0.15	Dress, light-weight	0.22 (a, b)
Waistcoat, heavy-weight	0.29	Dress, heavy-weight	0.70 (a, b)
Trousers, light-weight	0.26	Skirt, light-weight	0.10 (b)
Trousers, heavy-weight	0.32	Skirt, heavy-weight	0.22 (b)
Sweater, light-weight	0.20 (a)	Slacks, light-weight	0.26
Sweater, heavy-weight	0.37 (a)	Slacks, heavy-weight	0.44
Jacket, light-weight	0.22	Sweater, light-weight	0.17 (a)
Jacket, heavy-weight	0.49	Sweater, heavy-weight	0.37 (a)
Ankle socks	0.04	Jacket, light-weight	0.17
Knee socks	0.10	Jacket, heavy-weight	0.37
Shoes	0.04	Stockings or tights	0.01
Boots	0.08	Sandals	0.02
		Shoes	0.04
		Boots	0.08

(a) Deduct 10% if sleeveless or short sleeved.
(b) Add 5% if below knee length; deduct 5% if above knee length.

It is reckoned by ASHRAE (1997) that total clo-values cannot be estimated to be better than 20 per cent accuracy and this should be borne in mind. ASHRAE (1997) quote clo-values that are a little less than the figures given in Table 4.2.

EXAMPLE 4.1

Estimate the insulation value of the clothing of a man dressed as follows:
(a) T-shirt, underpants, light-weight trousers, ankle socks and shoes,
(b) Sleeveless singlet, underpants, long-sleeved shirt, heavy-weight trousers, jacket and waistcoat, knee-length socks and shoes.

Answers

(i) Using Table 4.2 the sum of the individual items of clothing is 0.09 + 0.05 + 0.26 + 0.04 + 0.04 = 0.48 and by equation (4.5) we have 0.835 × 0.48 + 0.161 = 0.56 clo.
(ii) The sum of the individual items is 0.06 + 0.05 + 0.22 + 0.32 + 0.49 + 0.29 + 0.10 + 0.04 = 1.57 and by equation (4.5) we have 0.835 × 1.57 + 0.161 = 1.47 clo.

Similar calculations for the light and heavy extremes of women's clothing yield figures of about 0.6 and 1.3 clo. We might therefore conclude that 1 clo represents average clothing for a man and perhaps 0.9 clo for a woman. It is not surprising that achieving satisfactory conditions of comfort for an air conditioned room with a mixed population sometimes proves difficult. It has been suggested by Berglund (1980) that the comfort of a clothed individual corresponds to a decrease in ambient dry-bulb of about 0.5°C for each clothing increase of 0.1 clo but this can only be true over a limited range of temperature. For sedentary workers a realistic lower limit for a period of more than one hour is about 18.5°C, provided the air movement is imperceptible, as it might be in a room that was only heated and not mechanically ventilated or air conditioned. In air conditioned rooms with typical air change rates of 5 to 20 per hour air movement is not imperceptible at such a low temperature, no matter how well the air distribution system is designed. For air conditioned rooms the realistic lower limit is 20°C in the UK and even then it will be unsatisfactorily cool for some of the occupants.

The intensity of air turbulence (T_u) is relevant to a sensation of draught and is defined by ASHRAE (1997) and Fanger (1987) as

$$T_u = 100(V_{sd}/V) \tag{4.6}$$

where V_{sd} is the standard deviation of the local air velocity, measured by an omnidirectional anemometer with a time constant of 0.2 s, and V is the mean air velocity in m s^{-1}. The value of T_u is used in equation (4.7) to predict the percentage of people dissatisfied (PD) because of the presence of the draught:

$$PD = (34 - t_a)(V - 0.05)^{0.62}(0.37VT_u + 3.14) \tag{4.7}$$

where t_a is the dry-bulb temperature of the air.

The equation is relevant for 20°C < t_a < 26°C and for 0.05 < V < 0.5 m s^{-1}, according to ASHRAE (1997). As an example, this reference shows that, for an air temperature of 22°C with a mean air speed of 0.25 m s^{-1} and a turbulence intensity of 2 or 3 per cent, it is still likely that 15 per cent of the people will be dissatisfied. Fanger *et al.* (1988) have shown that discomfort depends also on the frequency of fluctuation of the draught, people being particularly sensitive to the range from 0.3 to 0.6 Hz.

4.5 Environmental influences on comfort

From the foregoing it can be inferred that the body maintains a thermal equilibrium with the environment by heat exchanges involving evaporation (about 25 per cent), radiation (about 45 per cent) and convection (about 30 per cent), referred to an environment of approximately 18.5°C and 50 per cent relative humidity for a normally clothed, sedentary person. The figure for evaporation covers respiration and insensible perspiration, without sweating. It further follows that there are four properties of the environment that influence comfort by modifying the contributions of these three modes of heat transfer:

(i)   dry-bulb temperature (affecting evaporation and convection),
(ii)   relative humidity (affecting evaporation only),
(iii)   air velocity (affecting evaporation and convection) and
(iv)   mean radiant temperature (affecting radiation only).

Items (i) and (ii) may be directly under the control of the air conditioning system although the influence of humidity at comfortable temperatures and air velocities is considered less important today than in the past. Item (iii) is a consequence of the design of the air distribution system and is most important in air conditioning because of the associated high air change rates. It is unusual for item (iv) to be regarded as controllable even though many systems have been designed, installed and successfully commissioned making use of a chilled ceiling in conjunction with a dehumidified auxiliary air supply, but these only give partial regulation of the mean radiant temperature because the ceiling capacity is commonly controlled from room air temperature. It is not possible therefore to specify a comfortable environment in terms of a single physical variable, such as dry-bulb temperature (although this may be the single most important factor): instead, account should be taken of as many as possible of the variables over which the system has a say, usually dry-bulb temperature and air velocity, plus humidity to a restricted extent.

In the UK dry-bulb temperatures of between 20°C and 23°C are often regarded as satisfactory for long-term occupancy and a sedentary activity. For shorter term, transient occupancy higher temperatures are sometimes adopted, up to the outside summer design value. In warmer climates, where outside temperatures may be 10°C or more higher than in the UK, inside dry-bulbs for sedentary occupations are 25°C to 26°C, on a long-term basis but even so values as low as 22°C, depending on the circumstances, are sometimes necessary, commercial considerations permitting. The human body is tolerant of considerable variation in humidity and Green (1979) has shown that humidities corresponding to dew points between 1.7°C and 16.7°C are satisfactory in terms of comfort, skin dryness, respiratory health and mould growth. Such dew points give humidities approximately between 30 per cent and 80 per cent saturation at 20°C and between about 26 per cent and 72 per cent at 22°C. ANSI/ASHRAE Standard 55–1992 relates to an environment where 80 per cent of sedentary or slightly active people are comfortable and adopts a slightly different view of Green's results. Account is taken of seasonal change of clothing in summer and winter and the 18°C and 20°C wet-bulb lines, rather than 16.7°C dew point, related to the upper bounds of winter and summer moisture content. Figure 4.11 illustrates a current view of ASHRAE (1997). Standard 55–1992 also considers that in order to decrease discomfort due to low humidity the dew point should not be less than 3°C. The Standard's view on an upper limit for humidity is controversial and suggests that dew points exceeding 20°C are associated with unacceptable air quality (see section 4.10). It points out that in warm conditions discomfort increases with humidity and skin wettedness may then become a problem, being related to sweat on the skin and clothing. An index of skin wettedness has been explored by Gagge *et al.* (1969a and 1969b) and by Gonzalez *et al.* (1978).

The influence of humidity is small when the body is in comfort equilibrium. Although the work of Yaglou and Miller (1925), leading to the original concept of effective temperature (see section 4.8), seemed to show that humidity was significant, it was based on the very short-term (a few minutes) exposure of subjects to comparative environments and Yaglou's later work (1947), and that of Koch *et al.* (1960), Nevins *et al.* (1966) and McNall *et al.* (1971) substantiate the current view that humidity is not very significant in a condition of thermal balance. However, high humidities may pose health risks. See section 4.11.

Air movement is very important and is related to air temperature and the part of the body blown upon. Air distribution terminals are generally selected to give a maximum centre-line velocity for the air jet of 0.25 m s^{-1} at the end of its throw when about to enter the occupied zone. According to Fanger (1972 and 1987) small changes in air velocity are important, especially between 0.1 and 0.3 m s^{-1}. For sedentary activities he quotes the relationships given in Table 4.3 for the comfort equilibrium of lightly clothed people.

Table 4.3   Air velocity and comfort

Air velocity, m s^{-1}	0.1	0.2	0.25	0.3	0.35
Dry-bulb temperature, 0°C	25	26.8	26.9	27.1	27.2

In the practical case of an air conditioned room in the UK some departure from these temperatures might be expected because of uncertainty as to what actually constitutes the part of the body relevant to the air velocity and the variation in the thermal insulation of the clothing worn in a mixed population, referred to earlier.

An air distribution performance index was proposed by Miller and Nash (1971) and Nevins *et al.* (1974) that relates comfort to measured values of effective draught temperature, t_{ed}, and air velocity within the occupied space:

$$t_{ed} = (t_x - \bar{t}_r) - 7.65(v_x - 0.152) \qquad (4.8)$$

t_x is the local air temperature, v_x a local air velocity and $\bar{t}_r$ the mean room dry-bulb temperature. The air distribution performance index is the percentage of positions in the occupied zone, to a specified standard of measurement, where t_{ed} is between –1.7°C and +1.1°C and v_x is less than 0.35 m s^{-1}. An 80 per cent value is regarded as very satisfactory.

There appears to be no minimum value of air movement necessary for comfort according to ASHRAE (1989) but an upper practical limit is 0.8 m s^{-1}, above which loose papers start to fly. Small fluctuations in air velocity, with a cool temperature, may cause complaints of draught, as the earlier mention of draught intensity (equation (4.6)) defined.

Non-uniformity in the environment seems most important for the head-to-feet temperature gradient and this should not normally exceed 1.5 K and should never be more than 3 K from ISO 7730 (1995). Discomfort at the feet depends on the footwear as well as the floor temperature. Nevins *et al.* (1964) and Springer *et al.* (1966) found that values up to 29°C with a three hour exposure were acceptable to walking and sitting subjects. (The author's experience is that 26°C may cause complaint in an ambient dry-bulb of 22°C, particularly for people who are standing.) For cold floors, a lower limit of 17°C to 18°C has been suggested by Nevins and Flinner (1958).

During winter, when heating, the mean radiant temperature should be higher than the dry-bulb but, in summer, the reverse is true and a lower mean radiant temperature is pleasant with a higher dry-bulb temperature. People are more tolerant of lateral than vertical asymmetry in thermal radiation and so excessive radiation is best not directed on the tops of heads. Asymmetry in thermal radiation can be expressed in terms of the plane radiant temperature difference. See McIntyre (1976). The plane radiant temperature is defined as the mean radiant temperature of half the room, with respect to a small, one-sided elemental plane area. If there is a difference between opposing half-room, mean radiant temperatures of this sort it is termed the 'radiant temperature asymmetry'. According to McIntyre (1976) and Olsen and Thorshauge (1979), the limiting differences for comfort with asymmetric radiation are 5 K vertically and 10 K horizontally. The maximum difference between two opposite plane radiant temperatures is sometimes called the vector radiant temperature. In practical terms the three most common sources of radiant asymmetry provoking local discomfort are: cooling near windows in winter (complicated by natural infiltration and convective downdraught and sometimes loosely termed cold radiation), heating from overhead lighting and short-wave solar radiation through unshaded glazing.

The presence and proximity of other people can be regarded as an environmental influence on comfort. Fanger (1972) reported experimental evidence that crowding had little effect on the physiological response of subjects occupying floor areas and volumes as little as 0.8 m^2 and 2.0 m^3, respectively, reciprocal radiation being unimportant under warm conditions but, in cooler circumstances, the mean radiant temperature was increased enough to require a reduction in the dry-bulb in a crowded room. In conditions of extreme crowding the boundary layers of people intermingle (e.g. in rush-hour transport), suppressing heat transfer by convection and needing a lower dry-bulb for comfort.

4.6   Other influences on comfort

Fanger (1972) has investigated non-environmental effects and concluded the following, for similar standards of clothing and activity. National and geographical influences are negligible and there are no seasonal effects. There is, apparently, no difference between the comfort requirements of the elderly and young adults, but it must be remembered that elderly people are less active and this implies higher temperatures are needed. It seems that women prefer a temperature 0.3 K greater than men but on his experimental evidence this was not statistically significant at the 5 per cent level. On the other hand, women seem more sensitive to small variations from an optimum temperature and, of course, are more lightly dressed, suggesting the need for a higher temperature in the practical case. Bodily build has little significance for sedentary activities beyond fat adolescence, young adults prefer 0.2 K less whilst the menstrual cycle has no significance for comfort temperatures. The little work that has been done on ethnic influences suggests they are insignificant but the intake of food does have a bearing: the optimum comfort temperature could be decreased by as much as 1 K for some hours after a heavy, protein-rich meal. Although the deep tissue temperatures have a natural cycle of variation over 24 hours of between 0.3 K and 0.5 K with a maximum before sleeping and a minimum some while before waking, this has no significant effect on comfort.

Thermal transients do not have much influence provided the mean value of temperature is reasonably constant at a comfortable value. Furthermore, it has been well established by Glickman *et al.* (1947) and (1949), Hick *et al.* (1953) and Inouye *et al.* (1953) and (1954) that when a person walks from a hot outside environment into a room conditioned at a comfortable but relatively low temperature there is no adverse physiological effect. In other words, there is no such thing as 'shock effect'. The human body is resilient and acclimatisation is rapid.

Psychological influences have been examined by Rohles (1980), who has suggested that these are very short-term effects: the presence of warmer looking furnishings, for instance, would not give a long-term influence on comfort, neither would the setting of a thermostat that implied an air temperature different from that actually present. Colour is of no significance according to Berry (1961). Psychological trivia of this kind can be totally disregarded as significant factors for sustained thermal comfort.

4.7 Fanger's comfort equation

The comfort equation developed by Fanger (1972) was established from experimental work with North American subjects and allows the calculation (by computer or, with difficulty, manually using diagrams) of the combination of activity, clothing and environmental factors that will produce thermal comfort (see section 4.1). It was subsequently checked by applying it to Danish subjects and excellent correlation found. Fanger concludes his equation is appropriate within the temperate zone for adults, regardless of sex. The thermal balance of the body for comfort is expressed in simple terms by

$$H - E_d - E_{sw} - E_{re} - L = R + C \qquad (4.9)$$

where H is the internal rate of bodily heat production (related to the activity), E_d is the heat loss by vapour diffusion through the skin (insensible perspiration and not subject to thermo-regulatory control), E_{sw} is the heat lost by the evaporation of sweat, E_{re} is the latent heat loss by respiration, L is the sensible heat loss by respiration, R is the radiation loss from a clothed person and C is the similar convective loss. Fanger's comfort equation then defines comfort for the case when a function of the relevant variables equals zero, namely:

$$f\left(\frac{H}{A_D}, I_{cl}, t_a, T_{rm}, \phi_a, v \right) = 0 \qquad (4.10)$$

where I_{cl} is the insulating value of the clothing, t_a is the ambient dry-bulb temperature and ϕ_a the relative humidity. Fanger finds the effect of relative humidity is not great for persons in comfort balance and that for such people a change in humidity from 0 per cent to 100 per cent can be compensated by a temperature decrease of about 1.5°C to 3.0°C. Clothing and activity are very significant: an increase in clothing from 0 clo to 1.5 clo corresponds to a decrease of 8°C in the dry-bulb for sedentary work (115 W) but to 19°C for more strenuous activity (345 W).

Using his comfort equation Fanger (1972) has derived an index of thermal sensation, making it possible to predict a mean comfort response (predicted mean vote, abbreviated as PMV), on a standard scale from –3 to +3, in a large group of persons for any combination of the four environmental variables, activity and clothing. Tables and diagrams are given to simplify the intrinsically complicated mathematical approach to this. The predicted percentage dissatisfied (abbreviated PPD) for an indoor climate is perhaps more meaningful. The PPD is determined from the PMV for several positions in a room. After measurement of the environmental variables the lowest possible percentage of dissatisfied persons (abbreviated LPPD) attainable by altering the temperature can be established. To quote Fanger: 'The magnitude of the LPPD is an expression for the non-uniformity of the thermal environment and is therefore suitable for characterising the heating or air conditioning system ...'. Table 4.4 is based on Fanger's work and quotes some dry-bulb temperatures and associated air velocities in an environment of 50 per cent relative humidity, with the

Table 4.4 Temperature and air velocity for zero PMV at various clo-values for sedentary workers

clo-value	0	0.5			1.0		1.25		1.5
Temp (°C)	29	28	27	26	25	24	24	22	22
Velocity (m s^{-1})	0.5 to 0.15	0.5 to 1.0	0.2 to 0.3	0 to 0.1	0.5 to 1.0	0.2 to 0.3	1.0 to 1.5	0.1 to 0.15	0.5 to 1.0

mean radiant temperature equal to the dry-bulb, when the PMV passes through zero for sedentary workers and therefore indicates when they are most comfortable.

In experiments with sedentary subjects clothed at 0.6 clo those expressing 'warm dissatisfaction' and those expressing 'cold dissatisfaction' gave equal votes at 25.6°C, equal to the optimum temperature predicted by the comfort equation, implying that, for a given activity and clothing there is one comfortable temperature for all people, regardless of age, sex, etc., as outlined earlier. A further important conclusion of Fanger's work is that a value of less than 5 per cent for the PPD value is not achievable for similarly clothed people engaged in the same activity, no matter how perfect the environmental system: a few complaints must not necessarily be interpreted as an indication that the system is defective or the controls wrongly set. The system should be set up to suit the comfort of the majority and not subsequently tampered with to cater for the few who will invariably complain, see also BS EN ISO 7730 (1995).

4.8   Synthetic comfort scales

Many attempts have been made, with mixed success, to correlate the four environmental factors (dry-bulb, velocity, mean radiant temperature and relative humidity) that contribute to the comfort of humans by influencing bodily thermal equilibrium. Several scales of comfort have been proposed and used in air conditioning: effective temperature, new effective temperature, standard effective temperature, and resultant temperature.

Yaglou (1923, 1947) and Yaglou and Miller (1924, 1925) using a survey of the responses of subjects to a relatively shortterm exposure in different environments founded the scale of effective temperature, defined as the temperature of still, saturated air that gives a feeling of comfort similar to that of another combination of the three relevant environmental variables. It was considered to over emphasise the influence of humidity and more recently Rohles and Nevins (1971), Rohles (1973, 1974) and Gagge *et al.* (1971) it has been recognised that with a longer term occupancy of an environment the importance of humidity is less when the body is in comfortable thermal equilibrium. A new scale of effective temperature has been proposed by ASHRAE (1997), expressed in terms of operative temperature (see equation (4.13)) and combining the effects of dry-bulb temperature, mean radiant temperature and humidity into a single index. Two such combinations could be compared, but only if they had the same air velocity. Furthermore, clothing and activity must be defined. These difficulties have prompted ASHRAE (1997) to offer a Standard Effective Temperature, defined as 'the equivalent temperature of an isothermal environment at 50 per cent relative humidity in which a subject, while wearing clothing standardised for the activity concerned, has the same heat stress (skin temperature) and skin wettedness as in the actual environment'.

Figure 4.1 illustrates the ASHRAE summer and winter comfort zones, based on ASHRAE Standard 55–1992. It differs from the original in that the abscissa in the figure is expressed in terms of dry-bulb temperature, instead of operative temperature. The justification for this is that operative temperature is not used in the UK. The two temperatures are related by equation (4.13).

Figure 4.1 refers to clothing having insulation values of between 0.5 and 0.9 clo for summer and winter conditions, respectively, with a sedentary activity. The nearly vertical temperature lines, separating the winter and summer comfort zones, correspond to lines of constant standard effective temperature within which it is seen how comfort is affected by humidity. The lower level of comfort is defined by a dew-point of 2°C. The upper levels

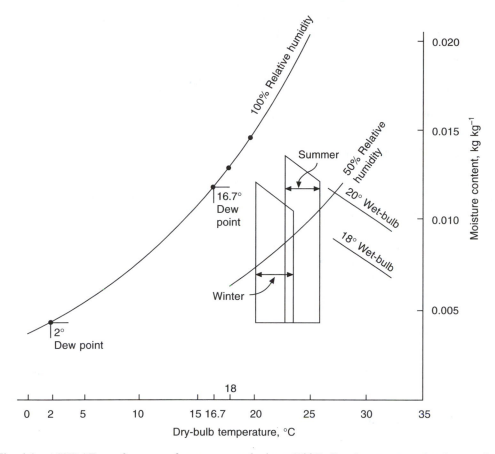

Fig. 4.1 ASHRAE comfort zones for summer and winter (1997). People are assumed to be wearing suitable summer or winter clothing for the appropriate season. The original ASHRAE figure uses operative temperature as the abscissa but the above figure uses dry-bulb temperature because operative temperature, although similar to resultant temperature, is not used in the UK. See equations (4.11), (4.12), and (4.13).

of humidity are theoretical and based on limited data but Nevins *et al.* (1975) recommended that relative humidity should not exceed 60 per cent, to prevent warm discomfort. The shape of the upper bounds of the comfort zones are corroborated by experimental data for an 80 per cent comfort acceptability level.

Dry resultant temperature (resultant temperature), developed by Missenard (1933 and 1935), is commonly used in Europe and, for most practical purposes, is defined by

$$t_{res} = [(T_{rm} - 273) + t_a]/2 \qquad (4.11)$$

where t_{res} is the dry resultant temperature (°C) and T_{rm} is the mean radiant temperature (K).

More exactly it is obtained from the reading of a thermometer at the centre of a blackened copper sphere of 100 mm diameter or, if the mean radiant temperature is known, by the equation

$$t_{res} = \frac{(T_{rm} - 273) + t_a \sqrt{10v}}{1 + \sqrt{10v}} \qquad (4.12)$$

in which v is the air velocity (m s^{-1}) and t_a is the dry-bulb temperature (°C).

Operative temperature, t_o, is defined by

$$t_o = [h_r(T_{rm} - 273) + h_c t_a]/(h_r + h_c) \qquad (4.13)$$

in which h_r and h_c are the radiative and convective heat transfer coefficients, respectively. In virtually still air, by putting $v = 0.1$ m s^{-1}, equation (4.12) degenerates to equation (4.11), the form commonly used.

4.9   Measuring instruments

To assess comfort conditions in an air conditioned room the following instruments may be used: mercury-in-glass thermometer, sling or aspirated psychrometer, Kata thermometer, stop watch, globe thermometer, and hot wire anemometer.

Dry-bulb temperature is measured by a mercury-in-glass thermometer, properly shielded from radiation, wiped dry beforehand and given time to settle down to a steady-state reading. The accuracy from a recent calibration should be known and the scale should be apt for the range of values to the measured.

The sling psychrometer comprises dry- and wet-bulb thermometers and has a reservoir which should be filled with clean, fresh water at a temperature about equal to the wet-bulb reading anticipated. The wick on the wet-bulb should be clean and free of scale or other solid deposits. Avoid handling the wick as this can cause contamination. Both thermometer stems and the bulb of the dry-bulb instrument must be wiped free of moisture. When the wick is completely wetted by capillary action from the reservoir the sling should be whirled rapidly for about 10 seconds to eliminate radiation errors (see section 2.17). As soon as whirling is stopped the wet-bulb reading is taken and then the dry-bulb immediately afterwards. The wet-bulb must be read first because the reduction of the evaporation rate and the onset of radiation error when whirling ceases causes the indicated temperature to rise rapidly. A similar rise in the dry-bulb occurs but this is less rapid. This procedure is repeated several times, ensuring each time that the wick is properly wetted, until three sets of successive readings show close agreement. Failure to ensure the wick is properly wetted, whirling the sling too slowly, or being late in reading the wet-bulb, will give high and unrealistic wet-bulb temperatures that imply the humidity is much greater than is really the case. Accurate measurements of wet-bulb temperature are quite difficult to obtain. An aspirated version uses a chromium-plated cylinder, having a clockwork or battery-driven fan at the top which induces air at an adequate velocity over the thermometer and gives stable readings free from radiation errors.

The Kata thermometer is not in very common use today because of the lengthy, elaborate procedure involved (requiring simultaneous readings of the dry-bulb and a significant time taken for the fluid within the Kata to fall between two markings on the stem as it cools). A thermos bottle of hot water is also needed to raise the temperature of the Kata above that corresponding to the upper markings on the stem, in the first place. Instead of using the Kata thermometer to measure very small air velocities, the hot-wire anemometer is adopted today. This instrument is basically a Wheatstone bridge, one resistance element of which is the anemometer velocity sensor. When an electric current flows through the sensor element its temperature is raised to a value at which its rate of heat loss to the environment is stable, being a function of ambient air velocity and dry-bulb. Since electrical resistance depends on the temperature of a conductor an indication can be obtained of air velocity. The instrument is easy to use down to a velocity of 0.05 m s^{-1} although there may be problems with local transient air movement. Temperature as well as air velocity may be measured simultaneously and both can be recorded.

The globe thermometer is an ordinary mercury-in-glass thermometer inserted in a blackened copper sphere so as to have its bulb at the centre. It is affected by dry-bulb temperature, air velocity and mean radiant temperature but in still air reads the mean radiant temperature exactly. A simultaneous dry-bulb reading must be taken and about 20 minutes allowed for the globe to settle down. Mean radiant temperature, T_{rm} in kelvin, is then determined from the following equation

$$T_{rm}^4 \times 10^{-9} = T_g^4 \times 10^{-9} + 1.442(T_g - T_a)\sqrt{v} \qquad (4.14)$$

where T_g is the globe reading in kelvin and T_a the ambient dry-bulb also in kelvin.

4.10   Outdoor air requirements

When resting quietly with a metabolic rate of 88 W an average man's oxygen consumption is only 0.006 litres s^{-1} but to provide this the breathing rate must be about 0.09 litres s^{-1} according to Haldane and Priestly (1905). Oxygen consumption is directly proportional to metabolic rate so for exhausting effort and a metabolic rate of 1055 W the oxygen consumption is 0.07 litres s^{-1} and the breathing rate reaches 1.15 litres s^{-1}. The amount of fresh air needed to sustain life is thus very small, extending from about 0.1 litres s^{-1} to 1.2 litres s^{-1}.

It is highly desirable to keep the quantity of fresh air handled by an air conditioning plant to a minimum for economic reasons and to conserve energy but this is not dictated by the requirement for breathing purposes. Enough air from outside must be supplied to reduce the accumulation of body odours, other smells and the pollution from smoking and other sources to a socially acceptable level. It is further necessary to dilute the concentration of carbon dioxide to a maximum acceptable value of 0.1 per cent by the introduction of fresh air with an average concentration of about 0.031 to 0.035 per cent (see section 2.2), but this is not as significant as the need to deal with odours when establishing the minimum fresh air required. As little as 5 litres s^{-1} for each person may be acceptable for a banking hall or a church where there is a large volume with no smoking but as much as 43 litres s^{-1} may be wanted in rooms where smoking is heavy. This topic is also dealt with in chapter 16.

In the past it has been thought by Dessauer (1931), Yaglou *et al.* (1932) and Gagge and Moriyama (1935) that air containing molecules with a high negative ion content exhibited indefinable qualities of freshness that some subjects found desirable, but subsequent work by Brandt (1933), Herrington (1935), Herrington and Kuh (1938) and Yaglou (1935) and earlier Yaglou *et al.* (1933), has failed to support this. Current, reputable, scientific evidence indicates that the ion content of air has no measurable effect on human beings in buildings. If any influence is produced by the ions present it must be exceedingly subtle.

The current outside air supply rates recommended by the CIBSE (1999) are given in Table 4.5.

An addendum to the ASHRAE Standard 62n–1999 (Ventilation for Acceptable Indoor Air Quality) proposes what appear to be smaller rates, modified according to the floor area and the use of the occupied space.

The proposals by Fanger (1988), to calculate the pollution balance in an occupied space (see section 16.1), have not found general acceptance in Europe or the UK. Results have been inconsistent and the supply rates of outside air so determined have proved unrealistically large.

Table 4.5   Recommended outdoor air supply rates for sedentary occupants

Level of smoking	Proportion of occupants who smoke (%)	Outdoor air supply rate (litre s^{-1} person^{-1})
No smoking	0	8
Some smoking	25	16
Heavy smoking	45	24
Very heavy smoking	75	36

Reproduced by kind permission of the CIBSE from Guide A, Environmental Design (1999).

4.11   Indoor air quality

Definition

There is no strict definition of indoor air quality (IAQ) and no instruments measure it. ANSI/ASHRAE Standard 62–1989 states: 'For comfort, indoor air quality may be said to be acceptable if not more than 50% of the occupants can detect any odour and not more than 20% experience discomfort, and not more than 10% suffer from mucosal irritation, and not more than 5% experience annoyance, for less than 2% of the time.'

Pollutants in occupied rooms

These come from six sources:

(1) The occupants, who emit carbon dioxide, water vapour, solid particles (lint from clothing, flakes of skin, hair), odours and biological aerosols.
(2) Smoking, which produces carbon monoxide, carbon dioxide, gases and vapours, solid particles, liquid droplets, volatile organic compounds (VOCs) and odours. Many of the emissions are short-term and long-term risks to health.
(3) Building materials, furnishings and their emissions (structural components, surface finishes, furniture, adhesives used in furniture, upholstery, carpets, dusts, house mites).
(4) The activities of people and the equipment they use (printers, copiers, fax machines, computers) give off gaseous and solid emissions and produce odours. The surface coatings of equipment emit pollutants and all electrical equipment may become hot, burn its coating of dust and emit odours. VOCs may also be produced.
(5) Air supplied from outside for ventilation may be contaminated. Hence outdoor air quality should be assessed if necessary. The direct measurement of the concentration of carbon monoxide in the outside air is the best index because CO is well correlated with the presence of oxides of nitrogen, aromatic hydrocarbons and other urban pollutants.
(6) The air handled by the plant and supplied to the occupied space through the ducting may be contaminated. Dirty air filters, particularly when wet, and silencers are potential sources of fungi. Air cooler coils and their condensate trays can provide bacteria, viruses and fungi. If self-draining and properly and regularly cleaned they are safe. Humidifiers using spray water or wetted surfaces are a worse problem because of the continuous presence of liquid water. They must never be used, but dry steam is safe for humidification. The ducts may harbour dust and give contamination if not properly cleaned. Other possible pollutants are: toxicity, radioactivity, materials with infectious

or allergenic potentials, the presence of irritants, extreme thermal conditions and objectionable odours.

Effects of humidity

Berglund (1998) observes that humidity affects the evaporation of water from the mucosal and sweating bodily surfaces, influencing its diffusion through the skin. Low humidities, with dew points less than 2°C, tend to give a dry nose and throat, and eye irritation. A dusty environment can exacerbate low humidity skin conditions. Liviana *et al.* (1988) found eye irritation increased with the passage of time in dew points less than 2°C. High humidities support the growth of pathogenic and allergenic organisms, certain fungi, mycotoxins and house mites. Their growth is enhanced by the presence of high cellulose materials such as fibreboard, dust, lint, skin particles and dandruff (even when these have a low nitrogen content). Fungal contamination is likely when the humidity exceeds 70 per cent.

Ventilation effectiveness (Ev)

This is defined as the fraction of the air introduced from outside that reaches the occupied zone. Some systems of air distribution are poorly designed and their effectiveness can be as low as 0.5. Since the occupants sometimes breathe only 1 per cent of the outside air introduced, poor ventilation effectiveness is bad. The CIBSE (1999) suggest values of ventilation effectiveness for different air distribution arrangements.

Odours

ASHRAE (1997) state that, for tobacco smoke, an increase in humidity at constant temperature lowers the intensity level of an odour but an increase in temperature at constant moisture content lowers the odour level only slightly. Initial adaptation to an odour is fairly rapid and perception decreases with time. On the other hand irritation of the eyes and nose generally increases as time passes and is greatest at low humidities. To minimise eye irritation humidity should be from 45 per cent to 60 per cent at normal temperatures. The minimum perception of odours generally appears to be in the humidity band 45 per cent to 65 per cent, at normal temperatures.

Perceived indoor air quality

Cooler, drier air is perceived as being freer of contaminants, even in clean, non-odourous, well-ventilated spaces, and perceived air freshness decreases with increasing temperature and humidity according to Berglund and Cain (1989). Fang *et al.* (1996) found similar results for air contaminated with emissions from building materials, even though temperature and humidity had no effect, or minimal effect, on the emission rates from the materials. Fanger (1998) observed that people preferred rather dry and cool air, the effect of temperature and humidity being combined as the enthalpy of the air related to the perceived air quality.

Oresczcyn *et al.* (1999) argue that, when people fill in questionnaires, different interpretations of IAQ may be due to:

(1) Perception or belief that the outdoor air quality is poor.
(2) Lack of air movement.
(3) High air temperature. See Berglund and Cain (1989) and Fang *et al.* (1996).
(4) High relative humidity. See Berglund and Cain (1989) and Fang *et al.* (1996).

(5) The combined effect of air movement, temperature and humidity leading to a perception of 'stuffiness'. (This does not agree with the original definition of stuffiness by Bedford (1964).)
(6) Visual appearance of the work environment (e.g. dirty supply air diffusers).
(7) Perception of poor IAQ used as a surrogate for poor working conditions or management, or for their lack of control over their job, workplace and/or environment.

Standard solutions to such expressed views are to increase the provision of fresh air or to remove sources of pollution but, as Oreszczyn *et al.* (1999) say, this may not be dealing with the actual cause, which could be 1, 6 or 7, above.

Oresczcyn *et al.* (1999) reported the results of a study of eight air conditioned buildings and found no correlation between outside air ventilation rate and perceived indoor air quality. Parine (1996) and Bluyssen *et al.* (1995) obtained similar results.

4.12   The choice of inside design conditions

For a person to feel comfortable it appears that the following conditions are desirable:

(1)   The air temperature should be higher than the mean radiant temperature in summer, but lower in winter.
(2)   The average air velocity in the room should probably not exceed 0.15 m s^{-1} in an air conditioned room but higher velocities may be acceptable with air temperatures greater than 26°C.
(3)   Relative humidity should desirably lie between about 45 per cent and 60 per cent.
(4)   Relative humidity should never exceed 70 per cent.
(5)   The dew point should never be less than 2°C.
(6)   The temperature difference between the feet and the head should be as small as possible, normally not exceeding 1.5°C and never more than 3°C.
(7)   Floor temperatures should not be greater than 26°C when people are standing and probably not less than 17°C.
(8)   The radiant temperature asymmetry should not be more than 5°C vertically or 10°C horizontally.
(9)   The carbon dioxide content should not exceed about 0.1 per cent.

As explained earlier, not all these variables are directly amenable to regulation and no air conditioning system is able to achieve control over all of them. The two most important are dry-bulb temperature and air velocity, with mean radiant temperature of slightly less importance. Of these a comfort air conditioning system can only exercise direct automatic control over the dry-bulb temperature. A suitable choice of air velocity may be achieved by proper attention to the system of air distribution and acceptable values of mean radiant temperature should result from co-operation between the design engineer and the architect, aiming to eliminate objectionable radiant effects from sunlit windows in summer, cold windows in winter, cold exposed floors or walls, and excessive radiation from light fittings.

The choice of inside comfort design conditions for an air conditioned room or building depends on the physiological considerations already debated and on economic factors. The designer then examines the outside design state, the clothing worn by the occupants, their rate of working and the period of occupancy. An air temperature of 22°C or $22\frac{1}{2}$°C with about 50 per cent saturation (the same as relative humidity as far as comfort is concerned) is a comfortable choice for long-term occupancy by normally clothed, sedentary people in the UK but the humidity can be allowed to rise to 60 per cent (or even a little more) or to fall towards 40 per cent, under conditions of peak summer heat gains if psychrometric, commercial or other practical considerations warrant it. (It is possible that as high as 23°C dry-bulb may be quite tolerable in this country if the clothing worn is appropriate, when there is sufficient air movement and the mean radiant temperature is less than the air temperature. In the author's view more than 23°C is usually too high as a design choice in the UK. As summer passes and heat gains diminish to become heat losses under winter design conditions outside, the inside dry-bulb may be allowed to fall to 20°C and the humidity also towards a value corresponding to 2°C dew point. (There is some evidence, according to Green (1981), that the incidence of the common cold, and absenteeism, diminishes in winter if people live or work in buildings where the relative humidity is kept in the vicinity of 50 per cent.)

There is economic sense in not adopting too low a temperature inside during summer when air conditioning a space such as a shop, or the foyer of a theatre, where the occupancy (by the patrons) is relatively short term. For example, a foyer at 25°C when it is 28°C outside gives an immediate impression (fading with time) of comfort upon entry. On the other hand, an auditorium should be at 22°C or $22\frac{1}{2}$°C because the audience is present for several hours during which 25°C would be noticeably too high with English customs of dress. There is a lower limit to the temperature of a conditioned space, set by winter design weather and the activity and clothing of the occupants. Thus in the UK the inside temperature should be 20°C (or perhaps 21°C) while the temperature outside rises from its winter design value to about 20°C. As the outside temperature rises from 20°C to about 28°C, the inside temperature should follow from 20°C or 21°C to 22°C or $22\frac{1}{2}$°C. Apart from the change in outside temperature, variation in the inside temperature between 20°C and $22\frac{1}{2}$°C (or even 23°C) may occur because of changes in the heat gains from internal sources or by solar radiation through windows. Such temperature variations should be quite acceptable and be under thermostatic control by the system. It is quite unnecessary to specify tight tolerances over automatic temperature control for comfort conditioning, apart from being needlessly expensive. Less than 20°C dry-bulb will usually be uncomfortably cool in an air conditioned building, at any time, because of the fairly high air change rates (5 to 20 per hour) that prevail.

In a warm or tropical climate, it is neither comfortable nor economically desirable to provide temperatures as low as those used in a temperate climate. It is customary in such circumstances to select a dry-bulb about 5° to 20°C less than the outside design value, with a preferred inside maximum of 25°C dry-bulb, coupled with a humidity of about 50 per cent to 60 per cent. However, there are occasions, even in such climates, where dry-bulbs of 22° to 23°, with humidities of 50 per cent to 60 per cent, are maintained for long-term occupancy.

4.13 Design temperatures and heat gains

The choice of inside and outside summer dry-bulb design temperatures affects the heat gains and hence influences the capital cost of the installation and its running cost, the latter implying the energy consumption of the system. Any change in the room design condition, particularly dry-bulb temperature, will have an effect on the comfort of the occupants. Similarly, any change in the chosen outside dry-bulb temperature will influence the system performance and the satisfaction given. Thus a relaxation of the outside design dry-bulb temperature to a lower value may give a small reduction in the sensible heat gains and the capital cost but this must be balanced against the fact that the system will only be able to maintain the inside design conditions for a shorter period of the summer. The relative merits of any such decisions must be carefully considered and the client advised. Example 7.19 considers this.

Exercises

1.   The human body adjusts itself, within limits, to maintain a relatively constant internal temperature of 37.2°C.

(*a*) How does the body attempt to compensate for a cool environment which tends to lower the internal temperature?

(*b*) How does the body attempt to compensate for a warm environment approaching body temperature or exceeding it?

(*c*) State, giving your reasons, whether an increased air motion as provided by a large rate of air change or the action of a 'punkah' fan is likely to be beneficial to comfort in a room at 29.5°C, 75 per cent relative humidity.

2.   It is required to visit a site to obtain measurements of comfort conditions at a point 1.5 m above floor level in the centre of a particular room. The measurements required are dry-bulb and wet-bulb temperatures, globe temperature and mean air speed. State what simple instruments and ancillary aquipment should be taken to site in order to obtain the required measurements. Briefly describe the instruments, with the aid of sketches and explain how to use each one, mentioning any precautions which should be taken to ensure accuracy of the values obtained.

3.   (*a*) Write down an equation expressing the thermal balance between the human body and its environment.

(*b*) Under what conditions is the temperature of the deep tissues of the human body going to change? Discuss the physiological mechanisms which the body employs to adjust such an imbalance. How can the air conditioning engineer, through an appropriate manipulation of the environment, assist the body in feeling comfortable?

4.   List the factors in the environment which affect the body's feeling of comfort and describe how they influence the rate of heat loss from the body.

5.   Briefly discuss the influence of clothing on human comfort. How is the effect of clothing expressed? Why would you expect a mixed group of people in an air conditioned room to have different attitudes to thermal comfort?

6.   Define plane radiant temperature and explain how it is used to describe the effect of asymmetrical radiation on comfort. State limiting values for vertical and horizontal radiant asymmetry.

7.   Explain briefly how Fanger's equation defines thermal comfort, stating the physical variables used for this purpose. Explain also the meaning of the terms: predicted mean vote, predicted percentage dissatisfied and lowest possible percentage of dissatisfied persons.

8.   Discuss the difference between the original concept of effective temperature and that of standard effective temperature.

9. State the conditions of the indoor environment that should be satisfied for a person to feel comfortable. Which of these conditions are under the control of the air conditioning system? How can the designer arrange for the other conditions to have values likely to achieve comfort?

10. Quote suitable inside design conditions for the provision of comfort conditions in summer and winter for sedentary workers in the UK. Explain how different design conditions would be chosen for summer in a tropical environment and suggest indoor design conditions for an outdoor state of 40°C dry-bulb, 30°C wet-bulb.

Notation

Symbol	Description	Unit
A_D	Du Bois bodily surface area	m^2
C	bodily rate of heat loss by convection	W or W m^{-2}
E	bodily rate of heat loss by evaporation	W or W m^{-2}
E_d	heat loss by vapour diffusion through the skin	W
E_{re}	latent heat loss by respiration	W
E_{sw}	heat loss by sweating from the skin	W
E_v	ventilation effectiveness	–
H	internal rate of bodily heat production	W
h	height of a person	m
h_c	convection heat transfer coefficient	W m^{-2} K^{-1}
h_e	evaporative heat transfer coefficient at the clothing surface	W m^{-2} kPa^{-1}
h_r	radiation heat transfer coefficient	W m^{-2} K^{-1}
I_{cl}	insulating value of clothing	m^2 K W^{-1}
I_{clo}	thermal insulation of the total clothing ensemble	m^2 K W^{-1}
I_{clui}	effective insulation of garment i	m^2 K W^{-1}
i_m	moisture permeability index for clothing	–
L	sensible heat loss by respiration	W
M	metabolic rate for a particular activity	W
m	body mass	kg
PD	percentage of people dissatisfied	%
PMV	predicted mean vote	–
PPD	predicted percentage of people dissatisfied	%
LPPD	lowest possible percentage of people dissatisfied	%
p_{es}	saturated vapour pressure at temperature t_{ef}	kPa
p_s	vapour pressure of humid air	Pa
p_w	vapour pressure of water	Pa
R	bodily rate of heat loss by radiation	W or W m^{-2}
S	bodily rate of heat storage	W or W m^{-2}
T_a	air dry-bulb temperature	K
T_g	globe temperature	K
T_{rm}	mean radiant temperature	K
T_u	intensity of air turbulence	–
t_a	ambient air dry-bulb temperature	°C
t_{ed}	effective draught temperature	°C

t_{ef}	standard effective temperature	°C
t_o	operative temperature	°C
$\bar{t}_r$	mean room dry-bulb temperature	°C
t_{res}	dry resultant temperature	°C
t_x	local air dry-bulb temperature	°C
V	mean air velocity	m s^{-1}
V_{sd}	standard deviation of the local air velocity	m s^{-1}
v	relative air velocity	m s^{-1}
v_x	local air velocity	m s^{-1}
W	rate of working	W or W m^{-2}
w	fraction of the skin surface that is wetted	–
ϕ_a	ambient air relative humidity	%

References

ASHRAE Handbook 1989: Fundamentals.

ANSI/ASHRAE Standard 62–1989: Ventilation for acceptable indoor air quality.

ANSI/ASHRAE Standard 55–1992: Thermal Environmental Conditions for Human Occupancy.

ASHRAE Handbook 1997: Fundamentals.

Bedford, T. (1964): *Basic Principles of Ventilation and Heating*, 2nd Edition, H.K. Lewis Publishers.

Belding, H.S. and Hatch, T.F. (1955): Index for evaluating heat stress in terms of resulting physiological strains, *Heating, Piping and Air Conditioning* **207**, 239.

Berglund, L.G. (1980): New Horizons for 55–74; implications for energy conservation and comfort, *ASHRAE Trans.* **86**, Part 1.

Berglund, L.G. and Cain, W. (1989): Perceived air quality and the thermal environment, Proc. IAQ 89, *The Human Health Equation*, 93–99.

Berglund, L.G. (1998): Comfort and Humidity, *ASHRAE Journal*, Aug. 1998, 35–41,

Berry, P.C. (1961): Effect of coloured illumination upon perceived temperature, *Journal of Applied Psychology* **45**, 248–250.

Bluysen, L. *et al.* eds (1995): European Audit Project to Optimise Indoor Air Quality and Energy Consumption of Office Buildings, Commission of the European Communities Joule II Programme, TNO Felf, The Netherlands.

Brandt, A.D. (1933): The influence of atmospheric ionisation upon the human organism, *J. Industr. Hyg.* **15**, 354.

BS EN ISO 7730 (1995): Moderate thermal environments. Determination of the PMV and PPD indices and specification of the conditions for human comfort (London: British Standards Institution).

CIBSE Guide (1999): *A1, Environmental criteria for design.*

Dessauer, F. (1931): Zehn Jahre Forschung auf dem Physikalische-Meizinischen Grenzgebeit, Leipzig, Georg Thieme.

Du Bois, D. and Du Bois, E.F. (1916): A formula to estimate approximate surface area if height and weight are known, *Archives of Internal Medicine* **17**, 863.

Fang, L., Clausen, G. and Fanger, P.O. (1996): The Impact of Temperature and Humidity on Perception and Emission of Indoor Air Pollutants, Yoshizawa, S. *et al.* eds, Proc. 7th International Conference on Indoor Air Quality and Climate, *Indoor Air '96* **4**, 349–354.

Fanger, P.O. (1972): *Thermal Comfort Analysis and Applications in Environmental Engineering*, McGraw-Hill Book Company.

Fanger, P.O., Melikov, A., Hanzana, H. and Ring, J. (1987): Air turbulence and sensation of draught, *Energy and Buildings* **12**, 21–39.

Fanger, P.O. (1998): Buildings and people, new evidence for air conditioning, *New Developments in Air Conditioning*, eds W.P. Jones and A.F.C. Sherratt, 1–4, Mid Career College Press.

Gagge, A.P. and Moriyama, I.M. (1935): The annual and diurnal variation of ions in an urban community, *Terrest. Mag. and Atmospheric Electricity* **40**, 295.

Gagge, A.P. *et al.* (1938): The influence of clothing on physiological reactions of the human body to varying environmental temperatures, *American Journal of Physiology* **124**, 30.

Gagge, A.P., Stolwiljk, A.J. and Saltin, B. (1969a): Comfort and thermal sensation and associated physiological responses during exercise at various ambient temperatures, *Environmental Research* **2**, 201.

Gagge, A.P., Stolwiljk, A.J. and Nishi, Y. (1969b): The prediction of thermal comfort when thermal equilibrium is maintained by sweating, *ASHRAE Trans.* **75**(2), 108.

Gagge, A.P., Stolwijk, J. and Nishi, I. (1971): An effective temperature scale based on a simple model of human physiological regulatory response, *ASHRAE Trans.* **77**(1), 247–262.

Glickman, N. *et al.* (1947): Physiological adjustments of human beings to sudden changes in environment, *ASHVE Trans.* **55**, 27.

Glickman, N. *et al.* (1949): Physiological adjustment of normal subjects and cardiac patients to sudden changes in environment, *ASHVE Trans.* **55**, 27.

Gonzalez, R.R., Berglund, L.G. and Gagge, A.P. (1978): Indices of thermo-regulatory strain for moderate exercise in heat. *J. Appl. Physiol.* **44**, 889.

Green, G.H. (1979): The effect of indoor relative humidity on colds, *ASHRAE Trans.* **85**, Part 1.

Green, G.H. (1981): Report in the New Scientist, 15 Jan. **89**, 134.

Haldane, J.S. and Priestly, J.G. (1905): The regulation of lung ventilation, *J. Physiol.* **32**, 225.

Herrington, L.P. (1935): The influence of ionised air upon normal subjects, *J. Clin. Invest.* **14**, 70.

Herrington, L.P. and Kuh, C. (1938): The reaction of hypertensive patients to atmospheres containing high concentrations of heavy ions, *J. Industr. Hyg. and Toxicol.* **20**, 179.

Hick, F.K. *et al.* (1953): Physiological adjustments of clothed human beings to sudden changes in environment—first hot moist and later comfortable conditions, *ASHVE Trans.* **59**, 189–198.

Inouye, T. *et al.* (1953): A comparison of physiological adjustment of clothed women and men to sudden changes in environment, *ASHVE Trans.* **59**, 35–48.

Inouye, T. *et al.* (1954): Physiological responses to sudden changes in atmospheric environment, *ASHVE Trans.* **60**, 315–328.

Koch, W. *et al.* (1960): Environmental Study II—Sensation responses to temperature and humidity under still air conditions in the comfort range, *ASHRAE Trans.* **66**, 264.

Liviana, J.E., Rohles, F.H. and Bullock, O.D. (1988): Humidity, comfort and contact lenses, *ASHRAE Trans.* **94**(1):, 3–11.

Love, M. (1998): The secret life of the microbe, *HAC Journal*, July, 34–37.

Mackean, D.G. (1977): *Introduction to Biology*, John Murray (Publishers) Ltd., London.

McCullough, E.A. and Jones, B.W. (1984): A comprehensive database for estimating clothing insulation. IER Technical Report, 84–101, Institute for Environmental Research, Kansas State University, KS.

McIntyre, D.A. (1976): *Overhead radiation and comfort*, Electricity Council Research Centre, R 918, Capenhurst.

McNall, P.E. *et al.* (1971): Thermal comfort (thermally neutral) conditions for three levels of activity, *ASHRAE Trans.* **77**, Part II.

Miller, P.L. and Nash, R.T. (1971): A further analysis of room air distribution performance, *ASHRAE Trans.* **77**, Part II.

Missenard, F.A. (1933): La température résultante d'un milieu, *Chal. Ind.*, July.

Missenard, F.A. (1935): Théorie simplifié du thermometre résultante, *Chauffage et Ventilation* **12**, 347.

Nevins, R.G., Michaels, K.B. and Feyerherm, A.M. (1964): The effect of floor surface temperature on comfort: Part I, college age males, *ASHRAE Trans.* **70**, 29.

Nevins, R.G. and Flinner, A.O. (1958): Effect of heated floor temperatures on comfort, *ASHRAE Trans.* **64**, 175.

Nevins, R.G., Michaels, K.B. and Feyerherm, A.M. (1964): The effect of floor surface temperature on comfort: Part II, college-age females, *ASHRAE Trans.* **70**, 37.

Nevins, R.G., Rohles, F.H. Jr., Springer, W.E. and Feyerherm, A.M. (1966): Temperature–humidity chart for thermal comfort of seated persons, *ASHRAE Trans.* **72**, Part 1, 283–291.

Nevins, E.G. and Miller, P.L. (1974): ADPI—An index for design and evaluation, *Australian Refrigeration, Air Conditioning and Heating* **28**, No. 7, 26.

Nevins, R., Gonzalez, R.R., Berglund, L.G. and Gagge, A.P. (1975): Effect of changes in ambient temperature and level of humidity on comfort and thermal sensation, *ASHRAE Trans.* **81**(2).

Olsen, B.W. and Thorshauge, J. (1979): *Differences in comfort sensations in spaces heated by different methods*, Indoor Climate, Danish Building Research Institute, Copenhagen.

Oreszczyn, T., Parine, N. and O'Sullivan, P. (1999): Air Conditioning, Health and Indoor Air Quality, *Air conditioning system design and application*, eds W.P. Jones and A.F.C. Sherrat, Mid Career College Press, 115–120.

Parine, N. (1996): Sensory Evaluation of Indoor Air Quality by Building Occupants versus Trained and Untrained Panels, *Indoor Built Environment* **5**, 34–43.

Rohles, F.H. and Nevins, R.G. (1971): The nature of thermal comfort for sedentary man, *ASHRAE Trans.* **77**, Part I, 239.

Rohles, F.H. (1973): The revised model comfort envelope, *ASHRAE Trans.* **79**, Part 2, 52.

Rohles, F.H. (1974): The measurement and prediction of thermal comfort, *ASHRAE Trans.* **80**, Part 2, 98.

Rohles, F.H. (1980): Temperature and temperament: a psychologist looks at thermal comfort, *ASHRAE Trans.* **86**, Part I.

Springer, W.E. *et al.* (1966): The effect of floor surface temperatures on comfort, Part III, the elderly, *ASHRAE Trans.* **72**, Part I, 292.

Whitehouse, A.G.R. *et al.* (1932): The osmotic passage of water and gases through human skin, *Proc. Roy. Soc.* **111**, 412.

Yaglou, C.P. and Miller, W.E. (1924): Effective temperature applied to industrial ventilation problems, *ASHVE Trans.* **30**, 339–364.

Yaglou, C.P. and Miller, W.E. (1925): Effective temperature with clothing, *ASHVE Trans.* **31**, 88–99.

Yaglou, C.P. *et al.* (1932): Changes in ionic content of air in occupied rooms ventilated by natural and by mechanical methods, *ASHVE Trans.* **38**, 191.

Yaglou, C.P. *et al.* (1933): Observations on a group of subjects before, during and after exposure to ionised air, *J. Industr. Hyg.* **15**, 341.

Yaglou, C.P. (1935): Observations of physiological efficiency in ionised air, *J. Industr. Hyg.* **17**, 280.

Yaglou, C.P. (1947): A method for improving the effective temperature index, *ASHVE Trans.* **53**, 307.

5
Climate and Outside Design Conditions

5.1 Climate

The variations in temperature, humidity and wind occurring throughout the world are due to several factors the integration of which, for a particular locality, provides the climate experienced. There is, first, a seasonal change in climatic conditions, varying with latitude and resulting from the fact that, because the earth's axis of rotation is tilted at about 23.5° to its axis of revolution about the sun, the amount of solar energy received at a particular place on the earth's surface alters throughout the year. The geography of the locality provides a second factor, influential in altering climate within the confines imposed by the seasonal variation.

Figure 5.1 illustrates the geometrical considerations which show that, at a particular latitude, the earth receives less solar radiation in winter than it does in summer. The importance of this in its effect on the seasons stems from the fact that, for all practical purposes, the sun is the sole supplier of energy to the earth.

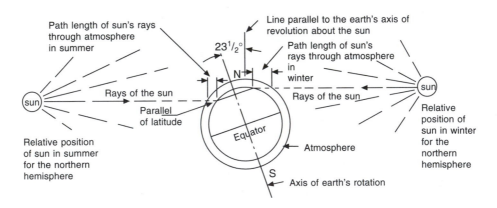

Fig. 5.1 Earth–sun geometrical relationships in summer and in winter: the path length of the solar radiation through the atmosphere is less in summer and the altitude of the sun is greater.

The geography of a place determines how much solar energy is absorbed by the earth, how much is stored and how readily it is released to the atmosphere. The atmosphere is comparatively transparent to the flux of radiant solar energy (termed 'insolation') but the land masses which receive the energy are opaque to it and are fairly good absorbers of it, although this depends on the reflectivity of the surface. This means that the thermal energy from the sun warms up land surfaces on which it falls. Some of this energy travels inwards and is stored in the upper layers of the earth's crust; some is convected to the atmosphere and some is re-radiated back to space, but at a longer wavelength (about 10 micrometres), since its mean surface temperature is very much less than that of the sun. Four-fifths of the earth's surface is water, not land, and water behaves in a different fashion as a receiver of insolation, being partially transparent to it; consequently, the energy is absorbed by the water in depth, with the result that its surface temperature does not reach such a high value during the daytime. On the other hand, at night, heat is lost from the land to the sky much more rapidly since less was stored in its shallow upper crust than was absorbed and stored in the deeper layers of water. The result is that land-surface temperatures tend to be lower at night than are water-surface temperatures. It is evident from this that places in the middle of large land masses will tend to have a more extreme annual variation of temperature than will islands in a large sea. Thus, the climate of places on the same latitude can vary enormously. To realise this we have but to compare the temperate seasons experienced in the British Isles with the extremes suffered in Central Asia and Northern Canada at about the same latitude. The exchanges of radiant energy cited above as responsible for the differences in maritime and continental climates are complicated somewhat by the amount of cloud. Cloud cover acts as an insulating barrier between the earth and its environment; not only does it reflect back to outer space a good deal of the solar energy incident upon it but it also stops the passage of the low-frequency infra-red radiation which the earth emits. In addition, the quantity of carbon dioxide in the atmosphere reduces the emission of infra-red radiation. Mountain ranges also play a part in altering the simple picture, presented above, of a radiation balance.

The effect of the unequal heating of land and sea is to produce air movement. This air movement results in adiabatic expansions and compressions taking place in the atmosphere, with consequent decreases and increases respectively in air temperature. These temperature changes, in turn, may result in cloud formation as values below the dew point are reached.

One overall aspect of the thermal radiation balance is prominent in affecting our weather and in producing permanent features of air movement such as the Trade Winds and the Doldrums. This is the fact that, for higher latitudes, the earth loses more heat to space by radiation than it receives from the sun but, for lower latitudes, the reverse is the case. The result is that the lower latitudes heat up and the higher ones cool down. This produces a thermal up-current from the equatorial regions and a corresponding down-current in the higher latitudes. While this is true for an ideal atmosphere, the fact that the earth rotates and that other complicating factors are present means that the true behaviour is quite involved and not yet fully understood.

5.2 Winds

It might be thought that wind flowed from a region of high pressure to one of low pressure, following the most direct path. This is not so. There are, in essence, three things which combine to produce the general pattern of wind flow over the globe. This pattern is complicated further by local effects such as the proximity of land and sea, the presence of mountains and so on. However, the three overall influences are:

(i)   the unequal heating of land and sea;
(ii)   the deviation of the wind due to forces arising from the rotation of the earth about its axis;

(iii)   the conservation of angular momentum—a factor occurring because the linear velocity of air at low latitudes is less than at high latitudes.

The general picture of wind distribution is as follows. Over equatorial regions the weather is uniform; the torrid zone is an area of very light and variable winds with frequent calms, cloudy skies and violent thunderstorms. These light and variable winds are called the 'Doldrums'. Above and below the Doldrums, up to 30° north and south, are the Trade Winds, which blow with considerable steadiness, interrupted by occasional storms. Land and sea breezes (mentioned later) also affect their behaviour.

Above and below 30° of latitude, as far as the sub-polar regions, the Westerlies blow. They are the result of the three factors mentioned earlier, but their behaviour is very much influenced by the development of regions of low pressure, termed cyclones, producing storms of the pattern familiar in temperate zones. This cyclonic influence means that the weather is much less predictable in the temperate zones, unless the place in question is in a very large land mass—for example in Asia or in North America. In temperate areas consisting of a mixture of islands, broken coastline and sea, as in north-western Europe, cyclonic weather is the rule and long-term behaviour difficult to forecast. The whole matter is much complicated by the influence of warm and cold currents of water.

5.3   Local winds

Exceptions exist to all the generalisations made in the preceding section. One important local effect occurs when, at coastlines, land and sea breezes result from the unequal heating. In the daytime, air rises over the hot land, and cool air comes in from the sea to displace it as the day advances. At night, the rapid cooling of the land chills the air in the vicinity of its surface which then moves out to sea to displace warmer air, which rises. This effect results in an averaging out of coastal air temperatures. The diurnal variation in dry-bulb temperature is less near the coast than it is further inland.

The seasonal variation of temperature (as well as the diurnal variation) is greatest over the middle of large land masses. This results in seasonal winds, called monsoons, blowing in from the sea to the land in the summer and outwards from land to sea in winter. India, Asia and China have distinct monsoon effects. Tornadoes, the origin of which is obscure, are instances of the highest wind speeds encountered. They are vortices, a few hundred metres across, which move in a well-defined path. They can occur all over the world but, fortunately are common only in certain areas. Their lives are short but violent. Wind speeds of 480 km h^{-1} or more are likely and pressures at their centre have reached values as low as about 800 millibars. (The lowest recorded barometric pressure in the United Kingdom is 925 mbar and the highest is 1055 mbar. In London the values are 950 mbar and 1049 mbar, respectively.)

At higher altitudes, the comparative absence of dust particles and the reduction in the amount of water vapour in the atmosphere mean that radiation outwards to space is less impeded at night. The consequence is that the surface of the earth on high ground cools more rapidly at night than it does at lower altitudes. The air in contact with the colder surface becomes chilled (and hence denser) and slides down mountain sides to low-lying ground where it is apparent as a gentle wind, termed a katabatic or gravity wind. Examples are common and a typical case is the Mistral, a katabatic wind into the Mediterranean from the high plateau of south and eastern France.

Other local winds result from the passage of air in front of an advancing depression, across a hot desert—the Sahara, for example, which gives a dry wind along the coast of north Africa. (The same wind changes its character in crossing the Mediterranean, becoming a warm moist wind to southern Europe.)

There is another local wind which results from a process of adiabatic expansion and cooling as air rises. Air striking the windward side of rising slopes expands as it ascends, cooling occurring meanwhile. A temperature below the dew point is eventually reached, clouds form and, in due course, descend. The same air then flows over the highest point of the slope and rain falls on the lee side, suffering adiabatic compression and an increase in temperature as it does so. It appears as a dry, warm wind on the lee side of the high ground. An example of this is the Chinook wind, blowing down the eastern slopes of the Rocky mountains.

One last instance of a local wind of importance occurs in West Africa during winter in the northern hemisphere. The Sahara desert cools considerably at night in the winter, causing a progressive drop in air temperature as the season advances. This cool, dry air flows westwards to the African coast, displacing the warmer and lighter coastal atmosphere.

5.4 The formation of dew

At nightfall, the ground, losing heat by radiation, undergoes a continued fall in temperature and the air in contact with the ground also suffers a fall in temperature, heat transfer between the two taking place by convection. Eventually, the temperature of the ground drops below the dew point and condensation forms. Although the rate of heat loss from solid surfaces is roughly constant, depending on the fourth power of the absolute temperature of the surface, all solid objects do not fall in temperature at the same rate. A good deal of heat is stored in the upper layers of the earth, and heat flows outwards to the surface to make good radiation losses therefrom. Thus, good conductors of heat in good thermal contact with the ground will fall in temperature at a rate much the same as that of the main surface mass of the earth nearby. Bad conductors, or insulated objects, however, will not be able to draw heat from the earth to make good their losses by radiation and so their temperature will fall more rapidly and dew will tend to be deposited first on such objects. Examples of these two classes are rocks, which are good conductors and are in intimate contact with the earth and grasses which are poor conductors. Dew tends to form on grass before it does on rocks.

5.5 Mist and fog

For condensation to occur in the atmosphere the presence is required of small, solid particles termed condensation nuclei. Any small solid particle will not do; it is desirable that the particles should have some affinity for water. Hygroscopic materials such as salt and sulphur dioxide then, play some part in the formation of condensation. The present opinion appears to be that the products of combustion play an important part in the provision of condensation nuclei and that the size and number of these nuclei vary tremendously. Over industrial areas there may be several million per cubic centimetre of air whereas, over sea, the density may be as low as a few hundred per cubic centimetre.

These nuclei play an important part in the formation of rain as well as fog, but for fog to form the cooling of moist air must also take place. There are two common sorts of fog: advection fog, formed when a moist sea breeze blows inland over a cooler land surface, and radiation fog. Radiation fog forms when moist air is cooled by contact with ground which has chilled as the result of heat loss by radiation to an open sky. Cloud cover discourages such heat loss and, inhibiting the fall of surface temperature, makes fog formation less likely. Still air is also essential; any degree of wind usually dissipates fog fairly rapidly. Fog has a tendency to occur in the vicinity of industrial areas owing to the local atmosphere being rich in condensation nuclei. Under these circumstances, the absence of wind is helpful in keeping up the concentration of such nuclei. The dispersion of the nuclei is further impeded by the presence of a temperature inversion, that is, by a rise in air temperature with increase of height, instead of the reverse. This discourages warm air from rising and encourages the products of combustion and fog to persist, other factors being helpful.

Long-wave thermal radiation, $I_{LW,}$ directed to the sky at night can be considered as radiation to a black body at a temperature of absolute zero. This is modified by a correction factor to account for variations in the absorption by water vapour in the lower reaches of the atmosphere. This changes as the amount of cloud cover and the area of the sky seen by the surface varies. According to Brunt (1932) the absorption effect can be accounted for by a vapour correction factor, K, expressed by

$$K = 0.56 - 0.08\sqrt{p_s} \tag{5.1}$$

where p_s is the vapour pressure of the air in millibars. The amount of sky seen by a surface is given by an angle factor, B,

$$B = 0.5(1 + \cos \delta) \tag{5.2}$$

where δ is the acute angle between the surface and the horizontal. Thus for a flat roof $\delta = 0$ and $B = 1$, whereas for a wall $\delta = 90°$ and $B = 0.5$. Cloud cover can play a part as well and if we assume it is wholly effective in suppressing loss from the surface to outer space, a cloud cover factor, C, can be introduced. Table 5.1 gives typical, approximate, average cloud cover factors for Kew, based on data recorded for the amount of bright sunshine received between sunrise and sunset in relation to the maximum amount of sunshine that could be received during the same period of the day.

Table 5.1   Typical, approximate cloud cover factors for Kew

Month	Jan	Feb	Mar	Apr	May	Jun	Jul	Aug	Sep	Oct	Nov	Dec
C	0.2	0.2	0.3	0.4	0.4	0.4	0.4	0.4	0.4	0.3	0.2	0.2

Hence the long-wave thermal radiation from a surface to the sky can be expressed by

$$I_{LW} = 5.77B(1 - C)K\left[\frac{273 + t_s}{100}\right]^4 \tag{5.3}$$

The surface temperature of a building could be calculated if steady-state heat flow is assumed but if the building has appreciable thermal inertia, as is usually the case, it becomes complicated.

EXAMPLE 5.1

Calculate the emission from a flat roof to the night sky in January, at Kew, under steady-state conditions, when the temperature of the room beneath the roof is 20°C and the outside air is at $-1°C$, assuming (*a*) 20 per cent cloud and (*b*) no cloud cover. Assume the *U*-value of the roof is 0.25 W m^{-2} K^{-1} and the outside surface film resistance is 0.0455 m^2 K W^{-1}.

Answer

(*a*) If $U = 0.25$ W m^{-2} K^{-1} then the overall thermal resistance is $1/0.25 = 4.0$ m^2 K W^{-1} and the roof surface temperature is

$$t_s = -1 + (0.0455/4) \times (20 + 1) = -0.76°C$$
$$\delta = 0 \text{ therefore, by equation (5.2), } B = 1.0$$

For an outside temperature as low as $-0.76°C$ the air is likely to be saturated in the UK and hence, from CIBSE psychrometric tables:

$$p_s = 0.5297 \text{ kPa (5.297 mbar) and, by equation (5.1),}$$
$$K = 0.56 - 0.08\sqrt{5.297} = 0.376$$

and hence, by equation (5.3),

$$I_{LW} = 5.77 \times 1 \times (1 - 0.2) \times 0.376[(273 - 0.76)/100]^4 = 95.3 \text{ W m}^{-2}$$

(*b*) $C = 0$, $B = 1$, $K = 0.376$

$$I_{LW} = 5.77 \times 1 \times (1 - 0) \times 0.376[(273 - 0.76)/100]^4 = 119.2 \text{ W m}^{-2}$$

5.6   Rain

Upward air currents undergo adiabatic expansion and cooling. The adiabatic lapse rate of air temperature which thus occurs is about 1 K for every 100 m of increase in altitude. Comparing this with the normal fall of temperature, 0.65 K per 100 m (see equation (2.2)), it can be seen that in due course the rising air, which was originally, perhaps, at a higher temperature than ambient air, is eventually at a lower temperature than its environment. At this point, upward motion ceases.

If, during the adiabatic cooling process, the air temperature in the rising current fell below its dew point, condensation would occur in the presence of adequate condensation nuclei. The liquid droplets so formed would tend to fall under the influence of the force of gravity but the rate of fall would be countered by the frictional resistance between the droplet and the rising current of air. Whether the cloud, formed as the result of the condensation, starts to rain or not depends on the resultant of the force of gravity downwards and the frictional resistance upwards. The rate at which rain falls depends on the size of the drops formed.

5.7   Diurnal temperature variation

The energy received from the sun is the source of heat to the atmosphere and so the balance of heat exchanges by radiation between the earth and its environment, which is reflected in changes of air temperature, must vary according to the position of the sun in the sky. That is to say, there will be a variation of air temperature against time.

The surface of the earth is at its coolest just before dawn, having had, in the absence of cloud cover, an opportunity of losing heat to the black sky during the whole of the night. Accordingly, it is usual to regard the lowest air temperature as occurring about one hour before sunrise. As soon as the sun rises, its radiation starts to warm the surface of the earth, and as the temperature of the ground rises heat is convected from the surface of the earth to the layers of air immediately above it. There is thus a progressive increase in air temperature as the sun continues to rise, and also for some little time after it has passed its zenith, because of the fact that some of the heat received by the ground from the sun and stored in its upper layers throughout the morning escapes upwards and is lost by convection in the early afternoon. It is, therefore, usual to find the highest air temperatures between 14.00 h and 16.00 h (sun-time). In fact, between about 13.00 h and 17.00 h one would not expect very great changes of temperature.

On an average basis, it is not unreasonable to suppose that there is some sort of rough sinusoidal relationship between sun-time and air dry-bulb temperature. The curve would not be wholly symmetrical since the time between lowest and highest temperatures would not necessarily equal that between highest and lowest.

For the month of June, sunrise is at about 04.00 h and sunset at about 20.00 h. The time of lowest temperature is at about 03.00 h and the time of highest temperature is at 15.00 h. There is, thus, a lapse of 12 hours while the temperature is increasing. Since the night period for unrestricted cooling is only about 7 hours (20.00 h to 03.00 h) the curve will be broader in daytime than at night. The reverse would be the case in December.

If we assume the outside temperature, t_θ, varies sinusoidally with time, θ, and its maximum value, t_{15}, occurs at 15.00 h sun-time we can write:

$$t_\theta = t_{15} - \frac{D}{2}\left[1 - \sin\frac{(\theta\pi - 9\pi)}{12}\right] \tag{5.4}$$

where D is the difference between the mean daily maximum and minimum temperatures, otherwise termed the diurnal range. Records are kept of weather data at hourly intervals for many locations in the UK and overseas. Such information is commonly stored electronically and, typically, includes: dry-bulb, screen wet-bulb, dew point, wind direction and speed, atmospheric pressure, type of cloud and cloud cover in several layers at different altitudes, solar radiation and so on. Some details for 17 July 1967, the hottest day of that year, for Wethersfield, near Braintree in Essex, have been taken from meteorological weather data and values of dry-bulb temperature and dew point temperature are plotted against time in Figure 5.2. The difference between the maximum (27.8°C at 14.00 h sun-time) and minimum (15.6°C at 03.00 h) is 12.2°C and inserting this for the value of D in equation (5.4) yields the sine curve also shown in Figure 5.2. We see that a sinusoidal assumption is not unreasonable for warm summer weather, at least for Wethersfield. Equation (5.4) is useful for estimating outside air temperatures when heat gains must be calculated for some time other than 15.00 h, in summer. If the meteorological records show that the maximum occurs at a time different from 15.00 h, equation (5.4) should be used assuming that the maximum is at 15.00 h and the answers then shifted by the number of hours necessary to give the maximum at the correct time, See example 5.2.

Equation (5.4) is useful for the determination of temperatures in the vicinity of the time of maximum daily temperature, in summertime. If there is a marked difference between the period from the mean daily maximum to the mean daily minimum and that from the mean daily minimum to the mean daily maximum, then the technique proposed by CIBSE (1999) should be used.

5.8   Diurnal variation of humidity

If the relative humidity is known in conjunction with the dry-bulb the moisture content can be established. Meteorological data usually include a measurement of relative humidity in

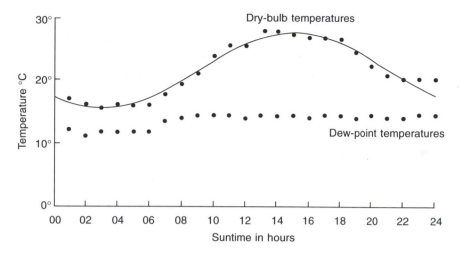

Fig. 5.2   Temperature variations on 17 July 1967 at Wethersfield, Essex.

the afternoon, at a time near that of the maximum temperature, even when the information is not more comprehensively stored. If the time of humidity measurement does not coincide with that of the highest mean daily temperature (about 15.00 h) equation (5.4) may be used to estimate the temperature at the same time as the humidity measurement. This will then establish the moisture content of the outside air at that time. In the absence of weather changes it might be supposed that the moisture content would stay constant throughout the day until the air cooled to its dew point. Although this is not strictly true, the dew point plotted in Figure 5.2 does remain fairly constant, providing a partial verification. In fact, it is difficult to measure dew point with high accuracy, except in the laboratory, and the apparent inconsistency of the values in Figure 5.2 may, perhaps, be attributed to measurement error. During hot weather, vegetation releases a good deal of its moisture into the air, becoming dehydrated in the process. This causes an increase in the moisture content of the air, even though no change of weather has occurred. However, assuming a constant moisture content during most of the day, the relative humidity of the air will rise as the afternoon and evening pass and night falls, because of the drop in dry-bulb temperature. As cooling of the air continues the dew point is reached and since the air is saturated any further fall in temperature will cause dew or mist to form, reducing the moisture content of the atmosphere. With the dawn and sunrise, temperatures increase and the relative humidity of the air falls somewhat, dew evaporating as the morning proceeds. Eventually all the dew disappears and the moisture content is back at its assumed constant value of the day before.

EXAMPLE 5.2

At a meteorological station in the Midlands the mean daily maximum and minimum temperatures are recorded as 20°C (assumed to occur at 1500 h sun-time) and 11.7°C (assumed to occur at 0300 h sun-time), respectively, in August. The mean relative humidity in August is recorded as 65 per cent (at 1300 h sun-time). Plot the typical diurnal variation of temperature and humidity in August.

Answer

First determine the mean outside moisture content and dew point in August. Since the maximum dry-bulb temperature occurs at about 1500 h sun-time, a typical temperature can be determined at 1300 h sun-time (the time at which the humidity is measured) by using equation (5.4).

$$t_{13} = 20 - [(20 - 11.7)/2][1 - \sin(13\pi - 9\pi)/12]$$
$$= 20 - [8.3/2][1 - \sin(4\pi/12)]$$
$$= 20 - 4.15[1 - \sin 60]$$
$$= 20 - 4.15[1 - 0.866]$$
$$= 20 - 0.56$$
$$= 19.4°C$$

Thus, at 1300 h sun-time the dry-bulb is 19.4°C and the relative humidity is 65 per cent. Remembering that relative humidity is not quite the same as percentage saturation, a typical moisture content is established as 0.009156 kg kg^{-1} dry air and the corresponding dew point as 12.7°C, by interpolation in CIBSE Guide, Psychrometric Table (1988).

The results are plotted in Figure 5.3 and extended over 24 hours to show how temperature and humidity change and how dew may form.

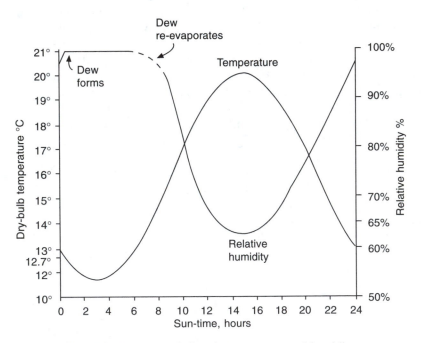

Fig. 5.3   Diurnal variations in temperature and humidity.

5.9   Meteorological measurement

Stations for measuring various atmospheric properties are set up all over the world by the meteorological authorities in different countries. The coverage of the surface of the earth is not complete and so there is a wealth of information for some areas and a scarcity of it for others.

Measurements are made of temperatures and humidity within louvred boxes, positioned in the open air. The louvres permit the ready circulation of outside air by natural means over the instruments but shield them from rain and sunshine. Since the air movement over any wet-bulb thermometer mounted in such a screened housing is by natural means, the velocity of airflow will be too low (less than 4.5 m s^{-1}) to minimise interference from radiation effectively (see section 2.17) and so, 'screened' wet-bulb values will be about $\frac{1}{2}$ degree higher than 'sling' wet-bulb values. Whereas wet-bulb values on the psychrometric chart published by the CIBSE are sling values, data taken from meteorological tables are always based on screened values, and due allowance must be made for this when using such data.

At modern meteorological stations hourly measurements are made and recorded of dry-bulb temperature, humidity and other relevant climatic variables. The highest and lowest values of temperature for each month are averaged over a suitably recent, relevant period of years and quoted as the mean daily maximum and mean daily minimum temperatures for a particular month. These values represent typical high and low temperatures in the month.

The extreme high and low values of temperature in each month are also noted, and averages of these are calculated over the number of years for which observations are made. This yields mean monthly maximum and minimum values of dry-bulb temperature which are typical of warm and cold spells of weather. Other records of temperature are kept but these are largely irrelevant to a study of air conditioning. In underdeveloped countries records may not be so extensive but the principles remain the same.

As regards records of relative humidity, the picture is not so comprehensive. Records are not always kept of wet-bulb temperature, but instead, measurements of relative humidity may be made twice daily—one at about 09.00 h and another at about 13.00 h. The readings are all taken at a height of about 1.5 m above ground level. Typical meteorological data are shown in Table 5.2.

Hourly values of dry-bulb and wet-bulb temperatures are recorded at many meteorological stations and the data processed to show their frequency of occurrence. The CIBSE (1999) covers both cold and warm weather and gives information for several places in the UK with data from Heathrow corresponding to London. The percentage of the total hours in the four summer months, June to September, inclusive, is presented in two ways: as a combination of coincident dry-bulb and wet-bulb values and, separately, as a percentage of the total hours in the summer months that quoted dry-bulb and wet-bulb values are not exceeded. Tables 5.3 and 5.4 list some values.

EXAMPLE 5.3

Making use of the data in Tables 5.3 and 5.4 determine the total number of hours in the four summer months that (*a*) the dry-bulb has exceeded 28°C, (*b*) the wet-bulb has exceeded 20°C and (*c*) the combination of dry- and wet-bulb temperatures has been within the limits 26°–28° dry-bulb and 18°–20° wet-bulb, simultaneously.

Answer

The total number of days in the four summer months is 30 + 31 + 31 + 30 = 122 days. The total number of hours is 24 × 122 = 2928. Hence the answers required are:

Table 5.2 Some typical meteorological data from observations made at Kew (now closed and transferred to Bracknell) during the period 1931 to 1960 but now out of date. Up-to-date observations are similar but warmer. For example, according to Levermore and Keeble (1997) the annual mean temperature at Heathrow in 1960 was 9.95°C but 11.25°C in 1995. They extrapolate a value of 11.7°C for 2010

Month	Temperature Mean daily max	Mean daily min	Mean monthly max	Mean monthly min	Relative humidity at 15.00 h
Jan	6.3	2.2	11.7	−4.3	77
Feb	6.9	2.2	12.1	−3.6	72
Mar	10.1	3.3	15.5	−2.3	64
Apr	13.3	5.5	18.7	0.1	56
May	16.7	8.2	23.3	2.7	57
Jun	20.3	11.6	25.9	6.9	58
Jul	21.8	13.5	26.9	9.3	59
Aug	21.4	13.2	26.2	8.5	62
Sep	18.5	11.3	23.4	5.4	65
Oct	14.2	7.9	18.7	0.4	70
Nov	10.1	5.3	14.4	−1.4	78
Dec	7.3	3.5	12.2	−3.2	81

Table 5.3 Percentages of the total hours in the months June to September, inclusive, that combinations of dry-bulb and wet-bulb temperatures lie within quoted ranges of values, at Heathrow for the period 1976–95

Dry-bulb temperatures	Wet-bulb temperatures 16°–18°	18°–20°	20°–22	22°–24°
24°–26°	1.37	1.32	0.25	
26°–28°	0.42	0.77	0.32	
28°–30°	0.10	0.31	0.25	0.01
30°–32°	0.01	0.12	0.14	0.02

Reproduced by kind permission of the CIBSE from *Guide A, Environmental Design* (1999).

Table 5.4 Percentages of the total hours in the months June to September, inclusive, that dry-bulb and wet-bulb temperatures exceed the stated value, at Heathrow, for the period 1976–95

Dry-bulb temperature	24°	25°	26°	27°	28°	29°	30°	31°
Precentage of the hours exceeded	5.73	3.88	2.53	1.60	0.98	0.62	0.29	0.12
Wet-bulb temperature	16°	17°	18°	19°	20°	21°	22°	
Percentage of the hours exceeded	20.64	12.03	6.16	2.89	1.09	0.19	0.03	

Reproduced by kind permission of the CIBSE from *Guide A, Environmental Design* (1999).

(a)   The dry-bulb exceeded 28°C for (0.98/100) × 2928 = 28.7 h.

(b)   The wet-bulb exceeded 20°C for (1.09/100) × 2928 = 31.9 h.

(c)   The outside state lay within the limits of 26°–28° dry-bulb and 18°–20° wet-bulb for (0.77/100) × 2928 = 22.5 h.

5.10   The seasonal change of outside psychrometric state

When up-to-date meteorological information is available for mean monthly maximum and minimum temperatures, mean daily maximum and minimum temperatures, and observations of relative humidity for each month of the year, it is possible to obtain an approximate, overall view of the psychrometric envelope within which the climate of the particular place lies. Remembering that humidity observations are made on a daily basis the method adopted in Example 5.2 can be used to link temperatures with moisture contents. This has to be done when meteorological data is not readily available on an hourly basis in terms of frequency of occurrence of dry- and wet-bulb temperatures.

The frequency of occurrence data for Heathrow, partly shown in Table 5.3, is available more extensively in the *CIBSE Guide A2* (1999) and can be used to provide a better psychrometric picture of the climate. Figure 5.4 shows values of coincident dry-bulb and wet-bulb temperatures with their frequencies of occurrence as percentages of the total hours in the four summer months, taken from the data published in the *CIBSE Guide A2* (1999). The wet-bulb temperatures recorded at hourly intervals and published in the meteorological data are screen readings. The CIBSE psychrometric chart shows aspirated (sling) values of wet-bulb temperature because these are stable values (see section 2.17), corrected for radiation error. The difference between a screen and a sling wet-bulb value varies according to the humidity and dry-bulb temperature.

At 100% saturation the two wet-bulbs are the same but the difference between them increases as the humidity falls, the sling value always being the lower. For instance, at 22°C dry-bulb, the *CIBSE Guide C1* (1988) gives the difference between the screen and the sling values as 0.4 K at 50 per cent saturation and 1.3 K at 0 per cent. At 32°C dry-bulb the corresponding differences are 0.4 K and 1.5 K. The humidity in the UK seldom falls below 30 per cent saturation and for most of the year it exceeds 50 per cent. Hence to obtain an adequate picture of the area on the psychrometric chart within which the climate lies it is reasonable to ignore the difference between sling and screen wet-bulbs. Figure 5.4 does this.

5.11   The choice of outside design conditions

There are three ways of establishing outside summer design states:

(1)   Use mean daily and monthly meteorological data in the following way:

(i)   Choose the highest mean monthly maximum temperature as the design dry-bulb value.

(ii)   For the same month used in (i) couple the mean daily maximum temperature and the afternoon record of relative humidity, if necessary using equation (5.4) to adjust the mean daily maximum to coincide with the time of the humidity value.

(iii)   With the temperature and humidity coupled in (ii) determine the corresponding moisture content from psychrometric tables or a chart.

(iv)   Use the moisture content from (iii) with the mean monthly maximum temperature to establish the corresponding screen or sling wet-bulb temperature.

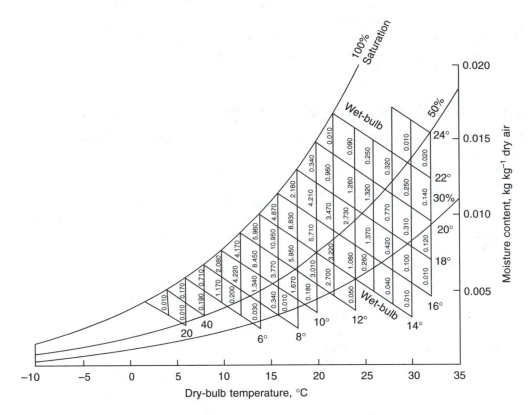

Fig. 5.4 Frequency of occurrence of coincident dry-bulb and wet-bulb temperatures over the total hours in the months June–September at Heathrow. (Reproduced by kind permission of the CIBSE from *Guide A, Environmental Design* (1999).)

(v)   Select the mean monthly maximum temperature (i) and wet-bulb obtained from (iv) as the outside summer design state.

(2)   Choose outside dry- and wet-bulb values in terms of their frequency of occurrence in the four summer months in the UK. Table 5.3 and Figure 5.4 should be referred to. In the past, it has been common to select an outside summer design state on a basis of a 1 per cent occurrence in the four summer months over the nine occupied hours in the day, rather than over the full 24 hours. This has yielded 27°C dry-bulb and 19.5°C wet-bulb (screen), referred to a non-urban location, such as the former meteorological station at Kew. Both Table 5.3 and Figure 5.4 refer to a frequency of occurrence over 24 h. The likelihood of warm temperatures occurring during the occupied period of about eight or nine hours during the day is three times the percentage quoted for the full 24 hours. Hence we should be looking for a figure of about 0.33 per cent, in the vicinity of high dry- and wet-bulb temperatures, as a basis for our outside design state. Possibilities are: 0.31 per cent for 28°C to 30°C dry-bulb with 18°C to 20°C wet-bulb (screen), or 0.32 per cent for 26°C to 28°C dry-bulb with 20°C to 22°C wet-bulb (screen).

   Making the actual choice needs further consideration. Global warming forecasts vary but something like 0.25 K per decade seems to be a possible rise in outside air temperature in southern England over the next few years. Accountants appear to allow 60 years for the life of an office block and it is very likely that an air conditioning system would be near the end of its useful life after 20 to 25 years, bearing in mind normal wear and obsolescence. This suggests that half a degree should be added to the dry-bulb with perhaps a similar increase for the wet-bulb, to cater for future possible increases in system performance over the life of the building. However, the choice is open to further thought: choosing a lower outside air temperature has an effect on the capital cost of the proposed design and this must be weighed against some measure of dissatisfaction that might result. Example 7.19 examines some of the consequences of choosing different inside and outside design conditions.

The foregoing leads to a view that, after correcting for the difference between screen and sling wet-bulbs, suitable outside choices for an open area near London might be 29°C dry-bulb, 18.3°C wet-bulb (sling) or 27°C dry-bulb, 20.7°C wet-bulb (sling). These states have corresponding moisture contents of 0.008616 kg kg^{-1} and 0.012 64 kg kg^{-1}. An addition of 0.5 K should be made to cover global warming over the life of the plant to give dry-bulbs of 29.5°C and 27.5°C. It is unlikely that the moisture content of the outside air will be constant because vegetation will dehydrate as air temperatures rise. A decision on this is speculative but one might suppose that the wet-bulb temperatures could increase by about half the rise in dry-bulb temperatures because, as we see on the psychrometric chart, the scale of wet-bulb temperature is about twice that of dry-bulb temperature, in the vicinity of 30°C dry-bulb and 20°C wet-bulb. The suggestion is that the wet-bulbs be increased by 0.3 K and a suitable outside design state for an open area near London could be 29.5°C dry-bulb with 18.6°C wet-bulb (sling), or 27.5°C dry-bulb with 21°C wet-bulb (sling), the corresponding moisture contents being 0.008 746 and 0.012 84 kg kg^{-1}. The matter does not end here because, as we shall see, the presence of buildings in London makes a further rise in the outside design dry-bulb necessary.

(3)   Adopt what is chosen by common custom, locally. For London, in the past, a value of 28°C dry-bulb, 19.5°C wet-bulb (sling) has been commonly adopted. This is no longer likely to be satisfactory over the next few decades.

In conurbations the mass of the building absorbs radiant solar heat during the day and releases this into the atmosphere later on. Cities are thus warmer areas than the surrounding countryside (often referred to as 'heat islands'). According to Chandler (1965) the influences at work are: the climate of the region, local morphology, the thermal properties of the congregation of buildings and surfaced roads: apparently each building in an urban complex has its own micro-climate which, because of local influences, may be significantly different from that of its neighbours. In the middle of London mean annual temperatures are about 1 K to 1.5 K warmer than those in the surrounding countryside. Dry-bulb temperatures after sunset in central London may be 5 K above those in the outlying country because of the heat released from the buildings in calm weather.

Bearing in mind the foregoing, the two dry-bulb temperatures considered suitable for an open area near London should be increased by 0.5 K to 30°C and 28°C to make them appropriate for London itself. It is argued that the higher dry-bulb is not likely to raise the moisture content because of the much smaller amount of vegetation in the conurbation, compared with open country. So the two outside states to be considered are 30°C dry-bulb, 0.008 746 kg kg^{-1}, 18.7°C wet-bulb (sling) and 28°C dry-bulb, 0.012 84 kg kg^{-1}, 20.8°C wet-bulb (sling). It is suggested that an arithmetic mean be adopted and a suitable outside summer design choice taken as 29°C dry-bulb, 20°C wet-bulb (sling), for which the moisture content is 0.010 80 kg kg^{-1}, the enthalpy is 56.76 kJ kg^{-1} and the specific volume is 0.8705 m^3 kg^{-1}.

When sizing air-cooled condensers and cooling towers, it is recommended that their selections are based on at least 30°C dry-bulb and 20.5°C wet-bulb (sling), respectively.

The selection of outside winter design temperatures is not made in the same way, the reason being that most people are at home in bed at the time of lowest temperature, about one hour before sunrise. For a building occupied during normal office hours, an outside design dry-bulb of –1°C or –2°C is often chosen and is satisfactory, although it should be noted that for purposes of sizing heater batteries that handle 100 per cent outside air, an air temperature of about –5°C should be adopted because the thermal inertia of the building plays no part in mitigating the effect on a heater battery of a drop in temperature below the design value, as it does for comfort temperatures within a building. The *CIBSE Guide A2* (1999) recommends choosing an outside design temperature for winter in terms of the frequency of occurrence of low temperatures and the thermal inertia of the building. Reference should be made to the *Guide* for the selection of winter design conditions.

According to Chandler (1965) records at Kew show that earth temperatures vary in London from a mean monthly minimum of 0.6°C in January to a maximum of 19.8°C in July, with annual averages in those months of 4.2°C and 17.8°C, at 300 mm below the surface. The corresponding values at a depth of 1200 mm are 5.4°C, 17.0°C (in August) and 15.8°C (in August).

Exercises

1. (*a*) Briefly explain the causes of outside air diurnal variation of temperature, accounting for the times at which maximum and minimum values occur.

(*b*) Briefly explain what differences there might be and why they occur, in the daily and annual temperature ranges between two places on the same latitude, one place being classified as a dry tropical region and the other a humid tropical region. Similarly, account for the differences between a temperate coastal region and a temperate inland region.

(*c*) Tabulate factors affecting the choice of outside design condition for air conditioning a modern building.

2. Briefly describe a simple method of estimating summer outside air conditions for the design calculations of an air conditioning scheme. If a more accurate assessment of heat gains to a building is required, explain what further meteorological information is wanted.

Notation

Symbol	Description	Unit
B	angle factor	–
C	cloud cover factor	–
D	diurnal variation in outside dry-bulb temperature	K
I_{LW}	long-wave thermal radiation from a surface to the sky	W m^{-2}
K	vapour correction factor	–
p_s	vapour pressure	mbar
t_o	outside air dry-bulb temperature	°C
t_s	surface temperature	°C
t_{15}	outside dry-bulb temperature at 1500 h sun-time	°C

t_θ	outside dry-bulb temperature at time θ	°C
δ	acute angle between a surface and the horizontal	degrees
θ	sun-time	h

References

Brunt, D. (1932): *Quart. J. Roy. Meteorological Soc.* **58**, 389.

Chandler, T.J. (1965): *The Climate of London*, Hutchinson & Co. Ltd.

CIBSE Guide C1 (1988): Properties of Humid Air.

CIBSE Guide A2 (1999): External Design Data.

Levermore, G. and Keeble, E. (1997): Dry-bulb temperature analysis for climate change at 3 UK sites in relation to the new CIBSE Guide to Weather and Solar Data, CIBSE National Conference 1997.

Bibliography

1. W.H. Pick, *A Short Course in Elementary Meteorology,* 5th edn, revised. HMSO, 1938.
2. O.G. Sutton, *Understanding Weather,* 2nd edn. Penguin Books, 1974.
3. *ASHRAE Handbook* 1997 Fundamentals.
4. H.C. Jamieson, Meteorological Data and Design Temperatures. *JIHVE*, 1955, **22**, 465.
5. R. Harrison, Air Conditioning at Home and Abroad, *JIHVE*, 1958, **26**, 177.

6

The Choice of Supply Design Conditions

6.1 Sensible heat removal

If there is a continuous source of heat having an output of Q in a hermetically sealed room the temperature within the room, t_r, will rise until the flow of heat through the walls, of area A and thermal transmittance U, equals the output of the source:

$$Q = AU\ (t_r - t_o) \tag{6.1}$$

in which t_o is the outside air temperature.

It then follows that

$$t_r = t_o + Q/AU \tag{6.2}$$

and hence t_r will always exceed t_o.

EXAMPLE 6.1

Calculate the temperature maintained in a room of dimensions 3 m × 3 m × 3 m if the average U-value of the walls, floor and ceiling is 1.1 W m^{-2} K^{-1}, the outside temperature is 27°C and the heat gains within the room are 2 kW.

Answer

By equation (6.2):

$$t_r = 27 + 2000/(6 \times 3^2 \times 1.1) = 27 + 33.7 = 60.7°C$$

This is clearly too high for the sustained comfort of any of the occupants and the first possible remedy to consider is mechanical ventilation. If outside air with a density ρ and a specific heat capacity c is supplied to the room at a rate of n air changes per hour, it will absorb some of the heat gain as its temperature rises to the value of t_r before being discharged to waste. If we write V for the volume of the room a new heat balance can be established:

$$Q = \rho cnV(t_r - t_o)/3600 + AU(t_r - t_o) \tag{6.3}$$

and hence a new expression for room temperature is

$$t_r = t_o + 3600Q/(\rho cnV + 3600AU) \tag{6.4}$$

Although t_r still exceeds t_o the excess will not be as great since the denominator of the second term in equation (6.4) is larger than that in equation (6.2).

EXAMPLE 6.2

For the conditions of example 6.1 calculate t_r if the room is ventilated at a rate of 10 air changes per hour, the density of air being 1.15 kg m^{-3} and its specific heat 1.034 kJ kg^{-1} K^{-1}.

Answer

By equation (6.4):

$$t_r = 27 + 3600 \times 2000/(1.15 \times 1034 \times 10 \times 3^3 + 3600 \times 6 \times 3^2 \times 1.1)$$
$$= 27 + 13.5°C = 40.5°C$$

The answers to examples 6.1 and 6.2 are numerically correct but misleading. First, the supply air temperature rises as it passes through the fan used to deliver it to the room (see section 6.5) and, secondly, the thermal capacity of the room walls, floor and ceiling will act as a reservoir for the heat gain, storing some of it for a while before releasing it to the space and mitigating the temperature rise of the air (see sections 7.17 to 7.20). Equation (6.4) shows that increasing the air change rate of mechanical ventilation gives diminishing returns. Supplying more than about ten air changes per hour is seldom worthwhile, without the use of mechanical refrigeration.

EXAMPLE 6.3

For the conditions of example 6.2 calculate the value of the steady outside temperature which will result in an inside steady-state value of 22°C. Assume the air delivered rises in temperature by 0.5°C as it is handled by the supply fan but ignore the effect of the thermal capacity of the room.

Answer

By equation (6.4)

$$22° = t_o + 13.5 + 0.5$$
$$t_o = 8.0°C$$

The heat loss through the room fabric plus the cooling effect of 10 air changes per hour delivered at 8.0°C exactly balance the internal heat gain of 2 kW. It is clear that for 22°C to be maintained within the room, in the presence of 2 kW of heat gain, when the outside temperature is greater than 8.1°C the air must be artificially cooled. This can be generalised as follows:

Sensible heat gain = mass flow rate of supply air
$$\times \text{ specific heat capacity} \times \text{temperature rise}$$
$$= \dot{m} \times c \times (t_r - t_s) \tag{6.5}$$

where $\dot{m}$ is the supply mass flow rate and t_s is the supply air temperature. The volumetric flow rate of air at temperature t, that corresponds to the mass flow rate $\dot{m}$, is $\dot{v}_t$ and equals the quotient $\dot{m}/\rho_t$ where ρ_t is the air density at temperature t. We can now apply a Charles' law correction (see section 2.5) and write $\rho_t = \rho_o(273 + t_o)/(273 + t)$ where ρ_o is a standard air density at a standard temperature t_o. Hence,

Sensible heat gain $= [\dot{v}_t \times \rho_o \times (273 + t_o)/(273 + t)] \times c \times (t_r - t_s)$

If we choose $\rho_o = 1.191$ kg m^{-3} at 20°C dry-bulb and 50 per cent saturation, and put $c = 1.026$ kJ kg^{-1} then

$$\dot{v}_t = \frac{\text{sensible heat gains}}{(t_r - t_s)} \times \frac{(273 + t)}{358} \tag{6.6}$$

If the sensible heat gain is expressed in kW then $\dot{v}_t$ is in m^3 s^{-1} but if the gain is in W then $\dot{v}_t$ is in litres s^{-1}. Equation (6.6) can equally be adopted for a heat loss when t_s must then be greater than t_r. In using equation (6.6) it is important to note that it is based on the fundamental principle shown by equation (6.5). The reason why it is necessary to establish equation (6.6) is because it is so very useful: air distribution fittings, duct systems, air handling plants, fans and other items of equipment are all sized and selected in terms of volumetric flow rate, not mass flow rate. It is also very important to understand that, since it is based on a mass flow rate absorbing the sensible heat gains, the volumetric flow rate must be associated with the corresponding air density. Thus the value of t in the expression $(273 + t)$ is always the temperature at which the volumetric airflow rate, $\dot{v}_t$, is expressed. Very often t is also t_s, but not always.

EXAMPLE 6.4

A room measuring 3 m × 3 m × 3 m suffers sensible heat gains of 2 kW and is to be maintained at 22°C dry-bulb by a supply of cooled air. If the supply air temperature is 13°C dry-bulb, calculate

(a)   the mass flow rate of air which must be supplied in kg s^{-1},
(b)   the volumetric flow rate which must be supplied in m^3 s^{-1},
(c)   the volumetric flow rate which must be extracted, assuming that no natural infiltration or exfiltration occurs.

Take the specific heat of air as 1.012 kJ kg^{-1} K^{-1}.

Answer

(a) By equation (6.5)

$$\dot{m} = 2/[1.012 \times (22 - 13)] = 0.2196 \text{ kg s}^{-1}$$

(b) By equation (6.6)

$$\dot{v}_{13} = \frac{2}{(22 - 13)} \times \frac{(273 + 13)}{358}$$
$$= 0.1775 \text{ m}^3 \text{ s}^{-1} \text{ at } 13°C$$

It is important to observe that, for a given mass flow rate, a corresponding volumetric flow rate has meaning only if its temperature is also quoted. This principle is not always adhered to since it is often quite clear from the context what the temperature is.

(c) That a variation of calculable magnitude exists is shown by evaluating the amount of air leaving the room. Since no natural exfiltration or infiltration occurs, the mass of air extracted mechanically must equal that supplied.

By equation (6.6)

$$\dot{v}_{22} = \frac{2}{(22 - 13)} \times \frac{(273 + 22)}{358}$$
$$= 0.1831 \text{ m}^3 \text{ s}^{-1} \text{ at } 22°C$$

This is just the same as using Charles' law:

$$\dot{v}_{22} = \dot{v}_{13} \times \frac{(273 + 22)}{(273 + 13)}$$

$$= 0.1775 \times \frac{295}{286}$$

$$= 0.1831 \text{ m}^3 \text{ s}^{-1} \text{ at } 22°C$$

Equation (6.6) can be used then to express the volumetric flow rate at any desired temperature for a given constant mass flow rate. The equation is sometimes simplified. A value of 13°C is chosen for the expression of the volumetric flow rate. We then have

$$\text{flow rate (m}^3 \text{ s}^{-1}) = \frac{\text{sensible heat gain (kW)}}{(t_r - t_s)} \times \frac{(273 + 13)}{358}$$

$$= \frac{\text{sensible heat gain (kW)}}{(t_r - t_s)} \times 0.8 \tag{6.7}$$

This version suffers from the disadvantage that a Charles' law correction for temperature cannot be made and so the answers it gives are correct only for air at 13°C. The inaccuracy that may result from this is usually unimportant in most comfort conditioning applications.

6.2   The specific heat capacity of humid air

The air supplied to a conditioned room in order to remove sensible heat gains occurring therein, is a mixture of dry air and superheated steam (see chapter 2). It follows that these two gases, being always at the same temperature because of the intimacy of their mixture, will rise together in temperature as both offset the sensible heat gain. They will, however, offset differing amounts of sensible heat because, firstly, their masses are different, and secondly, their specific heats are different too.

Consider 1 kg of dry air with an associated moisture content of g kg of superheated steam, supplied at temperature t_s in order to maintain temperature t_r in a room in the presence of sensible heat gains of Q kW. A heat balance equation can be written thus:

$$Q = 1 \times 1.012 \times (t_r - t_s) + g \times 1.890 \times (t_r - t_s)$$

where 1.012 and 1.890 are the specific heats at constant pressure of dry air and steam respectively. Rearrange the equation:

$$Q = (1.012 + 1.89g)(t_r - t_s)$$

The expression $(1.012 + 1.89g)$ is sometimes called the specific heat of humid air.

Taking into account the small sensible cooling or heating capacity of the superheated steam present in the supply air (or its moisture content) provides a slightly more accurate answer to certain types of problem. Such extra accuracy may not be warranted in most practical cases but it is worthy of consideration as an exercise in fundamental principles.

EXAMPLE 6.5

Calculate accurately the weight of dry air that must be supplied to the room mentioned in example 6.4, given that its associated moisture content is 7.500 g kg^{-1} of dry air and that the specific heat at constant pressure of superheated steam is 1.890 kJ kg^{-1} K^{-1}.

Answer

By equation (6.5)

$$2 = \dot{m} \times (1.012 + 0.0075 \times 1.89) \times (22 - 13)$$
$$\dot{m} = 2/(1.026 \times 9) = 0.2166 \text{ kg dry air per second}$$

This should be compared with 0.2196 kg s^{-1} the answer to example 6.4(*a*). Note that, for the moisture content quoted, the specific heat of the humid air is 1.026 kJ kg^{-1} K^{-1}.

6.3   Latent heat removal

If the air in a room is not at saturation, then water vapour may be liberated in the room and cause the moisture content of the air in the room to rise. Such a liberation of steam is effected by any process of evaporation as, for example, the case of insensible perspiration and sweating on the part of the people present. Since it is necessary to provide heat to effect a process of evaporation, it is customary to speak of the addition of moisture to a room as kW of latent heat rather than as kg s^{-1} of water evaporated.

The heat gains occurring in a room can be considered in two parts: sensible gains and latent gains. The mixture of dry air and associated water vapour supplied to a room has therefore a dual role; it is cool enough initially to suffer a temperature rise up to the room dry-bulb temperature in offsetting the sensible gains, and its initial moisture content is low enough to permit a rise to the value of the room moisture content as latent heat gains are offset. Figure 6.1 illustrates this.

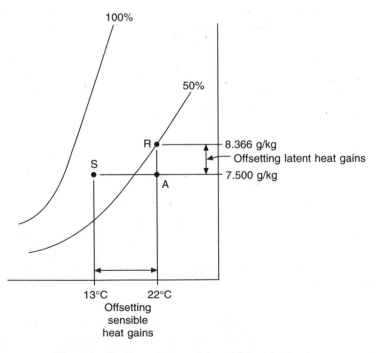

Fig. 6.1   Dealing with sensible and latent heat gains.

If $\dot{m}$ kg s^{-1} of dry air with an associated moisture content of g kg per kg of dry air is supplied at state S to a conditioned space wherein there are only sensible heat gains, then its temperature will rise from 13°C to 22°C as it diffuses through the room, offsetting the sensible gains. The resultant room condition would be typified by state point A. If there were then a latent gain in the room caused by some evaporation taking place (say from people's skin) without any further increase in sensible heat gain, then the state in the room would change from A to R, more or less up a line of constant dry-bulb temperature (see section 3.7). The exact amount that the moisture content of R exceeds g_a $(= g_s)$ would depend on the latent heat gain. If the mass of dry air supplied and its associated moisture content is known, then it is possible to calculate the rise in room moisture content corresponding to given latent heat gains:

Latent heat gain = mass flow rate of supply air in kg dry air per second
$$\times \text{ moisture pick-up in kg moisture per kg dry air}$$
$$\times \text{ latent heat of evaporation in kJ per kg moisture}$$
$$= \dot{m} \times (g_r - g_s) \times h_{fg}$$

where g_r and g_s are the moisture contents of the room and supply air, respectively, and h_{fg} is the latent heat of evaporation. As with the derivation of equation (6.6)

$$\dot{m} = \dot{v}_t \times \rho_o \times (273 + t_o)/(273 + t)$$

therefore

$$\text{Latent heat gain} = [\dot{v}_t \times \rho_o \times (273 + t_o)/(273 + t)] \times h_{fg} \times (g_r - g_s)$$

If $\rho_o = 1.191$ kg m^{-3} at 20°C dry-bulb and 50 per cent saturation and if $h_{fg} = 2454$ kJ kg^{-1} at 20°C then

$$\dot{v}_t = \frac{\text{Latent heat gain}}{(g_r - g_s)} \times \frac{(273 + t)}{856} \tag{6.8}$$

where $\dot{v}_t$ is in m^3 s^{-1}, the latent gain is in kW and $(g_r - g_s)$ is in g kg^{-1} dry air.

It is emphasised that the air quantity calculated in m^3 s^{-1} at temperature t is the same air quantity calculated by means of equation (6.6). We calculate the necessary air quantity by means *either* of equation (6.6), *or* equation (6.8), *or* by first principles. It is usual to use sensible gains to establish the required supply air quantity, by means of equation (6.6) first of all and then to establish by means of equation (6.8) the rise in moisture content which will result from supplying this air, with its associated moisture content (g_s) to the room. Alternatively, we may use equation (6.8) to establish what the moisture content of the supply air must be in order to maintain a certain moisture content in the room in the presence of known latent heat gains. The temperature in the room would then be the secondary consideration.

EXAMPLE 6.6

A room measures 20 m × 10 m × 3 m high and is to be maintained at a state of 20°C dry-bulb and 50 per cent saturation. The sensible and latent heat gains to the room are 7.3 kW and 1.4 kW, respectively.

(*a*) Calculate from first principles, the mass and volume of dry air that must be supplied at 16°C to the room each second. Also calculate its moisture content. Take the specific heats of dry air and superheated steam as 1.012 and 1.890 kJ kg^{-1} K^{-1}, respectively, the density of air as 1.208 kg m^{-3} at 16°C and the latent heat of evaporation as 2454 kJ kg^{-1} of water.

(*b*) Making use of equations (6.6) and (6.8), calculate the supply air quantity in m^{-3} s^{-1} and its moisture content in g kg^{-1}.

Answer

From CIBSE tables, the moisture content in the room is found to be 7.376 g kg^{-1} of dry air at 20°C dry-bulb and 50 per cent saturation.

(*a*) $$\dot{m} = \frac{7.3}{(20 - 16) \times (1 \times 1.012 + 0.007\,376 \times 1.89)}$$

$$= 1.779 \text{ kg dry air per second}$$

$$\dot{v}_{16} = \frac{1.779}{1.208} = 1.473 \text{ m}^3 \text{ s}^{-1} \text{ at } 16°C$$

Latent gain = (1.779 kg dry air per s) × (moisture pick-up in kg per kg dry air)

× (2454 kJ per kg moisture)

$$1.4 = 1.779 \times (0.007\,376 - g_s) \times 2454$$

$$g_s = 0.007\,055 \text{ kg per kg dry air}$$

(*b*) By equation (6.6)

$$\dot{v}_{16} = \frac{7.3}{(20 - 16)} \times \frac{(273 + 16)}{358}$$

$$= 1.473 \text{ m}^3 \text{ s}^{-1} \text{ at } 16°C$$

By equation (6.8)

$$g_s = 7.376 - \frac{1.4}{1.473} \times \frac{(273 + 16)}{856}$$

$$= 7.376 - 0.321$$

$$= 7.055 \text{ g per kg dry air}$$

6.4   The slope of the room ratio line

The room ratio line is the straight line, drawn on a psychrometric chart, joining the points representing the state maintained in the room and the initial condition of the air supplied to the room.

The slope of this line is an indication of the ratio of the latent and sensible heat exchanges taking place in the room, and the determination of its value plays a vital part in the selection of economical supply states.

Any supply state which lies on the room ratio line differs from the room state by a number of degrees of dry-bulb temperature and by a number of grams of moisture content. The values of this pair of differences are directly proportional to the mass of air supplied to the room for offsetting given sensible and latent gains or losses. Thus, in order to maintain a particular psychrometric state in a room, the state of the air supplied must always lie on the room ratio line.

EXAMPLE 6.7

For the room mentioned in example 6.6, calculate the condition which would be maintained therein if the supply air were not at a state of 16°C dry-bulb and 7.055 g kg^{-1} but at (*a*) 16°C dry-bulb and 7.986 g kg^{-1} and (*b*) 17°C dry-bulb and 7.055 g kg^{-1}.
   The heat gains are the same as in example 6.6.

Answer

(*a*) By equation (6.8)

$$g_r = 7.986 + 0.321$$
$$= 8.307 \text{ g per kg dry air}$$

The room state is therefore 20°C dry-bulb and 56 per cent saturation, by reference to CIBSE psychrometric tables.
   (*b*) By equation (6.6)

$$t_r = 17 + \frac{7.3}{1.473} \times \frac{(273 + 16)}{358} = 21°C$$

The room condition is therefore 21°C dry-bulb and 7.376 g kg^{-1}, to give 47 per cent saturation, by reference to CIBSE tables. Note that the Charles' law correction expression is still (273 + 16) because the volumetric flow rate used is 1.473 m^3 s^{-1}, which was evaluated at 16°C. Note also that the moisture content in the room remains at 7.376 g kg^{-1} because, by the calculation in example 6.6(*b*) the supply of 1.473 m^3 s^{-1} at a moisture content of 7.055 g kg^{-1} offsets a latent gain of 1.4 kW and suffers a moisture pick-up of 0.321 g kg^{-1}. In other words, this example assumes that some unmentioned device, such as a reheater battery, elevates the supply air temperature by 1°C but the supply air mass flow rate is unchanged.
   The calculation of the slope of the room ratio line is clearly a matter of some importance since it appears that one has a choice of a variety of supply air states. It seems that if any state can be chosen and the corresponding supply air quantity calculated, then the correct conditions will be maintained in the room. This is not so. Economic pressures restrict the choice of supply air state to a value fairly close to the saturation curve. It must, of course, still lie on the room ratio line.
   The calculation of the slope of the line can be done in one of two ways:

   (*a*)   by calculating it from the ratio of the sensible to total heat gains to the room,
   (*b*)   by making use of the ratio of the latent to sensible gains in the room.

   Method (*a*) merely consists of making use of the protractor at the top left-hand corner of the chart.

EXAMPLE 6.8

Calculate the slope of the room ratio line if the sensible gains are 7.3 kW and the latent gains are 1.4 kW.

Answer

Total gain = 7.3 + 1.4 = 8.7 kW

$$\text{Ratio:} \; \frac{\text{sensible gain}}{\text{total gain}} = \frac{7.3}{8.7} = 0.84$$

The value can be marked on the outer scale of the protractor on the chart and with the aid of a parallel rule or a pair of set squares the same slope can be transferred to any position on the chart. Figure 6.2 illustrates this. The line O O′ in the protractor is parallel to lines drawn through R_1 and R_2. The important point to appreciate is that the room ratio line can be drawn *anywhere* on the chart. Its slope depends only on the heat gains occurring in the room and *not on the particular room state*.

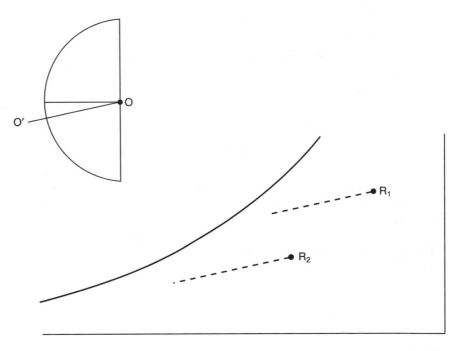

Fig. 6.2   The room ratio line.

The slope of the line can also be expressed in terms of the ratio of latent gain to sensible gain. This, in itself, is correct but a slightly inaccurate version of this is to calculate the ratio in terms of the moisture picked up by the air supplied to the room and the rise in dry-bulb temperature which it suffers.

Making use of equations (6.8) and (6.6), we can write,

$$\frac{\text{latent gain in kW}}{\text{sensible gain in kW}} = \frac{\text{flow rate (m}^3 \text{ s}^{-1}) \times (g_r - g_s) \times 856 \times (273 + t)}{\text{flow rate (m}^3 \text{ s}^{-1}) \times (t_r - t_s) \times 358 \times (273 + t)}$$

Hence

$$\frac{g_r - g_s}{t_r - t_s} = \frac{\text{latent gain in kW}}{\text{sensible gain in kW}} \times \frac{358}{856}$$

or

$$\frac{\Delta g}{\Delta t} = \frac{\text{latent gain}}{\text{sensible gain}} \times \frac{358}{856} \qquad (6.9)$$

This is slightly inaccurate because it expresses sensible changes in terms of dry-bulb temperature changes. Since the scale of dry-bulb temperature is not linear on the psychrometric chart, the linear displacement corresponding to a given change in dry-bulb temperature is not constant all over the chart. It follows that the angular displacement of the slope will vary slightly from place to place on the chart. This inaccuracy is small over the range of values of psychrometric state normally encountered and so the use of equation (6.9) is tolerated.

EXAMPLE 6.9

Calculate the slope of the room ratio line by evaluating $\Delta g/\Delta t$ for sensible and latent heat gains of 7.3 kW and 1.4 kW respectively.

Answer

$$\frac{\Delta g}{\Delta t} = \frac{1.4}{7.3} \times \frac{358}{856} \text{ by equation (6.9)}$$
$$= 0.0802 \text{ g kg}^{-1} \text{ K}^{-1}$$

The room ratio line does not necessarily have to slope downwards from right to left as in Figure 6.2. A slope of this kind indicates the presence of sensible and latent heat gains. Gains of both kinds do not always occur. In winter, for example, a room might have a sensible heat loss coupled with a latent heat gain. Under these circumstances, the line might slope downwards from left to right as in Figure 6.3(*a*). It is also possible, if a large amount of outside air having a moisture content lower than that in the room infiltrates, for the room to have a latent heat loss coupled with a sensible heat loss (or even with a sensible

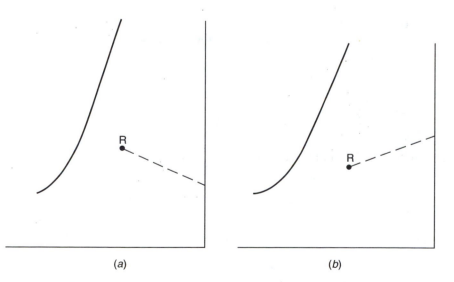

(*a*)   (*b*)

Fig. 6.3   Two possibilities for the slope of the room ratio line.

heat gain). The room ratio line for both latent and sensible losses would slope upwards from left to right, as in Figure 6.3(*b*).

EXAMPLE 6.10

If the room mentioned in example 6.6 suffers a sensible heat loss of 3 kW and a latent heat gain of 1.2 kW in winter, calculate the necessary supply state of the air delivered to the room. It is assumed that the supply fan handles the same amount of air in winter as in summer and that the battery required to heat the air in winter is positioned on the suction side of the fan.

Answer

Supply air quantity = 1.473 m³ s⁻¹.

   This volumetric flow rate is the same in winter even though its temperature will be at a much higher value than the 16°C used for its expression in summer because the fan handles a constant amount of air (see section 15.16).

   A diagram of the plant is shown in Figure 6.4(*a*) and of the psychrometry in Figure 6.4(*b*).

   Since a heat loss is to be offset, the air must be heated by the heater battery to temperature t_s and the value of t_s must exceed t_r the room temperature.

   From equation (6.6)

$$1.473 = \frac{3}{(t_s - 20)} \times \frac{(273 + t_s)}{358}$$

   There are two ways of solving this linear equation:

(1) $1.473t_s - 1.473 \times 20 = 3 \times (273 + t_s)/358$

whence

$$t_s = 21.68°C$$

   This method is tedious and prone to error.

(2)   Guess a value of t, say 25°C, and use in the expression $(273 + t)$.
   Then, from equation (6.6)

$$t_s = 20° + \frac{3}{1.473} \times \frac{(273 + 25)}{358}$$
$$= 20° + 1.70°$$
$$= 21.70°C$$

   Try again with $t = 22°C$ in the expression $(273 + t)$. Then

$$t_s = 20° + \frac{3}{1.473} \times \frac{(273 + 22)}{358}$$
$$= 21.68°C$$

   This method is preferred, provided that reasonable values are guessed for t.
   From equation (6.8)

$$g_s = 7.376 - \frac{1.2}{1.473} \times \frac{(273 + 21.68)}{856}$$

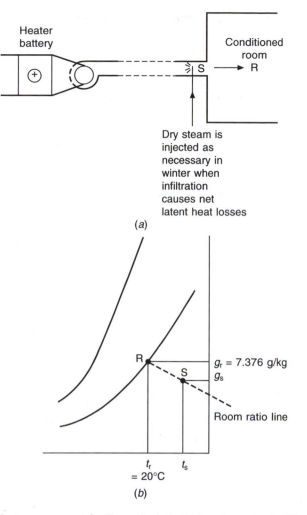

Heater battery

Conditioned room

S → R

Dry steam is injected as necessary in winter when infiltration causes net latent heat losses

(a)

R

S

g_r = 7.376 g/kg
g_s

Room ratio line

t_r t_s
= 20°C

(b)

Fig. 6.4   (a) Plant arrangement for Example 6.10. (b) Psychrometry for Example 6.10.

$$= 7.376 - 0.280$$
$$= 7.096 \text{ g per kg dry air}$$

This establishes the supply state, S, as 21.68°C dry-bulb and 7.096 g per kg, moisture content.

S must lie on the room ratio line. This is self-evident since we have just calculated

$$\Delta t = 1.68°C \text{ and } \Delta g = 0.28 \text{ g kg}^{-1}$$

Thus

$$\frac{\Delta g}{\Delta t} = \frac{0.28}{1.68}$$
$$= 0.167 \text{ g kg}^{-1} \text{ K}^{-1}$$

Strictly speaking, if the room ratio line is regarded as having a positive slope when both sensible and latent heat gains are occurring, its slope should be regarded as negative for this case. That is,

$$\frac{\Delta g}{\Delta t} = \frac{+0.28}{-1.68} = -0.167 \text{ g kg}^{-1} \text{ K}^{-1}$$

On the other hand, if both a latent heat loss and a sensible heat loss occurred, the slope would be positive again:

$$\frac{\Delta g}{\Delta t} = \frac{-0.28}{-1.68} = +0.167 \text{ g kg}^{-1} \text{ K}^{-1}$$

It is best not to adhere to a sign convention but to express the slope always as positive and to use one's understanding of fundamentals to decide on the way in which the line slopes.

It is just worth noting that the slope of the room ratio line is continuously changing as the heat gains vary. However, when designing a system, it is the slope of the room ratio line under summer design conditions that is relevant.

6.5   Heat gain arising from fan power

The flow of air along a duct results in the airstream suffering a loss of energy, and for the flow to be maintained a fan must make good the energy loss (see section 15.4). The energy dissipated through the ducting system is apparent as a change in the total pressure of the airstream and the energy input by the fan is indicated by the fan total pressure.

Ultimately, all energy losses appear as heat (although, on the way to this, some are evident as noise, in duct systems). So an energy balance equation can be formed involving the energy supplied by the fan and the energy lost in the airstream. That is to say, the loss of pressure suffered by the airstream as it flows through the ducting system and past the items of plant (which offer a resistance to airflow) constitutes an adiabatic expansion which must be offset by an adiabatic compression at the fan.

So, all the power supplied by the fan is regarded as being converted to heat and causing an increase in the temperature of the air handled, Δt, *as it flows through the fan.*

A heat balance equation can be written, accepting an expression for air power derived later in section 15.4.

$$\text{Air power} = \text{fan total pressure (N m}^{-2}) \times \text{volumetric flow rate (m}^3 \text{ s}^{-1})$$

The rate of heat gain corresponding to this is the volumetric flow rate $\times \rho \times c \times \Delta t$, where ρ and c are the density and specific heat of air, respectively. Hence

$$\Delta t = \frac{\text{fan total pressure (N m}^{-2})}{\rho c}$$

The air quantities have cancelled, indicating that the rise in air temperature is independent of the amount of air handled, and using $\rho = 1.2$ kg m^{-3} and $c = 1026$ J kg^{-1} K^{-1} we get

$$\Delta t = \frac{\text{fan total pressure (N m}^{-2})}{1231} \text{ in K} \tag{6.10}$$

Thus, the air suffers a temperature rise of 0.000 812 K for each N m^{-2} of fan total pressure.

The energy the fan receives is in excess of what it delivers to the airstream, since frictional and other losses occur as the fan impeller rotates the airstream. The power input to the fan shaft is termed the *fan power* (see section 15.4) and the ratio of the air power to the fan power is termed the *total fan efficiency* and is denoted by η. Not all the losses occur within the fan casing. Some take place in the bearings external to the fan, for example. Hence, for the case where the fan motor is not in the airstream, full allowance should not be made. It is suggested that a compromise be adopted.

If an assumption of 70 per cent is made for the fan total efficiency and if it is assumed that, instead of 30 per cent, only 15 per cent of the losses are absorbed by the airstream (since some are lost from the fan casing and the bearings) equation (6.10) becomes

$$\Delta t = \frac{\text{fan total pressure (N m}^{-2})}{1231 \times 0.85}$$

$$\Delta t = \frac{\text{fan total pressure (N m}^{-2})}{1045} \tag{6.11}$$

Thus almost one thousandth of a degree rise in temperature for each N m^{-2} of fan total pressure results from the energy input at the fan. In other words, a degree rise occurs for each kPa of fan total pressure.

When the fan and motor are within the airstream, as is the case with many air handling units, all the power absorbed by the driving motor is liberated into the airstream. Full account must then be taken of the motor inefficiency as well as all the fan inefficiency. Assuming a total fan efficiency of 70 per cent and a motor efficiency of 90 per cent the temperature rise of the airstream is

$$\Delta t = \frac{\text{fan total pressure (N m}^{-2})}{1231 \times 0.7 \times 0.9}$$

$$= \frac{\text{fan total pressure (N m}^{-2})}{776} \tag{6.12}$$

This represents a temperature rise of about 1.3 K for each kPa of fan total pressure.

6.6   Wasteful reheat

If an arbitrary choice of supply air temperature is made, it is almost certain that the resultant design will not be an economical one, either in capital or in running cost. The reason for this is that wasteful reheat may have to be offset by the refrigeration plant. The point is best illustrated by means of an example.

EXAMPLE 6.11

The sensible and latent heat gains to a room are 10 kW and 1 kW, respectively. Assuming that the plant illustrated in Figure 6.5(*a*) is used and that the cooler coil has a contact factor of 0.85, calculate the cooling load and analyse its make-up if the outside condition is 28°C dry-bulb, 20.9°C wet-bulb (sling) and the condition maintained in the room is 22°C dry-bulb and 50 per cent saturation. The supply air temperature is arbitrarily fixed at 16°C. Assume that the temperature rise due to fan power is 1 K.

Answer

From equation (6.6)

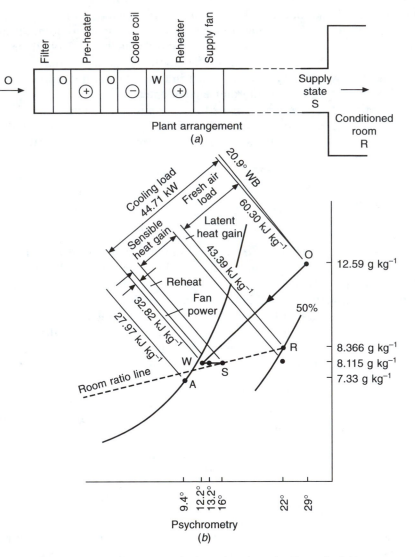

Fig. 6.5   Plant arrangement and psychrometry for Example 6.11.

$$\text{supply air quantity} = \frac{10}{(22° - 16°)} \times \frac{(273 + 16)}{358} = 1.345 \text{ m}^3 \text{ s}^{-1} \text{ at } 16°C$$

From equation (6.8):

$$\text{supply air moisture content} = g_r - \frac{1}{1.345} \times \frac{(273 + 16)}{856} = g_s$$

From CIBSE tables

$$g_r = 8.366 \text{ g kg}^{-1}$$
$$\therefore g_s = 8.366 - 0.251$$
$$= 8.115 \text{ g kg}^{-1}$$

The supply air state is thus 16°C dry-bulb and 8.115 g kg⁻¹. This is denoted by the letter S in Figure 6.5(*b*). The line joining the state point of the air entering a cooler coil to the state point of the air leaving the coil must always cut the 100 per cent relative humidity curve when suitably produced (see section 3.4). Since the line joining O and S does not cross the saturation curve when produced, the cooler coil will not give state S directly. The coil must therefore be chosen to give the correct moisture content, g_s, and reheat must be used to get the correct temperature, t_s.

Four things are now established for the behaviour of the cooler coil:

(i)   The state of the air entering the coil is known.
(ii) It has a contact factor of 0.85.
(iii) The moisture content of the air leaving the coil must be 8.115 g kg⁻¹.
(iv) The straight line joining the state points of the air entering and leaving the coil cuts the saturation curve at the apparatus dew point.

Making use of the above information and employing the definitions of contact factor expressed by equations (3.1) and (3.3), the state of air leaving the cooler coil W can be worked out.

Denote the apparatus dew point by A. Then, on a moisture content basis

$$0.85 = \frac{g_o - g_w}{g_o - g_a}$$

$$= \frac{12.59 - 8.115}{12.59 - g_a}$$

whence

$$g_a = 7.33 \text{ g kg}^{-1}$$

Thus, A is located at 100 per cent relative humidity and 7.33 g kg⁻¹.

W can now be marked in on the line joining O to A at a moisture content of 8.115 g kg⁻¹.

The temperature of A, from tables (or from a chart), is 9.4°C and its enthalpy is 27.97 kJ kg⁻¹.

Using the definition of contact factor once more, but in terms of differences of enthalpy:

$$0.85 = \frac{h_o - h_w}{h_o - h_a} = \frac{60.3 - h_w}{60.3 - 27.87}$$

where h_o has been determined from tables or a chart as 60.3 kJ kg⁻¹. Hence

$$h_w = 32.82 \text{ kJ kg}^{-1}$$

Using the approximate definition of contact factor we have:

$$0.85 = \frac{28° - t_w}{28° - 9.4°}$$

Whence

$$t_w = 12.2°C$$

Since the supply airflow of 1.345 m³ s⁻¹ is at 16°C and 8.115 g kg⁻¹ its specific volume is 0.8294 m³ kg⁻¹, from tables, and because we know the enthalpies at states O (on the cooler coil) and W (off the cooler coil) we can calculate the cooling load as:

$$\text{Cooling load} = \frac{1.345}{0.8294} \times (60.30 - 32.82) = 44.56 \text{ kW}$$

It is instructive to analyse the load:

$$\text{fresh-air load} = \frac{1.345 \times (60.30 - 43.39)}{0.8294} \qquad = 27.42 \text{ kW}$$

sensible heat gain $\qquad\qquad\qquad\qquad\qquad\qquad = 10.00 \text{ kW}$

latent heat gain $\qquad\qquad\qquad\qquad\qquad\qquad\quad = 1.00 \text{ kW}$

$$\text{fan power} = \frac{1.345 \times 1.026 \times (13.2° - 12.2°)}{0.8294} \qquad = 1.66 \text{ kW}$$

$$\text{reheat wasted} = \frac{1.345 \times 1.026 \times (16° - 13.2°)}{0.8294} \qquad = 4.66 \text{ kW}$$

$$\text{Total} = 44.74 \text{ kW}$$

(This is to be compared with 44.56 kW, computed by using the difference of enthalpy across the cooler coil.)

The analysis is illustrated in Figure 6.5(*b*). It can be seen that about 4.66 kW plays no useful thermodynamic part. The fact that reheat was needed arose from the arbitrary decision to make the supply air temperature 16°C.

A more sensible approach would be to dispense with reheat entirely and to use the lowest practical temperature for the supply air consistent with the contact factor of the cooler coil. The temperature difference between t_r and t_s would then be increased and less supply air would be needed. The consequence of this would be to reduce the cooling load because, although the sensible and latent heat gains are unaffected, the fresh-air load and the cooling dissipated in offsetting fan power are reduced, since less air is handled. This is clarified in the next section.

6.7   The choice of a suitable supply state

The choice of the supply air state is constrained by four practical issues:

(i)   The desire to minimise the air quantity handled

The supply air temperature should be as low as possible, consistent with providing draught-free air distribution in the occupied space and the required conditions of temperature and humidity. In practice this means that the supply air temperature is usually about 8° to 11° less than the room temperature, although larger differences are sometimes used. (With special attention to air distribution in the conditioned room, supply air temperatures as much as 15° less than the room air temperature have been successfully used in order to minimise the airflow rate.)

Equation (6.6) shows that, to deal with a given sensible heat gain, the supply airflow rate is inversely related to the difference between the room and supply air temperatures. Choosing a lower supply air temperature gives a smaller airflow rate, smaller air distribution system and smaller air handling plant. Fan power will be less with lower running costs, less space will be occupied by ducts and the system will be easier to install.

(ii) Cooler coil selection

The effectiveness of a cooler coil is expressed in terms of its contact factor (see sections 3.4 and 10.4) and this depends on the number of rows of tubes and other factors. An effective cooler coil must therefore be chosen that achieves the design cooling and dehumidification duty (see Figure 6.6) in an economical and practical manner. Air handling plants in the UK commonly use four or six rows for this purpose but, in a hotter and more humid climate, six or eight rows are generally needed. More than eight rows is unnecessary for comfort conditioning because the cooling duty of additional downstream rows progressively decreases, cooler coils being piped for counterflow heat exchange.

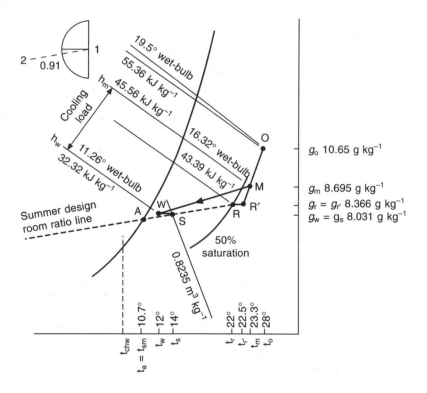

Fig. 6.6 The choice of supply air state (see also Example 6.12).

A practical contact factor is between 0.8 and about 0.95, depending principally on the number of rows, although the fin spacing and the face velocity of airflow also play a part (see section 10.4). Reference to Figure 6.6 shows that the contact factor is the ratio of the distance M–W to the distance M–A. W will lie closer to the saturation curve if the contact factor is larger.

(iii) Fan power and duct gain

If the supply fan is located on the downstream side of the cooler coil, as in a so-called draw-through plant, there is a temperature rise after the cooler coil because of fan power. The air will also rise in temperature, before it reaches the conditioned room, because of heat gain to the supply air ducting. Referring to Figure 6.6 it is seen that the supply air temperature, t_s, is greater than the temperature leaving the cooler coil, t_w, for these two reasons. Similar considerations apply when the fan blows air through a cooler coil, except that fan power then warms the air before it enters the coil and duct gains warm it after leaving the coil.

The supply fan total pressure depends on the velocity of air distribution in the ducts. Conventional, low velocity systems are likely to have fan total pressures of about 600 Pa to 750 Pa, medium velocity about 1000 Pa and high velocity about 2000 Pa, with corresponding temperature rises, as given by equations (6.11) and (6.12).

Duct heat gains often receive inadequate attention and can be surprisingly large (see section 7.21). It is difficult to quote typical values for temperature rise, except to say several degrees are possible.

The air extracted from the conditioned room can also rise in air temperature (t_r to $t_{r'}$ in Figure 6.6) and this will affect the cooler coil performance if, as is usually the case, some of the extracted air is recirculated. The temperature rise can be caused by air exhausted from the room through extract-ventilated luminaires (see section 7.23) and also by the extract fan power. A typical temperature rise through extract-ventilated light fittings is about 1° but such fittings are not always desirable with some types of fluorescent tube. Extract duct systems are invariably simpler than supply systems and airflow is usually at low velocity. There are generally no plant items for air treatment. Extract fan total pressure is therefore likely to be about 200 to 300 Pa. Hence a typical temperature rise in the recirculated air is about 0.2° to 1.3°K.

Heat gains to extract ducts can usually be ignored.

(iv) *The freezing temperature of water occurring at 0°C*

When chilled water is used as the cooling medium the behaviour of the coil depends on the flow temperature of the water and the temperature rise as it flows through the tubes. The mean coil surface temperature of the coil, t_{sm}, is the same as the temperature of the apparatus dew point, t_a (see sections 3.4 and 10.3). It follows that the chilled water flow temperature, t_{chw}, will be less than the mean coil surface temperature, as Figure 6.6 shows. A refrigeration plant must therefore be selected that can provide the chilled water flow temperature required, in a stable and safe manner, *under all conditions of load*. If reciprocating plant, controlled by cylinder unloading from return chilled water temperature, is used (see section 12.10), then a suitable chilled water flow temperature is 6.5°C. If a centrifugal, screw, or other machine is used, controlled from its flow temperature under proportional, plus integral, plus derivative control (see section 13.11), then a suitable flow temperature is 5°C. Manufacturers' advice should be sought. The lowest, practical dry-bulb temperature leaving a cooler coil is about 10°C to 11°C, depending on the chilled water flow temperature available. If chilled brine is used, lower temperatures are possible but the use of brine or glycol introduces significant complications, with a possible reduction in heat transfer coefficient.

Figure 6.6 shows the performance of a cooler coil. The ratio of the sensible to total heat gains in the conditioned room is known, so the slope of the design room ratio line can be drawn on the protractor and then transferred to run through the room state, R. If a supply state, S, is to be chosen in conformance with the four principles outlined above, the following procedure is suggested. It is usually carried out for the chosen summer design conditions, at 15.00 h sun time in July.

(1) Knowing the occupancy of the conditioned room, calculate the design minimum supply rate of outside air.

(2) Identify summer design state points O and R on the psychrometric chart.

(3) Knowing the sensible and latent heat gains for the conditioned room, calculate the ratio of the sensible to total heat gain. This is the slope of the summer design room ratio line.

(4) Mark the value of the slope on the psychrometric chart protractor and, using a parallel rule or a pair of set squares, draw the design room ratio line to run through the room state, R. Referring to Figure 6.6, the line 1–2 on the protractor is parallel to the room ratio line through R.

(5) Make a reasonable estimate of the temperature rise due to extract fan power and the use of any extract-ventilated luminaires. Identify the state point R′ on the chart, with $g_r = g_{r'}$. Join the points R′ and O by a straight line.

(6) Make a reasonable estimate of the temperature rise due to supply fan power and supply duct heat gain.

(7) Make a first, arbitrary choice of state S and identify this on the design room ratio line. Clearly, S must be to the right of the saturation curve and should be a reasonable choice, bearing in mind the four considerations mentioned earlier. The degree of arbitrariness will depend on experience, but the temperature of S is likely to be 8° or 9° less than the room dry-bulb temperature.

(8) Using equation (6.6) calculate the supply airflow rate.

(9) Knowing the minimum outside airflow rate and the supply airflow rate, determine the proportions of fresh and recirculated air and hence identify the mixture state, M, on the straight line O–R′.

(10) Knowing the temperature rise due to supply fan power and duct heat gain identify the point W on the chart, with $g_w = g_s$.

(11) Join the points M and W by a straight line and extend this to cut the saturation curve at A.

(12) Calculate the contact factor by means of equation (3.3):

$$\beta = \frac{t_m - t_w}{t_m - t_{sm}}.$$

(13) If the contact factor is reasonable (in the vicinity of 0.85 if a four-row coil is to be used or 0.93 if a six-row coil is likely) accept the choice of S made in (7), above. If the contact factor is not reasonable, or if other possibilities are to be explored, go back to (7) and make another choice of S, using a dry-bulb temperature half a degree warmer or cooler, to suit the circumstances.

(14) When a satisfactory choice of S has been made, calculate the supply air moisture content, g_s, by means of equation (6.8). This establishes the moisture content of the air leaving the cooler coil, equal to the supply air moisture content.

EXAMPLE 6.12

The sensible and latent heat gains to a room are 10 kW and 1 kW, respectively. The occupancy of the room is 12 people and the minimum fresh air allowance is 12 litres s^{-1} for each person. The outside summer design state is: 28°C dry-bulb, 19.5°C wet-bulb (sling), 10.65 g kg^{-1} dry air, 55.36 kJ kg^{-1} dry air. The room state is: 22°C dry-bulb, 50 per cent saturation, 8.366 g kg^{-1} dry air, 43.39 kJ kg^{-1} dry air. Select a suitable supply air state, specify the states on and off the cooler coil and calculate the design cooling load. Analyse and check the cooling load. The temperature rise through the extract system is estimated as 0.5°C. A low velocity supply system is to be adopted and the temperature rise for supply fan power and duct heat gain is estimated at 2°.

Answer

Following the steps suggested above:

(1) Minimum fresh air = 12 persons × 12 litres s^{-1} each

$$= 144 \text{ litres s}^{-1}$$

For convenience later (in step 9) it is assumed that this is expressed at the supply air state.

(2) Identify O and R on the psychrometric chart (Figure 6.6).

(3) Sensible/total heat ratio = 10/(10 + 1) = 0.91.

(4) The room ratio line is located on the chart with this slope on the protractor and transferred to run parallel to this through state R.

(5) Extract system temperature rise = 0.5°C. The point R′ is located on the chart with a temperature of 22.5°C and a moisture content of 8.366 g kg^{-1}. O is joined to R′ on the chart.

(6) Supply system temperature rise = 2.0°C.

(7) Make a first arbitrary choice of 14°C dry-bulb for state S.

(8) By equation (6.6)

$$\dot{v}_{14} = \frac{10\,000}{(22 - 14)} \times \frac{(273 + 14)}{358} = 1002 \text{ litres s}^{-1} \text{ at } 14°C$$

(9)
$$\frac{\text{Fresh airflow rate}}{\text{Supply airflow rate}} = \frac{144}{1002} = 0.144$$

On-coil state = mixture state M:

$$t_m = 0.144 \times 28 + 0.856 \times 22.5 = 23.3°C$$
$$g_m = 0.144 \times 10.65 + 0.856 \times 8.366 = 8.695 \text{ g kg}^{-1}$$

By equation (2.24)

$$h_m = (1.007 \times 23.3 - 0.026) + 0.008\,695\,[(2501 + 1.84 \times 23.3)]$$
$$= 45.56 \text{ kJ kg}^{-1}$$

(10) Off-coil dry-bulb temperature:

$$t_w = 14° - \text{supply system temperature rise} = 12°C$$

(11) The process line M–W, when projected, cuts the saturation curve on the chart at the apparatus dew point, A, for which the mean coil surface temperature is read from the chart as 10.7°C.

(12) The contact factor = $\dfrac{23.3 - 12}{23.3 - 10.7} = 0.90$

(13) This is a practical contact factor, needing four or six rows of tubes, depending on the fin spacing and the face velocity of airflow. Hence accept the choice of state S, made in (7), above.

(14) By equation (6.8):

$$g_s = 8.366 - \frac{1000}{1002} \times \frac{(273 + 14)}{856} = 8.031 \text{ g kg}^{-1}$$

Cooling Load Calculation:

On coil state (M): 23.3°C dry-bulb, 16.32°C wet-bulb (sling), 8.695 g kg^{-1},
45.56 kJ kg^{-1}

Off coil state (W): 12°C dry-bulb, 11.26°C wet-bulb (sling), 8.031 g kg^{-1}, 32.32 kJ kg^{-1}
(determined from equation (2.24))

From psychrometric tables or, less accurately, from a chart, the specific volume at state S is 0.8235 m^3 kg^{-1}.

$$\text{Cooling load} = \frac{1.002}{0.8235} \times [45.56 - 32.32] = 16.11 \text{ kW}$$

Analysis and check

Sensible heat gain:	10.00 kW
Latent heat gain:	1.00 kW
Supply system gain: $\dfrac{1.002 \times 2 \times 358}{(273 + 14)}$	$=$ 2.50 kW
Extract system gain: $\dfrac{1.002 \times 0.856 \times 0.5 \times 358}{(273 + 14)}$	$=$ 0.53 kW
Fresh air load: $\dfrac{0.144}{0.8235} \times [55.36 - 43.39]$	$=$ 2.09 kW

Total = 16.12 kW

6.8 Warm air supply temperatures

Air density is inversely proportional to its absolute temperature (see section 2.5). As a consequence, if the temperature difference between the warm supply air and the room air is too great, stratification will occur. The layer of air beneath the ceiling will be higher than the room temperature, under heat loss conditions, and the air in the occupied part of the room will be uncomfortably cool. This places an upper limit on the supply air temperature when it is used to offset heat losses in winter. Although supply air temperatures of more than 60°C have been used in the past, with direct-fired warm air heaters, these were with high discharge air velocities (10 m s^{-1}) from the supply openings and were for industrial applications.

It is recommended that, for comfort air conditioning in commercial applications, the maximum supply air temperature should not normally exceed 35°C, when room temperatures are about 20°C. The absolute limit should be 40°C. The risk of stratification is made worse if the face velocity at the supply grille is low and also if the extract or recirculation grilles are at high level. Large floor-to-ceiling heights, with supply grilles at high level, increase the chances of cold discomfort from undesirable stratification.

A consequence of this is that achieving a boosted heating capacity during pre-heat periods with intermittently operated systems is not always easy when warm air is used, supplied at high level.

Exercises

1.   A room is to be maintained at a condition of 20°C dry-bulb and 7.376 g kg^{-1} moisture content by air supplied at 15°C dry-bulb when the heat gains are 7 kW sensible and 1.4 kW latent. Calculate the weight of air to be supplied to the room and the moisture content at which it should be supplied. Take the latent heat of evaporation at room condition as 2454 kJ kg^{-1}.

Answers

1.365 kg s^{-1} dry air, assuming a value of 1.026 kJ kg^{-1} K^{-1} for the specific heat of humid air; 6.958 g kg^{-1}.

2.   Briefly discuss the factors to be considered when selecting

(i) a design room temperature,
(ii) the temperature differential between supply and room air.

3.   An air conditioning plant comprising filter, cooler coil, fan and distributing ductwork uses only fresh air for the purpose of maintaining comfort conditions in summer. Using the information listed below, choose a suitable supply air temperature, calculate the cooler coil load and determine its contact factor.

>   Sensible heat gains to conditioned space: 11.75 kW.
>
>   Latent heat gains to conditioned space: 2.35 kW.
>
>   Outside design state: 28°C dry-bulb, 19.9°C wet-bulb (sling).
>
>   Inside design state: 21°C dry-bulb, 50 per cent saturation.
>
>   Temperature rise due to fan power and duct heat gains: 1°C. A psychrometric chart should be used.

Answers

Suitable supply air temperature is about 11°C: corresponding cooler coil load is 33.1 kW; corresponding contact factor is 0.88.

Notation

Symbol	Description	Unit
A	surface area	m^2
c	specific heat capacity of air	kJ kg^{-1} K^{-1}
		J kg^{-1} K^{-1}
g	moisture content	g per kg dry air
g_a	moisture content at the apparatus dew point	g kg^{-1}
g_o	moisture content of outside air	g kg^{-1}
g_r	room air moisture content	g kg^{-1}
g_s	supply air moisture content	g kg^{-1}
h_a	enthalpy of air at the apparatus dew point	kJ kg^{-1}
h_o	enthalpy of the outside air	kJ kg^{-1}
h_w	enthalpy of the air leaving the cooler coil	kJ kg^{-1}
h_{fg}	latent heat of evaporation	kJ kg^{-1}

$\dot{m}$	mass flow rate of dry air and its associated water vapour	kg s^{-1}
n	air change rate per hour	h^{-1}
Q	rate of production of heat	kW or W
T	absolute dry-bulb temperature	K
t	temperature	°C
t_{chw}	chilled water flow temperature	°C
t_o	outside dry-bulb temperature	°C
t_r	room dry-bulb temperature	°C
t_s	supply air temperature	°C
t_w	dry-bulb temperature of air leaving the cooler coil	°C
U	overall thermal transmittance	W m^{-2} K^{-1}
V	volume	m^3
v	specific volume	m^3 per kg dry air
$\dot{v}$	volumetric flow rate of air	m^3 s^{-1} or litres s^{-1}
$\dot{v}_t$	volumetric flow rate of air at temperature t	m^3 s^{-1} or litres s^{-1}
β	contact factor	–
ρ	density of air	kg m^{-3}
ρ_t	air density at temperature t	kg m^{-3}
ρ_o	standard air density	kg m^{-3}
η	total fan efficiency	%

7

Heat Gains from Solar and Other Sources

7.1 The composition of heat gains

Heat gains are either sensible, tending to cause a rise in air temperature, or latent, causing an increase in moisture content. In comfort air conditioning sensible gains originate from the following sources:

(i)   Solar radiation through windows, walls and roofs.
(ii)   Transmission through the building envelope and by the natural infiltration of warmer air from outside.
(iii) People.
(iv)   Electric lighting.
(v)   Business machines and the like.

Latent heat gains are due to the presence of the occupants and the natural infiltration of more humid air from outside.

In the case of industrial air conditioning there may be additional sensible and latent heat gains from the processes carried out.

All the above sources of heat gain are well researched but a measure of uncertainty is introduced by the random nature of some, such as the varying presence of people and the way in which electric lights are switched. The thermal inertia of the building structure also introduces a problem when calculating the sensible heat gain arising from solar radiation. It follows that a precise determination of heat gains is impossible. Nevertheless, it is vital that the design engineer should be able to calculate the heat gains with some assurance and this can be done when generally accepted methods of calculation are followed, supported by sound common sense. The following text discusses and describes such methods.

7.2 The physics of solar radiation

The sun radiates energy as a black body having a surface temperature of about 6000°C over a spectrum of wavelengths from 300 to 470 nm. Nine per cent of the energy is in the ultra-violet region but 91 per cent of the energy is in the visible part of the spectrum (380–780 nm) and in the infra-red. Figure 7.1 shows a typical spectral distribution of the energy reaching the surface of the earth. The peak intensity of the solar energy reaching the upper limits of the atmosphere of the earth is about 2200 W m^{-2} at 480 nm but the average total,

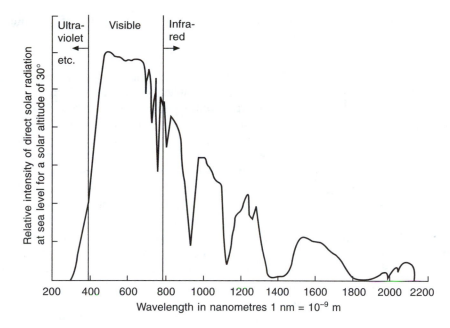

Fig. 7.1 A typical solar spectrum at sea level. The maximum irradiance (beyond the atmosphere) is 2130 W m^{-2} at 451 nm (in the green visible spectrum).

termed the solar constant, is 1367 W m^{-2}, according to Iqbal (1983). The orbit of the earth about the sun is an ellipse and the earth is slightly closer to the sun in January than it is in July. Consequently the solar irradiation has a maximum value of 1413 W m^{-2} in January and a minimum of 1332 W m^{-2} in July.

A total of only about 1025 W m^{-2} reaches the surface of the earth when the sun is vertically overhead in a cloudless sky. Of this figure, about 945 W m^{-2} is by radiation received directly from the sun, the remainder being solar radiation received indirectly from the sky.

7.3 Sky radiation

This is otherwise known as 'diffused' radiation or 'scattered' radiation. Its presence constitutes a heat gain to the earth, in addition to that from direct radiation, and it arises from the translucent nature of the atmosphere.

The atmospheric losses in direct radiation which produce the sky radiation stem from four principal phenomena:

(i)   A scattering of the direct radiation in all directions which occurs when the radiant energy encounters the actual molecules of the ideal gases (nitrogen and oxygen) in the atmosphere. This effect is more pronounced for the short wavelengths, and accounts for the blue appearance of the sky.
(ii)   Scattering resulting from the presence of molecules of water vapour.
(iii) Selective absorption by the ideal gases and by water vapour. Asymmetrical gaseous molecules such as ozone, water vapour, carbon dioxide etc. have a greater ability to

absorb (and hence to emit) radiation than do gaseous molecules of a symmetrical structure, such as nitrogen and oxygen.
(iv) Scattering caused by dust particles.

The losses incurred are largely due to the permanent gases and to water vapour. There is also some diffusion of the direct sunlight caused when direct (and indirect) radiation encounters cloud. Not all the solar energy removed from direct radiation by the processes of scattering and absorption reaches the surface of the earth. Some is scattered back to space, and the absorbed energy which is re-radiated at a longer wavelength (mostly from water vapour) is also partly lost to outer space. Some direct energy is additionally lost by reflection from the upper surfaces of clouds. However, a good deal of the energy which direct radiation loses does eventually reach the surface of the earth in a scattered form.

The amount of sky radiation varies with the time of day, the weather, the cloud cover and the portion of the sky from which it is received, the amount from the sky in the vicinity of the sun exceeding that from elsewhere, for example. Sky radiation cannot, however, be assigned a specific direction and, hence, it casts no directional shadow outdoors.

Equations are available (see section 7.15) for calculating the total scattered solar radiation received from the sky and the ground. The intensity of such radiation depends on seasonal variations of moisture content and the earth–sun distance (given by Allen (1973) as 149.5×10^6 km), the angular relation of the receiving surface with respect to the surroundings, and the reflectances of the relevant surfaces.

Values of monthly mean daily irradiance on inclined, plane surfaces, at various angles to the horizontal, for both direct and diffuse solar radiation are provided by the *CIBSE Guide A2* (1999). Several locations are given, including Bracknell, over various orientations. The intensity of diffuse radiation depends mostly on the solar altitude but also, to a lesser extent, on the orientation when the receiving surface is vertical.

7.4 Definitions

There are a few basic terms in common use to describe the attributes of direct solar radiation. These are defined below, without lengthy explanation. The full implications of their meaning will become clear later in the text.

Altitude of the sun (a). This is the angle a direct ray from the sun makes with the horizontal at a particular place on the surface of the earth. It is illustrated in Figure 7.2(a). For a given date and time, the sun's altitude is different at different places over the world.

Azimuth of the sun (z). This is the angle the horizontal component of a direct ray from the sun makes with the true north–south axis. It may be expressed in degrees west or east of south (in the northern hemisphere), as illustrated in Figure 7.2(b), but is more commonly given as an angular displacement through 360° from true north.

Wall-solar azimuth (n). This is the angle the horizontal component of the sun's ray makes with a direction normal to a particular wall. It is illustrated in Figure 7.2(c), where it can be seen that this angle, referring as it does to direct radiation, can have a value between 0° and 90° only. When the wall is in shadow (n > 90°) the value of n has no meaning.

Declination (d). This is the angular displacement of the sun from the plane of the earth's equator. The value of the declination will vary throughout the year between $+23\frac{1}{2}$° and $-23^1/_2$° because the axis of the earth is tilted at an angle of about $23\frac{1}{2}$° to the axis of the plane in which it orbits the sun. Declination is expressed in degrees north or south (of the equator). The geometry of declination is illustrated in Figure 7.3.

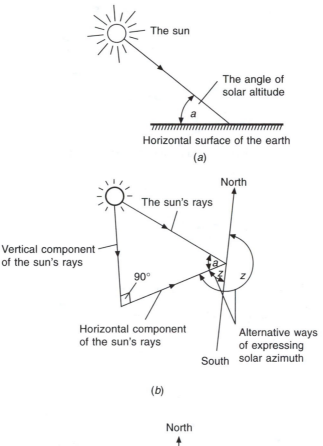

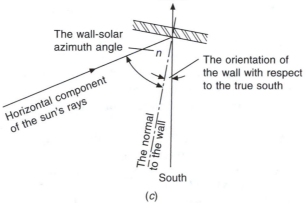

Fig. 7.2   (a) Definition of solar altitude; (b) Definition of solar azimuth; (c) Definition of wall-solar azimuth.

Latitude (*L*). The latitude of a place on the surface of the earth is its angular displacement above or below the plane of the equator, measured from the centre of the earth. It is illustrated in Figure 7.4. The latitude at London is about 51.5° North of the equator.

Longitude. This is the angle which the semi-plane through the poles, and a particular place on the surface, makes with a similar semi-plane through Greenwich. The semi-plane

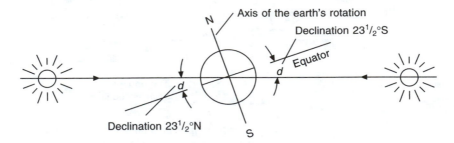

Fig. 7.3   Solar declination in June and December. See also equation (7.1).

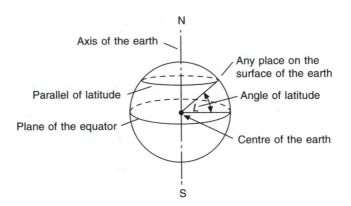

Fig. 7.4   The latitude of a place on the surface of the earth.

through Greenwich is an arbitrary zero, and the line it makes in cutting the earth's surface is termed the Greenwich meridian. Longitude is measured east or west of Greenwich and so its value lies between 0° and 180°. Latitude and longitude together are co-ordinates which locate any point on the surface of the earth.

Sun time (*T*). This is the time in hours, before or after noon, noon being defined as the time when the sun is highest in the sky. It is sometimes called local apparent time.

On the Greenwich meridian, sun time does not always exactly equal Greenwich Mean Time (or the local mean time for other longitudes). This is because the orbital velocity of the earth in its path about the sun is not a constant, since the orbit is elliptical, not circular. The variation, according to the *ASHRAE Handbook* (1997a), is the amount by which the sun time is different from the local mean time (for example from GMT on longitude 0°) and varies from nearly −14 minutes to just over +15 minutes, depending on the month. A good approximation is 4 minutes of time for each degree of longitude, as implied by the hour angle.

Hour angle (*h*). This is the angular displacement of the sun from noon. Thus,

$$h = \frac{360°}{24} \times T$$

or, put another way 1 hour corresponds to 15° of angular displacement.

7.5   The declination of the sun

Figure 7.5(*a*) shows the position of the earth, in the course of its orbit around the sun, seen slightly from above. If a section is taken through the earth at its two positions at mid-winter and mid-summer, the concept of declination is seen more clearly. Figure 7.5(*b*) is such a section, illustrating that the sun is vertically overhead at noon on a latitude of $23\frac{1}{2}°$S in mid-winter and also on $23\frac{1}{2}°$N at noon in mid-summer. These two values of latitude correspond to the sun's declination at those dates and are a consequence of the earth's pole being tilted at an angle of about $23\frac{1}{2}°$ to the axis of the plane in which it orbits the sun. Figure 7.5(*b*) also illustrates this.

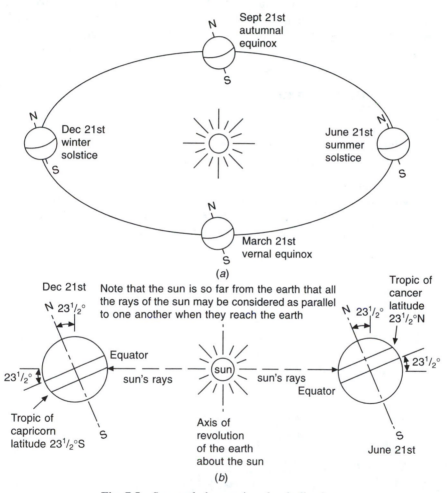

Fig. 7.5   Seasonal changes in solar declination.

The angular inclination of the pole does not vary substantially from this value owing to the gyroscopic effect of the earth's rotating mass. As a result of this, the sun can be vertically overhead only between the latitudes $23\frac{1}{2}°$N and $23\frac{1}{2}°$S. These two parallels are called the tropics of Cancer and Capricorn, respectively. They define the bounds of the Torrid Zone, often referred to as 'the Tropics'. It follows that, at dates other than the winter and summer solstices, the sun will be vertically overhead at noon in the tropics, at some latitude between $23\frac{1}{2}°$N and S. In fact, the sun is vertically overhead at noon on the equator itself on two dates in the year—the autumnal equinox and the vernal (spring) equinox. On these occasions the declination is zero. Referring to Figure 7.5 it can also be inferred that the latitude of the arctic circle is 66.5° north and that of the antarctic circle 66.5° south.

The dates of the solstices and equinoxes are not always the same because the earth's year is not exactly 365 days. An equation from which the declination can be determined is:

$$d = 23.45 \sin[360 \times (284 + N)/365] \qquad (7.1)$$

where N is the day number, starting from 1 on 1 January. Approximate values of declination are: 21 June, $23\frac{1}{2}°$N; 21 May and 21 July, $20\frac{1}{4}°$N; 21 April and 21 August, $11\frac{1}{2}°$N; 21 March and 21 September, 0°. Corresponding values for the other six months of the year have southerly, negative declinations.

7.6   The altitude of the sun

It is worth considering two cases:

(a)   the special case for the altitude at noon;
(b)   the general case for the altitude at any time of the day.

(a) The altitude at noon (a')
This is easily illustrated in geometrical terms, as in Figure 7.6, in which it can be seen that a very simple expression is derived from a consideration of the latitude of the place in question and the declination for the particular date:

$$a' = 90 - (L - d) \qquad (7.2)$$

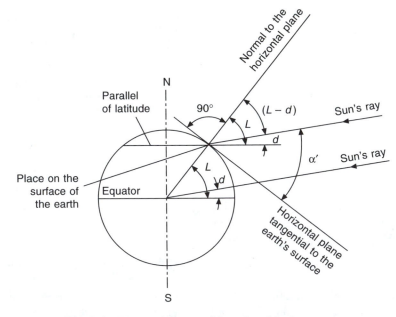

Fig. 7.6   The special case of the solar altitude at noon.

It should be borne in mind when considering Figure 7.6 that all parallels of latitude are parallel to the plane of the equator, and that, because the sun is so far away from the earth, all its rays parallel to one another when they strike the earth.

EXAMPLE 7.1

Calculate the maximum and minimum altitudes of the sun at noon, in London, given that the latitude of London is 51°N.

Answer

The greatest altitude occurs at noon on 21 June and the least at noon on 21 December. Hence, from equation (7.2), and knowing that the declinations are $23\frac{1}{2}$°N and S respectively in June and December, we can write

$$a'_{(max)} = 90° - (51° - 23\tfrac{1}{2}°) = 62\tfrac{1}{2}°$$
$$a'_{(min)} = 90° - (51° - (-23\tfrac{1}{2}°)) = 15\tfrac{1}{2}°$$

(*b*) The general case of the sun's altitude (*a*)

Although a geometrical illustration is possible, it is in three dimensions and consequently, being difficult to illustrate, is not shown. The result, obtained from such geometrical considerations, is that

$$\sin a = \sin d \times \sin L + \cos d \times \cos L \times \cos h \tag{7.3}$$

This equation degenerates, to give the special result obtained from equation (7.2), if a value of $h = 0$ is taken for noon.

EXAMPLE 7.2

Calculate the altitude of the sun in London on 21 June (*a*) at 13.00 h and (*b*) at 17.00 h (British Summer Time), making use of equation (7.3) in both cases.

Answer

(*a*)

$$d = 23\tfrac{1}{2}°N = +23\tfrac{1}{2}°$$
$$L = +51°$$
$$T = 0 \text{ (because sun time is British Summer Time minus 1 h)}$$

therefore

$$h = 0$$

Hence

$$\sin a = \sin 23\tfrac{1}{2}° \times \sin 51° + \cos 23\tfrac{1}{2}° \times \cos 51° \times \cos 0$$
$$= 0.399 \times 0.777 + 0.917 \times 0.629 \times 1.0 = 0.887$$

Therefore

$$a = 62\tfrac{1}{2}°$$

(*b*) At 17.00 h, $T = 16.00$ h, hence $h = 60°$ and

$$\sin a = \sin 23\tfrac{1}{2}° \times \sin 51° + \cos 23\tfrac{1}{2}° \times \cos 51° \times \cos 60°$$
$$= 0.399 \times 0.777 + 0.917 \times 0.629 \times 0.5 = 0.598$$

Therefore

$$a = 36\tfrac{3}{4}°$$

Note that using a more exact value of $51\tfrac{1}{2}$ °N for London gives a slightly different answer.

7.7   The azimuth of the sun

There is no special case that can easily be illustrated geometrically except, perhaps, the trivial one of the sun at noon, when the azimuth is zero by definition. The general case is capable of geometrical treatment but, since three dimensions are involved, it is preferred here to give the result:

$$\tan z = \frac{\sin h}{\sin L \cos h - \cos L \tan d} \tag{7.4}$$

EXAMPLE 7.3

Calculate the azimuth of the sun in London on 21 June at 17.00 h British Summer Time, making use of equation (7.4).

Answer

As before, L is 51°, d is $+23\tfrac{1}{2}$ ° and h is 60°, so by equation (7.4)

$$\tan z = 0.866/(0.777 \times 0.5 - 0.629 \times 435)$$
$$= 7.53$$

whence z is 82.5°W of S, or 262.5° from N.

   If the hour angle, h, is given a negative sign for times before mid-day and a positive one for times after, the sign of the azimuth is negative when it is east of south and positive when it is west of south.

EXAMPLE 7.4

Calculate the solar altitude and azimuth on 21 January for Perth, Western Australia, at 15.00 h sun time, given that the latitude is 32°S.

Answer

The day number is 21
Hence, by equation (7.1),

$$d = 23.45 \sin[360(284 + 21)/365]$$
$$= 23.45 \sin[300.82]$$
$$= -20.14°$$

$L = -32°$ and $h = +45°$. Hence, by equation (7.3)

$$\sin a = \sin(-20.14) \sin(-32) + \cos(-20.14) \cos(-32) \cos(45)$$

$$= (-0.3443)(\sim 0.5299) + (0.9389)(0.8480)(0.7071)$$
$$= 0.7454$$

whence

$$a = 48.2°$$

By equation (7.4)

$$\tan z = \sin(45)/[\sin(-32)\cos(45) - \cos(-32)\tan(-20.14)]$$
$$= 0.7071/[(-0.5299)(0.7071) - (0.8480)(-0.3667)]$$
$$= 0.7071/(-0.0637)$$
$$= -11.10$$

whence,

$$z = -84.9°$$

Note that, since the sun is to the north of the observer in the southern hemisphere, the solar azimuth is 84.9° west of north.

7.8   The intensity of direct radiation on a surface

If the intensity of direct solar radiation incident on a surface normal to the rays of the sun is I W m^{-2}, then the component of this intensity in any direction can be easily calculated. Two simple cases are used to illustrate this, followed by a more general case.

Case 1   The component of direct radiation normal to a horizontal surface (I_h), as illustrated in Figure 7.7(a).
   If the angle of altitude of the sun is a, then simple trigonometry shows that the component in question is $I \sin a$.

$$I_h = I \sin a \tag{7.5}$$

Case 2   The component of direct radiation normal to a vertical surface (I_v), as illustrated in Figure 7.7(b).
   The situation is a little more complicated since the horizontal component of the sun's rays, $I \cos a$, has first to be obtained and then in its turn, has to be further resolved in a direction at right angles to a vertical surface which has a particular orientation: that is, the wall-solar azimuthal angle n must be worked out. It then follows that the resolution of $I \cos a$ normally to the wall is $I \cos a \cos n$.

$$I_v = I \cos a \cos n \tag{7.6}$$

EXAMPLE 7.5

If the altitude and azimuth of the sun are $62\frac{1}{2}°$ and $82\frac{1}{2}°$W of S respectively, calculate the intensity of direct radiation normal to (a) a horizontal surface and (b) a vertical surface facing south-west.

Answer

Denote the intensity of direct radiation normal to the rays of the sun by the symbol I,

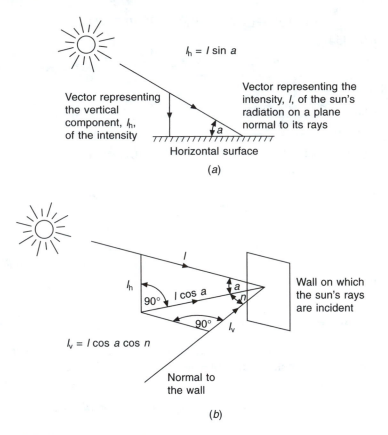

Fig. 7.7   The intensity of solar radiation on a horizontal and vertical surface.

having units of W m^{-2}, and denote its components normal to a horizontal and a vertical surface by I_h and I_v, respectively.

(*a*)

$$I_h = I \sin a$$
$$= I \sin 62\tfrac{1}{2}°$$
$$= 0.887I$$

(*b*) By drawing the diagram shown in Figure 7.8 it is evident that the wall-solar azimuthal angle, n, is $37\tfrac{1}{2}°$ and hence

$$I_v = I \cos a \cos n$$
$$= I \cos 62\tfrac{1}{2}° \cos 37\tfrac{1}{2}°$$
$$= 0.366I$$

Case 3   The component of direct radiation normal to a tilted surface ($I_δ$).

Figure 7.9 illustrates the case and shows, in section, a surface that is tilted at an angle $δ$ to the horizontal. The surface has an orientation which gives it a wall-solar azimuthal angle of n.

If the surface were vertical, then the resolution of I, normal to its surface, would be $I \cos$

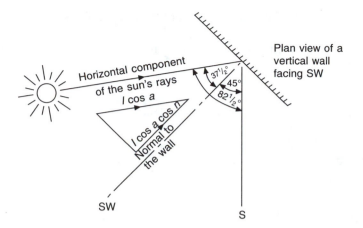

Fig. 7.8   Determining the wall-solar azimuth angle for Example 7.5.

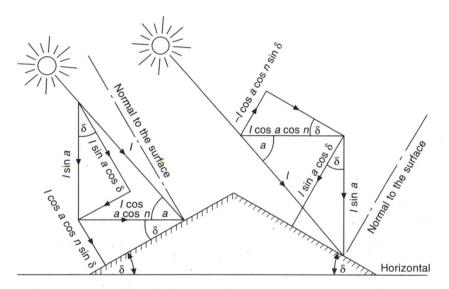

Fig. 7.9   The derivation of an expression for the intensity of solar radiation normally incident on a surface tilted at an angle δ to the horizontal.

$a \cos n$. The sun's rays also have a resolution normal to a horizontal surface of $I \sin a$. It follows that the rays normally incident on a tilted surface instead of a vertical one have a vector sum of $I \sin a$ and $I \cos a \cos n$, in an appropriate direction. The components of $I \sin a$ and $I \cos a \cos n$ in this appropriate direction are $I \sin a \cos \delta$ and $I \cos a \cos n \sin \delta$, respectively, both normal to the tilted surface. Their sum is the total value of the intensity of radiation normally incident on the tilted surface—the answer required. Hence,

$$I_\delta = I \sin a \cos \delta \pm I \cos a \cos n \sin \delta \qquad (7.7)$$

Equation (7.7) degenerates to equation (7.5) for a horizontal surface where δ is zero, and to equation (7.6) for a vertical surface where δ is 90°. The significance of the alternative signs is that if the tilted surface faces the sun, a positive sign should be used and vice versa if the surface is tilted away from the sun. Naturally enough, if the surface is tilted so much that it is in shadow, I is zero, since the angle of incidence of the radiation will be 90°.

Equation (7.7) can be written in a more general but less informative manner:

$$I_\delta = I \cos i \tag{7.8}$$

where i is the angle of incidence of the ray on the surface. This is true of all surfaces, but it must be borne in mind that the angle of incidence is between the incident ray and the *normal* to the surface. It is *not* the glancing angle.

All that equation (7.8) does is to state in a succinct form what equation (7.7) does more explicitly. This means that $\cos i$ is the same as $\sin a \cos \delta \pm \cos a \cos n \sin \delta$.

EXAMPLE 7.6

Calculate the component of direct solar radiation which is normally incident upon a surface tilted at 30° to the horizontal and which faces south-west, given that the solar altitude and azimuth are $62\frac{1}{2}°$ and $82\frac{1}{2}°$ west of south, respectively. The tilted surface is facing the sun.

Answer

For a wall facing south-west and an azimuth of $82\frac{1}{2}°$ west of south, the value of the wall solar azimuthal angle, n, is $37\frac{1}{2}°$; hence, by equation (7.7),

$$I_\delta = I \sin 62\frac{1}{2}° \cos 30° + I \cos 62\frac{1}{2}° \cos 37\frac{1}{2}° \sin 30°$$
$$= I(0.887 \times 0.866 + 0.462 \times 0.793 \times 0.5)$$
$$= 0.95I.$$

Doing the same calculation by means of the data published in the *CIBSE Guide A2* (1999) yields the same answer.

7.9   The numerical value of direct radiation

In order to evaluate the amount of solar radiation normally incident upon a surface, it is first necessary to know the value of the intensity, I, which is normally incident on a surface held at right angles to the path of the rays. The values of I can be established only by referring to experimental results for different places over the surface of the earth. These seem to suggest that I is independent of the place and that it depends only on the altitude of the sun. There is sense in this, since the amount of direct radiation reaching the surface of the earth will clearly depend on how much is absorbed in transit through the atmosphere, and the atmospheric path length is greater when the sun is lower in the sky; that is, when its altitude is less.

Numerical values for the intensity of direct solar radiation are given in Table 7.1, and Table 7.2 gives correction factors that account for the increased intensity with ascending altitude from the same source.

Figure 7.10 shows curves based on Curtis and Lawrence (1972) and ASHRAE (1993a) using equation (7.9):

$$I = A/\exp(B/\sin a) \text{ kW m}^{-2} \tag{7.9}$$

in which the constant A is the apparent solar radiation in the absence of an atmosphere and the constant B is an atmospheric correction factor. A and B (see Table 7.3) depend upon

Table 7.1   Intensity of direct solar radiation with a clear sky

Inclination and orientation of surface		Sun altitude (degrees)											
		5°	10°	15°	20°	25°	30°	35°	40°	50°	60°	70°	80°
		Intensity of basic direct solar radiation with a clear sky for a place 0–300 m above sea level (W m^{-2})											
1. Normal to sun		210	388	524	620	688	740	782	814	860	893	912	920
2. Horizontal roof		18	67	136	212	290	370	450	523	660	773	857	907
3. Vertical wall: orientation from													
sun in degrees	0°	210	382	506	584	624	642	640	624	553	447	312	160
(wall-solar	10°	207	376	498	575	615	632	630	615	545	440	307	158
azimuth angle)	20°	197	360	475	550	586	603	602	586	520	420	293	150
	30°	182	330	438	506	540	556	555	540	480	387	270	140
	40°	160	293	388	447	478	492	490	478	424	342	240	123
	45°	148	270	358	413	440	454	453	440	390	316	220	113
	50°	135	246	325	375	400	413	412	400	355	287	200	103
	55°	120	220	290	335	358	368	368	358	317	256	180	92
	60°	105	190	253	292	312	210	320	312	277	224	156	80
	65°	90	160	214	247	264	270	270	264	234	190	132	68
	70°	72	130	173	200	213	220	220	213	190	153	107	55
	75°	54	100	130	150	160	166	166	160	143	116	80	40
	80°	36	66	88	100	108	110	110	108	96	78	54	28

(Reproduced by kind permission from the *CIBSE Guide A2* (1986))

Table 7.2   Percentage increase in direct solar radiation at varying heights above sea level

Height above sea level	Solar altitude									
	10°	20°	25°	30°	35°	40°	50°	60°	70°	80°
1000 m	32	22	18	16	14	13	12	11	10	10
1500 m	50	31	26	23	21	18	16	15	14	14
2000 m	65	40	33	29	27	24	21	19	18	18
3000 m	89	52	43	37	34	31	27	24	23	22

Note that sky radiation *decreases* by approximately 30 per cent at 1000 m and by about 60 per cent at 1500 m above sea level.

seasonal variations in the earth–sun distance, the atmospheric moisture content and dust pollution according to Moon (1940). The equation does not give the maximum intensity but the value likely on an average cloudless day: the maximum intensity of direct radiation on a very clear day can be 15 per cent higher. The equation gives more accurate results than Moon.

CIBSE values for the intensity of direct radiation on a plane normal to the sun's rays, taken from Table 7.1, are in good agreement with the curve for May in Figure 7.10, for solar altitudes exceeding 10°, in northern latitudes.

If data which have been determined for the northern hemisphere are to be used for an application in the southern hemisphere, corrections must be applied to take account of the reduced sun–earth distance in the southern summer. Thus a solar intensity value (or a value

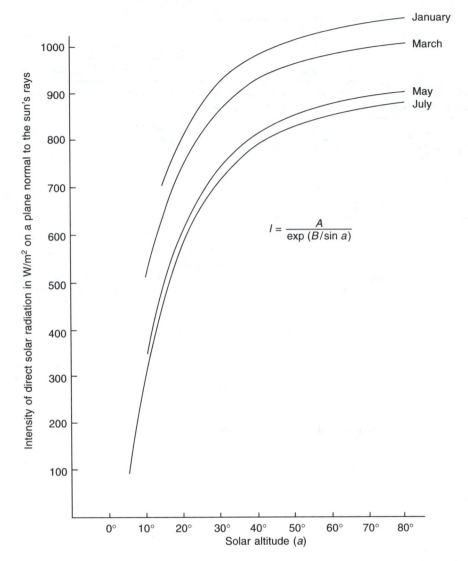

$$I = \frac{A}{\exp(B/\sin a)}$$

Fig. 7.10   Numerical values of direct solar radiation, incident on a surface at right angles to the sun's rays, at sea level in the northern hemisphere.

Table 7.3   Constants for determining the values of direct and scattered radiation at sea level, to be used in equations (7.9) and (7.14)

	Jan	Feb	Mar	Apr	May	Jun	Jul	Aug	Sep	Oct	Nov	Dec	Units
C	0.058	0.060	0.071	0.097	0.121	0.134	0.136	0.122	0.092	0.073	0.063	0.057	–
A	1.230	1.213	1.186	1.136	1.104	1.088	1.085	1.107	1.152	1.192	1.220	1.233	kW m^{-2}
B	0.142	0.144	0.156	0.180	0.196	0.205	0.207	0.201	0.177	0.160	0.149	0.142	–

of cooling load due to solar gain through windows) for the month of July in the northern hemisphere, should be multiplied by the ratio of the intensity in January to that in July, if it is to be used for the month of January in the southern hemisphere.

Correction factors for this purpose, based on the ratio of solar intensities outside the limits of the earth's atmosphere according to ASHRAE (1993a), for corresponding months, are given in Table 7.4.

Table 7.4 Correction factors for solar radiation data based on the northern hemisphere, when applied to the southern hemisphere

Month ratio	Dec/Jun	Jan/Jul	Feb/Aug	Mar/Sep	Apr/Oct
Correction factor	1.07	1.07	1.06	1.02	0.98

7.10 External shading

This is the best way of excluding the entry of direct solar radiation. If the shade is completely opaque, all the direct radiation is prevented from entering the window, although scattered radiation from the ground and the sky can still do so and must be taken into account. If the shade is partially opaque then some of the direct radiation is transmitted. The practical difficulty with motorised, external shading is that adequate access must be provided for maintenance.

7.11 The geometry of shadows

This is best illustrated by considering the dimensions of the shadow cast on a vertical window by a reveal and an overhanging lintel.

Figure 7.11(*a*) shows in perspective a window which is recessed from the wall surface by an amount R. The front elevation of the window, seen in Figure 7.11(*b*), shows the pattern of the shadow cast by the sunlight on the glass. The position of the point P' on the shadow, cast by the sun's rays passing the corner point P in the recess, is the information required. In other words, if the co-ordinates x and y of the point P' on the glass can be worked out, the dimensions of the shaded area are known.

Figure 7.11(*c*) shows a plan of the window. In this, the wall-solar azimuth angle is n and co-ordinate x is immediately obtainable from the relationship:

$$x = R \tan n \tag{7.10}$$

To see the true value of the angle of altitude of the sun, we must not take a simple sectional elevation of the window but a section in the plane of the sun's rays; that is, along A–A, as shown by Figure 7.11(*d*).

If the hypotenuse of the triangle formed by R and x in Figure 7.11(*c*) is denoted by M, then the value of M is clearly $R \sec n$. Reference to Figure 7.11(*d*) shows that the value of the co-ordinate y can be obtained from

$$y = M \tan a$$
$$= R \sec n \tan a \tag{7.11}$$

The *CIBSE Guide* publishes data in tabular form on the shading cast on a recessed window.

EXAMPLE 7.7

Calculate, by means of equations (7.10) and (7.11), the area of the shaded portion of a

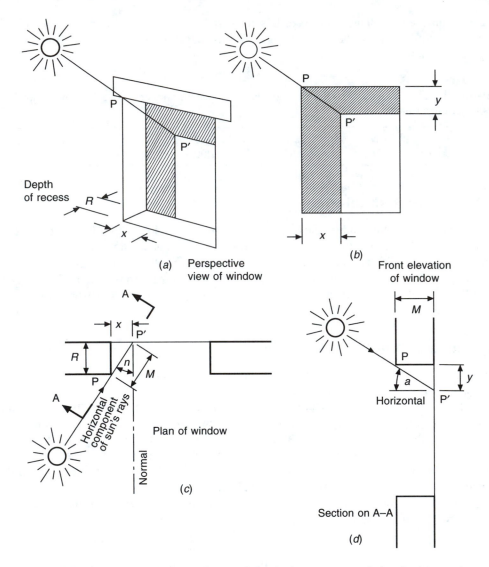

(a)   Perspective
       view of window

(b)   Front elevation
       of window

Plan of window

(c)

Section on A–A

(d)

Fig. 7.11   Determining the co-ordinates of the shadow cast on a window by its reveal.

window which is recessed by 50 mm from the surface of the wall in which it is fitted, given the following information:

Altitude of the sun	43°30′
Azimuth of the sun	66° west of south
Orientations of the window	facing south-west
Dimensions of the window	2.75 m high × 3 m wide

Answer

The angle *n* is 66° − 45° = 21° and so, by equation (7.10)

$$x = 50 \tan 21°$$
$$= 50 \times 0.384$$
$$= 19 \text{ mm}$$

By equation (7.11)

$$y = 50 \sec 21° \tan 43°30'$$
$$= 50 \times 1.071 \times 0.949$$
$$= 51 \text{ mm}$$

It is to be noted that the depth of the recess is the only dimension of the window which governs values of x and y. The actual size of the window is only relevant when the area of the shadow is required. It follows that equations (7.10) and (7.11) are valid for locating the position of any point, regardless of whether it is the corner of a recessed window or not. Thus, as an instance, the position of the end of the shadow cast by a flagpole on a parade ground could be calculated.

Continuing the answer to example 7.6,

$$\text{sunlit area} = (2750 - y) \times (3000 - x)$$
$$= 2699 \times 2981$$
$$= 8.05 \text{ m}^2$$

$$\text{total area} = 2.75 \times 3.00$$
$$= 8.25 \text{ m}^2$$

$$\text{shaded area} = 8.25 - 8.05$$
$$= 0.2 \text{ m}^2$$

This does not seem to be very significant, but it must be borne in mind that it is the depth of the recess which governs the amount of shading.

7.12   The transmission of solar radiation through glass

Of the energy which is incident upon the glass, some is reflected and lost, some is transmitted through the glass, and some is absorbed by the glass as the energy passes through it. This small amount of absorbed energy raises the temperature of the glass, and the glass eventually transmits this heat, by convection, partly to the room and partly to the exterior. The sum of the transmission coefficient (transmissivity), the absorption coefficient (absorptivity) and the reflection coefficient (reflectivity) is unity.

Table 7.5 gives figures for transmissibility and absorption but, in round figures, for angles of incidence between 60° and 0°, ordinary single window glass transmits about 85 per cent of the energy incident upon it. About 6 per cent is absorbed and the remaining 9 per cent is reflected. As the angle of incidence increases beyond 60°, the transmitted radiation falls off to zero, the reflected amount increasing. The absorption figure remains fairly constant at about 6 per cent for angles of incidence up to 80°.

For double glazing, the picture is more complicated but, in approximate terms, only about 90 per cent of what passes through single glazing is transmitted. Thus, the transmitted percentage for double glazing is about 76 per cent of the incident energy.

Table 7.5 Transmissivity and absorptivity for direct solar radiation through glass. (The absorptivity for indirect radition is taken as 0.06, regardless of the angle of incidence. The transmissivity for indirect radiation is taken as 0.79, regardless of the angle of incidence.)

	Angle of incidence						
	0°	20°	40°	50°	60°	70°	80°
*4 mm ordinary, clear glass**							
Transmissivity	0.87	0.87	0.86	0.84	0.79	0.67	0.42
Absorptivity	0.05	0.05	0.06	0.06	0.06	0.06	0.06
*6 mm plate or float, clear glass***							
Transmissivity	0.84	0.84	0.83	0.80	0.74	0.62	0.38
Absorptivity	0.08	0.08	0.08	0.08	0.10	0.09	0.09

*   Reproduced by kind permission from the *ASHRAE Handbook* (1993b).
** Based on data published by Pilkington Bros (1969).

EXAMPLE 7.8

For the window and information given in example 7.7 calculate the instantaneous gain from solar radiation to a room, given that the transmission coefficient for direct radition τ is 0.85, that the intensity of direct solar radiation on a surface normal to the rays is 832 W m^{-2}, that the intensity of sky radiation is 43 W m^{-2}, normal to the window, and that the transmission coefficient τ' for sky radiation is 0.80.

Answer

Total transmitted solar radiation is given by

Q = (transmitted direct radiation × sunlit area of glass)
     + (transmitted sky radiation × total area of glass)
   = $\tau I_\delta A_{sun} + \tau' \times 43 \times A_{total}$

By equation (7.6), for a vertical window

$$I_\delta = I \cos a \cos n$$
$$= 832 \cos 43°30' \cos 21°$$
$$= 832 \times 0.677$$
$$Q = 0.85 \times 832 \times 0.677 \times 8.05 + 0.8 \times 43 \times 8.25$$
$$= 4138 \text{ W}$$

As is seen later, not all of this energy constitutes an immediate load on the air conditioning system.

To summarise the procedure for assessing the instantaneous solar heat transmission through glass:

(1) For the given date, time of day and latitude, establish the solar declination, d, altitude, a, and azimuth, z, either by equations (7.1), (7.3) and (7.4) or by reference to tables.
(2) Determine the intensity of direct radiation, I, on a surface normal to the sun's rays for

the particular altitude, either by reference to tables or by means of equation (7.9). Apply the correction factor from Table 7.4, if needed.

(3) For the orientation of the window in question, calculate the wall-solar azimuth angle, n.

(4) Using equation (7.7) directly, or the simplified versions, equations (7.5) or (7.6), calculate the component of the direct radiation, I_δ, normally incident on the surface. If the surface is vertical, I_δ is I_v, and if it is horizontal, I_δ is I_h.

(5) For the particular angle of incidence on the glass, determine the transmission factor τ. If i is between 0° and 60°, take τ' as 0.80, in the absence of other information.

(6) Establish the dimensions of any shadows cast by external shading, hence, knowing the dimensions of the windows, calculate the sunlit and total areas.

(7) Knowing the altitude of the sun and the month of the year, refer to tables and find the value of the sky radiation normally incident on the glass. See Table 7.7. Take the transmission factor for sky radiation, τ, as 0.85, in the absence of other information.

(8) Calculate the direct transmitted radiation by multiplying τ, I_δ, and the sunlit area, and add to this the product of τ', the sky radiation and the total area. This is the total instantaneous solar radiation transmitted through the window.

7.13   The heat absorbed by glass

The amount of the solar energy absorbed by the glass during the passage of the direct rays of the sun through it depends on the absorption characteristics of the particular type of glass.

Ordinary glass does not have a very large coefficient of absorption, but certain specially made glasses absorb a good deal of heat. The heat absorbed causes an increase in the temperature of the glass, and heat then flows by conduction through the glass to both its surfaces. At the indoor and outdoor surfaces the heat is convected and radiated away at a rate dependent on the value of the inside and outside surface film coefficients of heat transfer, h_{si} and h_{so}.

If values are assumed for the temperature in the room, t_r, and for the temperature outside, t_o, a heat balance equation can be drawn up and a value calculated for the mean temperature of the glass. It is assumed in doing this that, because the glass is so thin, the surface temperatures are virtually the same as the mean.

Referring to Figure 7.12, taking the mean glass temperature as t_g and the absorptivity of the glass as α or α' the heat balance is

$$\alpha I_\delta + \alpha' I_s = (t_g - t_o)h_{so} + (t_g - t_r)h_{si}$$

whence

$$t_g = \frac{\alpha I_\delta + \alpha' I_s + h_{so}t_o + h_{si}t_r}{(h_{so} + h_{si})} \tag{7.12}$$

EXAMPLE 7.9

Given that the solar altitude is 43°30′, the solar azimuth is 66° west of south, the window faces south-west, the outside temperature is 28°C, the room temperature is 22°C, h_{so} is 22.7 W m^{-2} and h_{si} is 7.9 W m^{-2}, calculate the mean temperature of a single sheet of glass in July, for the following cases: (a) 6 mm clear float glass; (b) 6 mm heat-absorbing bronze glass.

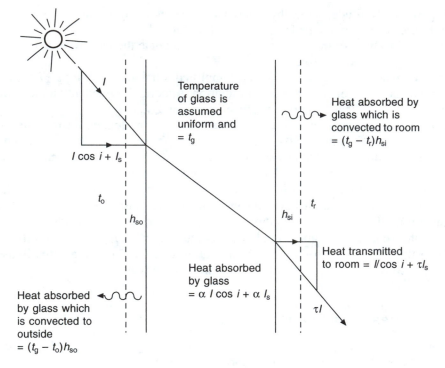

Temperature of glass is assumed uniform and $= t_g$

Heat absorbed by glass which is convected to room $= (t_g - t_r)h_{si}$

t_o

h_{so}

h_{si}

t_r

Heat transmitted to room $= I/\cos i + \tau I_s$

Heat absorbed by glass $= \alpha I \cos i + \alpha I_s$

τI

$I \cos i + I_s$

I

Heat absorbed by glass which is convected to outside $= (t_g - t_o)h_{so}$

Fig. 7.12   Heat absorbed by glass in sunlight.

Answer

The window faces 45° west of south hence the wall-solar azimuth is 66° – 45° = 21°. Refer to Table 7.1 and determine that the direct solar radiation on a plane normal to the rays of the sun is 830 W m^{-2} for a solar altitude of 43°30′. Then, for direct radiation, by equation (7.6):

$$I_v = 830 \cos 43°30′ \cos 21° = 830 \times 0.7254 \times 0.9336 = 564 \text{ W m}^{-2}$$

Further reference, to Table 7.7, shows that, for a solar altitude of 43°30′ the intensity of radiation scattered from the sky is 54 W m^{-2} and the intensity of radiation scattered from the ground is 66 W m^{-2}, for a vertical surface in July. Hence the additional, scattered radiation, normally incident on the vertical window, is (54 + 66) = 120 W m^{-2}.

Reference to Table 7.6 shows that the absorption coefficients for 6 mm clear float and 6 mm heat-absorbing bronze are 0.15 and 0.49, respectively. It is reasonable to assume that the coefficients refer to both direct and scattered solar radiation so we can now calculate the glass temperatures by equation (7.12):

(*a*) for 6 mm clear float glass

$$t_g = [0.15 \times 564 + 0.15 \times 120 + 22.7 \times 28 + 7.9 \times 22]/(22.7 + 7.9) = 29.8°C$$

(*b*) for 6 mm heat-absorbing bronze glass

$$t_g = [0.49 \times 564 + 0.49 \times 120 + 22.7 \times 28 + 7.9 \times 22]/(22.7 + 7.9) = 37.4°C$$

If no solar radiation is absorbed it can be verified by equation (7.12) that the glass temperature

Table 7.6   Transmission performance data for windows and shades. (Based on data from Pilkington (1991))

	Solar thermal radiation				Shading coefficients		
	light trans. %	absn. %	trans. %	total trans. %	radn.	conv.	total
Single unshaded glass							
Ordinary 4 mm glass	87	8	84	87	0.96	0.04	1.00
4 mm clear float	89	11	82	86	0.94	0.04	0.98
6 mm clear float	87	15	78	83	0.90	0.05	0.95
6 mm heat-absorbing bronze	50	49	46	62	0.53	0.19	0.72
6 mm heat-absorbing green	72	49	46	62	0.53	0.19	0.72
6 mm heat-reflecting bronze	10	73	6	24	0.07	0.20	0.27
6 mm heat-reflecting blue	20	64	15	33	0.17	0.21	0.38
Single glass + internal Venetian blinds							
Ordinary 4 mm glass	–	44	11	47	0.12	0.41	0.53
6 mm clear float	–	52	9	47	0.10	0.44	0.54
6 mm heat-absorbing bronze	–	80	5	42	0.06	0.42	0.48
6 mm heat-reflecting bronze	–	78	1	22	0.01	0.24	0.25
Double unshaded glass							
Ordinary glass							
4 mm inner							
4 mm outer	76	16	71	76	0.77	0.08	0.85
Clear float							
6 mm inner							
6 mm outer	76	28	61	72	0.70	0.12	0.82
Double heat-reflecting glass							
6 mm clear inner,							
6 mm bronze outer	9	74	5	16	0.06	0.12	0.18
6 mm clear inner,							
6 mm blue outer	18	67	12	24	0.14	0.13	0.27
Double glass + internal Venetian blinds							
Ordinary glass							
4 mm inner							
4 mm outer	–	–	–	48	0.12	0.35	0.55
Clear float							
6 mm inner							
6 mm outer	–	62	7	47	0.08	0.46	0.54
Double heat-reflecting glass + internal Venetian blinds							
6 mm clear float inner							
6 mm bronze outer	–	78	1	15	0.01	0.16	0.17
6 mm clear float inner							
6 mm blue outer	–	76	2	20	0.02	0.21	0.23
Double glass + Venetian blinds between the panes							
Ordinary glass							
4 mm inner							
4 mm outer	–	50	5	25	0.08	0.17	0.29

(Contd)

Table 7.6   *(Contd)*

	Solar thermal radiation				Shading coefficients		
	light trans. %	absn. %	trans. %	total trans. %	radn.	conv.	total
Clear float 4 mm inner 4 mm outer	–	54	7	25	0.08	0.21	0.29
Double heat-reflecting glass + Venetian blinds between the panes							
6 mm clear float inner 6 mm bronze outer	–	78	1	13	0.01	0.14	0.15
6 mm clear float inner 6 mm blue outer	–	76	2	16	0.02	0.16	0.18

Notes for Table 7.6

1. Except where stated otherwise, all sheets of glass are 6 mm thick.
2. The figures quoted for heat-absorbing and heat-reflecting glasses are for current proprietary brands, easily obtainable.
3. Venetian blinds are assumed to be light coloured with their slats at 45° and incident solar radiation at an angle of incidence of 30° to the glass. Their reflectivity is taken as 0.51, transmissivity as 0.12 and absorptivity as 0.37.
4. The external wind speed is assumed as about 2 m s^{-1}.
5. Of the heat absorbed by the glass, two-thirds is assumed to be lost to the outside and one-third to the room.
6. The total transmitted solar thermal radiation is the actual gain to the room and includes some of the heat absorbed by the glass (and blinds).
7. The convection shading coefficient is not influenced by storage effects for the building and therefore corresponds to an instantaneous load on the air conditioning system. The radiation shading coefficient is susceptible to building storage effects and so does not correspond to an instantaneous load. Hence, the total shading coefficient corresponds to a mixed load, part instantaneous and part delayed.
8. It is usually not worthwhile using Venetian blinds with heat-reflecting glass, although the large amount of heat absorbed by such glass can give it a surface temperature exceeding 40°C. Similarly, from the opposite point of view, it is not worth using heat-absorbing glass if internal Venetian blinds are to be used anyway. People cannot feel comfortable in strong direct sunlight in a conditioned room; so internal blinds or strongly reflective glass are essential for comfort.
9. To be worthwhile, reflective glass should have a total shading coefficient not exceeding 0.27, if internal shading is to be dispensed with. Never use internal shading, of any sort, with heat-absorbing or heat-reflecting glass unless the glass manufacturers have approved the arrangement. Refer to Pilkington (1980).

is 26.5°C and then, for the three cases considered, the heat transfer into the room may be calculated:

(*a*)

$$7.9(29.8 - 22) = 61.6 \text{ W m}^{-2}$$

(*b*)

$$7.9(37.4 - 22) = 121.7 \text{ W m}^{-2}$$

For no solar radiation:

$$7.9(26.5 - 22) = 35.6 \text{ W m}^{-2}$$

Table 7.7 The intensity of scattered radiation from a clear sky in W m^{-2}

Month	Surface	Radiation	5°	10°	15°	20°	25°	30°	35°	40°	50°	60°	70°	80°
	Horizontal	Sky	14	32	41	47	51	54	56	57	59	61	62	62
Jan	Vertical	Sky	7	16	21	24	26	27	28	29	30	30	31	31
	Vertical	Ground	3	13	22	32	42	52	61	69	84	96	106	109
	Horizontal	Sky	14	32	42	48	52	55	57	58	60	62	62	63
Feb	Vertical	Sky	7	16	21	24	26	27	28	29	30	31	31	32
	Vertical	Ground	3	12	22	32	42	51	60	68	83	95	104	110
	Horizontal	Sky	14	38	46	53	58	62	64	66	69	70	72	72
Mar	Vertical	Sky	7	19	23	27	29	31	32	33	34	35	36	36
	Vertical	Ground	3	13	21	31	40	50	58	66	81	93	102	107
	Horizontal	Sky	14	39	55	65	72	77	80	83	87	90	91	92
Apr	Vertical	Sky	7	20	27	32	36	38	40	42	44	45	45	46
	Vertical	Ground	3	11	20	30	39	47	56	64	77	89	97	102
	Horizontal	Sky	14	43	61	75	84	91	95	98	104	107	109	110
May	Vertical	Sky	7	21	31	38	42	46	47	49	52	53	54	55
	Vertical	Ground	2	10	19	29	38	46	55	62	75	87	95	100
	Horizontal	Sky	14	45	66	80	90	97	102	106	112	115	118	119
June	Vertical	Sky	7	22	33	40	45	48	51	53	56	57	59	59
	Vertical	Ground	2	10	19	29	37	46	54	62	75	86	94	99
	Horizontal	Sky	14	45	66	81	90	98	103	107	113	116	118	120
July	Vertical	Sky	7	22	33	40	45	49	52	53	56	58	59	60
	Vertical	Ground	2	10	19	28	37	46	54	61	75	86	94	99
	Horizontal	Sky	14	42	62	75	84	90	95	99	104	107	109	110
Aug	Vertical	Sky	7	21	31	38	42	45	48	50	52	53	55	55
	Vertical	Ground	2	10	19	29	38	46	54	62	76	86	95	100
	Horizontal	Sky	14	38	53	63	70	74	78	80	84	86	88	89
Sep	Vertical	Sky	7	19	27	32	35	37	39	40	42	43	44	44
	Vertical	Ground	3	11	20	30	39	48	56	64	78	90	98	104
	Horizontal	Sky	14	35	47	55	60	63	66	68	71	72	74	75
Oct	Vertical	Sky	7	17	24	27	30	32	33	34	35	36	37	38
	Vertical	Ground	3	12	21	31	40	50	58	67	81	93	102	107
	Horizontal	Sky	14	33	43	50	54	57	59	61	64	65	65	66
Nov	Vertical	Sky	7	16	22	25	27	29	30	31	32	32	33	33
	Vertical	Ground	3	12	22	32	42	51	60	69	84	96	104	110
	Horizontal	Sky	14	31	41	46	50	53	55	57	58	60	60	61
Dec	Vertical	Sky	7	16	20	23	25	26	27	28	29	30	30	30
	Vertical	Ground	4	13	23	33	42	52	61	69	85	97	106	112

We see that, for 4 mm clear float, the presence of solar radiation on the glass increases the heat transfer to the room by $(61.6 - 35.6) = 26.0$ W m^{-2}. If heat-absorbing glass is used the figure goes up to 86.1 W m^{-2}. It is evident that the heat absorbed by clear glass makes only a small contribution but, if heat-absorbing glass is used it can become significant.

Glass temperatures can rise to very high values (well over 60°C) when the incident solar radiation is high, the absorptivity is large and the heat transfer coefficients for the surfaces are small—as would be the case for glazing with a sheltered outside exposure and stratified temperature conditions on the inside. High glazing temperatures cause stresses that can be a risk if not considered. Reference to the manufacturers should be made in such cases as Pilkington (1980) shows.

7.14   Internal shading and double glazing

The effect of internal shading devices in reducing solar heat gain is considerable but not so great as the effect of external shades. In tropical and sub-tropical climates, fixed, horizontal, external shades are extensively used and are effective because the sun is high in the sky for most of the day, but in the temperate zones the lower solar altitudes that prevail for the majority of the time make them of much less use. Motorised, automatic, external shades are also used but they need access for maintenance and because of adverse winter weather such maintenance may be expensive. For this reason, internal shading is much adopted and is essential for sunlit windows if air conditioning is to be satisfactory. Table 7.6 indicates the effects of shading windows from solar radiation.

Internal shades are very often of the Venetian blind type and for these to give most benefit, they should be of a white or aluminium colour and should have polished surfaces. They should be adjusted so that they reflect the rays of the sun back to the outside and so that no direct rays pass between the slats. The blinds absorb some radiation, warm up, and convect and radiate heat into the room. The radiant emission is from a surface temperature that is several degrees above room temperature and is acceptable. In comparison, the radiation emitted by the sun is from a surface temperature of 6000°C and causes acute discomfort when received through unshaded windows. Blinds are most effective when fitted *between* sheets of double glazing.

The comparative effectiveness of various combinations of shades and glazing is expressed in terms of a shading coefficient, defined by ASHRAE (1997b) as the ratio of the solar heat gain coefficient (SHGC) of a glazing system for a particular angle of incidence and incident solar spectrum to the SHGC for clear, single pane glass (standard reference glazing in which $\tau = 0.86$, $\rho = 0.08$ and $\alpha = 0.06$ at normal incidence) with the same angle and spectral distribution. Glazing system refers to the combination of glass type and shade. The symbols τ, ρ and α are the transmissivity, reflectivity and absorptivity, respectively, of the glass. It is claimed by ASHRAE (1997b) that, defined in this way, the shading coefficient (SC) is independent of the solar spectral shape and the angle of incidence and that it can be used for single or double glass and for various types of tinted glass.

The shading coefficient for single, clear, 4 mm glass (see Table 7.6) is 1.0 and its transmissivity is 0.87.

Hence a simpler definition by ASHRAE (1997b) merely states that, for a particular glass and blind combination with a solar heat gain coefficient of SHGC, the shading coefficient, SC, is expressed by

$$SC = SHGC/0.87 \qquad\qquad (7.13)$$

This is the sense in which shading coefficient is generally used, expressing the heat gain of a particular combination of glass and shading in comparison with that of an unshaded, 4 mm clear glass window. It has proved mostly acceptable. However, Reilly *et al.* (1992) claim that this simplified approach may exaggerate the solar heat gain by as much as 35 per cent.

As Table 7.6 shows, the presence of two sheets of glass reduces the net solar transmitted heat by about 15 per cent only. It usually follows from this that the saving in capital and running costs of the air-conditioning system are not enough to pay for the extra cost of double glazing.

The main favourable feature of double glazing is the reduction in noise transmission, but this is effective only if the air gap between the sheets exceeds about 100 mm.

The obvious advantage that a higher relative humidity can be maintained in the room in winter (because the inside surface temperature of double glazing is higher than that of single glazing, thus permitting a higher room dew point) is often spurious; poorly fitting double glazing permits the air in the room to enter the space between the sheets, and objectionable condensation then takes place on the inside of the glass sheets, where it cannot readily be wiped away. On the other hand, a small amount of ventilation through the outer sheet and frame is sometimes deliberately arranged and is most successful in preventing condensation.

Reflective plastic films, if applied properly to the inside surface of clear glazing can be significantly useful, provided that the shading coefficient does not exceed 0.27. External application is not recommended because of potential degradation due to weathering. The use of an applied film with solar control glass is not recommended. The glass manufacturer's advice should be sought.

Certain types of curtain can also give a measure of solar control. The shading coefficient, according to ASHRAE (1997c), depends on the reflective nature of the fabric, the ratio of the open area between the fibres to the total area, and the fullness of the drape. Values can range from 0.87 to 0.37, with 3 m clear glass, depending on the factors mentioned. The necessary washing of curtains can spoil their effectiveness—particularly when they have had a reflective coating applied to the fibres, which may get washed away after a while.

7.15 Numerical values of scattered radiation

The total amount of solar radiation (I_t) normally incident on a surface is given by

$$I_t = I \cos i + I_s + I_r \qquad (7.14)$$

where I_s is sky radiation and I_r is radiation reflected from surrounding surfaces. ASHRAE (1993b) evaluates I_s and I_r according to

$$I_s + I_r = CIF_s + \rho I(C + \sin a)F_g \qquad (7.15)$$

The value of the dimensionless constant, C, varies through the year and is given in Table 7.3. The angle factor for the ground with relation to the particular surface, F_g, is complementary to F_s, the factor between the surface and the sky and is given by

$$F_g = 0.5(1 - \cos \delta) \qquad (7.16)$$

Note that the sum of all the angle factors between a surface and its surroundings is unity. Hence, $F_s = 1 - F_g$ for any surface seeing only the ground and the sky. δ is the angle between the ground and the horizontal.

The reflectivity, ρ, of the ground depends on solar altitude and on the type of surface. Over the range of solar altitudes from 30° to 60° it has values of about 0.32 for new concrete, 0.22 for old concrete, 0.23 for green grass and 0.1 for asphalt according to ASHRAE (1993c).

EXAMPLE 7.10

Calculate the total scattered radiation normally incident on a vertical surface in July for a solar altitude of 40°, a height above sea level of 2000 m and a ground reflectivity of 0.2.

Answer

From Table 7.1 or equation (7.9), I = 814 W m^{-2} (CIBSE figure) or 786 W m^{-2} (an

American-derived figure) at sea level and from Table 7.2 the increase in these values is 24 per cent at a height of 2000 m. Also, from Table 7.3, $C = 0.136$ and by equation (7.16),

$$F_g = 0.5(1 - \cos 90°) = 0.5$$

Hence, by equation (7.15), using the CIBSE figure,

$$
\begin{aligned}
I_s + I_r &= 1.24[0.136 \times 814 \times (1 - 0.5) + 0.2 \times 814 \times (0.136 + \sin 40) \times 0.5] \\
&= 1.24(55.4 + 63.4) \\
&= 147 \text{ W m}^{-2}
\end{aligned}
$$

Note that Table 7.7 quotes values of 53 and 61 W m^{-2} at sea level for sky and ground radiation in July with 40° solar altitude. These are less than the values calculated because they are based on $I = 786$ W m^{-2}, the American-derived figure, used when compiling the table.

The *CIBSE Guide* A2 (1999) uses solar data from measurements at Bracknell.

7.16   Minor factors influencing solar gains

There are five relatively minor factors that should be taken into account when calculating the instantaneous heat transmission resulting from solar radiation:

(1)   Atmospheric haze
(2)   The type of window frame
(3)   The height of the place above sea level
(4)   Variation of the dew point
(5)   The hemisphere

(1)   Atmospheric haze

This is most noticeable in industrial areas and, in this context, is regarded as resulting from contaminants emitted by traffic and industry, carried aloft by thermal up-currents. It is usually more pronounced in the afternoon, owing to the build up of the ground surface temperatures which produce the up-currents as the day progresses. Haze can reduce the value of I_δ, the direct radiation normally incident on a surface, so much that a diminution of 15 per cent in the total radiation received (scattered plus direct) may occur. A conservative factor of 0.95 may be applied for cities such as London.

(2)   The type of window frame

Tables of the actual heat gain occurring through windows are often published (see Table 7.9). Generally, tables of this sort are for wooden-framed windows. If, as in many modern buildings, the framework is metal, then because this is a much better thermal conductor than wood, the heat gains are increased by about 17 per cent.

A more convenient way of taking account of this effect is to apply the tabulated figure for the heat gain to the glass area, in wooden-framed windows, and to the area of the opening in the wall, in metal-framed windows.

(3)   The height of the place above sea level

There is some difference of opinion in published data as to the influence of increased height above sea level on the intensity of direct solar radiation. Two indisputable facts are that the intensity is about 1367 W m^{-2} on a surface normal to the sun's rays, at the limits of the atmosphere, but only about 1025 W m^{-2}, as a maximum, at the surface of the earth. Thus the intensity is reduced by about 25 per cent when the path length of the sun's rays is at minimum.

This minimum path-length occurs when the sun is at its zenith or, put another way, when the 'air-mass' is unity. The bulk of the earth's atmosphere is below 3000 m and hence the major part of the reduction in intensity occurs below this level. The reduction in intensity, being clearly bound up with the air mass, must also be dependent on the altitude of the sun, since this affects the path-length of the sun's rays. (It is approximately true to assume that the path-length is proportional to the cosecant of the angle of altitude of the sun for values of this from 90° to 30°.)

Experimental observations show that for altitude variations from 0 to 3000 m above sea level, the maximum intensity of direct radiation normal to the sun's rays alters from about 950 W m^{-2} to about 1170 W m^{-2}. That is to say, an average increase of about 2.5 per cent occurs for each 300 m of increase of height above sea level. It should be noted that as the intensity of direct solar radiation increases with height the intensity of sky radiation falls off. Table 7.2 refers to this.

(4)   *Variation of the dew point*

Although the dew point of the air falls with increasing height above sea level (see equation (2.10) and section 2.13), the effect of this lapse rate is already taken account of in the altitude correction mentioned above. That a variation in dew point has any effect at all stems from the fact that a change in dew point means a change in moisture content and this, in turn, means a change in the absorption capacity of the air–water vapour mixture which constitutes the lower reaches of our atmosphere.

However, dew point varies over the surface of the earth between places of the same height above sea level. Some variation in intensity is, therefore, to be expected on this count. The Carrier Air Conditioning Company (1965) suggests that an increase in the intensity of direct radiation of 7 per cent occurs for each 5 K reduction in dew point below 19.4°C.

(5)   *Hemisphere*

As has been mentioned in section 7.1, the sun is 3 per cent closer to the earth in January than it is in July. This results in an increase of about 7 per cent in the value of the intensity of radiation reaching the upper part of the earth's atmosphere in January. Hence, calculations carried out for summer in the southern hemisphere should take account of this increase.

EXAMPLE 7.11

Calculate the instantaneous transmission of solar radiation through a window recessed 300 mm from the outer surface of the wall in which it is set. Use the following data:

Single 4 mm ordinary clear glazing, facing south-west
Latitude	40°N
Outside air temperature (t_o)	32°C
Room air temperature (t_r)	24°C
Sun time	13.00 h on 23 July

Altitude of location   600 m above sea level
Moderate industrial haze
The window framework is steel and the size of the
opening in the wall is 3 m × 3 m. The surrounding
ground is covered with grass.

U glass = 5.86 W m^{-2} K^{-1}
h_{so} glass = 22.7 W m^{-2} K^{-1}
h_{si} glass = 7.9 W m^{-2} K^{-1}

Answer

It is convenient to divide the calculations into a number of steps, in order to illustrate
methods.

(a) Calculation of solar altitudes

Declination, d, is $20\frac{1}{4}°$ (see section 7.5) and hour angle, h, is $15°$ (see section 7.4). Then
by equation (7.3):

$$\sin a = \sin 20\tfrac{1}{4}° \sin 40° + \cos 20\tfrac{1}{4}° \cos 40° \cos 15°$$
$$= 0.3461 \times 0.6428 + 0.9382 \times 0.7660 \times 0.9659$$
$$= 0.9166$$

whence a is $66°26'$.
    The CIBSE table of solar altitude gives a similar result.

(b) Calculation of solar azimuth

$$\tan z = \frac{\sin 15°}{\sin 40° \cos 15° - \cos 40° \tan 20\tfrac{1}{4}°} \qquad \text{(see equation (7.4))}$$

$$= \frac{0.2588}{0.6428 \times 0.9659 - 0.7660 \times 0.3689}$$

whence z is $37°25'$, west of south, or $217°$ clockwise from north.

(c) Determination of wall-solar azimuth

Figure 7.13 shows that the wall-solar azimuth angle, n, is $45° - 37°25' = 7°35'$.

(d) Calculation of sunlit area

By equation (7.10)

$$x = 300 \tan 7°35' = 39.9 \text{ mm}$$

By equation (7.11)

$$y = 300 \sec 7°35' \tan 66°26'$$
$$= 300 \times 1.009 \times 2.292 \, 55 = 694 \text{ mm}$$

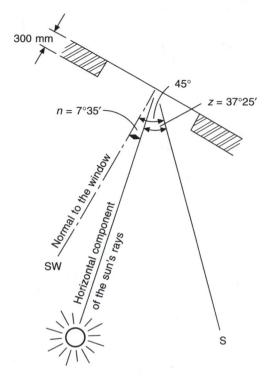

Fig. 7.13   Determining the wall-solar azimuth angle for Example 7.11.

Sunlit area = (3.0 − 0.0399)(3.0 − 0.694)
$$= 6.83 \text{ m}^2$$
Total area = 9 m^2

(e) Determination of the intensity of direct radiation

By equation (7.9)
$$I = A \exp(-B/\sin 66°26')$$
$$= 1.085 \exp(-0.207/0.9166)$$
$$= 0.866 \text{ kW m}^{-2}$$
This is normal to the sun's rays at sea level.
   Table 7.1 gives a value of 0.905 kW m^{-2}, by interpolation.

(f) Determination of the intensity of sky radiation

By interpolation, Table 7.7 (for the intensity of scattered solar radiation from a clear sky on a vertical surface) gives a value of 59 W m^{-2} at sea level. The footnote to Table 7.2 quotes a decrease of about 20 per cent, by extrapolation, for a height of 600 m above sea level where the value of sky radiation therefore becomes 47 W m^{-2}.
   There should be a correction applied to this figure to take account of the additional scattering resulting from industrial haze. However, data are scanty, the effect is small, and so it is ignored here.

(g) Determination of the effect of haze and height above sea level upon direct radiation

Table 7.2 also quotes an increase of 6 per cent for direct radiation and, by inference, for ground radiation also. A haze correction is 0.9 and so the total correction factor is $1.06 \times 0.9 = 0.954$.

(h) Calculation of the intensity of ground radiation

From Table 7.7, the intensity of radiation scattered from grass at sea level and for a solar altitude of $66°26'$ is 89 W m^{-2}, by interpolation. This intensity is normally incident on the window.

At 600 m the figure would be $1.06 \times 89 = 94$ W m^{-2}.

(i) Calculation of the intensity of direct radiation normal to the window

By equation (7.6)

$$I_v = 866 \cos 66°26' \cos 7°35'$$
$$= 343 \text{ W m}^{-2}, \text{ uncorrected for height and haze}$$

Table 7.1 gives 356 W m^{-2}, by interpolation.

(j) Direct radiation transmitted through the window

By equations (7.6) and (7.8)

$$\cos i = \cos a \cos n$$
$$= \cos 66°26' \cos 7°35' = 0.3963$$

whence $i = 66°39'$ and the transmissivity for this angle of incidence is 0.71, according to Table 7.5. Thus the transmission is 0.71×343, or 244 W m^{-2}, uncorrected for haze and height.

(k) Scattered radiation transmitted

Taking an approximate value of 0.79 for the transmissivity of scattered radiation, as recommended in Table 7.5, then, at sea level, the scattered radiation transmitted would be

$$0.79(59 + 91) = 118 \text{ W m}^{-2}$$

At an altitude of 600 m, ignoring the effect of haze, the transmission would be:

$$0.79(47 + 94) = 111 \text{ W m}^{-2}$$

(l) Direct plus scattered radiation transmitted

Through glass in sunlight, the radiation transmitted is due to direct and scattered sources and at sea level it is

$$= (244 + 118) = 362 \text{ W m}^{-2}$$

This value, even when corrected for altitude and haze, is of little use if only part of the window is in sunlight. Scattered radiation is transmitted through the entire window, but direct radiation is transmitted through the sunlit portion only.

Hence, transmitted direct radiation, corrected for haze and height

= 0.954 × 244 × sunlit area
= 0.954 × 244 × 6.83 = 1590 W

Transmitted scattered radiation, corrected for height but not for haze

= 111 × total area
= 111 × 9 = 999 W

Hence, the transmitted radiation is 1590 + 999 = 2589 W. As is seen later, this does not constitute an immediate load on the air conditioning system.

(m) Transmission due to absorbed solar radiation

Assume a value of 0.06 for the absorption of single glazing with an angle of incidence equal to 66°38′ in the case of both scattered and direct radiation (see Table 7.5).
By equation (7.12)

$$t_g = \frac{0.06 \times (0.954 \times 343 + 47 + 94) + 22.7 \times 32 + 7.9 \times 24}{30.6}$$

$$= 30.9°C$$

The convected and radiated heat gain to the interior by transmission and from heat absorbed by the glass is

$$6.83 \times 7.9(30.9 - 24) + (9 - 6.83) \times 5.86 \times (32 - 24) = 474 \text{ W}$$

Total instantaneous solar gain = 2589 + 474 = 3063 W.
Note that the gain because of solar energy absorbed by the glass is quite small. In the figure of 474 W the transmitted gain by virtue of air-to-air temperature difference, is 9 × 5.68 × (32 − 24), or 422 W, and so the gain by absorption is only 52 W, which is 1.7 per cent of the total gain of 3063 W and is evidently almost negligible for ordinary glass.

7.17   Heat gain through walls

The heat gain through a wall is the sum of the relatively steady-state flow (often simply termed 'transmission') that occurs because the inside air temperature is less than that outside, and the unsteady-state gain resulting from the varying intensity of solar radiation on the outer surface of the wall. The phenomenon of unsteady-state heat flow through a wall is complicated by the fact that a wall has a thermal capacity, and so a certain amount of the heat passing through it is stored, being released to the interior (or exterior) at some later time.

Two environmental factors are to be considered when assessing the amount of heat entering the outer surface of a wall:

(i)   the diurnal variation of air temperature, and
(ii)   the sinusoidal-type variation of solar intensity.

Figure 7.14 presents a simplified picture. At (*a*), under steady-state conditions, the graph of temperature through the wall is a straight line, the slope of which depends on the difference between the temperatures of the inner and outer surfaces and on the thickness of the wall. The calculation of heat gain under such circumstances is exactly the same as for the more familiar case of steady-state loss:

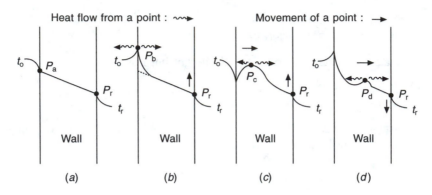

Fig. 7.14   Temperature gradients in a wall subjected to an unsteady heat input on the outside.

$$Q = U(t_r - t_o) \tag{7.17}$$

where Q is the rate of heat flow in W m^{-2}

U is the thermal transmittance coefficient in W m^{-2} K^{-1}

t_r and t_o are the air temperatures in the room and outside, respectively.

A heat gain occurs if t_o exceeds t_r.

The *CIBSE Guide* expresses the steady-state heat transfer through a wall in terms of the inside environmental temperature, instead of the inside air (dry-bulb) temperature, t_r. Environmental temperature is a hypothetical temperature that gives the same rate of heat transfer as that determined by more complicated methods. It is defined as one-third of the mean air temperature plus two-thirds of the mean radiant temperature in the room. Air conditioning systems control air temperature, not environmental temperature, hence equation (7.17) is adequate and is widely used.

Figure 7.14(*b*) shows the effect of raising the outer surface temperature. The temperature at the point P_b is greater than that at point P_a. Such an increase could be caused by the outer wall surface receiving solar radiation. Heat flows away from P_b in both directions because its temperature is higher than both the air and the material of the wall in its vicinity. If the intensity of solar radiation then diminishes, the situation in Figure 7.14(*c*) arises.

Since heat flowed away from P_b, the value of the temperature at P_c is now less than it was at P_b. When the surface temperature rises again, the situation is as shown at (*d*). It can be seen that the crest of the wave, represented by points P_a, P_b, P_c, P_d etc., is travelling to the right and that its magnitude is reducing. Further crests will follow the original one because a wave is being propagated through the wall as a result of an oscillation in the value of the outside surface temperature. Eventually the wave will reach the inner surface of the wall and will produce similar fluctuations in surface temperature. The inner surface temperature will have a succession of values corresponding to the point P_r as it rises and falls. Thick walls with a large thermal capacity will damp the temperature wave considerably, whereas thin walls of small capacity will have little damping effect, and fluctuations in outside surface temperature will be apparent, almost immediately, as similar changes in inner surface temperature.

It is possible for the inner surface temperatures to be less than room air temperatures, at certain times of the day, for walls of sufficiently heavy construction. A wide diurnal range of temperature can give this result. This outside surface temperature falls at night, by radiation to the black vault of the sky, and the effect of this is felt as a low inside surface temperature at some time later. At such a time, the air conditioning load will be reduced because of the heat lost into the wall from the room. Figure 7.15 presents a simplified picture of this.

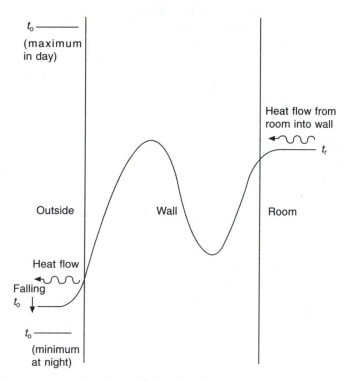

Fig. 7.15   Temperature gradient in a wall showing the possibility of heat flow outwards under unsteady state conditions.

7.18   Sol-air temperature

In the first instance, heat transfer through a wall depends on the rate at which heat enters its outer surface. The concept of 'Sol-air temperature' has been in use for some time as an aid to the determination of the initial rate of entry of heat. It is defined as the value of the outside air temperature which would, in the absence of all radiation exchanges, give the same rate of heat flow into the outer surface of the wall as the actual combination of temperature differences and radiation exchanges really does.

The sol-air temperature, t_{eo}, otherwise termed the outside environmental temperature, is used in the following equation

$$Q' = h_{so}(t_{eo} - t_{so}) \qquad (7.18)$$

where Q' = rate of heat entry into the outer surface, in W m^{-2}

h_{so} = outside surface heat transfer coefficient, in W m^{-2} K^{-1}

t_{eo} = sol-air temperature, in °C

t_{so} = outside surface temperature, in °C

Q' can be expressed in another way which does not involve the use of the sol-air temperature:

$$Q' = \alpha I_\delta + \alpha' I_s + h_{so}(t_o - t_{so}) + R \tag{7.19}$$

In this basic heat entry equation, α and α' are the absorption coefficients (usually about the same value) for direct, I_δ, and scattered, I_s, radiation which is normally incident on the wall surface. R is a remainder term which covers the complicated long wavelength heat exchanges by radiation between the wall and nearby surfaces. The value of R is difficult to assess; in all probability it is quite small and, if neglected, results in little error.

Equations (7.18) and (7.19) can be combined to yield an expression for t_{eo} in useful form:

$$t_{eo} = t_o + \frac{\alpha I_\delta + \alpha' I_s + R}{h_{so}} \tag{7.20}$$

If α' is made equal to α, and R is ignored, this expression becomes

$$t_{eo} = t_o + \frac{\alpha(I_\delta + I_s)}{h_{so}} \tag{7.21}$$

The *CIBSE Guide A5* (1999) quotes an equation virtually the same as this.

EXAMPLE 7.12

Calculate the sol-air temperature for a vertical, south-west wall at 13.00 h sun-time and latitude 40°N, given the following information:

$I_v = 343$ W m^{-2} $= I_\delta$ (see example 7.11)
$I_s = 118$ W m^{-2} (see example 7.11)
$\alpha = \alpha' = 0.9$
$h_{so} = 22.7$ W m^{-2} K^{-1}
$t_o = 32$°C

Answer

By equation (7.21)

$$t_{eo} = 32° + \frac{0.9(343 + 118)}{22.7}$$

$$= 50.3°C$$

Table 7.8 gives values of sol-air temperatures for average walls at different times and orientations at 51°33'N. From such a table the sol-air temperature for the above example is 42.8°C. So a difference of $11\frac{1}{2}$ degrees more southerly latitude and about 5 degrees more air temperature produces an increase of about $7\frac{1}{2}$ degrees in the value of t_e, in this particular comparison.

7.19   Calculation of heat gain through a wall or roof

The use of analytical methods to determine the heat transfer rate through a wall into a room, in the presence of unsteady heat flow into the outer wall surface, presents considerable difficulties. No exact solution is possible but the *CIBSE Guide A5* (1999) offers a method

Table 7.8   Sol-air temperatures at Bracknell (latitude 51°33′N) for the month of July. D and L signify dark and light coloured surfaces, respectively

Sun-time	Air temperature	Sol-air temperatures, °C									
		Horizontal		North		East		South		West	
		D	L	D	L	D	L	D	L	D	L
01	14.8	11.5	11.5	12.3	12.3	12.3	12.3	12.3	12.3	12.3	12.3
02	14.2	10.7	10.7	11.5	11.5	11.5	11.5	11.5	11.5	11.5	11.5
03	13.6	10.1	10.1	10.9	10.9	10.9	10.9	10.9	10.9	10.9	10.9
04	13.2	9.6	9.6	10.4	10.4	10.4	10.4	10.4	10.4	10.4	10.4
05	13.3	11.1	10.6	11.8	11.3	11.9	11.3	11.5	11.1	11.5	11.1
06	14.3	18.1	15.0	24.5	19.2	39.9	28.0	14.8	13.8	14.8	13.8
07	15.9	25.3	19.7	23.9	19.9	49.6	34.6	18.4	16.8	18.4	16.8
08	17.7	32.5	24.8	22.1	20.0	53.9	38.2	23.5	20.7	22.1	20.0
09	19.5	38.5	28.8	25.2	22.8	53.6	39.0	33.0	27.2	25.2	22.8
10	21.0	42.7	31.9	27.6	25.0	49.2	37.3	40.4	32.2	27.6	25.0
11	22.5	45.5	34.2	29.6	27.0	42.8	34.4	45.4	35.9	29.6	27.0
12	23.4	46.7	35.3	30.8	28.0	35.0	30.4	47.7	37.6	30.8	28.0
13	24.2	48.0	36.4	31.7	29.0	31.7	29.0	49.0	38.8	36.1	31.4
14	24.8	46.8	35.9	31.7	29.1	31.7	29.1	46.8	37.7	44.3	36.2
15	25.3	44.4	34.8	31.3	29.0	31.3	29.0	42.7	35.3	50.5	39.8
16	25.4	40.5	32.7	30.1	28.1	30.1	28.1	36.5	31.7	53.4	41.3
17	25.0	35.8	29.9	28.2	26.6	28.2	26.6	29.2	27.2	53.1	40.7
18	24.4	30.6	26.7	30.0	27.1	26.1	24.9	26.1	24.9	48.5	37.6
19	23.3	25.2	23.3	29.5	26.2	23.3	22.7	23.3	22.7	39.6	31.9
20	21.7	20.1	19.8	20.6	20.2	20.4	20.1	20.4	20.1	20.6	20.3
21	20.1	17.8	17.6	18.1	18.1	18.1	18.1	18.1	18.1	18.1	18.1
22	18.9	18.1	16.1	16.7	16.7	16.7	16.7	16.7	16.7	16.7	16.7
23	17.5	14.4	14.4	15.2	15.2	15.2	15.2	15.2	15.2	15.2	15.2
24	16.7	13.4	13.4	14.2	14.2	14.2	14.2	14.2	14.2	14.2	14.2
24 hr mean	19.6	27.4	22.6	22.4	20.7	27.8	23.8	25.8	22.6	26.5	23.0

Reproduced by kind permission of the CIBSE from their *Guide A5* (1999) Thermal response and plant sizing.

involving the use of sol-air temperatures and related to the concepts of environmental temperature, as well as dry resultant and air temperatures. On an average basis, the mean flow of heat, Q_m, through a wall into a room conditioned at a constant temperature, t_r, is given by

$$Q_m = AU(t_{em} - t_r) \qquad (7.22)$$

wherein t_{em} is the mean sol-air temperature over 24 hours. If thermal capacity effects were ignored the instantaneous heat gain to the room at time θ would be given by

$$Q_\theta = AU(t_{eo} - t_r) \qquad (7.23)$$

If thermal capacity is considered then a simplified picture of the heat flow into the room at some later time, $\theta + \phi$, where ϕ is the time lag, is shown by

$$Q_{\theta+\phi} = AU(t_{em} - t_r) + AU(t_{eo} - t_{em})f \qquad (7.24)$$

where t_{eo} is the sol-air temperature at time θ and in which f is a decrement factor (see Figure 7.16(b)).

Since most building materials have specific heats of about 0.84 kJ kg^{-1} K^{-1} their thermal capacities depend largely on their density and thickness. The *CIBSE Guide A5* (1999) derives equations for the calculation of decrement factor, time lag and admittance and tabulates values for a range of common building materials. Figures 7.16(a) and (b) show simple curves allowing an approximate determination of time lags and decrement factors

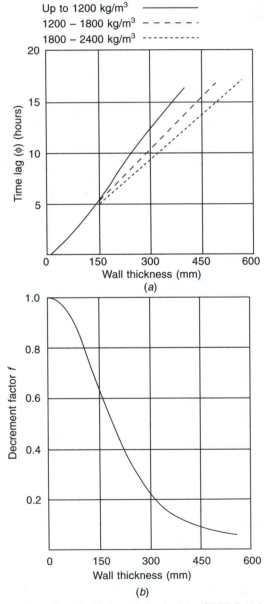

(Reproduced by kind permission from the CIBSE Guide)

Fig. 7.16 Approximate time and decrement factors for use with sol-air temperatures.

which may be used in equation (7.24) to yield answers of practical value.

The *CIBSE Guide* pursues a more complicated approach because a distinction is made between systems that are controlled to keep dry resultant and air temperatures constant. For all practical purposes of control dry resultant temperature can be ignored, although the *CIBSE Guide* does not do so. If, as is invariably the case for real-life air conditioning systems, the air temperature is held constant, then the *CIBSE Guide* introduces two dimensionless factors, F_{au} and F_{ay}, defined by

$$F_{au} = 4.5\Sigma(A)/[4.5\ \Sigma(A) + \Sigma(AU)] \tag{7.25}$$
$$F_{ay} = 4.5\Sigma(A)/[4.5\ \Sigma(A) + \Sigma(AY)] \tag{7.26}$$

in which 4.5 is a heat transfer coefficient and Y is the admittance of a surface.

When the air temperature is held constant in an air conditioned room the equation for the heat gain through a wall or a roof, exposed to the outside, becomes:

$$Q_{\theta+\phi} = F_{au}AU(t_{em} - t_r) + F_{ay}AU(t_{eo} - t_{em})f \tag{7.27}$$

In words, this could be stated as the mean heat transfer over 24 hours, plus the variation about the mean.

For the case of heat transfer through a glass window, which has no thermal inertia and where the heat gains by solar radiation are calculated separately (see section 7.20), equation (7.17) applies. Similarly, for the case of heat flow through a floor overhanging an open space, such as a car park, the same simple equation applies.

In equation (7.27) $Q_{\theta+\phi}$ is the heat gain into the room at the time $\theta + \phi$, t_{em} is the 24-hour mean sol-air temperature, t_r is the room temperature (dry-bulb) and is assumed to be held at a constant value, t_{eo} is the sol-air temperature at a time θ, when the heat entered the outside surface of the wall, ϕ is the time lag of the wall and f is its decrement factor.

Note: When using equations (7.25) and (7.26), ΣA is the sum of all the internal surface areas of the room. ΣAU refers to the surface under consideration through which heat is flowing, but $\Sigma(AY)$ refers to all the internal surfaces of the room.

Equations (7.24) and (7.27) show that $Q_{\theta+\phi}$ is the sum of the mean heat flow over 24 hours and the cyclic input from the inner surface of the wall to the room at the time $\theta + \phi$. Since this cyclic input depends on the external conditions at the earlier time, θ, its sum with the mean heat flow may be positive or negative, depending on the size and sign of the earlier heat flow into the outer surface. If the wall is very thick, say over 600 mm, with a density of more than about 1000 kg m^{-3}, the decrement factor will be very small and the influence of the second term in equations (7.24) and (7.27) slight. Equation (7.22) then gives a good approximation of the answer. If the wall is very thin and has negligible thermal capacity the heat gain will vary considerably over 24 hours because ϕ is small and f is large. For such a case (as with the roof of a factory of lightweight construction) ϕ *is best taken as zero and f as unity.*

EXAMPLE 7.13

Calculate the heat gain at 15.00 h sun-time in July through a light-coloured east-facing wall which is 2.4 m wide and 2.6 m high (floor-to-ceiling). The room is 6.0 m deep and is held at a constant temperature of 22°C dry-bulb. Make use of the following information:

Window: area 2.184 m^2, *U*-value 5.6 W m^{-2} K^{-1}, *Y*-value (admittance) 5.6 W m^{-2} K^{-1}

Wall: thickness 150 mm, density 1200 kg m^{-3}, U-value 0.45 W m^{-2} K^{-1}, Y-value
  3.7 W m^{-2} K^{-1}
Ceiling: Y-value 2.5 W m^{-2} K^{-1}
Floor: Y-value 2.5 W m^{-2} K^{-1}

Answer

From Figures 7.16(*a*) and (*b*), the time lag is 5 hours and the decrement factor is 0.65.
From Table 7.8 the sol-air temperature is 37.3°C at 10.00 h (five hours earlier than the time
at which the heat gain is to be calculated) and the 24-hour mean sol-air temperature is
23.8°C. The area of the wall, through which heat is flowing, is 2.4 × 2.6 − 2.184 =
4.056 m^2.

$$\Sigma A = (2.4 + 6.0 + 2.4 + 6.0)2.6 + (2.4 \times 6.0)2 = 72.48 \text{ m}^2$$
$$\Sigma(AU) = 4.056 \times 0.45 + 2.184 \times 5.6 = 14.06 \text{ W m}^{-2} \text{ K}^{-1}$$
$$\Sigma(AY) = 4.056 \times 3.7 + 2.184 \times 5.6 + (2.4 \times 6.0) \times 2 \times 2.5$$
$$+ [(6.0 + 2.4 + 6.0) \times 2.6] \times 3.7 = 237.77 \text{ W m}^{-2} \text{ K}^{-1}$$

By equation (7.25)

$$F_{au} = 4.5 \times 72.48/(4.5 \times 72.48 + 14.06) = 0.959$$

By equation (7.26)

$$F_{ay} = 4.5 \times 72.48/(4.5 \times 72.48 + 237.77) = 0.578$$

By equation (7.27)

$$Q_{10+5} = 4.056 \times 0.45[0.959(23.8 - 22) + 0.578(37.3 - 23.8)0.65] = 12.4 \text{ W}$$

Relative to the treated floor area of 14.4 m^2 this is 0.9 W m^{-2}. Typically, the maximum sum
of all sensible heat gains is about 80 or 90 W m^{-2} for the sort of office module considered
in example 7.13.

  Using the simplified version, namely equation (7.24) yields an answer of 19.3 W. Although
this is 56 per cent greater than the answer given by equation (7.27) the value of the
apparently greater accuracy is questionable. There is probably some uncertainty in the U-
values, decrement factor and time lag. There is much additional calculation involved and,
furthermore, the heat gain through the wall is usually only about one or two per cent of the
total sensible gain. The conclusion could be that the refinement introduced by the inclusion
of the factors F_{au} and F_{ay} is scarcely worth while and that the answers obtained by equation
(7.24) are good enough.

  For a roof of lightweight construction and of large plan area, as, for example, that of a
hypermarket, the heat gain can be a larger proportion of the total sensible heat gain and
hence of more significance. For such a case, a safer answer would be to assume a time lag
of zero and a decrement factor of unity. The heat gain is then instantaneous and equation
(7.24) simplifies to (7.23), which could be used.

  Note that if the design maximum air temperature is higher than the value tabulated with
the sol-air temperatures, the difference should be added to the tabulated sol-air temperatures.
The definition of sol-air temperature in equations (7.20) and (7.21) shows the dependence
on outside air temperature.

EXAMPLE 7.14

Calculate the heat gain through a light-coloured flat roof of negligible mass at 1300 h sun-time in July, given that the design outside air temperature is 28°C at 1600 h sun-time, the room temperature is 22°C and the *U*-value of the roof is 0.45 W m^{-2} K^{-1}.

Answer

Assume a decrement factor of 1.0 and a time lag of 0 h, because the roof has negligible mass. Add 2.6° to all sol-air temperatures taken from Table 7.8 because the design temperature of 28°C is 2.6° higher than the tabulated value of 25.4°C at 16.00 h. Hence t_{eo} at 13.00 h is 36.4° + 2.6° = 39.0°C and t_{em} is 22.6° + 2.6° = 25.2°C.
   By equation (7.23)

$$Q_{13} = 1 \times 0.45(39.0 - 22) = 7.65 \text{ W m}^{-2}$$

It is to be noted that the values of f and ϕ represented by the curves in Figure 7.16 are based on work done by Danter (1960) that take account of the characteristics of different wall and roof structures by adopting load curves of different shapes, as indicated by Figure 7.17.
   The early analytical work of Mackey and Wright (1944), modified and developed by Stewart (1948), is the basis of the current methods adopted by ASHRAE, involving the use of transfer functions (Mitelas and Stephenson (1967), Stephenson and Mitelas (1967) and Mitelas (1972)). A transfer function is a set of coefficients relating heat transfer into the outer surface of a wall or roof with that entering the room at a later time. ASHRAE propose several methods for the calculation of heat gain through a wall or roof, most being impractical without the use of a computer. One method in the *ASHRAE Handbook* (1993b), however,

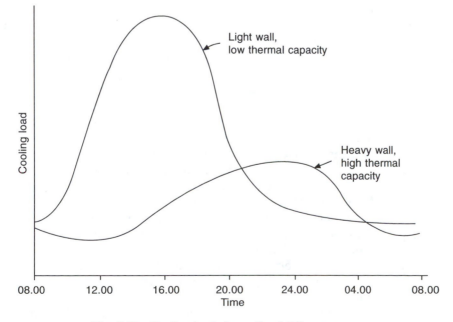

Fig. 7.17   Cooling loads for walls of different masses.

is amenable to manual calculation and expresses heat gains in terms of cooling load temperature differences. Such equivalent temperature differences are tabulated for various typical structures (according to American building practice) and would be applied, as factors, to the product of area and U-value to yield the heat flow into the room at a specified time.

7.20   Air conditioning load due to solar gain through glass

The solar radiation which passes through a sheet of window glazing does not constitute an immediate load on the air conditioning system. This is because

(*a*)   air is transparent to radiation of this kind, and
(*b*)   a change of load on the air conditioning system is indicated by an alteration to the air temperature within the room.

For the temperature of the air in the room to rise, solar radiation entering through the window must first warm up the solid surfaces of the furniture, floor slab and walls, within the room. These surfaces are then in a position to liberate some of the heat to the air by convection. Not all the heat will be liberated immediately, because some of the energy is stored within the depth of the solid materials. The situation is analogous to that considered in section 7.17 for heat gain through walls. There is, thus, a decrement factor to be applied to the value of the instantaneous solar transmission through glass, and there is also a time lag to be considered.

Figure 7.18 illustrates that, in the long run, all the energy received is returned to the room, but, because of the diminution of the peak values, the maximum load on the air conditioning system is reduced.

Modern buildings have most of their mass concentrated in the floor slab, which will, therefore, have a big effect on the values of the decrement factor and the time lag. Since the specific heat of most structural materials is about $0.84 \, \text{kJ} \, \text{kg}^{-1} \, \text{K}^{-1}$, the precise composition of the slab does not matter very much. Although most of the solar radiation entering through a window does strike the floor slab and get absorbed, the presence of furniture and floor coverings, particularly carpeting, reduces the influence of the slab. Wooden furniture has a smaller mass, hence any radiation received by it and absorbed will be subjected to only a small time lag and will be convected back to the room quite soon. The insulating effect of carpets means that the floor behaves as if it were thinner, resulting in a larger decrement factor. There is, thus, a tendency for a furnished carpeted room to impose a larger load on the air conditioning system, and to do so sooner than will an empty room.

Another factor of some importance is the time for which the plant operates. Figure 7.18 shows what happens if an installation runs continuously. Under these circumstances there is no, so-called, 'pull-down' load. If the plant operates for only, say, 12 hours each day, then the heat stored in the fabric of the building is released to the inside air during the night and, on start up next morning, the initial load may be greater than expected. This surplus is termed the pull-down load. Figure 7.19 illustrates the possible effect of such a surplus load. The importance of pull-down load is open to question: outside dry-bulb temperatures fall at night and, in the presence of clear skies, the building is then likely to lose a good deal of the stored heat by radiation. There is an initial load when the sun rises, but the major increase in load is unlikely to occur, in an office block for example, until people enter at 09.00 h and lights are switched on. This may swamp the effect of pull-down load and render its presence less obvious.

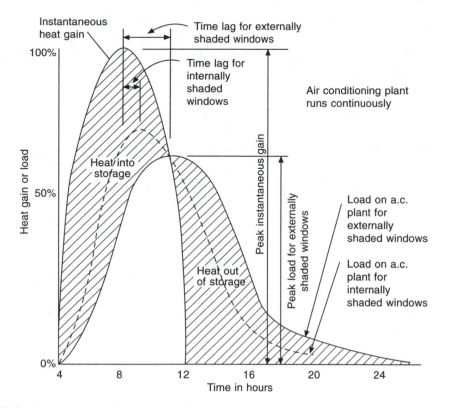

Fig. 7.18   Instantaneous solar heat gain through glass and the load on the air conditioning system.

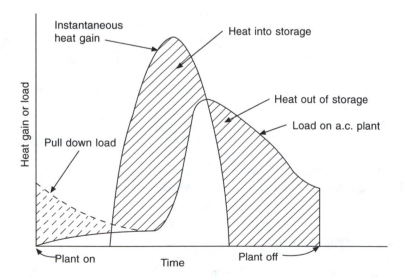

Fig. 7.19   The possible effect of pull-down load on the air conditioning system.

When the window has internal blinds these absorb part of the radiation and convect and re-radiate it back to the room. The remaining part is considered as direct transmission and so is susceptible to storage effects. The load imposed by convection and re-radiation is virtually instantaneous because the mass of the blinds is small and air is not entirely transparent to the long wavelength emission from the relatively low temperature blinds. The same argument holds for heat-absorbing glass.

For cases where the windows have internal Venetian blinds fitted, the air conditioning cooling load may be calculated directly by means of Tables 7.9 and 7.10.

Table 7.9 refers to what are commonly called lightweight buildings. The term lightweight is described in the CIBSE *Guide A5* (1999) as referring to buildings having demountable partitions and suspended ceilings, with supported, uncarpeted floors, or solid floors with a carpet. The thermal response factor, defined by *CIBSE Guide A9* (1986) as the ratio $(\Sigma(AY) + nV/3)/\Sigma(AU) + nV/3)$, should only be used with caution to describe the weight of a building structure when determining the load due to solar heat gain through glazing, because it may lead to the wrong conclusion. For the purposes of Table 7.9, a lightweight building is defined as one having a surface density of 150 kg m^{-2}. This is typical of most modern office blocks. Surface density is determined by calculating the mass of the room enclosure surfaces, using half the known thicknesses of the walls, floor, ceiling etc., applied to the relevant areas and densities. The mass of the glass is ignored. The sum of the calculation, in kg, is divided by the floor area of the room to yield its surface density in kg m^{-2}. When the floor slab is covered with a carpet, or provided with a supported false floor, its density is halved for purposes of the calculation.

EXAMPLE 7.15

Calculate the load arising from the solar heat gain through a double-glazed window, shaded by internal Venetian blinds, facing south-west, at 15.00 h sun-time, in June, at latitude 51.7°N, by means of Tables 7.9 and 7.10.

Answer

Reference to Table 7.9 shows that the load is 224 W m^{-2} for single-glazed windows shaded internally by Venetian blinds of a light colour. Reference to Table 7.10 gives a factor of 1.08 to be applied to the value of 224 W m^{-2} when the window is double glazed with ordinary glass. The load on the air conditioning system is, therefore, $1.08 \times 224 = 242$ W m^{-2}.

Note that if the blinds had been fitted between the sheets of glass, the factor would have been 0.55, and the load would then have been $0.55 \times 224 = 123$ W m^{-2}. Compare the simplicity of this with example 7.11.

For windows not fitted with internal Venetian blinds, where the direct use of Tables 7.9 and 7.10 is not appropriate, the air conditioning load may be determined by taking the maximum total solar intensity normal to a surface (Table 7.11) and multiplying this by factors for haze, dew point, altitude, hemisphere, storage (Table 7.12) and shading (Table 7.6).

EXAMPLE 7.16

Calculate the air conditioning load arising from solar gain through a window fitted with unshaded, single, heat-reflecting (bronze) glass, facing SW, at 15.00 h sun-time in June in London, for a floor slab density of 500 kg m^{-2}. Use the storage load factors in Table 7.12 and the maximum total solar intensity from Table 7.11.

Table 7.9 Solar air conditioning loads through windows in the UK only, for latitude 51.7°N for a room surface density of 150 kg m⁻² expressed in W m⁻²

Month	Exposure	Solar air-conditioning loads (W m⁻²) Sun time (hours)												
		06.00	07.00	08.00	09.00	10.00	11.00	12.00	13.00	14.00	15.00	16.00	17.00	18.00
June 21	N	0	6	22	25	28	28	28	28	28	32	32	32	32
	NE	136	186	177	142	88	60	47	41	38	32	28	28	16
	E	145	221	253	249	202	133	79	60	50	44	35	28	22
	SE	0	79	148	196	221	212	180	130	79	54	44	35	22
	S	19	38	79	114	139	155	158	143	101	88	44	28	19
	SW	6	9	16	19	22	60	123	177	212	224	205	158	69
	W	9	12	19	22	25	25	25	60	133	205	256	268	234
	NW	6	13	16	19	22	22	25	25	41	95	158	196	193
July 23 and May 21	N	0	6	19	22	25	25	25	25	25	25	25	28	28
	NE	127	174	164	133	82	54	44	38	35	28	28	25	16
	E	145	221	253	249	202	133	79	60	50	44	35	28	22
	SE	0	82	158	208	234	224	193	139	82	57	47	35	25
	S	22	44	88	130	158	177	180	167	114	101	50	32	22
	SW	9	13	16	19	25	63	130	186	224	237	218	167	73
	W	9	13	19	22	25	25	25	60	133	205	256	268	234
	NW	6	13	16	19	19	19	22	22	38	88	142	183	180
August 24 and April 20	N	0	6	16	19	19	19	19	19	19	22	22	22	22
	NE	101	139	132	104	68	44	35	32	28	22	22	19	13
	E	147	215	246	243	196	130	76	57	50	44	35	28	22
	SE	0	91	174	228	256	246	208	152	91	60	50	41	28
	S	25	57	114	167	205	231	234	218	146	133	63	44	28
	SW	9	13	19	22	28	69	142	205	246	262	240	183	79
	W	9	13	19	22	25	25	25	57	130	199	249	259	228
	NW	6	9	13	16	16	16	19	19	32	73	114	145	145
September 22 and March 22	N	0	3	13	13	13	13	16	16	16	16	16	16	16
	NE	63	85	82	66	41	28	22	19	16	16	13	13	6
	E	123	186	215	212	171	114	66	50	44	38	28	25	19
	SE	0	95	180	237	265	256	218	158	95	68	54	41	28
	S	32	63	133	193	237	262	268	249	171	152	78	50	35
	SW	9	13	19	22	28	73	148	212	256	271	249	189	32

(Contd)

Table 7.9 *(Contd)*

Month	Exposure	Solar air-conditioning loads (W m^{-2}) Sun time (hours)												
		06.00	07.00	08.00	09.00	10.00	11.00	12.00	13.00	14.00	15.00	16.00	17.00	18.00
	W	9	9	16	19	22	22	22	50	114	174	218	228	199
	NW	3	6	6	9	9	9	13	13	19	44	69	88	88
October 23 and February 20	N	0	3	6	6	9	9	9	9	9	9	9	9	9
	NE	32	44	41	32	19	13	9	9	9	6	6	6	3
	E	95	142	164	161	130	85	50	38	32	28	22	19	16
	SE	0	91	174	228	256	246	212	152	91	60	50	41	27
	S	32	69	139	205	249	278	284	265	180	161	79	50	35
	SW	9	13	19	22	28	69	142	199	246	262	240	183	79
	W	6	9	13	16	16	16	16	38	85	133	164	174	152
	NW	3	3	3	3	6	6	6	6	9	22	35	44	44
November 21 and January 21	N	0	3	6	6	6	6	6	6	6	6	6	6	6
	NE	9	13	13	9	6	3	3	3	3	3	3	3	3
	E	57	85	98	98	79	54	32	22	19	16	13	13	9
	SE	0	73	139	183	205	199	171	123	73	50	41	32	22
	S	28	63	127	186	228	256	262	243	167	148	73	47	32
	SW	6	9	15	16	22	57	117	164	199	212	196	148	63
	W	3	6	6	9	9	9	9	22	50	79	101	104	91
	NW	0	0	0	0	0	0	3	3	3	6	9	13	13
December 22	N	0	0	3	6	6	6	6	6	6	6	6	6	6
	NE	6	9	9	6	6	3	3	3	3	3	3	0	0
	E	41	63	73	73	57	38	22	16	16	13	9	9	6
	SE	0	66	127	167	189	183	155	114	66	44	38	28	19
	S	28	57	117	171	212	234	240	224	152	136	66	44	32
	SW	6	9	13	16	19	50	104	152	183	193	177	136	60
	W	3	3	6	6	6	6	6	16	38	60	73	79	66
	NW	0	0	0	0	0	0	0	3	3	6	9	9	9

Values are for single plate or float glass and, where correction is necessary for other types of glazing, should be multiplied by the factors given in Table 7.10. The area to be used is the opening in the wall for metal-framed windows and the area of the glass for wooden-framed windows. A haze factor of 0.9 has been allowed. It is assumed that shades are not provided on the windows facing north but that all other exposures have blinds which will be raised when the windows are not in direct sunlight. Scattered radiation is included and the storage effect of the building mass taken into account. Air-to-air transmission is excluded. (Reproduced by kind permission of Haden Young Ltd.)

Table 7.10   Correction factors for Table 7.9

Type of glazing and shading	Correction factor
Single glass + internal Venetian blinds	
Ordinary or plate glass	1.00
Heat-absorbing (bronze)	0.96
Heat-reflecting (gold)	0.49
Double glass + internal Venetian blinds	
Ordinary	1.08
Plate	1.00
Heat-absorbing (bronze)*	0.81
Double glass + Venetian blinds between the sheets	
Ordinary	0.55
Plate	0.49
Heat-absorbing (bronze)*	0.47

*Inner leaf ordinary or plate glass.
The above factors are obtained from the ratios of the shading coefficients
to that of ordinary plate glass from Table 7.6.
(Reproduced by kind permission of Haden Young Ltd.)

Table 7.11   Maximum total solar intensities normal to surfaces for latitude 51.7°N in W m^{-2}

Orientation	Dec	Jan/Nov	Feb/Oct	Mar/Sep	Aug/Apr	Jul/May	Jun
North	45	55	75	100	120	135	135
NE/NW	45	55	110	240	405	505	530
E/W	205	255	410	545	660	695	695
SE/SW	425	490	625	695	685	645	625
Horizontal	175	235	385	560	725	820	850

Note that the above figures should be increased by 7 per cent for every 5° by which the dew point is less
than 15° for an application other than in London.

Answer

The maximum total solar intensity normal to a SW surface in June is 625 W m^{-2}. Assume
a haze factor of 0.95 for London. The shading coefficient (Table 7.6) is 0.27 and the
storage factor (Table 7.12) is 0.62. Then

$$q_{ac} = 625 \times 0.95 \times 0.27 \times 0.62 = 99 \text{ W m}^{-2} \text{ of glass surface}$$

The data for cooling loads from heat gains through glazing for London, given in *CIBSE
Guide A5* (1999), are based on measurements of solar irradiances that were not exceeded
on more than 2.5 per cent of occasions at Bracknell (latitude 51°33′N), in the period 1976–
95. The same source provides a computer disc giving cooling loads, based on theoretical
predictions, for latitudes 0° to 60°N and 0° to 60°S. When using the data for the southern
latitudes the tabulated values must be increased because the sun is nearer to the earth in
their summer. See Table 7.4, section 7.16 and example 7.20.

Tables 7.13 and 7.14 give details of the CIBSE cooling loads through glazing for latitude 51°33′N and the related correction factors for that latitude.

Tables 7.13 and 7.14 assume that the system maintains a constant dry resultant temperature in the conditioned space and a correction is given in the CIBSE method to determine the load when the air temperature is held at a constant value, as is the invariable practical case. The answers are tabulated in the usual way for internally shaded glass and corrections are given to cover the case of heavy-weight buildings and various shading/glass combinations. The corrections to give loads in terms of a constant room temperature yield results that are less than those for a constant resultant temperature.

The calculation is then simple and is typified by the following equation:

$$Q_s = A_w F_b F_c q_s \tag{7.28}$$

where Q_s = cooling load due to solar gain through glass in W

A_w = area of the glass or of the opening in the wall in m^2

F_b = shading factor

F_c = air point control factor

q_s = specific cooling load due to solar gain through glass in $W\,m^{-2}$

The method used in the *ASHRAE Guide* calculates solar gain by the product of the glass area, a shading coefficient, the maximum solar heat gain and a cooling load factor generated by the use of transfer functions according to Mitelas and Stephenson (1967), Stephenson and Mitelas (1967), and Mitelas (1972). The method is slightly similar in appearance to that using storage load factors, as in Table 7.12 but is based on a sounder theoretical foundation.

The three methods mentioned in the foregoing for estimating the cooling load that occurs by solar gain through windows yield answers that are in approximate agreement, although sometimes with a difference of phase as well as amplitude. It is not possible to say that any one method is correct and the others wrong. However, the *CIBSE Guide A5* (1999) method yields solar cooling loads for glass that are somewhat larger than the other methods. See Example 7.18. Due to the large number of variables and imponderables concerned, particularly with regard to the thermal inertia of the building, it is probable that the sensible heat gain through the window of a real room could never be measured with enough accuracy to verify one method. For all its occasional inadequacy in coping with the moving shadows across a building face, it is worth noting that the Carrier method (using Tables 7.11 and 7.12) has been tested extensively throughout the world over many years and, in spite of its apparently inadequate theory, it seems to give answers that work. This is a worthy recommendation. It is also to be noted that air conditioning systems do not maintain constant dry resultant temperatures and calculations based on the assumption that they do are not realistic.

7.21   Heat transfer to ducts

A heat balance equation establishes the change of temperature suffered by a ducted airstream under the influence of a heat gain or loss:

$$Q = PLU\left\{\left(\frac{\mp t_1 \pm t_2}{2}\right) \pm t_r\right\} \tag{7.29}$$

Table 7.12   Storage load factors, solar heat gain through glass. 12 hour operation, constant space temperature

Exposure (north lat.)	Mass per unit area of floor kg m^{-2}	Sun time												
		6	7	8	9	10	11	N	1	2	3	4	5	
North and shade	500	0.98	0.98	0.98	0.98	0.98	0.98	0.98	0.98	0.98	0.98	0.98	0.98	
	150	1.00	1.00	1.00	1.00	1.00	1.00	1.00	1.00	1.00	1.00	1.00	1.00	
NE	500	0.59	0.68	0.64	0.52	0.35	0.29	0.24	0.23	0.20	0.19	0.17	0.15	
	150	0.62	0.80	0.75	0.60	0.37	0.25	0.19	0.17	0.15	0.13	0.12	0.11	
E	500	0.52	0.67	0.73	0.70	0.58	0.40	0.29	0.26	0.24	0.21	0.19	0.16	
	150	0.53	0.74	0.82	0.81	0.65	0.43	0.25	0.19	0.16	0.14	0.11	0.09	Internal shade
SE	500	0.18	0.40	0.57	0.70	0.75	0.72	0.63	0.49	0.34	0.28	0.25	0.21	
	150	0.09	0.35	0.61	0.78	0.86	0.82	0.69	0.50	0.30	0.20	0.17	0.13	
S	500	0.26	0.22	0.38	0.51	0.64	0.73	0.79	0.79	0.77	0.65	0.51	0.31	
	150	0.21	0.29	0.48	0.67	0.79	0.88	0.89	0.83	0.56	0.50	0.24	0.16	
SW	500	0.33	0.28	0.25	0.23	0.23	0.35	0.50	0.64	0.74	0.77	0.70	0.55	
	150	0.29	0.21	0.18	0.15	0.14	0.27	0.50	0.69	0.82	0.87	0.79	0.60	
W	500	0.67	0.33	0.28	0.26	0.24	0.22	0.20	0.28	0.44	0.61	0.72	0.73	
	150	0.77	0.34	0.25	0.20	0.17	0.14	0.13	0.22	0.44	0.67	0.82	0.85	
NW	500	0.71	0.31	0.27	0.24	0.22	0.21	0.19	0.18	0.23	0.40	0.58	0.70	
	150	0.82	0.35	0.25	0.20	0.18	0.15	0.14	0.13	0.19	0.41	0.64	0.80	
North and shade	500	0.81	0.84	0.86	0.89	0.91	0.93	0.93	0.94	0.94	0.95	0.95	0.95	
	150	1.00	1.00	1.00	1.00	1.00	1.00	1.00	1.00	1.00	1.00	1.00	1.00	
NE	500	0.35	0.45	0.50	0.49	0.45	0.42	0.34	0.30	0.27	0.26	0.23	0.20	
	150	0.40	0.62	0.69	0.64	0.48	0.34	0.27	0.22	0.18	0.16	0.14	0.12	
E	500	0.34	0.44	0.54	0.58	0.57	0.51	0.44	0.39	0.34	0.31	0.28	0.24	Bare glass or external shade
	150	0.36	0.56	0.71	0.76	0.70	0.54	0.39	0.28	0.23	0.18	0.15	0.12	
SE	500	0.29	0.33	0.41	0.51	0.58	0.61	0.61	0.56	0.49	0.44	0.37	0.33	
	150	0.14	0.27	0.47	0.64	0.75	0.79	0.73	0.61	0.45	0.32	0.23	0.18	
S	500	0.44	0.37	0.39	0.43	0.50	0.57	0.64	0.68	0.70	0.68	0.63	0.53	
	150	0.28	0.19	0.25	0.38	0.54	0.68	0.78	0.84	0.82	0.76	0.61	0.42	
SW	500	0.53	0.44	0.37	0.35	0.31	0.33	0.39	0.46	0.55	0.62	0.64	0.60	
	150	0.48	0.32	0.25	0.20	0.17	0.19	0.39	0.56	0.70	0.80	0.79	0.69	
W	500	0.60	0.52	0.44	0.39	0.34	0.31	0.29	0.28	0.33	0.43	0.51	0.57	
	150	0.77	0.56	0.38	0.28	0.22	0.18	0.16	0.19	0.33	0.52	0.69	0.77	
NW	500	0.54	0.49	0.41	0.35	0.31	0.28	0.25	0.23	0.24	0.30	0.39	0.48	
	150	0.75	0.53	0.36	0.28	0.24	0.19	0.17	0.15	0.17	0.30	0.50	0.66	

From *Handbook of Air Conditioning System Design*, by Carrier Air Conditioning Co., copyright 1965 by McGraw-Hill Inc. Used with permission of McGraw-Hill Book Company.

where Q = heat transfer through the duct wall in W,

P = external duct perimeter in m,

L = duct length in m,

U = overall thermal transmittance in W m^{-2} K^{-1},

Table 7.13   Solar cooling loads for SE England (Bracknell, latitude 51°33′N) for fast-responding buildings with the reference glazing type (single clear glass, internal blinds, used intermittently)

Month and façade	Solar cooling load at stated sun-time										
	07.30	08.30	09.30	10.30	11.30	12.30	13.30	14.30	15.30	16.30	17.30
May											
N	109	111	116	121	126	129	129	125	120	114	109
NE	263	227	107	163	148	140	139	135	130	122	111
E	344	355	325	255	116	163	150	140	136	127	116
SE	251	301	323	310	265	197	88	122	122	109	98
S	70	80	253	217	248	254	236	195	74	143	79
SW	80	93	104	117	205	222	280	312	310	280	223
W	96	109	120	128	133	141	269	280	334	354	330
NW	92	105	116	123	129	132	132	139	260	243	226
Horiz	349	293	347	382	396	398	390	368	324	269	124
June											
N	134	128	131	135	140	142	143	140	136	133	129
NE	303	255	124	182	161	152	152	150	146	138	129
E	388	385	343	268	128	173	160	153	149	142	132
SE	272	314	328	311	264	195	90	123	127	116	107
S	74	77	239	209	240	248	229	187	73	137	79
SW	85	97	106	117	198	220	277	309	309	283	229
W	107	119	128	136	140	148	276	289	345	369	350
NW	106	117	127	134	139	141	142	152	284	264	291
Horiz	251	311	360	391	404	406	397	375	334	280	218
July											
N	117	116	120	125	131	135	135	131	126	121	115
NE	272	233	113	169	153	144	145	140	136	128	116
E	350	354	319	251	119	165	153	144	139	131	120
SE	250	294	310	296	253	107	204	122	123	111	100
S	66	74	231	201	230	238	220	181	69	132	73
SW	80	92	102	115	114	328	267	296	297	271	219
W	96	109	119	127	133	142	265	271	325	346	327
NW	94	107	117	125	131	135	135	142	265	245	269
Horiz	355	294	342	373	385	389	382	360	321	269	126
August											
N	74	85	91	99	104	108	108	106	99	91	84
NE	225	193	88	130	126	121	122	120	112	104	91
E	314	332	303	238	104	148	135	127	120	111	99
SE	244	298	320	313	270	203	86	127	116	101	88
S	64	84	281	233	262	267	249	210	79	160	83
SW	67	81	92	108	212	224	278	309	299	253	187
W	75	88	99	108	113	122	248	257	300	298	254
NW	75	88	100	108	114	117	118	125	216	197	201
Horiz	278	250	301	340	356	357	346	323	276	214	96
September											
N	39	51	60	71	79	81	81	76	69	60	51
NE	180	175	158	87	100	97	97	93	86	77	66
E	294	318	288	219	88	127	113	102	95	86	75

(Contd)

Table 7.13 *(Contd)*

Month and façade	Solar cooling load at stated sun-time										
	07.30	08.30	09.30	10.30	11.30	12.30	13.30	14.30	15.30	16.30	17.30
SE	252	317	341	326	285	224	88	140	101	83	71
S	46	203	213	263	291	303	288	244	179	65	99
SW	51	63	75	96	250	248	304	328	315	357	180
W	52	64	75	87	95	103	226	240	284	272	220
NW	45	57	68	80	88	90	90	89	77	236	148
Horiz	105	305	244	278	296	305	295	264	217	87	171

Reproduced by kind permission of the CIBSE from their *Guide A5* (1999) Thermal response and plant sizing.

t_1 = initial air temperature in the duct in °C,
t_2 = final air temperature in the duct in °C,
t_r = ambient air temperature in °C.

Also, by equation (6.6):

$$Q = \text{flow rate (m}^3 \text{ s}^{-1}) \times (\mp t_1 \pm t_2) \times \frac{358}{(273 + t)}$$

The factor in equation (7.29), which is of considerable importance, is the U-value. This is defined in the usual way by

$$\frac{1}{U} = r_{si} + \frac{l}{\lambda} + r_{so} \tag{7.30}$$

In this equation the thermal resistance of the metal is ignored and the symbols have the following meanings:

r_{si} = thermal resistance of the air film inside the duct, in m² K W⁻¹
r_{so} = thermal resistance of the air film outside the duct, in m² K W⁻¹
l = thickness of the insulation on the duct, in metres
λ = thermal conductivity of the insulation in W m⁻¹ K⁻¹

The value of λ is usually easily determined but it is customary to take a value between 0.03 and 0.07. We have here selected a value 0.045 as typical. Small alterations in the value of λ are not significant within the range mentioned, but changing the thickness is, of course, very influential in altering the heat gain. Values of r_{so} are difficult to establish with any certainty; the proximity of the duct to a ceiling or wall has an inhibiting effect on heat transfer and tends to increase the value of r_{so}. A value of 0.1 is suggested.

The internal surface resistance, on the other hand, may be calculated with moderate accuracy if the mean velocity of airflow in the duct is known. Theoretical considerations suggest that the value of r_{si} is a function of the Reynolds number, and experimental evidence suggests that

$$r_{si} = 0.286 \frac{D^{0.25}}{v^{0.8}}, \text{ for circular ducts} \tag{7.31}$$

$$r_{si} = 0.286 \frac{[2AB/(A + B)]^{0.25}}{v^{0.8}}, \text{ for rectangular ducts} \tag{7.32}$$

Table 7.14 Correction factors for Table 7.13

Glazing/blind arrangement (inside-to-outside)	Correction factor for stated response	
	Fast	Slow
Absorbing/blind	0.52	0.55
Blind/clear	1.00	1.03
Blind/reflecting	0.69	0.71
Blind/absorbing	0.75	0.76
Blind/clear/clear	0.95	0.94
Blind/clear/reflecting	0.62	0.62
Blind/clear/absorbing	0.66	0.66
Blind/clear/clear/clear	0.86	0.86
Blind/clear/clear/reflecting	0.55	0.55
Blind/clear/clear/absorbing	0.57	0.56
Blind/low E/clear	0.92	0.92
Blind/low E/reflecting	0.59	0.60
Blind/low E/absorbing	0.63	0.62
Blind/low E/clear/clear	0.84	0.83
Blind/low E/clear/reflecting	0.53	0.53
Blind/low E/clear/absorbing	0.55	0.55
Clear/blind	0.73	0.81
Clear/blind/clear	0.69	0.72
Clear/clear/blind	0.57	0.61
Clear/blind/reflecting	0.47	0.48
Clear/reflecting/blind	0.37	0.38
Clear/blind/absorbing	0.50	0.51
Clear/absorbing/blind	0.38	0.40
Clear/clear/blind/clear	0.56	0.58
Clear/clear/clear/blind	0.47	0.51
Clear/clear/blind/reflecting	0.37	0.38
Clear/clear/reflecting/blind	0.30	0.33
Clear/clear/blind/absorbing	0.39	0.39
Clear/clear/absorbing/blind	0.32	0.34
Low E/clear/blind	0.56	0.58
Low E/reflecting/blind	0.36	0.39
Low E/absorbing/blind	0.39	0.41
Low E/clear/blind/clear	0.55	0.57
Low E/clear/clear/blind	0.48	0.55
Low E/clear/blind/reflecting	0.37	0.38
Low E/clear/reflecting/blind	0.32	0.35
Low E/clear/blind/absorbing	0.39	0.37
Reflecting/blind	0.50	0.53
Air point correction:		
internal blind	0.91	0.89
mid pane blind	0.87	0.83
external blind	0.88	0.85

Reproduced by kind permission of the CIBSE from their *Guide A5* (1999) Thermal response and plant sizing.

where D = internal duct diameter in m,

v = mean air velocity in a duct in m s^{-1},

A, B = internal duct dimensions in m.

The change of temperature suffered by the air as it flows through the duct is of prime importance and ASHRAE (1993d) gives an expression for this which yields a positive answer for the case of duct heat loss but a negative answer for duct heat gain:

$$t_2 = \frac{t_1(y-1) + 2t_r}{(y+1)} \qquad (7.33)$$

where

$y = 503\rho Dv/UL$ for circular ducts $\qquad\qquad (7.34)$

$y = 2010\rho Av/UPL$ for rectangular ducts $\qquad\qquad (7.35)$

in which ρ is the density of air in kg m^{-3} and A is the internal cross-sectional area of the duct in m^2.

The value of r_{si} is not sensitive to changes of air velocity for the range of duct sizes and velocities in common use; for example, r_{si} is 0.0284 m^2 K W^{-1} for 8 m s^{-1} in a 75 mm diameter duct and 0.0243 m^2 K W^{-1} for 20 m s^{-1} and 750 mm. Hence the U-value of a lagged duct is almost independent of the air velocity. From equation (7.31) the U values are 1.47, 0.81 and 0.56 W m^{-2} K^{-1} for lagging with thicknesses of 25, 50 and 75 mm, respectively, assuming r_{si} is 0.026 m^2 K W^{-1} and λ is 0.045 W m^{-1} K^{-1}. Then by equation (7.34) y has values of 400 DV, 726 DV and 1046 DV, respectively, for the three thicknesses mentioned, if we take ρ = 1.165 kg m^{-3} (as typical for air at 30°C) and L = 1 metre.

Equation (7.33) can be rewritten to give the temperature drop or rise per metre of duct length: Δt

$$\Delta t = t_1 - t_2 = \frac{2(t_1 - t_r)}{(y+1)} \qquad (7.36)$$

Then, for most practical purposes:

$\Delta t = (t_1 - t_r)/200\ Dv$ for 25 mm lagging $\qquad\qquad (7.37)$

$\Delta t = (t_1 - t_r)/363\ Dv$ for 50 mm lagging $\qquad\qquad (7.38)$

$\Delta t = (t_1 - t_r)/523\ Dv$ for 75 mm lagging $\qquad\qquad (7.39)$

These equations show that it is important to assess properly the ambient temperature, t_r, and that the temperature drop along the duct is inversely proportional to Dv, being independent of the method of duct sizing adopted.

The equations are easy to use and they indicate that the rate of temperature drop is considerable once the value of Dv falls below about 1.5 m^2 s^{-1} It is usually impractical to attempt to keep Dv above this value by sizing or by increasing the airflow; it follows that the last few lengths of ductwork will suffer a considerable temperature drop—particularly a difficulty with any system using warm air for heating. The only sure way of virtually stopping the drop is to use 75 mm of lagging. Figure 7.20 illustrates the relative merits of different thicknesses of insulation.

EXAMPLE 7.17

Calculate the air temperature rise in a 10 m length of 500 mm diameter duct, lagged with

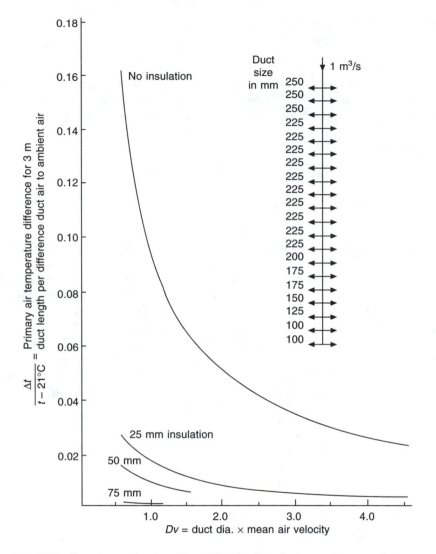

Fig. 7.20   Duct heat gain: the effect of lagging in reducing temperature change.

a thickness of 25 mm, when conveying air at (*a*) 10 m s^{-1} and (*b*) 5 m s^{-1}, given that the air within the duct is initially at 12°C and the ambient air is at 22°C.

Answer

Using equation (7.37) with temperatures reversed because it is a heat gain:
(*a*) *Dv* is 0.5 × 10 = 5.0 and

$$\Delta t = 10 \times (22 - 12)/200 \times 0.5 \times 10 = 0.1°$$

(*b*) *Dv* is 0.5 × 5 = 2.5 and

$$\Delta t = 10 \times (22 - 12)/200 \times 0.5 \times 5 = 0.2°$$

These are not insignificant temperature rises over 10 m of duct. For a long length of duct, the calculations should be done in short increments of, say 10 m, because the temperature difference, inside to outside the duct, changes with respect to the duct length.

7.22   Infiltration

Outside air filters into a conditioned space, even though the space may be slightly pressurised by an excess of air supplied over air extracted. Infiltration is principally due to:

(i)   wind pressure, particularly on tall buildings,

(ii)   stack effect, and

(iii) the entry of the occupants of the building, who also introduce dirt.

It is generally almost impossible to calculate, in advance of construction, the infiltration rate for a building. The only practical way of establishing the natural rate of infiltration for a building is to measure it, after construction. Even then the results will only be relevant to the particular set of circumstances prevailing at the time of measurement.

The product of the area of an opening, the mean theoretical velocity of airflow through it, and a factor to take account of frictional resistance, ought to give the rate of air flow through the opening. The theoretical velocity, v in m s^{-1}, depends on the pressure difference across the opening, Δp in Pa (see section 15.5), given by

$$\Delta p = 0.6v^2 \qquad (7.40)$$

In the case of a building this is complicated by two factors. First, the pressure is developed by two independent influences, wind effect and stack effect. Secondly, the air that enters through one opening must leave the building through another, so airflow through two openings in series must be considered. There is the further difficulty that, for a real building, the actual areas of inlet and outlet openings are not known with any certainty since they depend on the quality of the building structural components and the workmanship in construction.

Pressure differences across a building, due to wind effect, are tabulated and a method is presented in the *CIBSE Guide A4* (1999) for determining the natural infiltration due to wind effect. The difficulty is in deciding on the areas available for infiltration through the building fabric.

A theoretical equation for the natural airflow rate through a building by stack effect is derivable from first principles. Two columns of air having a common height of h metres are considered. The column within the air conditioned building is cooler (at temperature t_r) and therefore heavier, and the column outside the building is warmer (at temperature t_o) and therefore lighter. This is satisfactory for small buildings having clearly defined, simple inlets and outlets, but it is of little use for high-rise buildings because of the complication introduced by the presence of multiple potential inlets/outlets in the form of windows, one above the other.

In the United Kingdom, where reasonably good standards of building construction prevail, the common practice is to assume half an air change per hour for natural infiltration in summer and one air change in winter for office buildings in a city. Some designers assume no infiltration at all in summer. This is not very scientific but appears to be adequate because the sensible gains that result from these assumptions do not represent a very large proportion of the total (about 2 per cent or 3 per cent). The latent gains are larger and account for about 25 per cent of the total latent gains but this is usually not critical, in the

UK, for commercial office block air conditioning because of the comparative unimportance of relative humidity in human comfort.

For industrial air conditioning it is possible that infiltration could be significant and assumptions of the sort mentioned above should not be made. This is also particularly true for air conditioning in tropical climates where the outside moisture contents are often very high in summer design conditions. A wrong choice of infiltration rate could then give a significant error in the latent heat gain calculation. For such applications reference should be made to the *CIBSE Guide A4* (1999) and to the *ASHRAE Handbook*, Fundamentals (1997a).

Infiltration air change rates are much more critical in winter and the recommendations of the CIBSE *Guide* should be followed for the UK. This is notably true for entrance halls and lobbies, where infiltration rates can be very high in winter.

Heat gains by natural infiltration may be calculated using the following equations:

$$Q_{si} = 0.33nV(t_o - t_r) \tag{7.41}$$

$$Q_{li} = 0.8nV(g_o - g_r) \tag{7.42}$$

where Q_{si} = sensible heat gain by natural infiltration   W
$\quad\quad Q_{li}$ = latent heat gain by natural infiltration   W
$\quad\quad n$ = air change rate by natural infiltration h^{-1}
$\quad\quad V$ = room volume m^3
$\quad\quad t_o$ = outside air temperature   °C
$\quad\quad t_r$ = room air temperature   °C
$\quad\quad g_o$ = outside air moisture content $g\ kg^{-1}$
$\quad\quad g_r$ = room air moisture content $g\ kg^{-1}$

7.23   Electric lighting

Luminous intensity is defined, by international agreement, in terms of the brightness of molten platinum at a temperature of 1755°C, and the unit adopted for its expression is the candela. A point source of light delivers a flow of luminous energy which is expressed in lumens; the quotient of this luminous flux and the solid angle of the infinitesimal cone, in any direction, is the intensity of illumination expressed in candela. The density of luminous flux is the amount of luminous energy uniformly received by an area of one square metre. Thus, illumination, or density of luminous flux, is expressed in lumens per square metre, or otherwise termed lux.

Electric lighting is usually chosen to produce a certain standard of illumination and, in doing so, electrical energy is liberated. Most of the energy appears immediately as heat, but even the small proportion initially dissipated as light eventually becomes heat after multiple reflections and reactions with the surfaces inside the room.

The standard of illumination produced depends not only on the electrical power of the source but also on the method of light production, the area of the surfaces within the room, their colour and their reflective properties. The consequence is that no straightforward relation exists between electrical power and standard of illumination. For example, fluorescent tube light fittings are more efficient than are tungsten filament lamps. This means that for a given room and furnishings, more electrical power, and hence more heat dissipation, is involved in maintaining a given standard of illumination if tungsten lamps are used. Table 7.15 gives some approximate guidance on power required for various intensities of illumination.

Table 7.15   Typical total heat emissions for various illuminances and luminaires

	Watts liberated per m² of floor area, including power for control gear							
	Filament lamps		Discharge lighting		65 W white fluorescent			Poly-phosphor fluorescent tube 58 W (1.5 m)
			MBF	SON				
Illuminance in lux	Open industrial reflector	General diffusing fitting	Open industrial reflector		Enamel plastic trough	Enclosed diffusing fitting	Louvred ceiling panel	
150	19–28	28–36	4–7	2–4	4–5	6–8	6–8	4–8
200	28–36	36–50	—	—	6–7	8–11	9–11	6–10
300	38–55	50–69	7–14	4–8	9–11	12–16	12–17	10–16
500	66–88	—	13–25	7–14	15–25	24–27	20–27	14–26
750	—	—	18–35	10–20	—	—	—	—
1000	—	—	—	—	32–38	48–54	43–57	30–58

Notes: The larger figure in the range quoted is for small rooms which normally need from $^1/_3$ to $^1/_2$ more energy because of losses in reflection. The heat liberated by polyphosphor tubes depends on the type of fitting used. Gaps in the illuminances may be covered by interpolation but extrapolation is more risky.
MBF Mercury fluorescent high pressure.
SON Sodium high pressure.

The table is based on a room of dimensions 9 m × 6 m, in plan, with light fittings mounted 3 m above floor level, for tungsten lamps and fluorescent fittings, and of dimensions 15 m × 9 m × 4 m high, for mercury lamps. Light-coloured decorations and a reasonably clean room are assumed in each case. Generally speaking, larger rooms require fewer W m^{-2} than smaller rooms do, for the same illumination.

The efficiency of fluorescent lamps deteriorates with age. Whereas initially a 40 W tube might produce 2000 lux with the liberation of 48 watts of power, the standard of illumination might fall to 1600 lux with the liberations of 48 watts at each tube after 7500 hours of life.

A further comment on fluorescent fittings is that the electrical power absorbed at the fitting is greater than that necessary to produce the light at the tube. A fitting which has 80 watts printed on the tube will need 100 watts of power supplied to it; the surplus 20 watts is liberated directly from the control gear of the fitting as heat into the room.

As a rough guide it can be assumed that a typical modern lighting standard of 500 lux in an office involves a power supply of about 14 to 20 W m^{-2} of floor area. Thus an office measuring 5 m × 6 m will require about 600 W at its fluorescent light fittings to produce an illuminance of 500 lux. Six tubes each of 80 W (liberating 100 W) will be needed.

The heat liberated when electric lights are switched on is not felt immediately as a load by the air conditioning system since the heat transfer is largely by radiation. As with solar radiation, time must pass before a convective heat gain from solid surfaces causes the air temperature to rise. With an air conditioning system running for 12 hours a day and a floor slab of 150 kg m^{-2}, the decrement (storage) factor, according to the Carrier Air Conditioning Company (1965), to be applied to the heat gain from lights, is about 0.42 immediately and 0.86 one hour after the lights have been turned on, for projecting light fittings. If the light fittings are recessed, the corresponding factors are about 0.4 and 0.81. The decrement becomes 0.93 after 4 hours.

This can be advantageous if the lights are recessed and the ceiling space above is used as part of the air extract system. Under these circumstances the storage factors are 0.34

initially, 0.72 after one hour and 0.97 after 8 hours. If the light fitting itself is used as an air outlet, for the extraction of air from the room, then the effect is a long-term one and a permanent allowance can be made for the heat liberated at the light and carried away by the extracted air, provided adequate information is available from the manufacturers of the light fitting. If all the extracted air is discharged to the atmosphere then full value is credited for the reduction of the air conditioning load. However, if as is more likely, a good deal of the extracted air is recirculated, the full effect is not felt and due allowance must be made by increasing the temperature of the mixture air (outside air plus the recirculated air). Since the maximum gain is usually some time after the lights are first switched on, the storage factor for lights should usually be taken as unity, when calculating the sensible heat gains.

When the ceiling void is not used as an extract air path the full emission from the electric lighting and its control gear is a heat gain to the conditioned space. If the ceiling void is used as a return air path then, for the older type of fluorescent tubes, Westinghouse (1970) argue that only 60 per cent to 70 per cent of the heat is liberated into the conditioned room. If the luminaire is an extract-ventilated type then 50 per cent to 60 per cent of the heat from the lights and control gear is given to the room. These figures take account of the fact that the ceiling void gets warmer and some heat is retransmitted through the suspended ceiling into the treated space and some flows upwards to the room above. No improvement is achieved by connecting ducts directly to the extract luminaires but there must be a basic extract duct system to convey the air back to the plant, for recirculation or discharge to waste. Such a basic duct system should have a dampering arrangement that ensures reasonably uniform extract air balance among the extract ventilated luminaires. In principle, no dampered extract duct spigot, in such a basic duct system, should be further than about 18 m from any extract luminaire.

Before using extract-ventilated luminaires it is essential to consult the manufacturers of the fluorescent tubes and obtain their approval.

7.24   Occupants

As was mentioned in section 4.1, human beings give off heat at a metabolic rate which depends on their rate of working. The sensible and latent proportions of the heat liberated for any given activity depend on the value of the ambient dry-bulb temperature: the lower the dry-bulb temperature the larger the proportion of sensible heat dissipated.

Typical values of the sensible and latent liberations of heat are given in Table 7.16.

The figures for eating in a restaurant include the heat given off by the food.

Deciding on the density of occupation is usually a problem for the air conditioning designer. A normal density for an office block is 9 m^2 per person, as an average over the whole conditioned floor area. The density of occupation may be as low as 20 m^2 per person in executive offices or as high as 6 m^2 per person in open office areas.

Some premises may have much higher densities than this; for restaurants, 2 m^2 per person is reasonable, but for department stores, at certain times of the year, densities may reach values of 4.3 to 1.7 m^2 per person, even after allowance has been made for the space occupied by goods. In concert halls, cinemas and theatres, the seating arrangement provides the necessary information but in dance halls and night clubs estimates are open to conjecture. Occupation may be very dense indeed. A figure of 0.5 m^2 per person is suggested tentatively.

Table 7.16   Heat emissions from people

Activity	Metabolic rate W	Heat liberated in W Room dry-bulb temperature (°C)							
		20°		22°		24°		26°	
		S	L	S	L	S	L	S	L
Seated at rest	115	90	25	80	35	75	40	65	50
Office work	140	100	40	90	50	80	60	70	70
Standing	150	105	45	95	55	82	68	72	78
Eating in a restaurant	160	110	50	100	60	85	75	75	85
Light work in a factory	235	130	105	115	120	100	135	80	155
Dancing	265	140	125	125	140	105	160	90	175

7.25   Power dissipation from motors

Deciding on the heat liberated from motors hinges on the following:

(i)   The frequency with which the motors will be used, if there is more than one in the conditioned space, so that the maximum simultaneous liberation may be assessed
(ii)   The efficiency of the motor
(iii)   Whether the motor and its driven machine are in the conditioned space

All the power drawn from the electricity mains is ultimately dissipated as heat. If both the motor and the machine are in the conditioned space then the total amount of power drawn from the mains appears as a heat gain to the room. If only the driven machine is in the room, the motor being outside, then the product of the efficiency (motor plus drive) and the power drawn from the mains is the heat liberated to the space conditioned. Similarly, when only the driving motor is within the room, 100 minus the efficiency is the factor to be used.

Control of motor speed should be done by an efficient method, such as varying the frequency of the input electrical supply. If inefficient methods of speed control are adopted energy is not saved and the inefficiency of the method appears as heat gain in the vicinity of the speed control mechanism.

7.26   Business machines

In offices, the presence of personal computers with peripheral devices and the common-place use of other electrically energised equipment, gives a significant contribution to the sensible heat gains. Although power dissipations as high as 80 W m^{-2} have been quoted for dealers' rooms in applications such as merchant bankers and stockbrokers offices, such figures are exceptional and are not representative of small power in ordinary offices. The actual rate of energy dissipation is generally less than the nameplate power on the items of equipment. Furthermore, machines are used intermittently and the power consumed when idling is less than when operating at full duty. The *CIBSE Applications Manual AM7* (1992) gives useful details of suitable allowances for design purposes and quotes half-hour average powers in relation to nameplate powers.

Examples of some values are given in Table 7.17, based on an office total floor area of 583.2 m² (see Figure 7.21) and using data from *CIBSE AM7* (1992).

Table 7.17 Examples of powers for business machines. (For an office floor area of 583.2 m²)

Item	Nameplate power W			Half-hour average power W		
27 colour monitors	27 ×	185 =	4995	27 ×	120 =	3240
27 local memories	27 ×	30 =	810	27 ×	30 =	810
1 electrostatic plotter	1 ×	960 =	960	1 ×	557 =	557
25 dot matrix printers	25 ×	400 =	10000	25 ×	58 =	1450
2 laser printers	2 ×	1680 =	3360	2 ×	824 =	1648
2 fascimile machines	2 ×	820 =	1640	2 ×	260 =	520
2 electric typewriters	2 ×	105 =	210	2 ×	53 =	106
1 photocopier	1 ×	2400 =	2400	1 ×	1259 =	1259
2 coffee machines	2 ×	3000 =	6000	2 ×	600 =	1200
Totals (W)			30375			10790

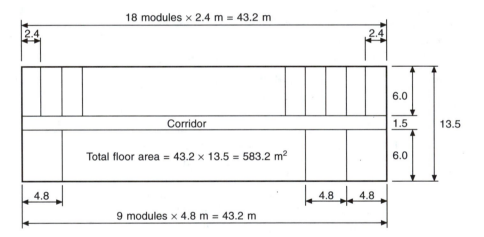

Fig. 7.21 A notional office plan area, used to establish the typical power dissipation from office machines (Table 7.17).

From the above table the specific totals, per unit of floor area, are 52.1 W m⁻² for nameplate powers and 18.5 W m⁻² for half-hour averages. Note that, for the above example, the coffee machines make a very large contribution to the total nameplate power but a good deal less to the half-hour average. For other cases the figures would be different.

Diversity factors that may be applied to the nameplate power to represent the load on the air conditioning system for the whole building are difficult to establish with any certainty. From Table 7.17 the ratio of total half-hour average power to total nameplate power gives a value of 10 790/30 375 = 0.36. From *CIBSE AM7* (1992) such ratios can be deduced and some are given in Table 7.18.

The variation in the values quoted depends on the assumptions made for the time that the machines are running or idling.

Table 7.18   Ratios of half-hour average powers to nameplate powers according to *CIBSE AM7* (1992)

Item	Half-hour average power / Nameplate power
Monochrome visual display terminal	0.60 to 0.65
Colour visual display terminal	0.60 to 0.79
Personal computers (inc. monochrome monitors)	0.56 to 0.61
Personal computers (inc. colour monitors)	0.59 to 0.65
Mini computer work-stations	0.25
Small graphics plotter, plotting	0.71
Small graphics plotter, idling	0.31
Continuous roll electrostatic, plotting	0.81
Continuous roll electrostatic, idling	0.35
Dot matrix printer	0.15
Laser printer (desk mounted)	0.27
Laser printer (floor mounted)	0.49
Facsimile machine (large)	0.32
Facsimile machine (small)	0.19
Photocopier (large)	0.52
Photocopier (small)	0.28

Reproduced by kind permission of the CIBSE, from *AM7* (1992). See also CIBSE Guide A2 (1999): Internal heat gains, 6.4.

A cautious conclusion is that a typical allowance for small power for business machines is from 10 W m^{-2} to 20 W m^{-2}, in the absence of firmer information to the contrary. As regards a diversity factor to be applied when calculating the total maximum sensible heat gain for the whole building (in order to determine the size of refrigeration plant) a value of 0.65 to 0.70 seems reasonable. For the purpose of calculating the maximum sensible heat gain for a particular room or module (to determine the amount of supply air necessary or the size of air conditioning unit to select), a diversity factor of unity must be applied to the heat gain from business machines, people and electric lights.

EXAMPLE 7.18

The dimensions of a west-facing module in a lightweight building (150 kg m^{-2}) are: 2.4 m width × 2.6 m floor-to-ceiling height × 6.0 m depth. The floor-to-floor height is 3.3 m. A double glazed, 6 mm clear glass window of 2.184 m^2 area is in the only exterior wall. Rooms on the other five sides of the module are air conditioned to the same temperature. The building is at latitude 51.7°N, approximately, and the window is protected by internal Venetian blinds, assumed to be drawn by the occupants when the window is in direct sunlight. The *U*-value of the window is 3.0 W m^{-2} K^{-1}. The time lag of the wall is 5 h and its decrement factor is 0.65.

Outside state: 28°C dry-bulb, 19.5°C wet-bulb (sling), 10.65 g kg^{-1}.
Room state: 22°C dry-bulb, 50 per cent saturation, 8.366 g kg^{-1}.

(a)   Making use of Tables 7.8, 7.9 and 7.10 calculate the sensible and latent heat gains at 1500 h sun-time in July. When calculating the heat gain through the wall use the floor-to-floor height for determining the wall area.

(*b*)   Repeat the calculation of the sensible heat gain using the data from Tables 7.11 and 7.12 for the solar load through glass. Take a haze factor of 0.95 and a blind factor of 0.54 (Table 7.6).

(*c*)   Repeat the calculation for the sensible heat gain using the data from Tables 7.13 and 7.14 for the solar load through glass.

(*d*)   Repeat the calculation using the solar heat gain data in Table A9.15 in the *CIBSE Guide A9* (1986).

(*e*)   Compare the results.

Answer

(*a*) First the relevant sol-air temperatures must be established from Table 7.8 for the month of July and any necessary corrections made:

Design outside air temperature at 15.00 h sun time:	28°C
Tabulated outside air temperature at 15.00 h sun time:	25.3°C
Correction to be applied to tabulated value:	+2.7 K
Time of heat gain to room:	15.00 h
Time lag of wall:	5 h
Time of relevant sol-air temperature	10.00 h
Tabulated sol-air temperature at 10.00 h	25.0°C
Correction:	+2.7 K
Corrected sol-air temperature at 10.00 h	27.7°C
Tabulated 24 h mean sol-air temperature:	23.0°C
Correction:	+2.7 K
Corrected 24 h mean sol-air temperature	25.7°C

Use the simplified equation (7.24) for the calculation of heat gain through the wall.

From Tables 7.9 and 7.10 the following is determined for a lightweight building with blinds closed (because the window is in direct sunlight), 12 h plant operation and air temperature control.

Solar air conditioning load through a west window at 15.00 h sun time in July: 205 Wm^{-2}.
Correction factor for a double glazed, clear, plate glass window: 1.00.
Sensible heat gain calculation:

	W
Glass transmission (equation (7.17)):	
$2.184 \times 3.0 \times (28 - 22) =$	39
Wall (equation (7.24)):	
$(3.3 \times 2.4 - 2.184) \times 0.45[(25.7 - 22) + 0.65(27.7 - 25.7)]$	= 13
Infiltration (equation (7.41)):	
$0.33 \times 0.5 \times (2.4 \times 2.6 \times 6.0)(28 - 22) =$	37
Solar gain through glass:	
$2.184 \times 1.00 \times 205 =$	448
People: $2 \times 90 =$	180
Lights: $17 \times 2.4 \times 6.0 =$	245
Business machines: $20 \times 2.4 \times 6.0 =$	288

Sensible heat gain =	1250

Latent heat gains:

Infiltration by equation (7.42):

$$0.8 \times 0.5 \times (2.4 \times 2.6 \times 6.0) \times (10.65 - 8.366) =$$ 34 W

People: $2 \times 50 =$ 100 W

Latent heat gain =   134 W

(b) From Table 7.11: maximum solar intensity: 695 W

From Table 7.12: storage load factor: 0.67

From Table 7.6 the shading coefficient: 0.54

Solar load: $0.95 \times 0.54 \times 0.67 \times 695 = 239$ W m^{-2}

Total sensible heat gain from (a):	1250 W
Solar gain through glass from (a):	448 W
Remaining other sensible gain:	802 W
Solar gain: $2.184 \times 239 =$	522 W
Sensible heat gain:	1324 W

(c) From Table 7.13:

Solar cooling load $(271 + 325)/2 = 298$ W m^{-2}

From Table 7.14, the correction factor for shading is 0.95 and the air-point control factor (because the tabulated data are for the control of dry resultant temperature) is 0.91

Remaining other sensible gain:	802 W
Solar gain: $2.184 \times 298 \times 0.95 \times 0.89$	550 W
Sensible gain:	1352 W

(d) From CIBSE Table A9.15 (1986) the cooling load due to solar gain through glass is 270 W m^{-2}, the shading factor is 0.74 and the air-point control factor is 0.91.

Remaining other sensible gain:	802 W
Solar gain: $2.184 \times 270 \times 0.74 \times 0.91$	397 W
Sensible gain:	1199 W

(e) Comparing the results:

	Sensible gain		Glass solar gain	
(d) Table A9.15 (CIBSE 1986)	1199 W	100%	397 W	100%
(a) Tables 7.9 and 7.10 (Haden Young)	1250 W	104%	448 W	113%
(b) Tables 7.11 and 7.12 (Carrier Air Conditioning Company 1965)	1324 W	110%	522 W	131%
(c) Tables 7.13 and 7.14 (CIBSE A2 1999)	1352 W	113%	550 W	139%

Note that the approximate percentages of the components of the sensible gain to a typical office module are: glass transmission 3 per cent, wall transmission 1 per cent, infiltration 3 per cent, solar gain through glass 36 per cent, people 14 per cent, lights 20 per cent and business machines 23 per cent. It is also worth noting that the approximate specific sensible gain is about 80 to 90 W m^{-2}, referred to the treated modular floor area.

The effects of changing the summer outside design state and the state maintained in the room are worth consideration. Tables 5.3 and 5.4 give details of the percentages of the hours in the summer that outside dry-bulbs and wet-bulbs were exceeded at Heathrow for the period 1976–95. If lower outside design temperatures are chosen the heat gains will be reduced but the chances of the air conditioning system not giving full satisfaction are increased. Similarly, raising the dry-bulb temperature maintained in the room, with the risk of less comfort, may be a consideration. Further, if the outside wet-bulb or the room enthalpy are altered there will be an effect on the fresh air load (see sections 8.1 and 8.2). All these matters have implications for comfort, the satisfaction of the client, capital costs and running costs. The benefits of any measure considered must be balanced against its disadvantages, by considering Table 5.3 or 5.4.

EXAMPLE 7.19

(*a*) Ignoring heat gains through the wall, which are trivially small, investigate the effect on the sensible heat gain of changing the outside and inside summer design states for the module used in example 7.18. Use Tables 7.9 and 7.10 for calculating the solar heat gains through glass.

Answer

The sensible heat gains from solar gain through glass, lights and business machines are unchanged at: $448 + 245 + 288 = 981$ W. The sensible gain from people reduces and the latent gain increases as the room temperature rises (see Table 7.16). The sensible gain from infiltration reduces for a given room condition, as the outside dry-bulb goes down. The sensible heat gains are considered as follows.

Sensible heat gain:

t_o	28°	27°	26°
t_r	22°	22°	22°
Glass	39	33	26
Infiltration	37	31	25
People	180	180	180
S + L + M	981	981	981
Total	1237	1225	1212

If we assume that the fresh air allowance is 1.4 litres s^{-1} m^{-2} over the modular floor area of 14.4 m^2 the supply is 20.16 litres s^{-1} of fresh air with a specific volume, v, expressed at the room state. Assuming the room is at 50 per cent saturation, the latent heat gain by infiltration can be calculated by equation (7.42) and we can also calculate the fresh air load (see example 8.1) by means of equation (7.43):

$$Q_{fa} = (v_t/v_t(h_o - h_r))$$

where Q_{fa} is the fresh air load in kW, v_t is the volumetric flow rate of fresh air at temperature

t in $m^3 s^{-1}$, v_t is the specific volume of the fresh air at temperature t, h_o is the enthalpy of the outside air in $kJ kg^{-1}$ and h_r is the enthalpy of the room air in $kJ kg^{-1}$.
   The following table is compiled.

t_o (°C)	28	27	27	26	26
t_o' (°C)	20	19	18	19	18
t_r (°C)	22	22	22	22	22
g_o (g kg^{-1})	10.65	9.719	8.354	10.18	8.859
g_r (g kg^{-1})	8.366	8.366	8.366	8.366	8.366
People latent gain, see Table 7.16 (W)	100	100	100	100	100
Infiltration latent gain, see equation (7.42) (W)	34	20	0	27	7
People + infiltration latent gain (W)	134	120	100	127	107
h_o (kJ kg^{-1})	55.36	51.96	48.47	52.10	48.74
h_r (kJ kg^{-1})	43.39	43.39	43.39	43.39	43.39
v_r ($m^3 kg^{-1}$)	0.847	0.847	0.847	0.846	0.847
Fresh air load (see equation (7.43) and section 6.7) (W)	285	204	121	207	127
Sensible gain + latent gain + fresh air load (W)	1636	1537	1446	1530	1442

Interpreting the data in Table 5.3 is difficult. One way might be to use mid-values of the bands and the averages of the limits of the bands of dry-bulb and wet-bulb temperatures quoted:

For 28°C dry-bulb and 20°C wet-bulb the average is $(0.77 + 0.32 + 0.31 + 0.25)/4 = 0.41$ per cent of the four summer months, namely 12.0 hours. For 27°C dry-bulb and 19°C wet-bulb it is 0.77 per cent or 22.5 hours. For 27°C dry-bulb, 18°C wet-bulb it is $(0.42 + 0.77)/2 = 0.60$ per cent or 17.6 hours. For 26°C dry-bulb and 19°C wet-bulb it is $(1.32 + 0.77)/2 = 1.04$ per cent or 30.4 hours and for 26°C dry-bulb and 18°C wet-bulb it is $(1.37 + 1.32 + 0.42 + 0.77)/4 = 0.97$ per cent which is 28.4 hours. The results are summarised in the following table.

1. Dry-bulb	28°	27°	27°	26°	26°
2. Wet-bulb	20°	19°	18°	19°	18°
3. Per cent exceeded	0.41%	0.77%	0.60%	1.04%	0.97%
4. Hours exceeded	12.0 h	22.5 h	17.6 h	30.4 h	28.4 h
5. Sensible gain	100%	99%	99%	98%	98%
6. Latent gain	100%	95%	83%	97%	90%
7. Fresh air load	100%	72%	43%	73%	45%
8. Totals	1636 W	1537 W	1446 W	1530 W	1442 W
	100%	94%	88%	94%	88%

Conclusions

1. The sensible gains do not change very much and the reduction is probably within the accuracy of their calculation. Any effect they have is most likely to be on the size of the air handling plant and duct system, although this depends very much on the type of system adopted.

2. For the case of an outside state of 27° dry-bulb, 18° wet-bulb, the large drop in the latent gain to 83 per cent must be disregarded since this is entirely due to the reduction of the infiltration gain to zero, because the outside moisture content is less than the room moisture content. The allowance of half an air change per hour for infiltration is very much open to question and the reduction must be ignored.

The reduction of the latent gains to 95 per cent and 90 per cent for the cases of 27° dry-bulb, 19° wet-bulb and 26 per cent dry-bulb, 18° wet-bulb look interesting but the latent gains are a fairly small proportion of sensible plus latent gain, namely, about 10 per cent. Any benefit in reducing the capital or running cost is likely to be in the size of the refrigeration duty. The impact of the reduction in latent gains on this is smaller still, at about 7 per cent.

3. If, as an approximation, the cooling load is taken as the sum of the sensible gain, the latent gain and the fresh air load, we can see that this is probably the most significant factor. There is an established correlation between the cooling load and capital cost. The running cost is also likely to correlate although other factors, such as system choice and the quality of maintenance, intervene. It must be remembered that this example is based on the heat gains and cooling load for a single, west-facing module and diversity factors for people, lights, and business machines have not been considered, whereas they would be for an entire building. Neither has the natural diversity in the solar load through glazing, as the sun moves round the building during the day, been taken into account. Nevertheless, designing the system for an outside state of 27°C dry-bulb with 19°C wet-bulb (screen) gives a 6 per cent reduction in the cooling load (implying a 6 per cent reduction in capital cost). This is at the expense of a failure in apparant comfort satisfaction for 22.5 h in the four summer months. The decision is commercial and the client must be fully aware of the risks involved for the benefit obtained.

The view of the author is that, with global warming likely to become increasingly important over the next 20 years, prudence is to be recommended. Air conditioning systems are likely to be required to do more than they were initially designed for in the immediate future.

EXAMPLE 7.20

Using the modular details from example 7.18 calculate the sensible and latent heat gains at 1500 h sun time on 21 January, assuming the location to be Perth, Western Australia. Take the same values for heat gains from lights and business machines, assume two people present and half an air change of infiltration. The window is single glazed ($U = 5.6$ W m^{-2} K^{-1}) and fitted with internal Venetian blinds. The latitude of Perth is 32°S. Make the following design assumptions:

Outside state: 35°C dry-bulb, 19.2°C wet-bulb (screen), 5.876 g kg^{-1}, 50.29 kJ kg^{-1}, 0.8809 m^3 kg^{-1}.
Room state: 25°C dry-bulb, 40 per cent saturation, 8.063 g kg^{-1}, 45.69 kJ kg^{-1}.

The load due to sensible heat gain through single glass by solar radiation for a fast response building with internal blinds, used intermittently when the sun shines on the west face of the building, is 241 W m^{-1}, based on data from the *CIBSE Guide A2* (1999). This should be increased by 7 per cent because the earth–sun distance is 3.5 per cent less in December than in June and the intensity of solar radiation follows an inverse square law with respect to distance. See Table 7.4 and section 7.16.

Answer

First the relevant sol-air temperatures should be calculated. Sol-air temperature is given by equation (7.21). Using the results of example 7.4, equation (7.9) and the data in Tables 7.3 and 7.4, the calculation of the sol-air temperature at a particular time is possible, but calculating the heat gain by equation (7.24) requires a knowledge of the 24 hour mean sol-air temperature. This is not possible here, without detailed weather data, on an hourly basis, for January at Perth. However, the heat gain through a wall usually amounts to about 1 per cent of the whole (see example 7.18) and is ignored in this case.

Sensible heat gain calculation:	W
Glass: $2.184 \times 5.6 \times (35 - 25)$	122
Infiltration: $0.33 \times 0.5 \times (2.4 \times 2.6 \times 6.0) \times (35 - 25)$	62
Solar through glass: $2.184 \times 241 \times 1.07$	563
People*: 2×75	150
Lights: $17 \times 2.4 \times 6.0$	245
Business machines: $20 \times 2.4 \times 6.0$	288
Total sensible gain	1430

*Refer to Table 7.16
Specific sensible gain per unit of floor area:

$$1430/14.4 = 99.3 \text{ W m}^{-2}$$

Latent heat gain calculation:	
Infiltration (equation (7.42)):	W
$0.8 \times 0.5 \times 2.4 \times 2.6 \times 6.0(5.876 - 8.063)$	−33
People:	
2×65	130
Total latent gain	97

Since any rate of natural infiltration is uncertain, it might be prudent to take the latent gain as 130 W. However, local custom should be considered as well as meteorological data (see section 5.11). There is an on-shore breeze from the Indian Ocean that often occurs in the afternoons during summer months in Perth, which reduces the dry-bulb temperature and increases the moisture content. This could make the assumed outside design state temporarily irrelevant. The contribution of any infiltration in the latent heat gain calculation would then be positive with the total latent gain exceeding 130 W.

Exercises

1.   (*a*) Why do the instantaneous heat gains occurring when solar thermal radiation passes through glass not constitute an immediate increase on the load of the airconditioning plant? Explain what sort of effect on the load such instantaneous gains are likely to have in the long run.

  (*b*) A single glass window in a wall facing 30° west of south is 2.4 m wide and 1.5 m high. If it is fitted flush with the outside surface of the wall, calculate the instantaneous heat gain due to direct solar thermal radiation, using the following data:

Intensity of direct radiation on a plane normal to the sun's rays	790 W m^{-2}
Altitude of the sun	60°
Azimuth of the sun	70° west of south
Transmissivity of glass	0.8

Answers

(*b*) 870 W.

2.   A window 2.4 m long × 1.5 m high is recessed 300 mm from the outer surface of a wall facing 10° west of south. Using the following data, determine the temperature of the glass in sun and shade and hence the instantaneous heat gain through the window.

Altitude of sun	60°
Azimuth of sun	20° east of south
Intensity of sun's rays	790 W m^{-2}
Sky radiation normal to glass	110 W m^{-2}
Transmissivity of glass	0.6
Reflectivity of glass	0.1
Outside surface coefficient	23 W m^{-2} K^{-1}
Inside surface coefficient	10 W m^{-2} K^{-1}
Outside air temperature	32°C
Inside air temperature	24°C

Answers

37.9°C, 29.6°C, 983 W.

3.   An air conditioned room measures 3 m wide, 3 m high and 6 m deep. One of the two 3 m walls faces west and contains a single glazed window of size 1.5 m by 1.5 m. The window is shaded internally by Venetian blinds and is mounted flush with the external wall. There are no heat gains through the floor, ceiling, or walls other than that facing west and there is no infiltration. Calculate the sensible and latent heat gains which constitute a load on the air conditioning system at 16.00 h in June, given the following information.

Outside state	28°C dry-bulb, 19.5°C wet-bulb (sling)
Inside state	22°C dry-bulb, 50% saturation
Electric lighting	33 W per m^2 of floor area

Number of occupants	4
Heat liberated by occupants	90 W sensible, 50 W latent
Solar heat gain through window with	
Venetian blinds fully closed	258 W m^{-2}
U-value of wall	1.7 W m^{-2} K^{-1}
U-value of glass	5.7 W m^{-2} K^{-1}
Time lag for wall	5 hours (= ϕ)
Decrement factor for wall	0.62 (= f)

Diurnal variations of air temperature and sol-air temperature are as follows:

Sun-time	09.00	10.00	11.00	12.00	13.00	14.00	15.00	16.00
Air temperature (°C)	20.6	22.0	23.3	24.7	25.8	26.8	27.5	28.0
Sol-air temperature (°C)	23.7	25.3	26.8	28.3	39.4	47.3	53.6	57.0

The mean sol-air temperature over 24 hours is 29.9°C (= t_{em}).
The heat gain through a wall, $q_{(\theta+\phi)}$, at any time $(\theta + \phi)$, is given by the equation:

$$q_{(\theta+\phi)} = UA(t_{em} - t_i) + UA(t_e - t_{em})f$$

where t_e is the sol-air temperature
     t_i is the inside air temperature
     θ is the time in hours.

Answers

1662 W, 200 W.

4. (*a*) Derive an expression for sol-air temperature.
  (*b*) Using your derived expression determine the sol-air temperature for a flat roof if the direct radiation, normal to the sun's rays, is 893 W m^{-2} and the intensity of scattered radiation normal to the roof is 112 W m^{-2} Take the absorption coefficient of the roof for direct and scattered radiation as 0.9, the heat transfer coefficient of the outside surface as 22.7 W m^{-2}, the outside air temperature as 28°C and the solar altitude as 60°C.
  (*c*) Given that the time lag of the roof structure is zero and its decrement factor is unity, calculate the heat gain to the room beneath the roof referred to in part (*b*) if the *U*-value of the roof is 0.5 W m^{-2} K^{-1} and the room temperature is 22°C. The mean sol-air temperature over 24 hours is 37°C.

Answers

(*b*) 63.1°C, (*c*) 20.6 W m^{-2}.

5. Repeat the calculation of the sensible gain for the module used in example 7.18 but for an east-facing module, with a wall *U*-value of 0.6 W m^{-2} K^{-1} and an infiltration rate of 2 air changes per hour.

Answer

1111 W.

Notation

Symbol	Description	Unit
A	apparent solar radiation in the absence of an atmosphere	W m^{-2}
	solar radiation constant	–
	area	m^2
	surface area of a structural element in a room	m^2
	internal duct dimension	m
A_f	floor area	m^2
A_w	area of glass or area of opening in a wall	m^2
	angular movement of the sun	medians
a	altitude of the sun	degrees
a'	altitude of the sun at noon	degrees
B	atmospheric correction factor	–
	internal duct dimension	m
C	dimensionless constant	–
c	specific heat capacity	$\text{kJ kg}^{-1}\,\text{K}^{-1}$
D	internal duct diameter	m
d	declination of the sun	degrees
F_b	shading factor	–
F_c	air-point control factor	–
F_g	angle factor for the ground	–
F_s	angle factor for the sky	–
F_{au}	room conduction factor with respect to the air node	–
F_{ay}	room admittance factor with respect to the air node	–
f	decrement factor	–
g_o	outside air moisture content	g kg^{-1}
g_r	room air moisture content	g kg^{-1}
h	hour angle	degrees
	coefficient of heat transfer	$\text{W m}^{-2}\,\text{K}^{-1}$
h_{si}	inside surface film coefficient of heat transfer	$\text{W m}^{-2}\,\text{K}^{-1}$
h_{so}	outside surface film coefficient of heat transfer	$\text{W m}^{-2}\,\text{K}^{-1}$
h_o	enthalpy of the outside air	kJ kg^{-1}
h_r	enthalpy of the room air	kJ kg^{-1}
I	intensity of direct solar radiation on a surface normal to the rays of the sun	W m^{-2} or kW m^{-2}
I_h	component of direct solar radiation normal to a horizontal surface	W m^{-2}
I_r	intensity of radiation reflected from surrounding surfaces	W m^{-2}
I_s	intensity of diffuse (sky) radiation normally incident on a surface	W m^{-2}
I_t	intensity of total radiation on a surface	W m^{-2}
I_v	component of direct solar radiation normal to a vertical surface	W m^{-2}
I_δ	component of direct solar radiation normal to a tilted surface	W m^{-2}

i	angle of incidence of a ray on a surface	degrees
L	latitude of a place on the surface of the earth	degrees
	duct length	m
l	thickness of the insulation on the duct	m
M	dimension of a hypotenuse formed by R and x	m or mm
n	wall-solar azimuth	degrees
	number of air changes per hour	h^{-1}
P	external duct perimeter	m
Q	rate of heat flow	W m^{-2}
	heat transfer through a duct wall	W
Q'	rate of heat entry to an outer wall/roof surface	W m^{-2}
Q_{fa}	fresh air load	kW
Q_{li}	latent heat gain by natural infiltration	W
Q_m	mean rate of heat flow through a wall or roof	W
Q_s	cooling load due to solar gain through glass	W
Q_{si}	sensible heat gain by natural infiltration	W
Q_θ	rate of heat flow into a room at time θ	W
$Q_{\theta+\phi}$	rate of heat flow into a room at time $\theta + \phi$	W
q_{max}	maximum instantaneous sensible solar heat gain through glass	W m^{-2}
q_s	specific cooling load due to solar gain through glass	W m^{-2}
R	depth of a window recess	m or mm
	remainder term to cover long-wave radiation exchanges	W m^{-2}
r_{si}	thermal resistance of an inside surface air film	$\text{m}^2\text{ K W}^{-1}$
r_{so}	thermal resistance of an outside surface air film	$\text{m}^2\text{ K W}^{-1}$
SC	shading coefficient	–
SHGC	solar heat gain coefficient	–
T	sun time	h
	absolute temperature	K
t	dry-bulb temperature	°C
	time	h
t'	wet-bulb temperature	°C
t_a	air temperature	°C
t_{eo}	sol-air temperature	°C
t_{ei}	inside environmental temperature	°C
t_{em}	mean sol-air temperature over 24 hours	°C
t_g	mean glass temperature	°C
t_o	outside air dry-bulb temperature	°C
t_r	room air temperature	°C
	ambient air dry-bulb temperature	°C
t_{rm}	mean radiant temperature	°C
t_{si}	inside surface temperature	°C
t_{so}	outside surface temperature	°C
t_{sm}	mean inside surface temperature	°C
t_1	initial air temperature in duct	°C
t_2	final air temperature in duct	°C
U	overall thermal transmittance coefficient	$\text{W m}^{-2}\text{ K}^{-1}$
V	volume of a room	m^3

v	mean air velocity in a duct	m s^{-1}
	velocity of airflow through an opening	m s^{-1}
$\dot{v}_t$	volumetric flow rate of fresh air at temperature t	m^3 s^{-1}
v_t	specific volume of air at temperature t	m^3 kg^{-1}
x	horizontal co-ordinate	m or mm
Y	admittance of a surface	W m^{-2} K^{-1}
y	vertical co-ordinate	m or mm
	dimensionless parameter related to duct heat gain	–
z	azimuth of the sun	degrees
α	absorptivity of glass for direct solar radiation	–
α'	absorptivity of glass for scattered solar radiation	–
Δp	pressure drop through an opening	Pa
Δt	temperature change per metre of duct length	K m^{-1}
δ	angle between the ground and the horizontal	degrees
	angle of a surface with the horizontal	degrees
θ	time	h
λ	thermal conductivity of the insulation on a duct	W m^{-1} K^{-1}
ρ	density of air	kg m^{-3}
	glass reflection coefficient for direct solar radiation	–
	reflectivity of the ground	–
τ	glass transmission coefficient for direct radiation	–
τ'	glass transmission coefficient for sky radiation	–
ϕ	time lag	h

References

Allen, C.W. (1973): *Astrophysical Quantities*. The Athlone Press, University of London.
ASHRAE Handbook (1993a): Fundamentals, 27.4.
ASHRAE Handbook (1993b): Fundamentals 26.6.
ASHRAE Handbook (1993c): Fundamentals, 26.33–26.34.
ASHRAE Handbook (1993d): Fundamentals, 32.15.
ASHRAE Handbook (1997a): Fundamentals, 29.14.
ASHRAE Handbook (1997b): Fundamentals, 29.23–29.24.
ASHRAE Handbook (1997c): Fundamentals, 29.43–29.44.
Carrier Air Conditioning Company (1965): *Handbook of Air Conditioning Systems Design*, McGraw-Hill Inc.
CIBSE Guide A2 (1986): Weather and solar data.
CIBSE Guide A9 (1986): Estimation of plant capacity, Tables A19.9, A9–9.
CIBSE Guide A9 (1986): Table A9.15.
CIBSE Guide A2 (1999): Internal heat gains, 6.4.
CIBSE Guide A3 (1999): Thermal properties of building structures.
CIBSE Guide A4 (1999): Air infiltration and natural ventilation.
CIBSE Guide A5 (1999): Thermal response and plant sizing.
CIBSE Guide A2 (1999): External design data.
CIBSE Applications Manual AM7 (1992): Information technology in buildings, 24–25.
Curtis, D.M. and Lawrence, J.M. (1972): Atmospheric effects on solar radiation for computer analysis of cooling loads for buildings of various location heights, *JIHVE* **39**, 254.
Danter, E. (1960): Periodic heat flow characteristics of simple walls and roofs, *JIHVE* **28**, 136–146.

Iqbal, M. (1983): *An Introduction to Solar Radiation.* Academic Press, Toronto.

Mackey, C.M. and Wright, L.T. (1944): Periodic heat flow–homogeneous walls or roofs, *ASHVE Trans.* **50**, 293–312.

Mitelas, G.P. and Stephenson, D.G. (1967): Room thermal response factor, *ASHRAE Trans.* **73**, III–2.1.

Mitelas, G.P. (1972): Transfer function method of calculating cooling loads, heat extraction rate and space temperatures, *ASHRAE Journal*, No. 12, **14**, 52.

Moon, P. (1940): Proposed standard solar radiation curves for engineering use, *Journal of Environmental Science*, November, **230**, 583.

Pilkington Bros (1969): *Windows and Environment.*

Pilkington Flat Glass Ltd. (1980): *Glass and Thermal Safety*, June.

Pilkington Flat Glass Ltd. (1991): *Glass and Transmission Properties of Windows*, Febuary.

Reilly, M.S., Winkelmann, F.C., Arsateh, D.K. and Carroll, W. L. (1992): Modeling windows in DOE-2, *Energy and Buildings* **22**, 59–66.

Stephenson, D.G. and Mitelas, G.P. (1967): Cooling load calculations by thermal response factor method, *ASHRAE Trans.* **73**, III–1.1.

Stewart, J.P. (1948): Solar heat gain through walls and roofs for cooling load calculations, *ASHVE Trans.* **54**, 361–388.

Westinghouse Electric Corporation (1970): Thermolume Water Cooled Lighting Environmental Systems, *Applications Manual 61–300.*

8

Cooling Load

8.1 Cooling load and heat gains

Examples 6.11 and 6.12 showed how a cooling load is calculated for a simple system and how the load can be broken down into its component parts and the answer checked. In carrying out the check it is important to do so in a way that is as different as possible from the method used to determine the cooling load in the first place. This is straightforward for the sensible and latent heat gains because they were calculated in a different way prior to determining the cooling load. Equation (6.6) can be used to evaluate the contributions by the supply and extract fans, the supply duct gain, and any heat gain to the extracted air as it flows through ventilated luminaires. However it is not possible to assess the fresh air load except by using the enthalpy difference between the ouside air and the room air, applied to the fresh air flowrate. A less satisfactory way, because it is directly duplicating the method used to determine the original cooling load, is to use the difference in enthalpy between the mixed air and the room air, applied to the supply air flowrate. To appreciate the relative proportions of the various components in the cooling load, in a practical case, the design data and results in example 7.18 can be used. This example referred to the west-facing module on an intermediate floor of an office block under typical summer design conditions in London.

EXAMPLE 8.1

Using the assumptions and results of example 7.18(*a*) calculate the design cooling load. Check the answer and establish the relative percentages of the elements comprising the load. Assume a temperature rise of 2° for the supply fan power and duct gain, and 0.25° for the extract fan power.

Answer

Identify on a psychrometric chart the following state points:

>   *O* (outside air): 28°C dry-bulb, 19.5°C wet-bulb (sling)
>   *R* (room air): 22°C dry-bulb, 50 per cent saturation, 8.366 g kg^{-1}
>   *R'* (recirculated air): 22.25°C dry-bulb, 8.366 g kg^{-1}

The states *O* and *R'* on the chart are joined by a straight line.
  From example 7.18(*a*):

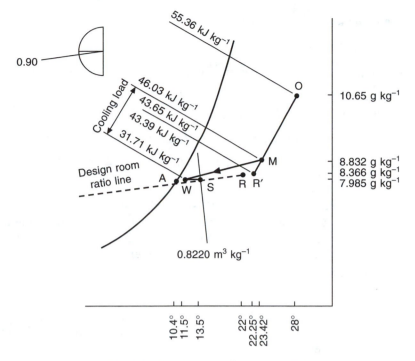

Fig. 8.1 Psychrometry for example 8.1.

Sensible heat gain: 1250 W

Latent heat gain: 134 W

Sensible/total heat gain ratio: 1250/(1250 + 134) = 0.90

Mark the value of the sensible/total ratio on the chart protractor and draw the room ratio line parallel to this through the room state, *R*.

Bearing in mind the methods outlined in chapter 6, an inspection of the psychrometric chart suggests that a supply air temperature of 13.5°C dry-bulb could be a satisfactory choice. On this basis the following calculations are carried out.

By equation (6.6):

$$\dot{v}_{13.5} = \frac{1250}{(22 - 13.5)} \times \frac{(273 + 13.5)}{358} = 117.7 \text{ litres s}^{-1}$$

The supply air moisture content is found from equation (6.8):

$$g_s = 8.366 - \frac{134}{117.7} \times \frac{(273 + 13.5)}{856} = 7.985 \text{ g kg}^{-1}$$

Identify the supply air state *S*, at 13.5°C dry-bulb and 7.985 g kg^{-1}, on the room ratio line.

Identify the off-coil state *W*, at 11.5°C dry-bulb and 7.985 g kg^{-1}, on the psychrometric chart.

Allow 12 litres s^{-1} of fresh air for each of the two persons assumed to be in a module. This represents a total of 24 litres s^{-1} of fresh air. It is convenient to assume that this is measured at the supply air state in order to simplify the calculation of the proportions of fresh and recirculated air handled. Hence:

Fresh air fraction $= \dfrac{24}{117.7} = 0.204$

Recirculated air fraction $= 0.796$

The mixture state, M, which is also the on-coil state, must now be determined:

$t_m = 0.796 \times 22.25 + 0.204 \times 28 = 23.42°C$

$g_m = 0.796 \times 8.366 + 0.204 \times 10.65 = 8.832 \text{ g kg}^{-1}$

From the chart or, more accurately, by equation (2.24):

$h_m = [(1.007 \times 23.42 - 0.026) + 0.008\,832(2501 + 1.84 \times 23.42)]$

$= 46.03 \text{ kJ kg}^{-1}$

The state M is identified on the chart and M is joined to W by a straight line, which is extended to cut the saturation curve at the apparatus dew point, A. The temperature of A is found to be 10.4°C. The practicality of the proposed choice of the supply state S is now established by determining the contact factor of the cooler coil, using equation (3.3):

$\beta = \dfrac{(23.42 - 11.5)}{(23.42 - 10.4)} = 0.92$

This implies a six-row coil, with 315 fins per metre (8 per inch) and a face velocity of about 2.5 m s^{-1} (see section 10.4). This is quite practical and so the choice of 13.5°C dry-bulb for the supply air temperature is accepted.

The specific volume at the supply state, S, is found from the chart or, more accurately, from psychrometric tables, to be 0.8220 m^3 kg^{-1}. Similarly, the enthalpy at the off-coil state, W, is determined as 31.71 kJ kg^{-1}. The design cooling load can now be calculated.

Design cooling load $= \dfrac{0.1177}{0.8220} (46.03 - 31.71) = 2.050 \text{ kW}$

The cooling load is analysed into its component parts and, by evaluating the contribution of each, a check on the answer is obtained:

	watts	*per cent*
Sensible heat gain:	1250	61.0
Latent heat gain:	134	6.6
Supply fan power and duct gain (equation (6.6))		
$= (2 \times 117.7)(358)/(273 + 13.5) =$	294	14.3
Recirculation fan power (equation (6.6))		
$= (0.25 \times 117.7 \times 0.796) \times 358/(273 + 13.5) =$	29	1.4
Fresh air load $= (0.024/0.8220)(55.36 - 43.39) \times 1000 =$	342	16.7
Total	2049	100.0

Although the sensible heat gain dominates the load a sizeable part is due to the fresh air component. This provides a good indication of the time of the day and month of the year for which the cooling load will be a maximum. The fresh air load is greatest when the outside air enthalpy is at its highest value and, in most countries in the northern hemisphere, this is at about 15.00 h, sun-time, in July. There are occasional exceptions but it is rare for the maximum load to be earlier than 13.00 h, or later than 17.00 h. Sometimes the maximum is in August.

The end of example 7.18 compared the solar gain through glass from four different sources and the above example used data from Tables 7.9 and 7.10. If, instead, solar gain data in Tables 7.13 and 7.14 from the *CIBSE Guide A2* (1999) is used, the sensible heat gain increases by about 8 per cent from 1250 W to 1352 W, requiring a similarly larger supply airflow rate of 127.3 litres s^{-1}. This has no effect on the fresh air load and the latent heat gain. Consequently, the cooling load does not increase as much, rising only by about 6 per cent in the simple case considered, because of the extra fan power needed for the increased airflow rate. Such an increase is not insignificant because it affects the selection of the refrigeration plant and its ancillaries, the size of air handling plants, duct systems and the amount of building space used. Capital and running costs will increase.

8.2 Cooling load for a whole building

To determine the size of the necessary refrigeration plant the cooling load should be calculated at about 15.00 h sun-time for the entire building. The design refrigeration load is defined by the following equation:

$$Q_r = [Q_s + Q_1 + Q_{fa} + Q_{sf} + Q_{sd} + Q_{ra} + Q_{rh}]f_p \qquad (8.1)$$

where

Q_r =	design refrigeration load	kW
Q_s =	sensible heat gain	kW
Q_1 =	latent heat gain	kW
Q_{fa} =	fresh air load	kW
Q_{sf} =	supply fan power	kW
Q_{sd} =	supply duct heat gain	kW
Q_{ra} =	return air load (recirculation fan power and heat gain through extract ventilated luminaires)	kW
Q_{rh} =	reheat load	kW
f_p =	factor to cover chilled water pump power and heat gain to pipes	

Provided that the type of air conditioning system used for the building can take advantage of the reduction in sensible heat gains, the calculation of Q_s, at 15.00 h in July, will involve the use ot diversity factors for people, lights and business machines. People will be absent for various reasons and not all the lights will be on, or all the business machines in use. Diversity in the solar gain through glass will be naturally accounted for because windows on the easterly side of the building will be in shade at 15.00 h in July. Similarly, the part of the latent gain due to people will also be subject to the use of a diversity factor.

The fresh air load is maximised by the choice of 15.00 h in July. Supply fan power is not subject to diversity, unless a variable air volume system is used and then only if an efficient method of controlling fan capacity is employed. There is no contribution from heat gain to recirculated air ducts because the air within is at about 22°C. With a well-designed system there should be no reheat load at the time of greatest cooling load. The factor f_p is used only if a chilled water system is adopted, as would be the case for larger installations. The contribution of pump power and heat gain to pipes is usually very small and f_p could have a value of 1.01 or 1.02. In such a case it might as well be taken as unity, since the calculation of the other elements in the load is not done with this degree of accuracy. The design engineer must exercise discretion.

8.3   Partial load

An air cooler coil works at partial load either when the enthalpy of the air entering the coil is less than the summer design value, or when the capacity of the coil is deliberately reduced by the use of motorised valves in the chilled water piping, or by face and by-pass dampers (see sections 10.7 and 8.6).

The majority of air conditioning systems operate at partial load for most of the time. As an example, an office block in London is likely to be at full cooling duty for a total of only 30 hours, in July and August, out of a total annual working year of about 2300 hours.

It is good practice to design systems to take advantage of the natural cooling capacity of the outside air. Figures 13.21(*a*) and (*b*) illustrate how this may be done. The dry-bulb temperature in London exceeds 10°C for about 1300 hours of the total working time. Since 10°C or 11°C is typical of the design dry-bulb temperature commonly required from a cooler coil, the refrigeration plant could be switched off for the remaining 1000 hours, in the case of an all-air system, such as variable air volume.

On the other hand, with air–water systems (such as fan coil), chilled water must be available for the terminal units all the time they are working and, to provide this, natural cooling is consequently related to the outside wet-bulb. In London, this exceeds a sling value of 5°C for about 1600 hours in a working year. It is possible to obtain a chilled water flow temperature of 11°C or 12°C, when the wet-bulb is less than about 5°C, by pumping primary chilled water through the cooler coil in the air handling plant, without running the refrigeration compressor, for the remaining 700 hours.

The behaviour of an air cooler coil at partial load is not easily predicted in a simple way, but section 10.7 deals with this and it is illustrated in Figure 10.7.

8.4   Cooling load offset by reheat

With a constant volume system, the room temperature will fall as the sensible heat gains diminish, unless the cooling capacity of the air supplied is also reduced: the sensible cooling capacity of the supply air must equal the sensible heat gain in the conditioned room. Similarly, with industrial applications, the moisture content of the supply air must be changed to achieve a match between the latent cooling capacity of the supply air and the latent gain in the room.

One simple, but inherently wasteful, method of achieving a match between sensible cooling capacity and sensible heat gain is to reheat the air leaving the cooler coil to a higher temperature, before it is supplied to the room.

EXAMPLE 8.2

If the sensible heat gains used in example 8.1 reduce to 500 W, other data remaining unchanged, calculate the necessary supply air temperature and reheat duty to maintain the room temperature at its design value.

Answer

The supply fan continues to handle 117.7 litres s^{-1} at a temperature of 13.5°C hence, by equation (6.6):

$$t_s = 22 - \frac{500}{117.7} \times \frac{(273 + 13.5)}{358} = 18.6°C$$

The reheat load should be calculated by using the conventional method (refer to Figure 8.2):

$$\text{Reheat duty} = \frac{0.1177}{0.8220} \times (38.95 - 33.74) = 0.746 \text{ kW}$$

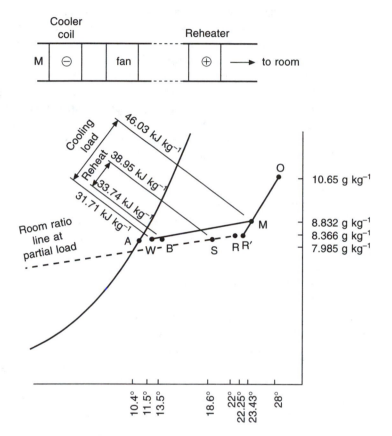

Fig. 8.2   Psychrometry for example 8.2.

Alternatively, equation (6.6) could be used:

$$\text{Reheat duty} = \frac{0.177 \times (18.6 - 13.5) \times 358}{(273 + 13.5)} = 0.750 \text{ kW}$$

(Note that it is really unnecessary to carry out a calculation in this case since, with a reheat system, the reduction in sensible gain is made good by reheat, hence the reheat duty is 1250 W − 500 W = 750 W.)

The process is illustrated in Figure 8.2. The cooling load is unchanged and the surplus sensible cooling capacity is cancelled by wasteful reheating. It is seen that the room ratio line has a steeper slope for the partial load condition because the sensible heat gains have fallen while the latent heat gains are unchanged.

A single air handling plant can deliver air through multiple reheaters to several rooms. The common cooler coil in the air handling unit cools and dehumidifies all the air to a nominally constant value and temperature control is achieved independently in each room by thermostatic control over the individual reheaters. If the latent heat gains change in the rooms then there will be some variation in the humidity because the mosture content delivered to each room will be at a nominally constant value. For comfort conditioning this will not matter since the dry-bulb temperature is under control and humidity is not very important for human comfort.

For industrial conditioning, on the other hand, control over humidity as well as temperature may be important. In this case it becomes necessary to provide independent dry steam humidifiers for the air supply to each treated room, as well as reheaters. The cooler coil in the central air handling plant cools and dehumidifies all the air down to the lowest necessary moisture content (for supply to the room with the largest latent heat gain). Dry steam humidification is then provided for rooms where this supply air moisture content is too low to deal with the latent heat gains. Meanwhile, the reheaters control room temperatures, independently. The air conditioning process is wasteful in two respects, humidity and temperature, but this does not matter because the industrial process is the prime concern.

Where dry steam humidification is provided for individual rooms the steam should be delivered into the air stream in a place where the air is not near to saturation. This is usually as close as possible to the treated room itself, after any reheater.

There are many ways in which the required supply air state can be achieved for a partial load condition and some of these are considered in the following sections. It might be thought that the cooler coil capacity could itself be reduced so that the air leaving the coil had the correct temperature. This is possible and can be done, but simple conclusions cannot be drawn because the behaviour of cooler coils is complicated. This is dealt with in chapter 10. Figure 8.3 illustrates one important aspect of cooler coil performance (also discussed in section 3.4). The process of cooling and dehumidification from state M to state W' in Figure 8.3(a) is not possible. For such a process to be possible the line MW', if extended, must cut the saturation curve. That is to say, there must be an apparatus dew point, A. Figure 8.3(b) shows a process that is possible.

Reheat and humidification duties must be worked out for the most exacting condition. If the reheater is the only source of heating for the room then it must be able to cope with the heat loss for the design winter state. The humidifier must be able to make good the latent loss that can arise from infiltration in winter, no benefit being allowed for the latent heat gain from people.

EXAMPLE 8.3

Determine the design reheat and dry steam humidification loads for the module forming the subject of examples 7.18 and 8.1. Assume that motorised, automatically controlled dampers are used to provide a mixed air temperature of 11.5°C in winter design conditions, the cooler coil not then being in operation. The winter inside design condition is 20°C dry-bulb, 50 per cent saturation, 7.376 g kg^{-1}, 38.84 kJ kg^{-1} and the outside design condition is –2°C saturated, 3.205 kJ kg^{-1}, 5.992 kJ kg^{-1}. A natural infiltration rate of one air change per hour is assumed.

Answer

See Figure 8.4. Using equations (7.17) and (7.41) the heat loss from the module is calculated as follows.

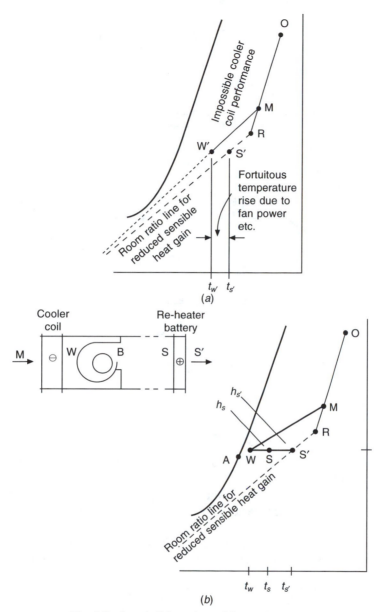

Fig. 8.3   Impossible and possible psychrometry.

Glass: $2.184 \times 3.0 \times (20 + 2)$ = 144 W
Wall: $(3.3 \times 2.4 - 2.184) \times 0.45 \times (20 + 2)$ =   57 W
Infiltration: $0.33 \times 1.0 \times (2.4 \times 2.6 \times 6.0) \times (20 + 2)$ = 272 W

Total: 473 W

The supply air temperature needed to offset this is determined from equation (6.6):

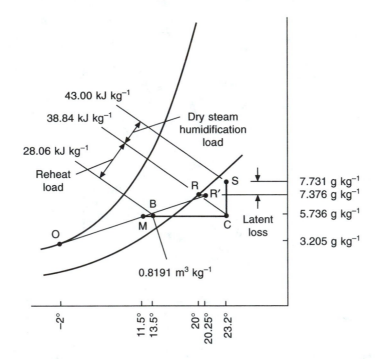

Fig. 8.4 Psychrometry for example 8.3. The enthalpy at *R* is almost the same as at *C* by coincidence.

$$t_s = 20° + \frac{473}{117.7} \times \frac{(273 + 13.5)}{358} = 23.2°C$$

The motorised mixing dampers in the air handling plant adjust the proportions of fresh and recirculated air to give a mixed temperature of 11.5°C. The fan power then causes this to rise to 13.5°C. This value is used for the temperature in the bracket in the above calculation because the reheater is after the fan. Hence the reheater must warm the air from 13.5°C to 23.2°C.

Referring to Figure 8.4, it is necessary to establish the enthalpies of the states *B* and *C*, and the specific volume at state *B* (for the airflow handled by the supply fan).

$$g_c = g_b = g_m = 3.205 + \frac{(7.376 - 3.205) \times (11.5 + 2)}{(20.25 + 2)}$$

$$= 5.736 \text{ g kg}^{-1}$$

Since the temperature of *B* is 13.5°C its enthalpy can be determined from tables, or from a chart, or by equation (2.24):

$$h_b = (1.007 \times 13.5 - 0.026) + 0.005\,736(2501 + 1.84 \times 13.5)$$
$$= 28.06 \text{ kJ kg}^{-1}$$

The temperature at state *C* is 23.2° and the enthalpy is established in a similar way:

$$h_c = (1.007 \times 24.1 - 0.026) + 0.005\,736(2501 + 1.84 \times 23.2)$$
$$= 38.84 \text{ kJ kg}^{-1}$$

By coincidence, this happens to be almost the same as the enthalpy of the air at the room state and we shall regard it as such, as a simplification.

From psychrometric tables or a chart the specific volume at B is 0.8191 m^3 kg^{-1}. This is for the volumetric airflow rate of 117.7 litres s^{-1}, handled by the fan.

The reheater duty is then

$$\frac{0.1177}{0.8191} \times (38.84 - 28.06) = 1.549 \text{ kW}$$

At an outside winter design condition of –2°C, saturated, the moisture content is 3.205 g kg^{-1}. One air change per hour of infiltrating air at this moisture content will represent a latent heat loss, it being necessary to assume the absence of people under the most exacting conditions. At the winter design state, the moisture content in the rooms is 7.376 g kg^{-1}.

Hence, the duty of the dry steam humidifier is to raise the moisture content of the supply air to a value higher than 7.376 g kg^{-1}, in order to counter the latent loss, which must be calculated.

By equation (7.42)

$$\text{Latent loss} = 0.8 \times 1 \times (2.4 \times 2.6 \times 6.0) \times (7.376 - 3.205)$$
$$= 125 \text{ W}$$

Then, by equation (6.8), the necessary supply air moisture content is

$$g_\text{s} = 7.376 + \frac{125}{117.7} \times \frac{(273 + 13.5)}{856} = 7.731 \text{ g kg}^{-1}$$

The moisture content of the mixed air, at 11.5°C in the air handling plant, has been determined as 5.736 g kg^{-1}. Hence the duty of the dry steam humidifier is to raise the moisture content of air supplied to the room from 5.736 g kg^{-1} to 7.731 g kg^{-1}.

Assuming that dry steam humidification occurs up a dry-bulb line (see section 3.7), the state of the supply air, S, is 23.2°C dry-bulb and 7.731 g kg^{-1}. Its enthalpy can be found from psychrometric tables, or from the psychrometric chart, or by equation (2.24):

$$h_\text{s} = (1.007 \times 23.2 - 0.026) + 0.007\ 731(2501 + 1.84 \times 23.2)$$
$$= 43.00 \text{ kJ kg}^{-1}$$

The dry steam humidification duty is

$$\frac{0.1177}{0.8191} (43.00 - 38.84) = 0.598 \text{ kW}$$

8.5   The use of by-passed air instead of reheat

It is clearly wasteful to cool and dehumidify air and then to cancel some of the sensible cooling capacity to deal with a reduced sensible heat gain in a room. One alternative approach, that avoids some reheating, is to by-pass a portion of the recirculated air around the cooler coil. Figure 8.5 shows the plant arrangement and the psychrometry. If a state S were required for supply to a room it could be achieved by cooling and dehumidifying all the air from a mixture state M to an off-coil state W, and then reheating this to state S. Alternatively, without the use of wasteful reheat, part of the recirculated air could be diverted around the cooler coil, the remaining part mixing with fresh air and forming a new off-coil state, W'. State W' then mixes with state R to give the desired state S.

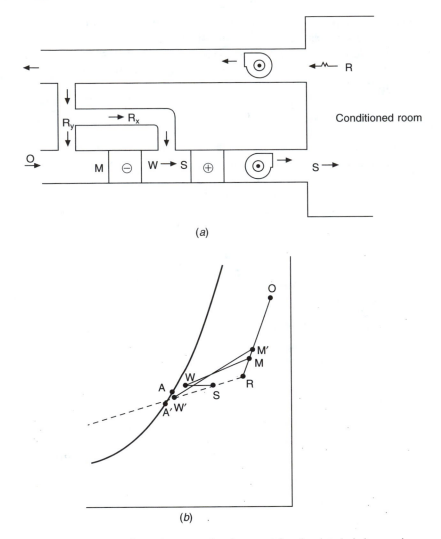

(a)

(b)

Fig. 8.5 Plant arrangement and psychrometry for the case of recirculated air by-passing a cooler coil, ignoring the temperature rise across the supply air fan for simplicity of illustration.

The actual performance is more complicated than this. When some of the recirculated air is by-passing the cooler coil there is less recirculated air available to mix with the fresh air so the mixture state, M', is not the same as the mixture state M. Further, when air is by-passing the cooler coil, less air is flowing through it and this affects performance (see sections 8.6, 10.4 and 10.7).

Although the method has been considered in the past it has not proved popular, for several reasons. The ducting arrangement round the air handling unit to accommodate the by-passed airflow is a complication in plantroom layout and occupies too much room. The capital cost is consequently greater than for other methods. Commissioning the correct airflow proportions may sometimes not be easy. In any case, a reheater is needed for winter operation. Other techniques, such as the use of face and by-pass dampers, have given better results.

8.6   Face and by-pass dampers

This is a method used to vary the output of a cooler coil in order to respond to changes in the sensible heat gain in a conditioned room, without using reheat and without using an automatic motorised valve in the chilled water supply to the coil.

Figure 8.6 illustrates the plant and shows the psychrometry for the simplifying assumption that 100 per cent fresh air is handled. Under design conditions the by-pass dampers are fully closed and all the airflow is over the face of the coil, the dampers of which are fully open. In the usual way, air is cooled and dehumidified from state O to state W. The apparatus dew point is A and the corresponding mean coil surface temperature is t_{sm}. Fan power and duct heat gain cause the air temperature to rise from t_w to t_s. Air at this temperature is supplied to the room in order to maintain the design state R therein.

If the sensible heat gains in the room reduce, the room thermostat sends a signal to the motorised dampers on the cooler coil which causes the face dampers partly to close and the by-pass dampers partly to open. Less air flows over the coil. The two consequences of this are:

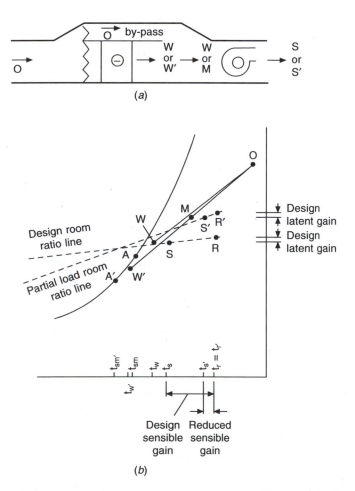

Fig. 8.6   Plant and psychrometry for face and by-pass dampers. 100 per cent fresh air is shown to simplify the psychrometry. The system works equally well with recirculated air.

(i)   The cooling load reduces and so the rise in temperature of the chilled water flowing inside the tubes of the coil is smaller. Hence the mean coil surface temperature falls from t_{sm} to $t_{sm'}$, in Figure 8.6(*b*), and the apparatus dew point, *A*, slides down the saturation curve to become *A'*.

(ii)   With the reduced airflow rate the face velocity over the coil is less and hence the contact factor improves. See section 10.4.

The outcome of this is that the air leaving the face of the coil is drier under partial load conditions, with face and by-pass dampers, than it is under design conditions.

Air at state *W'*, leaving the face of the coil, mixes with air at state *O*, by-passing the coil, to form a mixture state, *M*. Fan power and duct gain cause a temperature rise and the air is finally supplied to the room at a state *S'*. Figure 8.6 shows that state *S'* has the correct temperature to deal with the reduced sensible heat gain without the use of reheat. The humidity in the room will probably rise above the design condition but this is seldom important for comfort conditioning. The fact that the air leaving the coil face is drier than under design operating conditions is helpful in countering the tendency for the room humidity to rise.

EXAMPLE 8.4

If a room suffers sensible gains of 11.70 kW and latent gains of 3.15 kW when 22°C dry-bulb with 50 per cent saturation is maintained therein during outside design conditions of 28°C dry-bulb with 19.5°C wet-bulb (sling), calculate the inside conditions if the sensible gain diminishes to 5 kW, the latent gain remaining at 3.15 kW and the outside condition staying at the design value.

The plant comprises air intake, pre-heater, cooler coil with face and by-pass dampers (used for thermostatic control of room temperature), supply fan handling 1.26 kg per s of fresh air, and the usual distribution ducting. Assume that the performance of the cooler is unaltered by variations in air flow across it and that the temperature rise due to fan power etc. is 3°C.

Answer

It can be calculated that the supply state under design conditions is 13°C dry-bulb with 7.352 g kg^{-1}. The supply temperature for dealing with a sensible gain of 5 kW is approximately 18°C dry-bulb, and so the air temperature needed after the face and by-pass dampers must be 15°C at partial load.

The psychrometric changes involved are illustrated in Figure 8.7.

Assuming that the dry-bulb scale is linear, a good approximate answer is as follows:

$$g_{w'} = g_w + \frac{15° - 10°}{28° - 10°} \times (g_o - g_w)$$

$$= 7.352 + \frac{5}{18} \times (10.65 - 7.352)$$

$$= 7.352 + 0.916$$

$$= 8.268 \text{ g kg}^{-1}$$

$\therefore$ $g_{r'} = g_{w'}$ + moisture pick-up for the design latent load

$$= 8.268 + (8.366 - 7.352)$$

$$= 9.282 \text{ g kg}^{-1}$$

The room relative humidity is about 56 per cent.

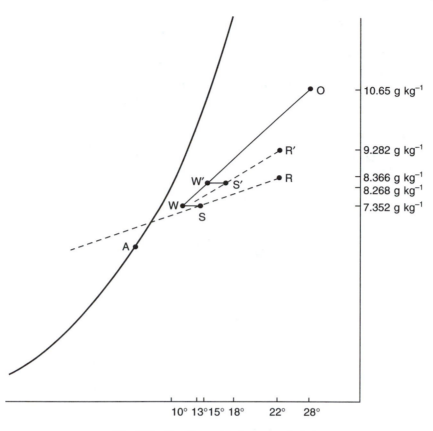

Fig. 8.7 Psychrometry for example 8.4.

8.7 Cooling in sequence with heating

When it is necessary to control only temperature in the conditioned space, reheat can be avoided by arranging that the heater battery is used only when the cooler coil is off. Figure 8.8(*a*) illustrates the control operation: Ra and Rb are three-port motorised mixing valves which are controlled from a room thermostat C, in sequence. As the room temperature falls, C progressively reduces the capacity of the cooler coil, the by-pass port of Ra opening until eventually no chilled water flows through the coil. Upon further fall in room temperature, C progressively increases the capacity of the heater, the by-pass port of Rb closing. Thus, the two work in sequence, and since the heater is never on when the cooler is also on, none of the cooling capacity is ever wasted by cancellation with reheat.

The effect of a reduction in sensible heat gain, while the outside state remains constant, is shown in Figure 8.8(*b*). If a supply temperature $t_{s'}$ is required instead of t_s, then, presuming the latent gains to be unchanged, the relative humidity in the room will rise, state R′ being maintained.

If objectionable increases in humidity are to be prevented, a high limit humidistat is necessary. This overrides the action of C, partly closing the by-pass port of Ra for the purpose of doing some latent cooling. The temperature fall which would result is prevented by C partly closing the by-pass port of Rb. Thus, under these circumstances, the heater is behaving as a reheater rather than as a sequence heater.

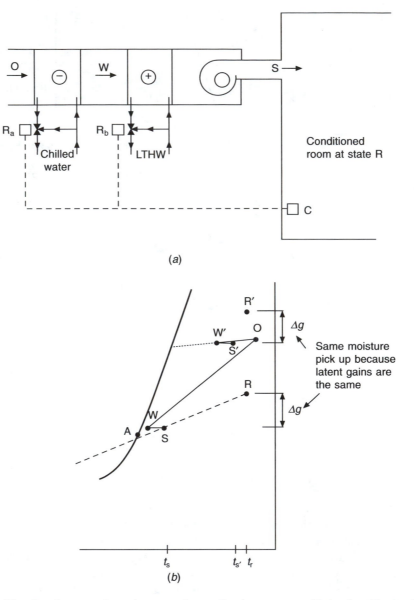

(a)

(b)

Fig. 8.8 The plant layout and psychrometry for cooling in sequence with heating. To simplify the illustration of the psychrometry it is assumed that 100 per cent fresh air is handled. The system works equally well when recirculated and fresh air are mixed, which is usually the case.

8.8 Hot deck–cold deck systems

Certain types of packaged air handling plant offer the facility of providing multiple zone control of temperature without the use of multiple reheaters. In some respects there is a similarity here with double-duct systems.

The type of plant is illustrated in Figure 8.9(*a*). Air is filtered and drawn through a fan which blows the air over a pair of coils, one heating and one cooling, arranged in parallel.

The downstream part of the plant is split by a horizontal dividing plate into two chambers, an upper 'hot deck' and a lower 'cold deck'. The outlets from these decks are dampered, and a number of ducts may accept air from the outlet. Any particular duct outlet can then have air at a controlled temperature by accepting the appropriate proportions of air from the two decks, through the motorised dampers. The particular damper group for one zone would be thermostatically controlled.

Figure 8.9(*b*) illustrates the psychrometry, the temperature rise due to fan power etc., being ignored. Air comes off the hot deck at state *H* and off the cold deck at state C. The two airstreams fed to zone 1 are dampered automatically and mix to form state S_1. This is the correct state to give condition R_1 in zone 1. Similarly H and C mix to give state S_2 for supply to zone 2, maintaining state R_2 therein. It is seen that differences in relative humidity are possible between zones.

As with sequence heating, there is no direct cancellation of cooling, hence account can be taken of cooling load diversity in assessing refrigeration load.

8.9 Double duct cooling load

Whether load diversity may be taken advantage of in estimating the required refrigeration capacity depends on which one of two systems is used.

The most common form of system (Figure 8.10(*a*)) has the cooler coil in the cold duct and the heater battery in the hot duct on the discharge side of the fan. The advantage is that no air is first cooled and then reheated. If a higher supply air temperature than for summer design conditions is required (say t_s rather than t_c), some hot air is mixed with cold air. No wasteful reheat is involved. Ultimately, all the air would be from the hot duct and none from the cold duct. The disadvantage of the system is that the moisture content of the supply air varies. With a supply state S, for the same moisture pick-up in the room, a condition R' rather than R is maintained, as shown in Figure 8.10(*b*).

The system is seldom used because it is expensive in capital cost, in running cost and in its use of building space.

8.10 The load on air–water systems

These systems take account of the diversity of sensible heat gains in the cooling load. Just how much depends on how they are designed; for instance, reheat on the air side can nullify some of the drop in sensible load caused by a fall in outside air temperature. The essence of such systems is that, by having individual chilled water cooler coils in units, such as fan coil, in each conditioned space, reductions in sensible load can be dealt with by throttling the flow of chilled water through local coils.

It must be remembered that the primary air delivered to the units has a sensible cooling capacity that must be regarded as offsetting part of the sensible gain.

8.11 Diversification of load

As was indicated earlier, if reductions in the sensible gain to a conditioned space are dealt with by elevating the supply air temperature through the use of reheat, no corresponding reduction in refrigeration load will occur. All the air supplied is cooled to a fixed dew point and the false load imposed by the reheat equals the reduction in sensible gain to the room.

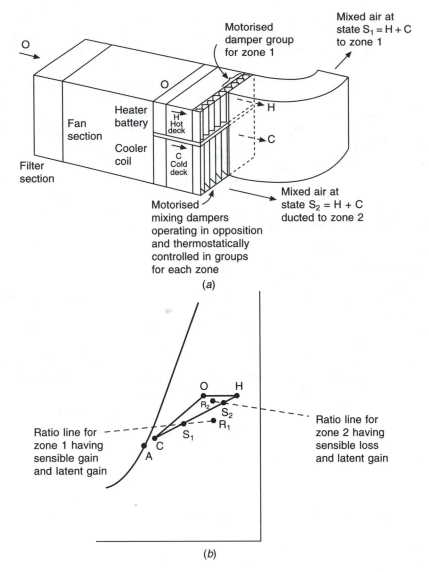

Fig. 8.9   The plant layout and psychrometry for a multi-zone unit.

If methods are adopted such as the use of face and by-pass dampers, which raise the supply air temperature as the sensible gain diminishes without wasting any cooling, then the refrigeration load will not necessarily involve the sum of the maximum sensible gains to two or more rooms conditioned.

If a plant which separates the dehumidifying and fresh-air load from the main bulk of the sensible load is used, then reductions in sensible gain are dealt with by reducing the secondary sensible cooling, rather than by cancelling cooling with reheat.

8.12   Load diagrams

Diagrams can be drawn to give a useful picture of seasonal load variations for multi-

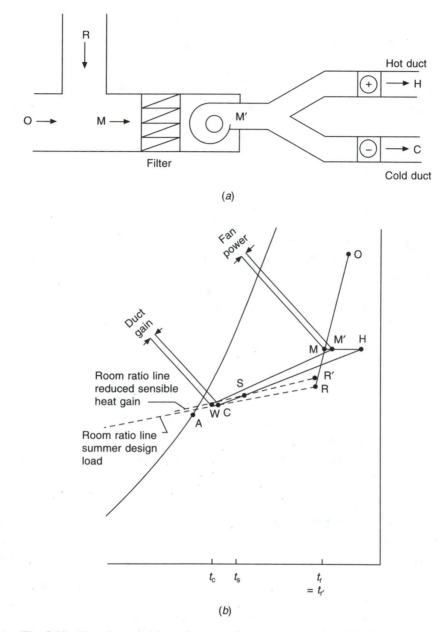

Fig. 8.10   Plant layout and psychrometry for a conventional double duct system.

roomed buildings with several orientations. The sensible heat gain to a room comprises: transmission (T) by air-to-air temperature difference, heat gain from lights (L), heat gain from people (P), heat gain from business machines (M) and solar heat gain through glass (S). For purposes of this simple analysis solar heat gains through walls are ignored. The construction of such load diagrams is conveniently illustrated by the two following examples.

EXAMPLE 8.5

Construct a load diagram for the west-facing module forming the subject of example 7.18, assuming that the natural infiltration rate stays constant at 0.5 air changes per hour throughout the year and that the design outside air temperature in January is –2°C. Take the room temperature as 22°C in July and 20°C in January. Make use of Tables 7.9 and 7.10, instead of CIBSE data, for solar gain through glass, Assume metal-framed single glazing, protected by internal Venetian blinds, and a building surface density of 150 kg m^{-2}. For simplicity, and without significant error, ignore the effects of solar gain through the wall and hence use equation (7.17) for both the glass and the wall.

Answer

Table 5.2 shows that the maximum temperature is 26.9°C in July at 15.00 h, sun-time, with a diurnal range of: 21.8 – 13.5 = 8.3°. Allow 1.1° for the heat island effect in the middle of London (see section 5.11). Hence take the temperature as 28°C at 15.00 h sun-time in July. On the other hand, Table 7.9 shows that the maximum solar gain through glass is at 17.00 h sun-time in July. This indicates that the maximum sensible gain will almost certainly be at 17.00 h sum-time in July. (This could be verified by separate calculation.) On this basis it is necessary to determine the outside air temperature at 17.00 h, using equation (5.4):

$$t_{17} = 28 - (8.3/2)[1 - \sin\{(17\pi - 9\pi)/12\}]$$
$$= 27.4°C$$

Hence the sensible heat gains are calculated at 17.00 h in July:

		watts
Glass (equation (7.17)): 2.184 × 3.0 × (27.4 – 22)	=	35
Wall (equation (7.16)): (3.3 × 2.4 – 2.184) × 0.45 × (27.4 – 22)	=	14
Infiltration (equation (7.41)): 0.33 × 0.5 × (2.6 × 2.4 × 6.0) × (27.4 – 22)	=	33
	$T =$	82
People: 2 × 90	$P =$	180
Lights: 17 × 2.4 × 6.0	$L =$	245
Machines: 20 × 2.4 × 6.0	$M =$	288
Solar (equation (7.41)): 2.184 × 1.0 × 268	$S =$	585
	$T + P + L + M + S =$	1380

This allows the points 1, 2, 3, 4 and 5 to be plotted on Figure 8.11.

From T we can determine that $\Sigma(AU) = 82/5.4 = 15.2$ W K^{-1}. Hence the heat loss in January is 15.2 × (20 + 2) = 334 W. This establishes the point 6 on the figure. (Note that the heat loss is less than was calculated in the answer to example 8.3 because that was based on an infiltration rate of one air change per hour in winter. As a simplification, the infiltration is kept constant at half an air change per hour in this example.) A line joining the points 1 and 6 is drawn and this represents the transmission load, T, against outside air temperature. The contributions of P, L and M are constants and load lines representing $T + P$, $T + P + L$ and $T + P + L + M$ are drawn parallel to the line for T.

The contribution of the solar gain to the diagram needs some thought. The diagram is intended to show seasonal rather than diurnal variations of maximum load and hence the peak solar gains for each month of the year are relevant. The mean monthly maximum temperatures are not directly related to solar gains and they are not indicative of winter conditions, but they do offer a way of associating solar gains with outside air temperature and so drawing a line on the figure for the solar contribution. Reference to Tables 5.2 (for temperatures plus 1.1° for the heat island effect) and 7.9 leads to the following tabulation, using equation (5.4) to determine t_{17}.

Month	Jan	Feb	Mar	Apr	May	Jun	Jul	Aug	Sep	Oct	Nov	Dec
q_s	104	174	228	259	268	268	268	259	228	174	104	79
S	227	380	498	566	585	585	585	566	498	380	227	173
t_{max}	12.8	13.2	16.6	19.8	24.4	27.0	28.0	27.3	24.5	19.8	15.5	13.3
D	4.1	4.7	6.8	7.8	8.5	8.7	8.3	8.2	7.2	6.3	4.8	3.8
t_{17}	12.5	13.0	16.1	19.3	23.8	26.4	27.4	26.8	24.0	19.4	15.2	13.0

where q_s = specific solar gain (Table 7.9), W m^{-2}

S = solar gain at 17.00 h sun-time for west-facing glass, W

t_{max} = mean monthly maximum temperatrue (Table 5.4), plus an allowance of 1.1° for the heat island effect, °C

D = diurnal range (Table 5.2), K

t_{17} = outside air temperature at 17.00 h (equation (5.4)), °C

Plotting these values of the solar component at 17.00 h for a west-facing window gives the points through which a curved broken line runs in Figure 8.11. This is unsatisfactory. To deal with the difficulty it is suggested that the solar gain for January should be associated with the outside design temperature for January, namely, –2°C. Referring as necessary to the answer of example 8.5, the following establishes the point 7 in Figure 8.11.

$$P + L + M = 180 + 245 + 288 = \quad 713$$
$$T \text{ (Jan)} = 15.2 \times (-2 - 20) \quad = -334$$
$$S \text{ (Jan)} \qquad\qquad\qquad = \quad 227$$
$$T + P + L + M + S \text{ (Jan)} \quad = \quad 606 \text{ W net heat gain.}$$

The points 5 and 7 are joined to give the load line for $T + P + L + M + S$.

Although clear skies and sunshine are associated with cold, dry weather in winter, the altitude of the sun is low and cloud cover prevails for much of the time, making an accurate forecast of the solar gain impossible. For this reason, the simple straight lines in Figure 8.11 are proposed as a reasonable load diagram.

It is to be noted that there are three reasons why there is a significantly high sensible heat gain in January:

(i)   The U-values for the walls and glass are low.
(ii)   The infiltration rate is only 0.5 air changes per hour (as might be expected with a tight building).
(iii)   Business machines are liberating 20 W m^{-2}. (This could easily be greater.)

EXAMPLE 8.6

Construct a load diagram for the module in example 8.5, assuming a southerly orientation.

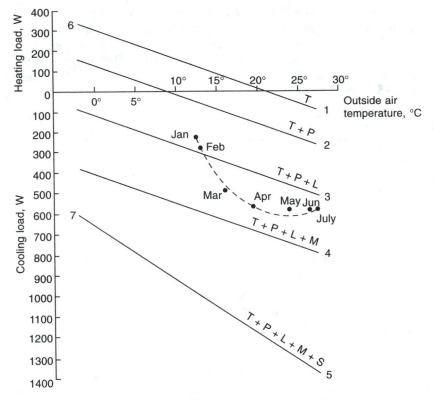

Fig. 8.11   Load diagram—western orientation.

Answer

Reference to Table 7.9 shows that the peak solar gain for a southern aspect is 284 W m⁻², occurring at 12.00 h sun-time in October/February. In fact, for a lightweight building, the maximum solar gain nearly always occurs at mid-day, for a south-facing window. This complicates the construction of a load diagram and an investigation must be made of the heat gains by transmission (T) and solar gain (S). Since the room temperature is allowed to fall from 22°C in July to 20°C in January, account must be taken of this and it is proposed that a simple proportional change is used. Knowing that $\Sigma(AU)$ for the module is 15.2 W m⁻² and adding 1.1° to the values of mean monthly maximum temperature taken from Table 5.2, the following tabulation is drawn up, using equation (5.4) as necessary to determine temperatures at mid-day.

Month	t_r °C	Mean monthly max. temp at 15.00 h °C + 1.1°	t_o at 12.00 h °C	T W	q_s at 12.00 h W m⁻²	S W	$T + P + L + M + S$ W
Dec	20.3	13.3	12.7	−116	240	524	1121
Jan	20	12.8	12.2	−119	262	572	1166
Feb	20.3	13.2	12.5	−119	284	620	1214
Mar	20.7	16.6	15.6	−78	268	585	1220
Apr	21	19.8	18.7	−35	234	511	1189
May	21.3	24.4	23.2	+29	180	393	1135
Jun	21.7	27.0	25.7	+61	158	345	1119
Jul	22	28.0	26.8	+73	180	393	1179

The points 1, 2, 3 and 4 are plotted in Figure 8.12, in the same way as before. The maximum value for $T + P + L + M + S$ is 1179 W in July and this identifies the point 5, against an outside air temperature of 26.8°C at 12.00 h. If the contribution by $P + L + M$ is smaller, the solar gain is more dominant and the maximum value for $T + P + L + M + S$ is in February or March, when the solar gain, S, is a larger proportion of the total. This is seen in Figure 8.12 by the plot of solar gains alone, emphasised by the curved broken line.

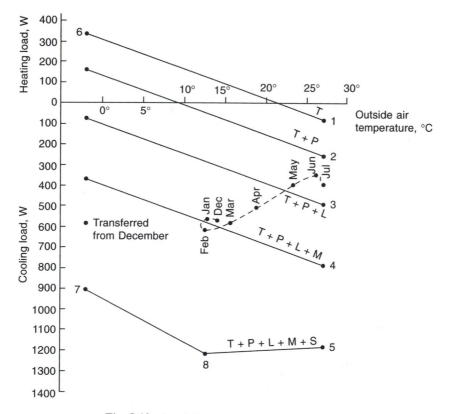

Fig. 8.12   Load diagram—southern orientation.

The lowest winter solar gain is in December. If it is assumed that, on a cold clear day in this month, the outside air temperature is −2°C, then the December solar gain can be added to the load line for $T + P + L + M$ at −2°C, to give the point 7 on the diagram.

The peak value for the load diagram will occur when the maximum value of the solar gain is added to $T + P + L + M$. This is done in Figure 8.12 by adding the value of S, in February, to the value of $T + P + L + M$ against an outside air temperature of 12.5°C (for February). In Figure 8.12 this identifies the point 8 (with a cooling load of 1214 W) and the load line for $T + P + L + M + S$ is established by joining the points 5, 8 and 7.

Exercises

1.   Air enters a cooler coil at a state of 32°C dry-bulb, 20.7°C wet-bulb (sling) and leaves at 11°C dry-bulb, 10.3°C wet-bulb (sling). The airflow rate at entry to the coil is 4.72 m³ s⁻¹. Determine: (*a*) the cooling load in kW, (*b*) the rate of moisture condensation in litres per minute, (*c*) the airflow rate leaving the coil in m³ s⁻¹.

Answers

(*a*) 155.2 kW, (*b*) 0.942 litres per minute, (*c*) 4.37 m³ s⁻¹.

2. An air conditioning plant supplies air to a room at a rate of 1.26 kg s⁻¹, the supply air being made up of equal parts by weight of outdoor and recirculated air. The mixed air passes through a cooler coil fitted with a by-pass duct, face-and-by-pass dampers being used for humidity control. The air then passes through an after-heater and supply fan. In a test on the plant, the following conditions were measured:

	Dry-bulb °C	Moisture content g kg⁻¹	Enthalpy kJ kg⁻¹
Outdoor air	27	12.30	58.54
Room air	21	7.86	41.08
Supply air	16	7.07	33.99
Air leaving cooler coil	8	6.15	23.51

Make neat sketches showing the plant arrangement and the psychrometric cycle and identify corresponding points on both diagrams. Show on the plant arrangement diagram the controls required to maintain the desired room condition.

Ignoring any temperature rise through the supply fan calculate the mass flowrate of air through the cooler coil, and the cooler coil and heater battery loads, in kW.

Answers

0.963 kg s⁻¹, 25.3 kW, 5.43 kW.

3. The air in a room is to be determined at 22°C dry-bulb and 50 per cent saturation by air supplied at a temperature of 12°C. The design conditions are as follows:

Sensible heat gain   6 kW
Latent heat gain   1.2 kW
Outside condition   32° dry-bulb, 24° wet-bulb (sling)

The ratio of recirculated air to fresh air is fixed at 3:1 by weight. The plant consists of a direct expansion cooler coil, a reheater and a constant-speed fan. Allowing 1°C rise for fan power etc., calculate:

(*a*)   The supply air quantity in m³ s⁻¹ and its moisture content in g kg⁻¹.
(*b*)   The load on the refrigeration plant in kW of refrigeration.
(*c*)   The cooler coil contact factor.

Answers

(i) 0.478 m³ s⁻¹, 7.53 g per kg, (ii) 11.96 kW, (iii) 0.86.

4. A room is air conditioned by a system which maintains 22°C dry-bulb with 50 per cent saturation inside when the outside state is 28°C dry-bulb with 19.5°C wet-bulb (sling), in the presence of sensible and latent heat gains of 44 kW and 2.9 kW respectively. Fresh air flows over a cooler coil and is reduced in state to 10°C dry-bulb and 7.046 g per kg. It is then mixed with recirculated air, the mixture being handled by a fan, passed over another cooler coil and sensibly cooled to 13°C dry-bulb. The air is then delivered to the conditioned room.

If the fresh air is to be used for dealing with the whole of the latent gain and if the effects of fan power and duct heat gain are ignored, determine the following.

(a) The amount of fresh air handled in $m^3 s^{-1}$ at the outside state and in kg s^{-1}.
(b) The amount of air supplied to the conditioned space in $m^3 s^{-1}$ at the supply state and in kg s^{-1}
(c) The dry-bulb temperature, enthalpy and moisture content of the air handled by the fan.
(d) The load, in kW, involved in cooling and dehumidifying the outside air.
(e) The load in kW of refrigeration on the sensible cooler coil.

Answers

(a) 0.7725 $m^3 s^{-1}$, 0.891 kg s^{-1}, (b) 3.91 $m^3 s^{-1}$, 4.76 kg s^{-1}, (c) 19.8°C, 40.46 kJ kg^{-1}, 8.118 g per kg, (d) 24.5 kW, (e) 34.1 kW.

Notation

Symbol	Description	Unit
A	area	m^2
D	diurnal range	K
f_p	factor for pump power and chilled water pipe heat gain	–
g_b	moisture content at state B	g kg^{-1}
g_c	moisture content at state C	g kg^{-1}
g_m	moisture content at state M	g kg^{-1}
g_o	moisture content at state O	g kg^{-1}
g_r	moisture content at state R	g kg^{-1}
$g_{r'}$	moisture content at state R'	g kg^{-1}
g_s	moisture content at state S	g kg^{-1}
g_w	moisture content at state W	g kg^{-1}
$g_{w'}$	moisture content at state W'	g kg^{-1}
h_b	enthalpy at state B	kJ kg^{-1}
h_c	enthalpy at state C	kJ kg^{-1}
h_m	enthalpy at state M	kJ kg^{-1}
h_s	enthalpy at state S	kJ kg^{-1}
L	sensible heat gain from lighting	W
M	sensible heat gain from business machines	W
P	sensible heat gain from people	W
Q_{fa}	fresh air load	W
Q_l	latent heat gain	kW
Q_r	design refrigeration load	kW
Q_{ra}	return air load	kW
Q_{rh}	reheat load	kW
Q_s	sensible heat gain	kW

Q_{sd}	supply duct heat gain	kW
Q_{sf}	supply fan power	kW
q_s	specific solar heat gain	W m^{-2}
S	sensible heat gain from solar radiation through glass	W
T	transmission heat gain through the building envelope, including infiltration, by virtue of air-to-air temperature difference	W
t_c	temperature at state C	°C
t_m	temperature at state M	°C
t_{max}	mean monthly maximum temperature plus 1.1 K for the heat island effect in a built-up area	°C
t_o	outside air temperature at state O	°C
t_r	room temperature at state R	°C
$t_{r'}$	room temperature at state R'	°C
t_s	supply air temperature at state S	°C
$t_{s'}$	supply air temperature at state S'	°C
t_{sm}	mean coil surface temperature for state A	°C
$t_{sm'}$	mean coil surface temperature for state A'	°C
t_w	off-coil temperature for state W	°C
$t_{w'}$	off-coil temperature for state W'	°C
U	overall thermal transmittance coefficient	W m^{-2} K^{-1}
$\dot{v}_t$	volumetric airflow rate at a temperature t	m^3 s^{-1} or litres s^{-1}
β	contact factor of a cooler coil	–

9

The Fundamentals of Vapour Compression Refrigeration

9.1 The basis of vapour compression refrigeration

If a liquid is introduced into a vessel in which there is initially a vacuum and whose walls are kept at a constant temperature it will evaporate at once. In the process the latent heat of vaporisation will be abstracted from the sides of the vessel. The resulting cooling effect is the starting point of the refrigeration cycle, which is to be examined in this chapter.

As the liquid evaporates, the pressure inside the vessel will rise until it reaches a certain maximum value for the temperature—the saturation vapour pressure (see section 2.8). After this, no more liquid will evaporate and, of course, the cooling effect will cease. Any further liquid introduced will remain in liquid state in the bottom of the vessel. If we now remove some of the vapour from the container, by connecting it to the suction of a pump, the pressure will tend to fall, and this will cause more liquid to evaporate. In this way, the cooling process can be rendered continuous. We need a suitable liquid, called the *refrigerant*; a container where the vaporisation and cooling take place, called the *evaporator*, and a pump to remove the vapour, called, for reasons which will be apparent later, the *compressor*.

The system as developed so far is obviously not a practical one because it involves the continuous consumption of refrigerant. To avoid this it is necessary to convert the process into a cycle. To turn the vapour back into a liquid it must be cooled with whatever medium is on hand for the purpose. This is usually water or air at a temperature substantially higher than the temperature of the medium being cooled by the evaporator. The vapour pressure corresponding to the temperature of condensation must, therefore, be a good deal higher than the pressure in the evaporator. The required step-up in pressure is provided by the pump acting as a *compressor*.

The liquefaction of the refrigerant is accomplished in the *condenser*, which is, essentially, a container cooled externally by air or water. The hot high-pressure refrigerant gas from the compressor is conveyed to the condenser and liquefies therein. Since there is a high gas pressure in the condenser, and the liquid refrigerant there is under the same pressure, it is easy to complete the cycle by providing a needle valve or other regulating device for injecting liquid into the evaporator. This essential component of a refrigerant plant is called the *expansion valve*.

This basic vapour compression refrigeration cycle is illustrated in Figure 9.1 where it is shown applied to a water-chilling set.

Figure 9.2 illustrates the changes in the state of the refrigerant as the simple, basic

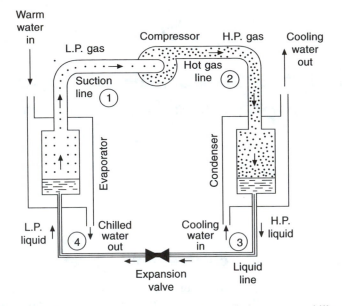

Fig. 9.1   Basic vapour compression cycle applied to a water chiller.

vapour compression cycle takes place. The co-ordinates of the diagram are absolute pressure and enthalpy but, in order to explain the cycle and describe the diagram, it is first necessary to consider the thermodynamics of the subject.

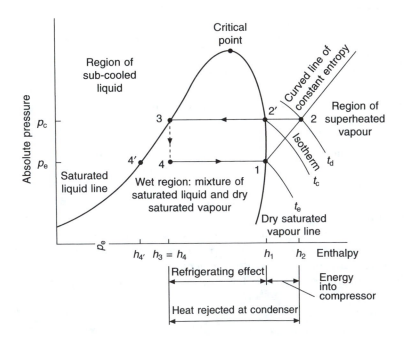

Fig. 9.2   Simple saturation refrigeration cycle on a pressure–enthalpy diagram.

9.2   Thermodynamics and refrigeration

The subject of thermodynamics was dealt with in part, in chapter 2, where it referred to the physics of air–water vapour mixtures. It is necessary to reconsider the topic here, as it refers to the behaviour of refrigerants in vapour compression cycles of refrigeration, in order to deal with such cycles quantitatively. As a preamble to this, some principles and definitions must be reviewed.

(*a*) *Thermodynamics*. This is simply the study of changes involving energy but it is also defined by ASHRAE (1997) as the study of energy, its transformations, and its relation to states of matter.

(*b*) *Thermodynamic system*. This is defined by ASHRAE (1997) as '... a region of space or quantity of matter, bounded by a closed surface'. There are two sorts of systems to be considered: closed systems, where there is no interchange of matter with the surroundings, and open systems where there is such an interchange. With a closed system the mass within the boundary of the system is constant as, for example, in a hermetic refrigeration plant. With an open system there is a mass flow through the system boundaries, for example with a pumping process that introduces a fluid to the system from the surroundings.

(*c*) *The first law of thermodynamics*. Otherwise interpreted as the conservation of energy, this law states that energy can neither be created nor destroyed.

For an open system, under steady-state conditions and with unit mass flow rate of a pure substance, the first law is expressed by the steady flow energy equation:

$$\left(h + \frac{v^2}{2} + gz \right) + (Q - W) = 0 \tag{9.1}$$

where h = enthalpy   J kg^{-1}
        v = velocity   m s^{-1}
        g = local acceleration due to gravity   m s^{-2}
        z = elevation above a reference level   m
        Q = rate of heat transfer to the system   W
        W = rate of work done by the system   W

Note that equation (2.19) offers an alternative expression of enthalpy.

(*d*) *The second law of thermodynamics*. In simple terms, this states that heat only flows from a higher temperature to a lower temperature. More formally, Spalding and Cole (1961) state: it is impossible for a system working in a cycle to have, as its sole effect, the transfer of heat from a system at a low temperature to a system at a high temperature.

(*e*) *Heat*. Energy has been described by ASHRAE (1997) as a capacity for producing an effect and it can be in a stored form or in a transient form. Stored energy is exemplified by such concepts as potential energy and kinetic energy, whereas heat is a form of transient energy. Heat can be defined as an interaction between two systems of differing temperatures and the flow is always from the higher to the lower temperature.

(*f*) *Work*. This is an aspect of energy. Work is the application of a force through a distance, transferring energy across the boundary between two systems.

(*g*) *Entropy*. This is a concept that is of value when analysing the behaviour of a thermodynamic system. It is expressed in terms of change in entropy, defined as the quantity of heat crossing the boundary of a reversible system, divided by the absolute temperature of the system, and is given by

$$\Delta s = \int \frac{\mathrm{d}q}{T} \qquad (9.2)$$

where s = specific entropy in kJ kg^{-1} K^{-1}

  q = heat energy in kJ kg^{-1}

  T = absolute temperature in K

Entropy is a property of the system and it depends on the state of the substance. For a pure substance, its value can be established since it depends on two other independent properties, heat and absolute temperature. Since it is defined as a difference, in equation (9.2), an arbitrary zero must be adopted if it is to be tabulated. This is usually at a temperature of zero degrees absolute.

Entropy can also be considered in terms of the disorder of the molecules in a system: if they are disordered the entropy is greater than if they are in some sort of order. Alternatively, it can be regarded as the availability of a given amount of heat: in equation (9.2) it is seen that if the entropy change is small, the absolute temperature must be large, for the given change of heat. Hence, it is possible to construct an absolute temperature–entropy diagram (see Figures 9.4 and 9.5), in which areas represent heat.

A process which occurs at constant entropy is termed isentropic.

(*h*) *Reversibility*. A reversible process is one which, after completion, has returned both the system and its surroundings to their original states. This is not true for an irreversible process, an example of which is any process involving friction.

If there were no heat exchange with the surroundings and no internal frictional losses, a reciprocating machine could either act as a compressor or as an expansion engine. When acting as a compressor the supply of power to the machine would be used to compress the gas handled. When acting as an expansion machine the pressure difference across the inlet and outlet ports would allow the gas handled to expand, driving the machine. The machine would liberate to the surroundings the same power that was taken from them when the machine acted as a compressor. The processes of compression and expansion would then be reversible.

It is possible to prove that the efficiency of a reversible engine is always greater than that of an irreversible engine, operating between the same two heat reservoirs. Hence it is desirable that refrigeration compressors should execute reversible compression, as far as possible.

(*i*) *Adiabatic processes*. If a system is isolated from its surroundings as regards heat transfer, the processes performed are termed adiabatic. Thus, an adiabatic process is one in which no heat is supplied or rejected.

All reversible, adiabatic processes are isentropic. The converse may not be true: all isentropic processes are not necessarily reversible and adiabatic.

The line joining the points 1 and 2 in Figure 9.2 depicts a reversible, adiabatic, isentropic process of compression from an evaporating pressure, p_e, to a condensing pressure, p_c.

(*j*) *Saturated liquid*. This is a substance existing at its saturated temperature and pressure. If the pressure falls, the substance can no longer exist as a saturated liquid and some of it flashes to vapour, with a corresponding fall in the temperature of the parent liquid, until it is again a saturated liquid, existing at a lower saturated temperature and pressure.

(k) *Throttling expansion.* This is an irreversible, adiabatic process of expansion that occurs at constant enthalpy, no heat being supplied or rejected and no work being done. It is the process occurring when liquid refrigerant flows through an expansion valve: the loss of pressure resulting from the frictional resistance to fluid flow causes some of the saturated liquid to flash to vapour, with a corresponding fall in temperature. The broken line joining the points 3 and 4 in Figure 9.2 represents a process of throttling expansion from the condensing pressure, p_c, to the evaporating pressure, p_e. It is customary to show the process by a broken line because it is irreversible.

(l) *Sub-cooled liquid.* This is liquid existing at a temperature less than the saturation temperature for the prevailing pressure. Figure 9.2 shows the region of sub-cooled liquid.

(m) *Wet vapour.* This is a mixture of saturated liquid and saturated vapour. The quality of the mixture is expressed in terms of its dryness fraction, defined as the mass of saturated vapour divided by the total mass of saturated vapour and saturated liquid. The point 4 in Figure 9.2 represents wet vapour: it is a mixture of saturated liquid at state 4′ and dry saturated vapour at state 1. The state of the wet vapour entering the evaporator can be defined in terms of the relevant enthalpies and its dryness fraction, f, at state 4, is given by

$$f = \frac{h_4 - h_{4'}}{h_1 - h_{4'}} = \frac{h_3 - h_{4'}}{h_1 - h_{4'}} \tag{9.3}$$

(n) *Dry saturated vapour.* This is the fluid existing as a vapour, without the presence of any saturated liquid, at its saturation vapour pressure. The points 1 and 2′ in Figure 9.2 represent states of dry saturated vapour, at saturated vapour pressures of p_e and p_c, respectively.

(o) *Superheated vapour.* This is the fluid existing as a vapour at a temperature greater than the saturated temperature for the pressure prevailing. The point 2 in Figure 9.2 is a state of superheated vapour.

In the following examples (9.1–9.6) R134a is used as the refrigerant but any other, single substance refrigerant would be dealt with using similar principles. Example 9.9 repeats these examples using ammonia as a refrigerant which, because of the format of Table 9.1, is an easier calculation. The treatment is likely to be different when mixtures of refrigerants are used.

EXAMPLE 9.1

An air conditioning plant using Refrigerant 134a has evaporating and condensing temperatures of 0°C and 35°C, respectively. Determine the dryness fraction of the vapour entering the evaporator.

Answer

Using the notation of Figure 9.2, refer to Table 9.1 and interpolate as necessary to determine the saturated liquid leaving the condenser at state 3:

$t_c = 35°C$

$p_c = 887.11 \text{ kPa}$

$h_3 = 248.94 \text{ kJ kg}^{-1}$

Table 9.1   Saturated properties of R134a

Temperature °C	Absolute pressure kPa	Specific heats		Vapour volume m³ kg⁻¹	Liquid enthalpy kJ kg⁻¹	Vapour enthalpy kJ kg⁻¹	Vapour entropy kJ kg⁻¹ K⁻¹
		Liquid kJ kg⁻¹ K⁻¹	Vapour				
− 10	200.52	1.306	0.842	0.099 63	186.78	392.75	1.7337
− 8	216.84	1.312	0.850	0.092 46	189.40	393.95	1.7323
− 6	234.18	1.317	0.858	0.085 91	192.03	395.15	1.7310
− 4	252.57	1.323	0.866	0.079 91	194.68	396.33	1.7297
− 2	272.06	1.329	0.875	0.074 40	197.33	397.51	1.7285
0	292.69	1.335	0.883	0.069 35	200.00	398.68	1.7274
2	314.50	1.341	0.892	0.064 70	202.68	399.84	1.7263
4	337.55	1.347	0.901	0.060 42	205.37	401.00	1.7252
6	361.86	1.353	0.910	0.056 48	208.08	402.14	1.7242
8	387.49	1.360	0.920	0.052 84	210.80	403.27	1.7233
10	414.49	1.367	0.930	0.049 48	213.53	404.40	1.7224
12	442.89	1.374	0.939	0.046 36	216.27	405.51	1.7215
14	472.76	1.381	0.950	0.043 48	219.03	406.61	1.7207
16	504.13	1.388	0.960	0.040 81	221.80	407.70	1.7199
18	537.06	1.396	0.971	0.038 33	224.59	408.78	1.7191
20	571.59	1.404	0.982	0.036 03	227.40	409.84	1.7183
22	607.77	1.412	0.994	0.033 88	230.21	410.89	1.7176
24	645.66	1.420	1.006	0.031 89	233.05	411.93	1.7169
26	685.31	1.429	1.018	0.030 03	235.90	412.95	1.7162
28	726.76	1.438	1.031	0.028 29	238.77	413.95	1.7155
30	770.08	1.447	1.044	0.026 67	241.65	414.94	1.7149
32	815.30	1.457	1.058	0.025 16	244.55	415.90	1.7142
34	862.50	1.467	1.073	0.023 74	247.47	416.85	1.7135
36	911.72	1.478	1.088	0.022 41	250.41	417.78	1.7129
38	963.01	1.489	1.104	0.021 16	253.37	418.69	1.7122
40	1016.5	1.500	1.120	0.019 99	256.35	419.58	1.7115
42	1072.1	1.513	1.138	0.018 90	259.35	420.44	1.7108
44	1130.0	1.525	1.156	0.017 86	262.38	421.28	1.7101
46	1190.1	1.539	1.175	0.016 89	265.42	422.09	1.7094
48	1252.7	1.553	1.196	0.015 98	268.49	422.88	1.7086
50	1317.7	1.569	1.218	0.015 11	271.59	423.63	1.7078
52	1385.2	1.585	1.241	0.014 30	274.71	424.35	1.7070
54	1455.3	1.602	1.266	0.013 53	277.86	425.03	1.7061
56	1528.0	1.621	1.293	0.012 80	281.04	425.68	1.7051
58	1603.3	1.641	1.322	0.012 12	284.25	426.29	1.7041
60	1681.5	1.663	1.354	0.011 46	287.49	426.86	1.7031

Data reproduced with permission of the American Society of Heating, Refrigerating and Air-Conditioning Engineers from the (1997) *ASHRAE Handbook.*

For the mixture of saturated liquid and saturated vapour leaving the expansion valve at state 4:

$$t_e = 0°C$$
$$p_e = 292.69 \text{ kPa}$$
$$h_4 = h_3 = 248.94 \text{ kJ kg}^{-1}$$

Table 9.2   Superheated properties of R134a

Temperature °C	Absolute pressure kPa	Vapour volume m^3 kg^{-1}	Liquid enthalpy kJ kg^{-1}	Vapour enthalpy kJ kg^{-1}	Vapour entropy kJ kg^{-1} K^{-1}
Saturated					
– 10.07	200	0.099 90	186.69	392.71	1.7337
Superheated					
– 10	200	0.099 90		392.77	1.7339
– 5	200	0.102 36		396.99	1.7496
0	200	0.104 82		401.21	1.7654
5	200	0.107 12		405.47	1.7808
10	200	0.109 53		409.73	1.7961
Saturated					
8.94	400	0.051 23	212.08	403.80	1.7229
Superheated					
10	400	0.051 52		404.78	1.7263
15	400	0.052 86		409.39	1.7423
20	400	0.054 20		414.00	1.7583
25	400	0.055 50		418.60	1.7738
30	400	0.056 79		423.21	1.7892
Saturated					
21.58	600	0.034 33	229.62	410.67	1.7178
Superheated					
25	600	0.035 00		414.04	1.7290
30	600	0.035 98		418.97	1.7455
35	600	0.036 92		423.84	1.7614
40	600	0.037 86		428.72	1.7772
45	600	0.038 74		433.58	1.7924
50	600	0.039 67		438.44	1.8077
Saturated					
31.33	800	0.025 65	243.58	415.58	1.7144
Superheated					
35	800	0.026 24		419.40	1.7268
40	800	0.027 04		424.61	1.7437
45	800	0.027 80		429.73	1.7598
50	800	0.028 55		434.85	1.7758
Saturated					
39.39	1000	0.020 34	255.44	419.31	1.7117
Superheated					
40	1000	0.020 43		419.99	1.7139
45	1000	0.022 12		415.45	1.7310
50	1000	0.021 81		430.91	1.7482
Saturated					
46.32	1200	0.016 74	265.91	422.22	1.7092
Superheated					
50	1200	0.017 21		426.51	1.7226
55	1200	0.017 81		432.17	1.7398
60	1200	0.018 41		437.83	1.7571

Data reproduced with permission of the American Society of Heating, Refrigerating and Air-Conditioning Engineers from the *ASHRAE Handbook* (1997).

Table 9.3   Saturated properties of R717, ammonia (NH$_3$)

Temperature °C	Absolute pressure kPa	Liquid volume m^3 kg^{-1}	Vapour volume m^3 kg^{-1}	Liquid enthalpy kJ kg^{-1}	Vapour enthalpy kJ kg^{-1}	Liquid entropy kJ kg^{-1} K^{-1}	Vapour entropy kJ kg^{-1} K^{-1}
− 5	3.549	0.001 550	0.3468	158.0	1437.6	0.6297	5.4023
− 4	3.689	0.001 553	0.3343	162.6	1438.7	0.6467	5.3888
− 3	3.834	0.001 556	0.3224	167.2	1439.8	0.6637	5.3753
− 2	3.983	0.001 559	0.3109	171.8	1440.9	0.6806	5.3620
− 1	4.136	0.001 563	0.3000	176.4	1442.0	0.6975	5.3487
0	4.294	0.001 566	0.2895	181.1	1443.1	0.7143	5.3356
1	4.457	0.001 569	0.2795	185.7	1444.2	0.7312	5.3225
2	4.625	0.001 573	0.2698	190.3	1445.2	0.7479	5.3096
3	4.797	0.001 576	0.2606	194.9	1446.3	0.7646	5.2967
4	4.975	0.001 580	0.2517	199.6	1447.3	0.7813	5.2839
5	5.157	0.001 583	0.2433	204.2	1448.3	0.7980	5.2712
6	5.345	0.001 587	0.2351	208.9	1449.2	0.8146	5.2587
7	5.538	0.001 590	0.2273	213.6	1450.2	0.8311	5.2461
8	5.736	0.001 594	0.2198	218.2	1451.1	0.8477	5.2337
9	5.940	0.001 597	0.2126	222.9	1452.1	0.8641	5.2214
10	6.149	0.001 601	0.2056	227.6	1453.0	0.8806	5.2091
11	6.364	0.001 604	0.1990	232.3	1453.9	0.8970	5.1949
12	6.585	0.001 608	0.1926	237.0	1454.8	0.9134	5.1849
13	6.812	0.001 612	0.1864	241.7	1455.6	0.9297	5.1728
14	7.044	0.001 616	0.1805	246.4	1456.5	0.9460	5.1609
15	7.283	0.001 619	0.1748	251.1	1457.3	0.9623	5.1490
16	7.528	0.001 623	0.1693	255.8	1458.1	0.9785	5.1373
17	7.779	0.001 627	0.1641	260.6	1458.9	0.9947	5.1255
18	8.037	0.001 631	0.1590	265.3	1459.7	1.0109	5.1139
19	8.301	0.001 635	0.1541	270.0	1460.4	1.0270	5.1023
20	8.571	0.001 639	0.1494	274.8	1461.2	1.0432	5.0908
21	8.849	0.001 643	0.1448	279.6	1461.9	1.0592	5.0794
22	9.133	0.001 647	0.1405	284.3	1462.6	1.0753	5.0680
23	9.424	0.001 651	0.1363	289.1	1463.3	1.0913	5.0567
24	9.722	0.001 655	0.1322	293.9	1464.0	1.1073	5.0455
25	10.03	0.001 659	0.1283	298.7	1464.6	1.1232	5.0343
26	10.34	0.001 663	0.1245	303.5	1465.2	1.1391	5.0232
27	10.66	0.001 667	0.1208	308.3	1465.9	1.1550	5.0121
28	10.99	0.001 671	0.1173	313.1	1466.4	1.1708	5.0011
29	11.32	0.001 676	0.1139	318.0	1467.0	1.1867	4.9902
30	11.66	0.001 680	0.1106	322.8	1467.6	1.2025	4.9793
31	12.02	0.001 684	0.1075	327.7	1468.1	1.2182	4.9685
32	12.37	0.001 689	0.1044	332.5	1468.6	1.2340	4.9577
33	12.74	0.001 693	0.1014	337.4	1469.1	1.2497	4.9469
34	13.12	0.001 698	0.0986	342.2	1469.6	1.2653	4.9362
35	13.50	0.001 702	0.0958	347.1	1470.0	1.2810	4.9256
36	13.89	0.001 707	0.0931	352.0	1470.4	1.2966	4.9149
37	14.29	0.001 711	0.0905	356.9	1470.8	1.3122	4.9044
38	14.70	0.001 716	0.0880	361.8	1471.2	1.3277	4.8938
39	15.12	0.001 721	0.0856	366.7	1471.5	1.3433	4.8833
40	15.54	0.001 726	0.0833	371.6	1471.9	1.3588	4.8728
41	15.98	0.001 731	0.0810	376.6	1472.2	1.3742	4.8623

(Contd)

Table 9.3 *(Contd)*

Temperature °C	Absolute pressure kPa	Liquid volume m³ kg⁻¹	Vapour volume m³ mg⁻¹	Liquid enthalpy kJ kg⁻¹	Vapour enthalpy kJ kg⁻¹	Liquid entropy kJ kg⁻¹ K⁻¹	Vapour entropy kJ kg⁻¹ K⁻¹
42	16.42	0.001 735	0.0788	381.5	1472.4	1.3897	4.8519
43	16.88	0.001 740	0.0767	386.5	1472.7	1.4052	4.8414
44	17.34	0.001 745	0.0746	391.4	1472.9	1.4206	4.8310
45	17.81	0.001 750	0.0726	396.4	1473.0	1.4360	4.8206
46	18.30	0.001 756	0.0707	401.4	1473.2	1.4514	4.8102
47	18.79	0.001 761	0.0688	406.4	1473.3	1.4668	4.7998
48	19.29	0.001 766	0.0669	411.4	1473.3	1.4822	4.7893
49	19.80	0.001 771	0.0652	416.5	1473.4	1.4977	4.7789
50	20.33	0.001 777	0.0635	421.6	1473.4	1.5131	4.7684

Reproduced from *Thermodynamic Properties of Ammonia* by W.B. Gosney and O. Fabris, with the kind permission of the authors.

For saturated liquid at a temperature of 0°C and a pressure of 292.69 kPa, $h_{4'} = 200.00$ kJ kg⁻¹.

For dry saturated vapour at 0°C and 292.69 kPa, $h_1 = 398.68$ kJ kg⁻¹.

Hence, by equation (9.3)

$$f = (248.94 - 200.00)/(398.68 - 200.00) = 0.25$$

Thus 25 per cent, by weight, of the liquid refrigerant entering the expansion valve flashes to vapour as the pressure drop through the valve occurs. Since the volume of the vapour is much greater than that of the liquid, the space occupied by the vapour is significantly large. Consequently, the mixture of liquid and vapour leaving the expansion valve must be distributed uniformly into the evaporator before the vapour and liquid get a chance to separate. This is because it is the liquid in the mixture that has the ability to provide refrigeration, by absorbing heat through the surfaces of the evaporator and boiling to a saturated vapour at the evaporating pressure.

9.3 The refrigerating effect

In the basic vapour compression cycle of refrigeration shown in Figure 9.2 a mixture of relatively cold, saturated liquid and saturated vapour enters the evaporator. The liquid part of the mixture is boiled to a saturated vapour and leaves the evaporator at state 1 (in Figure 9.2). The refrigerating effect (q_r) is the enthalpy change across the evaporator. Using the notation of Figure 9.2, this is expressed by

$$q_r = (h_1 - h_4) \tag{9.4}$$

EXAMPLE 9.2

(*a*) Calculate the refrigerating effect for the plant used in example 9.1.

(*b*) If the duty is 352 kW of refrigeration, determine the mass flow rate of refrigerant.

(*c*) What is the volumetric flow rate under suction conditions?

Table 9.4   Superheated properties of R717, ammonia (NH₃)

$t - t_s$ K	$t_s = -3°C$ $p = 383.4$ kPa			$t_s = -2°C$ $p = 398.3$ kPa			$t_s = -1°C$ $p = 413.6$ kPa		
	v	h	s	v	h	s	v	h	s
0	0.3224	1439.8	5.3753	0.3109	1440.9	5.3620	0.3000	1442.0	5.3487
5	0.3302	1453.0	5.4237	0.3185	1454.2	5.4105	0.3073	1455.4	5.3973
10	0.3379	1465.9	5.4700	0.3259	1467.1	5.4567	0.3145	1468.3	5.4436

$t - t_s$ K	$t_s = 0°C$ $p = 429.4$ kPa			$t_s = 1°C$ $p = 445.7$ kPa			$t_s = 2°C$ $p = 462.5$ kPa		
	v	h	s	v	h	s	v	h	s
0	0.2895	1443.1	5.3356	0.2795	1444.2	5.3225	0.2698	1445.2	5.3096
5	0.2966	1456.5	5.3842	0.2863	1457.6	5.3712	0.2765	1458.8	5.3584
10	0.3035	1469.5	5.4306	0.2930	1470.7	5.4177	0.2830	1471.9	5.4049

$t - t_s$ K	$t_s = 3°C$ $p = 479.7$ kPa			$t_s = 4°C$ $p = 497.5$ kPa			$t_s = 5°C$ $p = 515.7$ kPa		
	v	h	s	v	h	s	v	h	s
0	0.2606	1446.3	5.2967	0.2517	1447.3	5.2839	0.2433	1448.3	5.2712
5	0.2670	1459.9	5.3456	0.2580	1461.0	5.3329	0.2493	1462.0	5.3203
10	0.2733	1473.1	5.3922	0.2641	1474.2	5.3796	0.2552	1475.4	5.3671

$t - t_s$ K	$t_s = 6°C$ $p = 534.5$ kPa			$t_s = 7°C$ $p = 553.8$ kPa			$t_s = 8°C$ $p = 573.6$ kPa		
	v	h	s	v	h	s	v	h	s
0	0.2351	1449.2	5.2587	0.2273	1450.2	5.2461	0.2198	1451.1	5.2337
5	0.2410	1463.1	5.3078	0.2330	1464.1	5.2954	0.2253	1465.1	5.2831
10	0.2467	1475.5	5.3546	0.2385	1477.6	5.3423	0.2307	1478.7	5.3301

$t - t_s$ K	$t_s = 9°C$ $p = 594.0$ kPa			$t_s = 10°C$ $p = 614.9$ kPa			$t_s = 11°C$ $p = 636.4$ kPa		
	v	h	s	v	h	s	v	h	s
0	0.2126	1452.1	5.2214	0.2056	1453.0	5.2091	0.1990	1453.9	5.1969
5	0.2179	1466.2	5.2708	0.2108	1467.1	5.2587	0.2040	1468.1	5.2466
10	0.2231	1479.8	5.3179	0.2159	1480.8	5.3509	0.2089	1481.9	5.2939

$t - t_s$ K	$t_s = 26°C$ $p = 1034.0$ kPa			$t_s = 27°C$ $p = 1066.0$ kPa			$t_s = 28°C$ $p = 1098.7$ kPa		
	v	h	s	v	h	s	v	h	s
0	0.1245	1465.2	5.0232	0.1208	1465.9	5.0121	0.1173	1466.4	5.0011
5	0.1279	1480.8	5.0749	0.1241	1481.5	5.0640	0.1205	1482.2	5.0531
10	0.1311	1495.8	5.1238	0.1273	1496.6	5.1130	0.1236	1497.4	5.1023
15	0.1343	1510.3	5.1703	0.1304	1511.2	5.1597	0.1266	1512.1	5.1491
20	0.1374	1524.4	5.2149	0.1334	1525.4	5.2044	0.1296	1526.3	5.1939
25	0.1404	1538.2	5.2578	0.1364	1539.2	5.2473	0.1325	1540.3	5.2369
30	0.1434	1551.7	5.2992	0.1393	1552.8	5.2887	0.1353	1553.9	5.2784
35	0.1463	1565.0	5.3392	0.1421	1566.2	5.3288	0.1380	1567.3	5.3185

(Contd)

Table 9.4   *(Contd)*

$t - t_s$ K	$t_s = 29°C$ $p = 1132.2$ kPa			$t_s = 30°C$ $p = 1166.5$ kPa			$t_s = 31°C$ $p = 1201.6$ kPa		
	v	h	s	v	h	s	v	h	s
0	0.1139	1467.0	4.9902	0.1106	1467.6	4.9793	0.1075	1468.1	4.9685
5	0.1170	1482.9	5.0423	0.1137	1483.6	5.0316	0.1104	1484.2	5.0210
10	0.1201	1498.2	5.0916	0.1166	1498.9	5.0811	0.1133	1499.6	5.0706
15	0.1230	1512.9	5.1386	0.1195	1513.8	5.1281	0.1161	1514.6	5.1177
20	0.1259	1527.3	5.1834	0.1223	1528.2	5.1731	0.1189	1529.1	5.1628
25	0.1287	1541.2	5.2265	0.1250	1542.2	5.2163	0.1215	1543.2	5.2060
30	0.1314	1554.9	5.2681	0.1277	1556.0	5.2579	0.1241	1557.0	5.2477
35	0.1341	1568.4	5.3083	0.1303	1569.5	5.2982	0.1267	1570.6	5.2881

$t - t_s$ K	$t_s = 32°C$ $p = 1237.4$ kPa			$t_s = 33°C$ $p = 1274.1$ kPa			$t_s = 34°C$ $p = 1311.6$ kPa		
	v	h	s	v	h	s	v	h	s
0	0.1044	1468.6	4.9577	0.1014	1469.1	4.9469	0.0986	1469.6	4.9362
5	0.1073	1484.8	5.0103	0.1043	1485.4	4.9998	0.1014	1486.0	4.9893
10	0.1101	1500.4	5.0601	0.1070	1501.1	5.0497	0.1041	1501.7	5.0393
15	0.1129	1515.4	5.1074	0.1097	1516.2	5.0971	0.1067	1516.9	5.0869
20	0.1155	1529.9	5.1525	0.1123	1530.8	5.1424	0.1092	1531.7	5.1323
25	0.1181	1544.1	5.1959	0.1148	1545.1	5.1858	0.1117	1546.0	5.1758
30	0.1207	1558.0	5.2377	0.1173	1559.0	5.2277	0.1141	1560.0	5.2177
35	0.1232	1571.7	5.2781	0.1198	1572.7	5.2681	0.1165	1573.8	5.2582

$t - t_s$ K	$t_s = 35°C$ $p = 1349.9$ kPa			$t_s = 36°C$ $p = 1389.0$ kPa			$t_s = 37°C$ $p = 1429.0$ kPa		
	v	h	s	v	h	s	v	h	s
0	0.0958	1470.0	4.9256	0.0931	1470.4	4.9149	0.0905	1470.8	4.9044
5	0.0985	1486.5	4.9788	0.0958	1487.1	0.9684	0.0931	1487.6	4.9580
10	0.1012	1502.4	5.0290	0.0984	1503.0	5.0188	0.0957	1503.7	5.0086
15	0.1037	1517.7	5.0767	0.1009	1518.4	5.0666	0.0981	1519.1	5.0565
20	0.1062	1532.5	5.1222	0.1033	1533.3	5.1122	0.1005	1534.1	5.0923
25	0.1086	1546.9	5.1658	0.1056	1547.8	5.1559	0.1028	1548.7	5.1461
30	0.1110	1561.0	5.2078	0.1079	1562.0	5.1980	0.1050	1562.9	5.1883
35	0.1133	1574.8	5.2484	0.1102	1575.9	5.2387	0.1072	1576.9	5.2290
40	0.1155	1588.4	5.2878	0.1124	1589.5	5.2781	0.1094	1590.6	5.2685
45	0.1178	1601.8	5.3260	0.1146	1603.0	5.3163	0.1115	1604.1	5.3068
50	0.1200	1615.1	5.3632	0.1167	1616.3	5.3536	0.1136	1617.4	5.3441
55	0.1222	1628.2	5.3995	0.1189	1629.4	5.3900	0.1157	1630.6	5.3805
60	0.1243	1641.1	5.4350	0.1210	1642.4	5.4255	0.1177	1643.7	5.4160

$t - t_s$ K	$t_s = 38°C$ $p = 1469.9$ kPa			$t_s = 39°C$ $p = 1511.7$ kPa			$t_s = 40°C$ $p = 1554.3$ kPa		
	v	h	s	v	h	s	v	h	s
0	0.0880	1471.2	4.8938	0.0856	1471.5	4.8833	0.0833	1471.9	4.8728
5	0.0906	1488.1	4.9477	0.0881	1488.6	4.9374	0.0857	1489.0	4.9371
10	0.0931	1504.3	4.9984	0.0905	1504.8	4.9883	0.0881	1505.4	4.9783
15	0.0954	1519.8	5.0465	0.0929	1520.5	5.0366	0.0904	1521.2	5.0267
20	0.0977	1534.9	5.0924	0.0951	1535.7	5.0826	0.0926	1536.4	5.0728

Table 9.4   *(Contd)*

$t - t_s$ K	$t_s = 38°C$ $p = 1469.9$ kPa			$t_s = 39°C$ $p = 1511.7$ kPa			$t_s = 40°C$ $p = 1554.3$ kPa		
	v	h	s	v	h	s	v	h	s
25	0.1000	1549.6	5.1363	0.0973	1550.4	5.1266	0.0947	1551.2	5.1169
30	0.1022	1563.9	5.1786	0.0995	1564.8	5.1690	0.0968	1565.7	5.1594
35	0.1043	1577.9	5.2194	0.1016	1578.9	5.2098	0.0989	1579.8	5.2003
40	0.1065	1591.6	5.2589	0.1036	1592.7	5.2494	0.1009	1593.7	5.2400
45	0.1085	1605.2	5.2973	0.1057	1606.3	5.2878	0.1029	1607.4	5.2784
50	0.1106	1618.6	5.3346	0.1076	1619.8	5.3252	0.1048	1620.9	5.3159
60	0.1126	1631.8	5.3710	0.1096	1633.1	5.3617	0.1067	1634.3	5.3524

$t - t_s$ K	$t_s = 41°C$ $p = 1597.9$ kPa			$t_s = 42°C$ $p = 1642.4$ kPa			$t_s = 43°C$ $p = 1687.8$ kPa		
	v	h	s	v	h	s	v	h	s
0	0.0810	1472.2	4.8623	0.0788	1472.4	4.8519	0.0767	1472.7	4.8414
5	0.0834	1489.4	4.9169	0.0812	1489.8	0.9067	0.0790	1490.2	4.8965
10	0.0857	1505.9	4.9682	0.0834	1506.5	4.9583	0.0812	1507.0	4.9483
15	0.0879	1521.8	5.0168	0.0856	1522.4	5.0070	0.0833	1523.0	4.9972
20	0.0901	1537.1	5.0631	0.0877	1537.9	5.0534	0.0854	1538.6	5.0438
25	0.0922	1552.0	5.1073	0.0898	1552.8	5.0978	0.0874	1553.6	5.0883
30	0.0943	1566.6	5.1499	0.0918	1567.4	5.1404	0.0894	1568.3	5.1310
35	0.0963	1580.8	5.1909	0.0937	1581.7	5.1815	0.0913	1582.7	5.1722
40	0.0982	1594.7	5.2306	0.0957	1595.8	5.2213	0.0932	1596.8	5.2120
45	0.1002	1608.5	5.2691	0.0976	1609.6	5.2599	0.0950	1610.6	5.2507
50	0.1021	1622.0	5.3066	0.0994	1623.2	5.2974	0.0968	1624.3	5.2882
60	0.1058	1648.7	5.3788	0.1031	1649.9	5.3697	0.1004	1651.1	5.3606

Reproduced from *Thermodynamic Properties of Ammonia* by W.B. Gosney and O. Fabris, with the kind permission of the authors.

Answer

(*a*) From Table 9.1, h_1, leaving the evaporator, is 398.68 kJ kg^{-1}. Since the value of h_4, entering the evaporator, is the same as that of h_3, leaving the condenser as a saturated liquid at a temperature of 35°C, the value of h_4 is found from Table 9.1 to be 248.94 kJ kg^{-1}. Hence by equation (9.4) the refrigerating effect is

$$q_r = (398.68 - 248.94) = 149.74 \text{ kJ kg}^{-1}$$

(*b*) The mass flow rate of refrigerant, $\dot{m}$, is given by

$$\dot{m} = Q_r/q_r \tag{9.5}$$

where Q_r is the refrigeration duty in kW.
Hence the mass flow rate of refrigerant handled is

$$\dot{m} = 352/149.74 = 2.351 \text{kg s}^{-1}$$

(*c*) Assuming that there is no pressure drop in the suction line and that there are no heat exchanges between the suction line and its surroundings, the state entering the compressor is the same as that leaving the evaporator, namely, 0°C saturated. From Table 9.1 the specific volume at this state is 0.06935 m^3 kg^{-1}. Hence the volumetric flow rate at the suction state is

$$\dot{v}_1 = 2.351 \times 0.06935 = 0.16304 \text{ m}^3 \text{ s}^{-1}$$

9.4 The work done in compression

In the ideal case, where there is no heat lost from the cylinders and no friction, the process of compression is adiabatic, reversible and isentropic. This is shown in Figure 9.2 by the line joining the points 1 and 2. It is assumed that there is no pressure drop in the pipeline between the outlet from the evaporator and the suction port on the compressor and that no heat gains or losses occur between the suction line and its environment. The compressor sucks dry saturated vapour from the evaporator and power must be provided to effect this process. The work done in compression (w_r) for this ideal case is given by the enthalpy increase across the compressor:

$$w_r = (h_2 - h_1) \tag{9.6}$$

Multiplying this by the mass flow rate of refrigerant handled, $\dot{m}$, yields the power needed for compression, w_r, in W or kW:

$$W_r = \dot{m}(h_2 - h_1) \tag{9.7}$$

State 1 (Figure 9.2) is known because, with the simplifying assumptions made, it is that of the dry saturated vapour leaving the evaporator. State 2 is that of superheated vapour leaving the compressor, which can be identified by two known facts for the ideal case considered:

(i) its pressure is the same as the condensing pressure and
(ii) it has the same entropy as that of the dry saturated vapour entering the suction port.

Hence h_1 and h_2 can be established.

EXAMPLE 9.3

Calculate the work done in compression and the corresponding compressor power, for the plant used in example 9.2.

Answer

Figure 9.3 illustrates the answer. From Table 9.1, the enthalpy of dry saturated vapour at 0°C leaving the evaporator is 398.68 kJ kg^{-1} and the entropy is 1.7274 kJ kg^{-1} K^{-1}. At state 2, leaving the compressor, the pressure of the superheated vapour is 887.11 kPa and the entropy is also 1.7274 kJ kg^{-1} K^{-1}. Unfortunately, Table 9.2 is not particularly user-friendly and reference to it shows that this entropy, for the relevant pressure, occurs somewhere between 800 kPa and 1000 kPa. A double linear interpolation is necessary to identify the enthalpy and temperature of the superheated vapour discharged from the compressor at state 2. First, values of enthalpy and temperature having entropies above and below 1.7274 kJ kg^{-1} K^{-1} must be determined for the pressures each side of the condensing pressure of 887.11 kPa:

$p = 1000$ kPa			$p = 800$ kPa		
s	h	t	s	h	t
1.7310	425.45	45	1.7437	424.61	40
1.7274	424.30	43.95	1.7274	419.58	35.18
1.7139	419.99	40	1.7268	419.40	35

It is seen that linear interpolations yield a pair of values of h and t at an entropy of 1.7274 kJ kg^{-1} K^{-1}. The values of h and t obtained for an entropy of 1.7274 kJ kg^{-1} K^{-1} are then interpolated between 1000 and 800 kPa to yield the required values with a pressure of 887.11 kPa, at compressor discharge.

p	s	h_2	t_2
1000	1.7274	424.30	43.95
887.11	1.7274	421.64	39.00
800	1.7274	419.58	35.18

At compressor discharge $h_2 = 421.64$ kJ kg^{-1} K^{-1} and, from example 9.1 and Figure 9.3, h_1 at compressor suction is 398.68 kJ kg^{-1}. Hence, by equation (9.6), the work done in compression, w_r, is $(421.64 - 398.68) = 22.96$ kJ kg^{-1} and, by equation (9.7), the corresponding compressor power, W_r, is $2.351 \times 22.96 = 53.98$ kW.

9.5   Heat rejected at the condenser

All the heat removed at the evaporator together with all the energy provided by the compressor must be rejected from the system. This heat is rejected at the condenser and is expressed by

$$q_c = (h_2 - h_3) \tag{9.8}$$

Heat rejection is possible, in accordance with the second law of thermodynamics, by arranging that the temperature of the condensing refrigerant is greater than the temperature of the air or water used for cooling purposes. For the examples considered earlier, the condensing temperature was 35°C and heat could conveniently be rejected to cooling water or air at a temperature of about 25°C to 30°C.

The higher the temperature (and hence pressure) chosen for condensing, the greater will be the work done in compression and hence the greater the compressor power. Referring to Figures 9.2 and 9.3 it is seen that choosing a higher condensing pressure will move the state 2 along a line of constant entropy further into the superheated zone and will increase its enthalpy.

EXAMPLE 9.4

Calculate the heat rejected, and the rate of heat rejection, at the condenser used for examples 9.2 and 9.3.

Answer

See Figure 9.3. From examples 9.2 and 9.3, h_3 is 248.94 kJ kg^{-1} and h_2 is 421.64 kJ kg^{-1}. Hence, by equation (9.8), the heat rejected at the condenser is:

$$q_c = (421.64 - 248.94) = 172.7 \text{ kJ kg}^{-1}$$

Alternatively, the energy used for compression could be added to the refrigerating effect:

$$q_c = 22.96 + 149.74 = 172.7 \text{ kJ kg}^{-1}$$

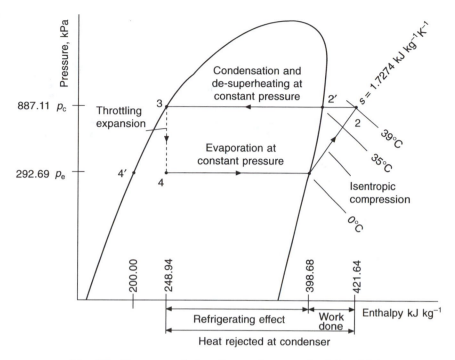

Fig. 9.3   Pressure-enthalpy diagram for examples 9.3 and 9.4.

The rate of heat rejection at the condenser is determined by

$$Q_c = \dot{m}q_c \qquad (9.9)$$

Hence, in this example,

$$Q_c = 2.351 \times 172.7 = 406 \text{ kW}$$

Alternatively, the compressor power could be added to the refrigeration duty:

$$Q_c = 53.98 + 352 = 405.98 = 406 \text{ kW}$$

9.6   Coefficient of performance

The *coefficient of performance* of a refrigeration machine is the ratio of the energy removed at the evaporator (refrigerating effect) to the energy supplied to the compressor. Thus, using the notation of Figure 9.2, we have the formula:

$$COP = \frac{(h_1 - h_3)}{(h_2 - h_1)} \qquad (9.10)$$

The best possible performance giving the highest *COP* would be obtained from a system operating on a Carnot cycle. Under such conditions the refrigeration cycle would be thermodynamically reversible, and both the expansion and compression processes would be isentropic. A Carnot cycle is shown on a *temperature-entropy diagram* in Figure 9.4.

Point 3 indicates the condition of the liquid as it leaves the condenser. Line 34 represents the passage of the refrigerant through an expansion engine, the work of which helps to

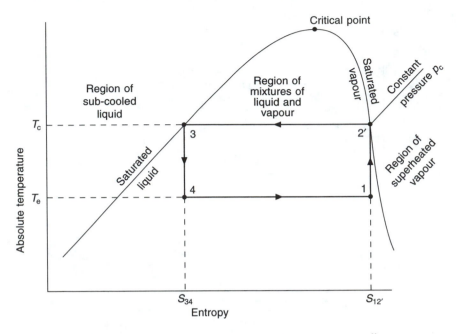

Fig. 9.4   Carnot refrigeration cycle on a temperature-entropy diagram.

drive the refrigeration machine. During the expansion process the pressure of the refrigerant drops from p_c to p_e, its temperature is reduced from T_c to T_e and part of the refrigerant vaporises. The expansion process, being reversible and adiabatic, takes place at constant entropy, with the result that the line 34 is vertical.

During the next part of the cycle, refrigerant is evaporated at constant pressure p_e and temperature T_e, this being represented on the diagram by the line 41. The process comes to an end at the point 1, where the refrigerant still consists of a liquid–vapour mixture but the proportion of liquid is small.

The mixture then enters another engine, the compressor, which is driven by some external source of power supplemented by the output of the expansion engine already described. In the compressor, the refrigerant is increased in pressure from p_e to p_c, the temperature rises from T_e to T_c and the remaining liquid evaporates, so that the refrigerant leaves the machine as a saturated vapour—point 2′ on the diagram. The compression process is adiabatic, reversible and, therefore, isentropic.

The final part of the cycle is shown by the line 2′3, which represents the liquefaction of the refrigerant in the condenser at constant pressure p_c and temperature T_c.

Since an area under a process line on a temperature-entropy diagram represents a quantity of energy (see equation (9.2)) the quantities involved in the Carnot refrigeration cycle are as follows:

> Heat rejected at condenser $T_c(s_{12'} - s_{34})$
> Heat received at evaporator $T_e(s_{12'} - s_{34})$
> Work supplied to machines $(T_c - T_e)(s_{12'} - s_{34})$

Since the coefficient of performance is the energy received at the evaporator divided by the energy supplied to the machine, we have for the Carnot cycle:

$$COP \text{ (Carnot)} = \frac{T_e}{T_c - T_e} \tag{9.11}$$

EXAMPLE 9.5

Using the data of examples 9.1 to 9.4 calculate the coefficient of performance. What *COP* would be obtained if a Carnot cycle were used?

Answer

From equation (9.10)

$$COP = (398.68 - 248.94)/(421.64 - 398.68)$$
$$= 149.74/22.96 = 6.52$$

and from equation (9.11)

Carnot $COP = (0 + 273)/[(35 + 273) - (0 + 273)] = 7.8$

It will be seen from this example that the simple saturation cycle under consideration had a coefficient of performance which was 83.7 per cent of that attainable with a Carnot cycle. Comparisons of this kind are sometimes used as an index of performance (see Table 9.6).

The simple, basic saturation cycle shown in Figure 9.2 is illustrated in Figure 9.5 as an absolute temperature-entropy diagram. The principal reasons why the coefficient of performance of this cycle is less than that of the corresponding Carnot cycle (as in Figure 9.4) are: the excursion of the compression process line into the superheated region and the irreversible nature of the method of throttling expansion used for the simple cycle.

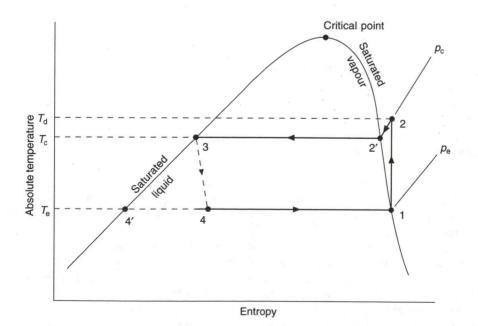

Fig. 9.5 Simple saturation refrigeration cycle on a temperature-entropy diagram.

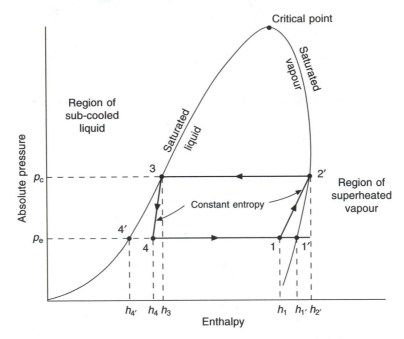

Fig. 9.6   Carnot refrigeration cycle on a pressure-enthalpy diagram.

9.7   Actual vapour-compression cycle

An ideal, simple, reversible vapour-compression cycle is not a practical proposition. The departure from reversibility arises from the irreversible nature of the throttling expansion process, pressure losses in the evaporator, condenser and pipelines, heat transfer through finite temperature differences and a measure of irreversibility in the compression process. For example, Figure 9.5 shows two of these irreversibilities: the throttling expansion (3–4) through the expansion valve and the desuperheating process (2–2′) when the vapour first enters the condenser. The condensation process (2′–3) and the evaporation process (4–1), on the other hand, are reversible.

   Liquid refrigerant is incompressible and if it entered the suction valve of a reciprocating compressor it could damage the cylinder head or the valve plates at top dead centre. To avoid this risk it is arranged that evaporation is completed before the vapour leaves the evaporator and, consequently, the state then entering the suction valve has several degrees of superheat.

   It is often also arranged that there is more than enough heat transfer surface in the condenser to change the refrigerant from a superheated vapour to a saturated liquid. In this case the liquid is sub-cooled to a temperature less than its saturated temperature for the prevailing pressure. A consequence of this is that any loss of position head (because the condenser might be at a lower level than the expansion valve), or any frictional pressure drop in the liquid line, is less likely to cause the liquid to flash to gas before it reaches the expansion valve. Figure 9.7 shows a pressure-enthalpy diagram of a simple cycle with superheat at the outlet from the evaporator and sub-cooling at the outlet from the condenser. Pressure drops through the piping, the evaporator and the condenser are ignored and isentropic compression is assumed. It can be seen that the dryness fraction is reduced and the refrigerating effect is improved, compared with the simple plant shown in Figure 9.1.

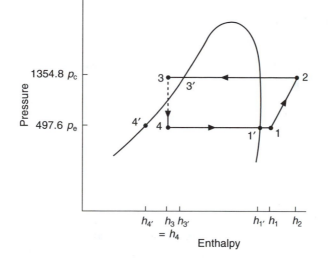

Fig. 9.7 Pressure-enthalpy diagram of a simple, actual, vapour-compression cycle showing superheat at evaporator outlet and sub-cooling at condenser outlet.

EXAMPLE 9.6

A plant using R134a evaporates at 0°C and condenses at 35°C, with 5 K of superheat at the evaporator outlet and 5 K of sub-cooling at the outlet from the condenser. The duty is 352 kW of refrigeration. Assuming isentropic compression and ignoring pressure drops in the piping, evaporator and condenser, calculate the following: (*a*) the dryness fraction at entry to the evaporator, (*b*) the refrigerating effect, (*c*) the mass flow rate of refrigerant, (*d*) the volumetric flow rate at the suction state, (*e*) the work done in compression, (*f*) the compressor power, (*g*) the temperature of the superheated vapour at discharge from the compressor, (*h*) the rate of heat rejection at the condenser, (*i*) the *COP*, (*j*) the corresponding Carnot *COP*, (*k*) the percentage Carnot cycle efficiency.

Answers

(*a*) Using equation (9.3), the notation of Figure 9.7 and referring to Tables 9.1 and 9.2 as necessary, it is established that $h_{4'} = 200.00$ kJ kg^{-1} kg^{-1} and $h_{1'} = 398.68$ kJ kg^{-1}. In Figure 9.7 it is seen that state 4 has the same enthalpy as state 3, which is 5 K cooler than state 3'. Table 9.1 quotes the enthalpy of saturated liquid at state 3' as 248.94 kJ kg^{-1}. Table 9.1 also gives the specific heat of saturated liquid at 35°C as the mean of the specific heats at 34°C and 36°C, namely, 1.4725 kJ kg^{-1} K^{-1} and as 1.447 kJ kg^{-1} K^{-1} at 30°C. The mean over the range 30°C to 35°C is then (1.4725 + 1.447)/2 = 1.460 kJ kg^{-1} K^{-1}. Hence $h_3 = h_4 = 248.94 - 1.460 \times 5 = 241.64$ kJ kg^{-1}. By equation (9.3)

$$f = (h_4 - h_{4'})/(h_{1'} - h_{4'}) = (241.64 - 200)/(398.68 - 200) = 0.21$$

(*b*) Equation (9.4) gives the refrigerating effect. In Table 9.1 the enthalpy of saturated vapour at 292.69 kPa and 0°C is given as 398.68 kPa kg^{-1} and the specific heat of saturated vapour as 0.883 kJ kg^{-1} K^{-1}. Assuming the specific heat does not alter very much for the small temperature change of 5 K at constant pressure, the enthalpy of state 1, leaving the evaporator, is calculated as $h_1 = 398.68 + 0.883 \times 5 = 403.10$ kJ kg^{-1}. Since $h_4 = h_3$ the refrigerating effect is

$$q_r = (h_1 - h_4) = (h_1 - h_3) = (403.10 - 241.64) = 161.46 \text{ kJ kg}^{-1}$$

Compare this with 149.74 kJ kg^{-1} obtained as the answer to example 9.2(*a*). The increase is about 8 per cent. The benefit of 5 degrees of superheat at entry to the compressor is $0.883 \times 5 = 4.415$ kJ kg^{-1}, which is less than 3 per cent of the refrigerating effect. Increasing the superheat to get more cooling capacity is seldom worthwhile.

(*c*) Equation (9.5) gives the mass flow rate of refrigerant:

$$\dot{m} = 352/161.46 = 2.180 \text{ kg s}^{-1}$$

(*d*) In the absence of pressure drops and heat exchanges with the surroundings, the state into the suction side of the compressor is the same as at the outlet from the evaporator, namely, 5°C and 292.69 kPa. Taking the specific volume of saturated vapour at 0°C as 0.069 35 m^3 kg^{-1} from Table 9.1 and making the assumption that, over a small temperature change near saturation, Charles' law (see section 2.5) may be applied without much error, gives the required specific volume at state 1 (Figure 9.7):

$$v_1 = 0.069\ 35 \times (273 + 5)/(273 + 0) = 0.070\ 62 \text{ m}^3 \text{ kg}^{-1}$$

This corresponds to a density of 14.160 kg m^{-3}. Reference to a pressure–enthalpy chart for R134a (Figure 9.10) verifies this. Hence the volumetric flow rate is $2.180 \times 0.070\ 62 = 0.153\ 95$ m^3 s^{-1}.

(*e*) Equation (9.6) gives the work done in compression. From Table 9.1 the enthalpy and entropy of dry saturated vapour at 0°C and 292.69 kPa are 398.68 kJ kg^{-1} and 1.7274 kJ^{-1} kg^{-1} K^{-1}, respectively. The enthalpy at 5 K of superheat and 292.69 kPa has already been established as 403.10 kJ kg^{-1}. Hence the enthalpy change from 0°C saturated to 5 K of superheat at a constant pressure of 292.69 kPa is $403.10 - 398.68 = 4.42$ kJ kg^{-1} and the mean temperature at which this enthalpy change occurs is 2.5°C. The principles discussed in section 9.2, leading to equation (9.2), allow a calculation of the corresponding change in entropy. Hence a reasonable calculation for the entropy at state 1, entering the compressor, is

$$s_1 = 1.7274 + 4.42/(273 + 2.5) = 1.7434 \text{ kJ kg}^{-1} \text{ K}^{-1}$$

Reference to a pressure–enthalpy diagram for R134a verifies this. This is also the entropy at state 2, leaving the compressor. We must now refer to Table 9.2 and interpolate between 800 and 1000 kPa to determine the enthalpy and temperature of the gas discharged at state 2. First values of *h* and *t* must be established at the correct entropy of the adiabatic compression process for the two relevant tabulated pressures:

p = 1000 kPa			*p* = 800 kPa		
s	*h*	*t*	*s*	*h*	*t*
1.7310	415.45	45	1.7268	419.40	35
1.7434	426.60	48.60	1.7434	424.52	39.91
1.7482	430.91	50	1.7437	424.61	40

Finally, values of h and t must be determined from the above interpolations at the correct pressure and entropy:

p	s	h	t
1000	1.7434	426.60	48.60
887.11	1.7434	425.43	43.69
800	1.7434	424.52	39.91

By equation (9.6) the work done in compression is

$$w_r = (425.43 - 403.10) = 22.33 \text{ kJ kg}^{-1}$$

(*f*) By equation (9.7) the power absorbed by the compressor is

$$W_r = 2.180 \times 22.33 = 48.68 \text{ kW}$$

(*g*) From the final interpolation in part (*e*), above, $t = 43.69°C$, leaving the compressor.

(*h*) The rate of heat rejection at the condenser is given by equation (9.9):

$$Q_c = 2.180 \times (425.43 - 241.64) = 400.7 \text{ kW}$$

This equals the total refrigeration duty plus the power absorbed by the compressor: 352 + 48.68 = 400.7 kW.

(*i*) Equation (9.10) gives the *COP*:

$$161.46/22.33 = 7.23$$

Such high *COP*s are academic and would not be realised in practice because they assume isentropic compression and ignore inefficiencies, pressure drops and heat gains or losses.

(*j*) The Carnot *COP*, according to equation (9.11), is:

$$(273 + 35)/(35 - 0) = 8.8$$

(*k*) Percentage Carnot efficiency = $(7.23 \times 100)/8.8 = 82.1$ per cent.

9.8   Pressure–volume relations

The gas being compressed in a refrigeration machine can sometimes be regarded as an ideal one where pressures and volumes are related as follows:

$$p_e V_e^n = p_d V_d^n \tag{9.12}$$

Three cases can then be distinguished:

(i)   The case of reversible adiabatic compression, in which the process is isentropic and we have for the value of n.

$$n = c_p/c_v = \gamma \tag{9.13}$$

Here c_p and c_v are the specific heats of the refrigerant gas at constant pressure and constant volume respectively.

(ii)   Reversible but non-adiabatic compression in which there is heat transfer between the refrigerant and its surroundings. Reciprocating compressors with cylinder cooling may approach this case and in such circumstances n will be less than γ.

(iii)   Irreversible but adiabatic compression, in which there are thermal effects due to gas friction and turbulence but no actual heat transfer between the refrigerant and its enclosing surfaces. Centrifugal machines approach this type of compression, in which case n will be greater than γ.

When $n \neq \gamma$ the process is called *polytropic compression*. With polytropic compression the value of n depends upon the process – it is *not* a property of the refrigerant. During polytropic compression the entropy of the gas increases in case (iii) and diminishes in case (ii). Thus, the process departs from the simple saturation cycle which we have hitherto considered.

If the compression is reversible, as in cases (i) and (ii), the work done on 1 kg mass of refrigerant in Nm during steady flow is given by the following expression, where pressures are in Pa absolute and volumes are in m^3.

$$\text{Work done} = \int_{p_e}^{p_d} V\,dp \tag{9.14}$$

Using equation (9.12) we have:

$pV^n = c$ (where c is a constant), or
$V = c^{1/n}p^{-1/n}$

Thus,

$$\text{work done} = c^{1/n} \int_{p_e}^{p_d} p^{-1/n}\,dp$$

$$= \frac{n}{(n-1)} \left[p^{(n-1)/n} \right]_{p_e}^{p_d} c^{1/n}$$

$$= \frac{n}{(n-1)} \left[\left(\frac{p_d}{p_e} \right)^{(n-1)/n} - 1 \right] p_e V_e \tag{9.15}$$

It should be noted that the work done on 1 kg mass of refrigerant in Nm is here equal to the head in metres against which the compressor is working. Developing the expression by using equation (9.5) we get the rate of work done in compression, in w, on m kg of refrigerant

$$W_r = \frac{\dot{m}n}{(n-1)} \left\{ \left(\frac{p_d}{p_e} \right)^{(n-1)/n} - 1 \right\} p_e V_e$$

$$= \frac{p_e V_e}{(h_{ve} - h_{lc})} \cdot \frac{n}{(n-1)} \left\{ \left(\frac{p_d}{p_e} \right)^{(n-1)/n} - 1 \right\} Q_r \tag{9.16}$$

where h_{ve} is the enthalpy of saturated vapour at the evaporating pressure and h_{1c} is the enthalpy of saturated liquid at the condensing pressure.

If the compression process is the isentropic one described as case (i) above, the equation becomes:

$$W_r = \left(\frac{p_e V_e}{(h_{ve} - h_{1c})}\right)\left(\frac{\gamma}{(\gamma - 1)}\right)\left\{\left(\frac{p_d}{p_e}\right)^{(\gamma-1)/\gamma} - 1\right\}Q_r \tag{9.17}$$

In case (iii), where the process is irreversible but adiabatic, equation (9.14) still gives the fluid output in Nm kg^{-1} of refrigerant, or the head in metres against which the machine is working, but it no longer represents the required input energy to the fluid.

This exceeds the quantity

$$\int_{p_c}^{p_d} V\,dp$$

and is calculated from the following expression:

$$W_r = \frac{p_e V_e}{h_{ve} - h_{1c}} \cdot \frac{\gamma}{\gamma - 1}\left\{\left(\frac{p_d}{p_e}\right)^{(n-1)/n} - 1\right\}Q_r \tag{9.18}$$

Comparing this with equation (9.17) we can define what is termed the isentropic efficiency (η_s) as

$$\eta_s = \frac{\text{theoretical work for case (i)}}{\text{theoretical work for case (iii)}}$$

$$= \frac{(p_d/p_e)^{(\gamma-1)/\gamma} - 1}{(p_d/p_e)^{(n-1)/n} - 1} \tag{9.19}$$

The quantity

$$\frac{n}{(n - 1)} \cdot \frac{(\gamma - 1)}{\gamma}$$

is called the polytropic efficiency and that represented by equation (9.15) is called the polytropic head. Polytropic head has to be divided by the polytropic efficiency to give the power required by case (iii) in Nm kg^{-1} of gas pumped.

Other forms of equations (9.15) to (9.18) are possible. Thus, for an ideal gas we can write RT_e in place of $p_e V_e$, R being the particular gas constant for the refrigerant.

It should be noted that although equation (9.7) is applicable to cases (i) and (iii), it cannot be applied to case (ii), since the quantity $h_{vd} - h_{ve}$ does not, in that case, represent all the work of compression. Instead, we must write for case (ii):

$$W_r = n[(h_{vd} - h_{ve}) + q_j] \tag{9.20}$$

Here, q_j is the heat lost to the jacket in J kg^{-1} of refrigerant in circulation. Its value can be calculated from

$$q_j = \left(\frac{n}{(n - 1)} - \frac{\gamma}{\gamma - 1}\right)\left\{\left(\frac{p_d}{p_e}\right)^{(n-1)/n} - 1\right\}p_e V_e \tag{9.21}$$

The temperature of the hot gas leaving the compressor can be calculated from the formula:

$$T_d = T_e \left(\frac{p_d}{p_e} \right)^{(n-1)/n}$$ (9.22)

EXAMPLE 9.7

A refrigerant which behaves as an ideal gas has a molecular mass of 64.06 and a specific heat ratio of 1.26 (equal to c_p/c_v). If the compression ratio (p_d/p_e) is 3.119 and the refrigerating effect is 322 kJ kg^{-1} with evaporation at 4.5°C, find the theoretical COP, the enthalpy gain during compression and the temperature of the discharge gas for the following cases:

(i)   isentropic compression,
(ii) polytropic compression with $n = 1.22$,
(iii) polytropic compression with $n = 1.30$.

Answer

(i) Write RT_e in place of $p_e V_e$ and set $R = 8314.41/64.06$ (see section 2.6).

Equation (9.17) expresses the work done in compression for a refrigeration duty of Q_r kW. Hence, re-arranging the equation:

$$\frac{W_r}{Q_r} = \frac{8314.41 \times (273 + 4.5) \times (1.26)}{64.06 \times 1000 \times 322 \times (1.26 - 1)} \times [3.119^{(0.26/1.26)} - 1] = 0.1434$$

$$COP = \frac{Q_r}{W_r} = 1/0.1434 = 6.97$$

By means of equation (9.10), the enthalpy gain during compression is determined as

$$h_{vd} - h_{ve} = \frac{322}{6.97} = 46.20 \text{ kJ kg}^{-1}$$

where h_{vd} is the enthalpy of superheated vapour at the discharge condition.

From equation (9.22) the temperature of the discharged gas is

$$T_d = (273 + 4.5) \times 3.119^{(1.26-1)/1.26} = 350.9 \text{ K}$$

which is equivalent to 77.9°C.

(ii) Here we proceed as in case (i) but re-arrange equation (9.16) instead of (9.17):

$$\frac{W_r}{Q_r} = \frac{8314.41 \times (273 + 4.5) \times 1.22}{64.06 \times 1000 \times 322 \times (1.22 - 1)} \times [3.119^{(0.22/1.22)} - 1] = 0.1412$$

$$COP = \frac{Q_r}{W_r} = 1/0.1412 = 7.08$$

Equation (9.5) yields the mass flow rate of refrigerant:

$$\dot{m} = \frac{Q_r}{322}$$

By equation (2.20)

$$W_r = \dot{m}\,[(h_{vd} - h_{ve}) + q_j]$$
$$= \frac{Q_r}{322}\,[(h_{vd} - h_{ve}) + q_j]$$

whence

$$(h_{vd} - h_{ve}) + q_j = 322\,\frac{W_r}{Q_r} = 322 \times 0.1412 = 45.47 \text{ kJ kg}^{-1}$$

Equation (9.21) gives the jacket loss:

$$q_j = \left[\frac{1.22}{(1.22 - 1)} - \frac{1.26}{(1.26 - 1)}\right][3.119^{(1.22-1)/1.22} - 1] \times \frac{(8314.41 \times (273 + 4.5))}{64.06}$$
$$= 5734.48 \text{ J kg}^{-1} = 5.73 \text{ kJ kg}^{-1}$$

The enthalpy gain during compression is then

$$(h_{vd} - h_{ve}) = 45.47 - 5.73 = 39.74 \text{ kJ kg}^{-1}$$

By equation (9.22) the temperature of the discharged gas is

$$T_d = (273 + 4.5) \times 3.119^{(1.22-1)/1.22} = 340.7 \text{ K}$$

which is equivalent to 67.7°C.

(iii) Proceeding as for case (i) but re-arranging equation (9.18) instead of (9.17) we have

$$\frac{W_r}{Q_r} = \frac{8314.41 \times (273 + 4.5) \times 1.26}{64.06 \times 322 \times 1000 \times (1.26 - 1)}\,[3.119^{(1.3-1)/1.3} - 1] = 0.1627$$

$$COP = \frac{Q_r}{W_r} = \frac{1}{0.1627} = 6.15$$

By equation (9.5)

$$\dot{m} = \frac{Q_r}{322}$$

By equation (9.7)

$$W_r = \frac{Q_r}{322}\,(h_{vd} - h_{ve})$$

whence the enthalpy gain during compression is

$$(h_{vd} - h_{ve}) = 322\,\frac{W_r}{Q_r} = 322 \times 0.1627 = 52.39 \text{ kJ kg}^{-1}$$

By equation (9.22) the temperature of the discharged gas is

$$T_d = (273 + 4.5) \times 3.119^{(1.3-1)/1.3} = 360.8 \text{ K}$$

which is equivalent to 87.7°C.

Summarising the results:

Case	n	COP	Enthalpy gain during compression kJ kg^{-1}	Discharge gas temperature °C
(i)	1.26	6.97	46.20	77.9
(ii)	1.22	7.08	39.74	67.7
(iii)	1.3	6.15	52.39	87.8

9.9 Volumetric efficiency

The capacity of a refrigeration compressor is a function of the mass flow rate of gas handled. However, the full volume of the cylinder swept by the piston is not available for pumping gas: some clearance volume must be left at the top of the cylinder, otherwise there would be a danger of the piston damaging the cylinder head or the valves as the stroke increased with wear on the crankshaft and bearings. This is illustrated in Figure 9.8. Since the actual amount of gas pumped is less than the theoretical amount, there is a concept of volumetric efficiency, defined by:

$$\eta_v = \frac{\text{actual volume of fresh gas}}{\text{swept volume}} \times 100$$

$$= \frac{V_a - V_d}{V_a - V_c} \times 100 \qquad (9.23)$$

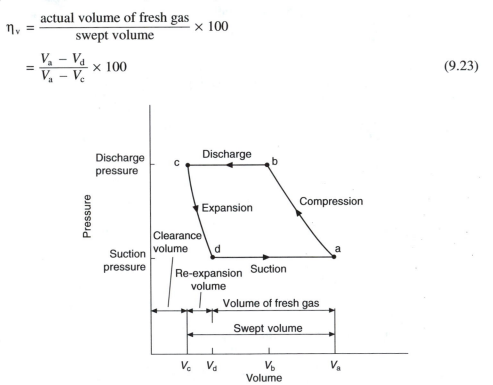

Fig. 9.8   Pressure-volume relationship for an ideal compression cycle.

The small amount of gas trapped in the clearance volume at the top of the stroke re-expands as the piston moves downward and reduces the volume available for the entry of fresh gas. Fresh gas cannot enter the cylinder until re-expansion has reduced the pressure within the cylinder to less than the suction pressure.

The ratio of the absolute condensing pressure to the absolute suction pressure is termed the compression ratio and this clearly has an effect on the volumetric efficiency. Table 9.5 gives some typical values of volumetric efficiency.

Table 9.5 Some typical, approximate volumetric efficiencies, according to *Trane Air Conditioning Manual* (1961)

Compression ratio	2.0	3.0	4.0	5.0	6.0
Volumetric efficiency	78%	74%	70%	66%	62%

Since the densities of refrigerants differ, the choice of refrigerant will affect the capacity of a machine, for a given piston displacement.

Two further reasons why the volume of fresh gas is always different from the swept volume are:

(i)   The cylinder walls are warmer than the gas leaving the evaporator and hence the gas in the cylinder expands and resists the entering flow through the suction port.

(ii) The density of the gas within the cylinder is less than that of the gas about to enter because of the pressure drop accompanying the flow through the suction valve. Hence the refrigerant used also has an effect on the volumetric efficiency.

The actual refrigeration capacity is defined as:

$$Q_{ra} = \eta_v \dot{m} q_r \tag{9.24}$$

Other efficiencies, that influence the power input needed for the process of compression, are:

Isentropic adiabatic efficiency

This is defined as the ratio of the work done for isentropic adiabatic compression of the gas, to the work input to the crankshaft.

Compression efficiency

This refers only to what happens within the cylinder and is the ratio of work done for isentropic compression to the measured work done for the actual compression of the gas.

Mechanical efficiency

This is the ratio of the measured work done for compression of the gas to the measured rate of work input to the crankshaft.

The product of the above three efficiencies, as fractions, would be divided into the ideal power required for compression (as related to the actual refrigeration duty obtained from equation (9.24)), to determine the actual power input to the crankshaft.

EXAMPLE 9.8

Using the results of example 9.6 as necessary, calculate the stroke and bore of an eight cylinder compressor running at 1425 rpm. Assume the dimensions of the stroke and bore are equal and obtain a suitable volumetric efficiency from Table 9.5.

Answer

Compression ratio = 887.11/292.69 = 3.031
    From Table 9.5 η_v = 73.9 per cent.
    Denoting the cylinder diameter by d, also equal to the bore, the required swept volume is

$$\frac{8 \times \pi \times d^3 \times 1425}{0.739 \times 4 \times 60} = 201.93 d^3 \text{m}^3 \text{ s}^{-1}$$

This can be equated to the volumetric flow rate at the suction state, to yield the value of d:

$$201.93 d^3 = 0.154$$

whence

$$d = 0.995 \text{ m} = 995 \text{ mm}$$

and is the diameter of the cylinders and their stroke.

EXAMPLE 9.9

A plant using R717 (ammonia) has a duty of 352 kW of refrigeration with a four cylinder compressor running at 1425 rpm. It evaporates at 0°C and condenses at 35°C, with 5 K of superheat at the evaporator outlet and 5 K of sub-cooling at the condenser outlet. The mean specific heat capacity of saturated liquid ammonia at 32.5°C is given as 4.867 kJ kg^{-1} K^{-1} (see *ASHRAE Handbook* (1997)). Assuming isentropic compression and ignoring pressure drops in the piping, the evaporator and condenser, determine the following, making use of Tables 9.3 and 9.4: (*a*) the dryness fraction at entry to the evaporator, (*b*) the refrigerating effect, (*c*) the mass flow rate of refrigerant, (*d*) the volumetric flow rate at the suction state, (*e*) the work done in compression, (*f*) the compressor power, (*g*) the temperature of the superheated vapour at discharge from the compressor, (*h*) the rate of heat rejection at the condenser, (*i*) the COP, (*j*) the corresponding Carnot COP, (*k*) the percentage Carnot efficiency, (*l*) the stroke and bore of the compressor cylinders, assuming these to be equal.

Answer

(*a*) Making use of the value given for the mean specific heat of saturated liquid refrigerant at 32.5°C the enthalpy of sub-cooled liquid at 30°C is 347.1 − 5 × 4.867 = 322.8 kJ kg^{-1} K^{-1}. Using the notation of Figure 9.7 and equation (9.3):

$$f = (322.8 - 181.1)/(1443.1 - 181.1) = 0.11$$

(*b*) By equation (9.4):

$$q_r = (1456.5 - 322.8) = 1133.7 \text{ kJ kg}^{-1}$$

(*c*) By equation (9.5):

$$m = 352/1133.7 \text{ kg s}^{-1} = 0.3105 \text{ kg s}^{-1}$$

(*d*) At compressor suction (5°C and 429.4 kPa) $v = 0.2966$ m^3 kg^{-1} and hence $v = 0.3105 \times 0.2966 = 0.0921$ m^3 s^{-1}.

(*e*) At compressor discharge the entropy, $s_2 = s_1 = 5.3842$ kJ kg^{-1} K^{-1}. Interpolating for this entropy in Table 9.4 at a pressure of 1350 kPa yields:

s	$t - t_s$	h
5.3632	50°	1615.1
5.3842	52.9°	1622.7
5.3995	55°	1628.2

whence $h_2 = 1622.7 \text{ kJ kg}^{-1}$

By equation (9.6):

$$w_r = (1622.7 - 1456.5) = 166.2 \text{ kJ kg}^{-1}$$

(*f*) By equation (9.7) the compressor power is

$$W_r = 0.3105 \times 166.2 = 51.60 \text{ kW}$$

(*g*) From the interpolation in part (*e*) the temperature of the vapour at discharge from the compressor is 35°C + 52.9 K = 87.9°C.

(*h*) Using the mass flow rate and equation (9.8):

Rate of heat rejection at the condenser

$$= 0.3105 \times (1622.7 - 322.8) = 403.6 \text{ kW}$$

(*i*) By equation (9.10):

$$COP = (1456.5 - 322.8)/(1622.7 - 1456.5) = 6.82$$

(*j*) By equation (9.11):

$$\text{Carnot } COP = (0 + 273)/[(35 + 273) - (0 + 273)] = 7.8$$

(*k*) Percentage Carnot efficiency = $6.82 \times 100/7.8 = 87.4$ per cent.

(*l*) The compression ratio is 1350/429.4 = 3.14 and from Table 9.5 the volumetric efficiency is 73.4 per cent. The swept volume is $(4 \times \pi \times d^3 \times 1425)/(0.734 \times 4 \times 60) = 101.65d^3$. This is equated to the volumetric flow rate at the suction state (part (*d*), above):

$101.65d^3 = 0.0921$ whence $d^3 = 0.000\ 906\ 05 \text{ m}^3$ and $d = 0.0968$ m.

The stoke and bore are each equal to 96.8 mm.

9.10 Thermosyphon cooling

Systems of air conditioning using fan coil units, induction units or chilled ceilings require a supply of chilled water throughout the year. As an alternative to operating a vapour compression system in the conventional manner during the winter, it is possible to switch the compressor off and to rely on a natural circulation of refrigerant between the evaporator and the condenser, under favourable operating conditions, the performance then being like that of one form of heat pipe. See Figure 9.9. The condenser is located above the evaporator and the compressor and expansion valve are temporarily by-passed. The water being chilled by the evaporator causes the liquid refrigerant to evaporate and migrate upwards to the condenser, under the pressure difference corresponding to the higher temperature in the evaporator (water chiller) and the lower temperature in the condenser. The vapour is condensed by cold airflow or waterflow over the outer heat transfer surface of the condenser and the liquid refrigerant flows by gravity back to the evaporator. The three-port, two-position valves at *A* and *B* must be switched to the appropriate positions.

Thermosyphon cooling was first developed in 1984, according to Pearson (1990) and

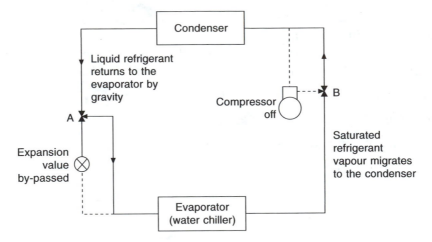

Fig. 9.9   Diagrammatic arrangement to achieve thermosyphon cooling. The broken line shows the path followed when normal vapour compression is taking place.

Blackhurst (1999), and has been successfully used with conventional refrigerants. Since 1997 it has also been applied for use with ammonia as the refrigerant, essential attention being paid to safety, in accordance with BS 4434, 1989.

Pearson (1998) points out that heat transfer at the surfaces of cupro-nickel tubes can be enhanced by providing a coating that increases the number of nucleation sites on the surface. It is claimed that overall heat transfer can be increased by a factor of over three. When properly designed, installed and commissioned, thermosyphon cooling provides a very efficient way of operating refrigeration plant in air conditioning systems. The compressor does not run when the local weather conditions allow free cooling and the changeover from using the compressor to free cooling can be automatic.

9.11   Refrigerants

The comparative performance of refrigerants commonly used in the vapour compression cycle is given in Table 9.6, the values of which were all obtained by calculation, using the methods of this chapter. Actual operating conditions will therefore be somewhat different, owing to the effects of the factors mentioned in section 9.2.

In general, it has been assumed that the vapour enters the compressor in a saturated condition at 5°C. In refrigerants 113 and 114, however, where saturated suction gas would result in condensation during compression, enough superheat has been assumed to ensure saturated discharge gas. This superheat has not been counted as part of the refrigerating effect.

A condensing temperature of 40°C has been taken for all refrigerants.

The performances have been calculated on the assumption that compression is isentropic.

A number of factors have to be taken into consideration when the thermodynamic characteristics listed in the table are being evaluated. For example, if operating pressures are high, then the materials comprising the refrigeration system will be heavy and the equipment expensive. The refrigerant containers will also be heavy, and this will increase the cost of transport. Sub-atmospheric operating pressures, on the other hand, mean that any leakage will result in air entering the system.

Table 9.6 Comparative performance of refrigerants evaporating at 5°C and condensing at 40°C

Refrigerant number	Suction temp. (°C)	Evaporating pressure (bar)	Condensing pressure (bar)	Compression ratio	Refrigerating effect (kJ kg⁻¹)	Specific vol. of vapour (m³ kg⁻¹)	Compressor displacement (1 s⁻¹ kW⁻¹)	Power in kW per kW refrigeration	% Carnot cycle efficiency
718	5	0.009	0.074	8.46	2370.0	147.0	62.0	0.1355	92.9
11	5	0.496	1.747	3.52	157.0	0.332	2.12	0.1395	90.2
717	5	5.160	15.55	3.01	1088.0	0.243	0.214	0.1456	86.4
114	12.7	1.062	3.373	3.18	106.2	0.122	1.14	0.1484	84.8
12	5	3.626	9.607	2.65	115.0	0.047	0.409	0.1502	83.8
113	10.4	0.188	0.783	4.16	129.5	0.652	5.03	0.1511	83.3
134a	5	3.451	10.032	2.91	145.0	0.0584	0.403	0.1516	83.0
22	5	5.838	15.34	2.63	157.8	0.040	0.255	0.1518	82.9
502	5	6.678	16.77	2.51	101.0	0.026	0.259	0.1631	77.1

718   Water, H_2O
11   Trichlorofluoromethane, CCl_3F
717   Ammonia, NH_3
114   Dichlorotetrafluoroethane, $CClF_2$–$CClF_2$
12   Dichlorodifluoromethane, CCl_2F_2
113   Trichlorotrifluoroethane, CCl_2FCClF_2
134a Tetrafluoroethane, CF_3CH_2F
22   Chlorodifluoromethane, $CHClF_2$
502   Azeotropic mixture (48.8% R22, 51.2% R115) for low temperature work

Note: An azeotrope is a mixture of refrigerants, the liquid and vapour phases of which, in equilibrium, have a constant boiling point.

In reciprocating compressors, the displacement (litre s^{-1} kW^{-1}) should be low so that the required performance can be achieved with a small machine. For a centrifugal compressor, on the other hand, large displacements are desirable in order to permit the use of large gas passages. The reduced frictional resistance to such passages improves the compressor efficiency.

The power required to drive the compressor is obviously very important because it affects both the first cost and the running cost of the refrigeration plant.

Although not listed in the table, the critical and freezing temperatures of a substance have also to be noted when assessing its suitability for use as a refrigerant.

There are, of course, factors other than thermodynamic ones which have to be taken into consideration when choosing a refrigerant for a particular vapour compression cycle. These include the heat transfer characteristics, dielectric strength of the vapour, inflammability, toxicity, chemical reaction with metals, tendency to leak and leak detectability, behaviour when in contact with oil, availability and, of course, the cost.

The general concern about the ozone depletion and global warming effects of the refrigerants commonly in use has stimulated searching for alternative, more acceptable refrigerants. Among those considered are: well-established refrigerants such as ammonia (provided proper safety measures are adopted), hydrocarbons (also requiring safety precautions) and mixtures of two or more refrigerants, which introduce unusual effects and are incompatible with conventional mineral oils for lubrication of the compressor.

Butler (1998) discussed refrigerant mixtures with components that boil at different temperatures, giving changes in the boiling points (termed 'glide') as the mixture evaporates or condenses. 'Bubble point' is the temperature at which a refrigerant liquid just starts to evaporate and 'dew point' is the temperature at which the vapour starts to condense. With a single refrigerant these points coincide and an azeotropic mixture having liquid and vapour phases in equilibrium also has a constant boiling point. On the other hand, for a zeotropic mixture of refrigerants, the constituents have different boiling points and the difference is termed the glide value for the mixture. Glide may lead to differential frosting temperatures across an evaporator, and condensers and evaporators tend to be larger than for a single refrigerant. Changes in heat transfer characteristics and refrigerant handling difficulties are possible with glide. Note that a refrigerant and a lubricant is a zeotropic mixture requiring special consideration. Whereas hydrocarbon refrigerants are compatible with conventional mineral oils, HFC refrigerants (such as R134a) and mixtures of them are not and a polyester lubricant is required.

9.12   Ozone depletion effects

The refrigerants that have been widely used in air conditioning and other applications throughout the world since the nineteen thirties, comprise molecular combinations of chlorine, fluorine, carbon and hydrogen. Reference to the footnotes in Table 9.6 verifies this. There is a significant turnover in the quantity of refrigerant used for topping-up purposes in plant, as leaks occur and maintenance and replacement work is carried out. Butler (1991) has suggested that this amounts to 80 per cent of the total usage of refrigerant for air conditioning purposes.

Fully halogenated refrigerants are termed chlorofluorocarbons (CFC). They are chemically very stable and exist for a long time in the atmosphere, after leaking from a refrigeration plant. In due course they rise into the stratosphere and, at a height of between six and thirty miles above the surface of the earth, the molecules break down in the presence of solar

Table 9.7 Ozone depletion factors, according to AAF Ltd. (1990)

Refrigerant	ODF	Proportion of chlorine by weight	Atmospheric half life in years
CFCs			
R11	1.0	77.4%	65
R12	1.0	58.6	146
R113	0.8	56.7	90
R114	0.8	41.5	185
R115	0.4	22.9	380
HCFCs			
R22	0.05	41.0	20
R142b	0.06	35.2	13
R141b	0.1(?)	60.6	10
R125	0.02	26.0	4
R123	0.02	46.3	2
HFCs			
R152a	0	0	1
R134a	0	0	6

Reproduced by courtesy of McQuay Chillers, a Snyder General brand.

radiation, losing an atom of chlorine. Two reactions then take place. First, the chlorine combines with a molecule of ozone to form chlorine monoxide and oxygen:

$$Cl + O_3 \rightarrow ClO + O_2$$

Secondly, the molecules of chlorine monoxide then combine with oxygen to form oxygen molecules and chlorine molecules.

$$2ClO + O_2 \rightarrow 2Cl + 2O_2$$

The chlorine molecules are once again free to combine with ozone and the process repeats up to 10 000 times, some CFCs having an atmospheric half life of the order of 100 years, according to AAF Ltd. (1990). The net result appears to be a steady depletion of the ozone content of the atmosphere at high altitudes. Since ozone is the agent in the upper atmosphere that prevents the entry of excessive, dangerous, ultraviolet radiation the consequences of ozone depletion are obvious and serious.

An ozone depletion factor (ODF) has been calculated for each refrigerant, in relation to a value of 1.0, assigned to R11, which is the worst offender because it contains the largest proportion of chlorine by weight.

If one of the chlorine atoms in the molecule of a CFC refrigerant is replaced by a hydrogen atom the stability of the molecule is much reduced and its ozone depleting effect becomes smaller. This has encouraged the use of such refrigerants, which are termed hydrochlorofluorocarbons (HCFC), as an interim measure, before the use of CFC refrigerants can be fully phased out. An example of an HCFC refrigerant is R22, which has been extensively used for many years in reciprocating compressors, but is no longer acceptable.

Refrigerants with no chlorine present have ODF values of zero. Such refrigerants are termed hydrofluorocarbons (HFC).

Of the refrigerants listed, R134a offers promise, being similar in many respects to R12 but with no ozone depletion effect.

9.13   Global warming

The British Refrigeration Association Specification (1996) describes a way of comparing the influence of different refrigeration systems on climate change. A value, termed the Total Equivalent Warming Impact (TEWI) is determined. This estimates a direct effect, which is the total release of refrigerant and an indirect effect, which is an estimate of the total energy consumption, both expressed in equivalent tonnes of CO_2 over the expected life of the system. The sum of the two effects is the TEWI value. Apparently, the indirect effect of energy consumption usually dwarfs the direct effect of refrigerant release. Miro and Cox (1999) have pointed out that HFC refrigerants, such as R134a, have a much greater global warming potential *on a molecule-to-molecule basis* than CO_2 but their overall contribution is trivially small when the enormous abundance of CO_2 is considered. HFCs contain no chlorine. See Table 9.7.

9.14   Other methods of refrigeration

There are several other forms of refrigeration, the most important of which is vapour absorption, dealt with in chapter 14. Other methods, considered by Butler and Perry (1999), are as follows.

(1) Water vapour systems. The vapour pressures to achieve effective cooling are very low, making the method impractical. With a steam jet as the source of energy the *COP* is less than 0.3.

(2) Stirling cycle. In principle, this should have a higher *COP* than conventional vapour compression. However, hot and cold heat transfer areas are very small, which introduces difficulties. The method has been used in small domestic refrigerators.

(3) Acoustic cooling. Alternate compressions and expansions, causing heating and cooling, are produced in a closed tube by a sound generator. Efficiencies appear to be poor and the plant is comparatively large.

(4) Magnetic cooling. Some metals heat up when magnetised and cool when demagnetised. A prototype plant at the US Department of Energy Ames Laboratory, in Iowa, has rotated a metal disc continuously through strong and weak magnetic fields with heat transferred by water or an anti-freeze fluid to heat exchangers. Initial efficiencies are claimed to be 20 per cent better than conventional methods but with higher capital costs and a simple pay back of 5 years.

(5) Pulse tube cooling. This is similar to acoustic cooling but with a compressor instead of a sound generator, and an inert gas in the tube.

(6) Thermo-electric cooling. This makes use of the Peltier effect and is commercially available with cooling capacities up to 500 W. It has a high capital cost and very low *COP*s.

(7) Thermionic cooling. Cooling occurs at the negative electrode when high energy electrons flow between two electrodes in a vacuum. A better *COP* than vapour compression gives is possible if semiconductor materials are used. It is potentially cheap with future development.

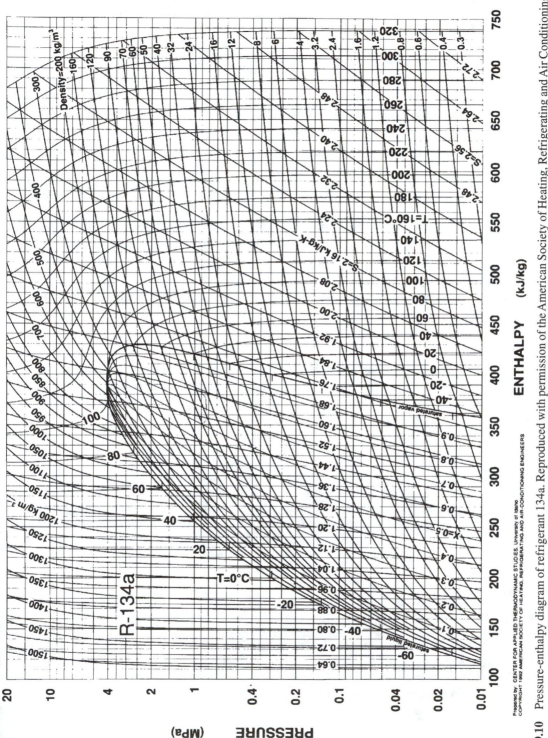

Prepared by: CENTER FOR APPLIED THERMODYNAMIC STUDIES, University of Idaho
COPYRIGHT 1992 AMERICAN SOCIETY OF HEATING, REFRIGERATING AND AIR-CONDITIONING ENGINEERS

Fig. 9.10 Pressure-enthalpy diagram of refrigerant 134a. Reproduced with permission of the American Society of Heating, Refrigerating and Air Conditioning Engineers from the (1977) ASHRAE Handbook.

(8) Air cycle cooling. This is well established for aircraft, but with high capital cost and low efficiency. *COP*s are in the range 0.35 to 0.57. Air is compressed and expanded without a phase change of refrigerant.

9.15 Safety

Whichever refrigerant is used it must be safe. This is dealt with in BS 4434: 1989, covering the design, construction and installation of refrigeration plant and systems. Refrigerants are classified in three groups:

1. These are non-inflammable in vapour form at any concentration in air at standard atmospheric pressure and 20°C. They have a low toxicity although when in contact with a flame or a hot surface toxic products of decomposition may form. Examples are: R11, R12, R13, R22, R113.

2. Toxicity is the dominant feature with these refrigerants and it is almost impossible to avoid a toxic concentration if an escape of refrigerant occurs. An example is: R717 (ammonia).

3. These are inflammable and are an explosive hazard, although with a low order of toxicity. Examples are: R170 (ethane), R290 (propane), R600 (butane).

  Group 3 refrigerants should not be used for institutional or residential buildings, or those buildings used for public assembly.

Exercises

1.   A refrigeration machine works on the simple saturation cycle. If the difference between the enthalpies of saturated liquid at the condensing and evaporating pressures is 158.7 kJ kg^{-1} and the latent heat of vaporisation under evaporating conditions is 1256 kJ kg^{-1}, find the dryness fraction after expansion. Calculate the refrigerating effect.

Answers

0.1265 and 1097.1 kJ kg^{-1}.

2.   Calculate the displacement of a compressor having 176 kW capacity if the refrigerating effect is 1097 kJ kg^{-1} and the volume of the suction gas is 0.2675 m^3 kg^{-1}. Assuming a volumetric efficiency of 75 per cent, what cylinder size will be needed if the speed is to be 25 rev s^{-1} and there are to be 6 cylinders with equal bore and stroke?

Answers

0.0429 m^3 s^{-1} and 78.6 mm.

3.   Find the change in entropy when a liquid refrigerant whose specific heat is 4.71 kJ kg^{-1} K^{-1} cools from 35.5°C to 2°C. If 12.64 per cent liquid then vaporises and the latent heat of that process is 1256 kJ kg^{-1}, what further change in entropy occurs? State the total change in entropy per kg of refrigerant for the cooling and partial vaporisation.

Answers

−0.541 kJ kg^{-1} K^{-1}, +0.577 kJ kg^{-1} K^{-1} and +0.036 kJ kg^{-1} K^{-1}.

4. A 4-cylinder 75 mm bore × 75 mm stroke compressor runs at 25 rev s^{-1} and has a volumetric efficiency of 75 per cent. If the volume of the suction gas is 0.248 m^3 kg^{-1} and the machine has an operating efficiency of 75 per cent, what power will be required on a simple saturation cycle when the difference between the enthalpies of the suction and discharge gases is 150 kJ kg^{-1}? If the refrigerating effect is 1087 kJ kg^{-1} what is the output of the machine in kW of refrigeration? State the coefficient of performance.

Answers

20 kW, 108.7 kW of refrigeration and *COP* = 7.25.

5. Water is used in a simple saturation vapour compression refrigeration cycle and the evaporating and condensing temperatures and absolute pressures are 4.5°C with 0.8424 kPa and 38°C with 6.624 kPa, respectively. Assume that water vapour behaves as an ideal gas with c_p/c_v = 1.322 and calculate the discharge temperature if compression is isentropic. Find also the kW/kW if the refrigerating effect is 2355 kJ kg^{-1}.

Answers

(i) 186°C, (ii) 0.1454 kW per kW of refrigeration.

Notation

Symbol	Description	Unit
COP	coefficient of performance	–
c	constant in equation $pV^n = c$ (see section 9.8)	–
c_1	specific heat of liquid	kJ kg^{-1} K^{-1}
c_p	specific heat of gas at constant pressure	kJ kg^{-1} K^{-1}
c_v	specific heat of gas at constant volume	kJ kg^{-1} K^{-1}
d	diameter	mm or m
f	dryness fraction	–
g	local acceleration due to gravity	m s^{-2}
h	enthalpy	kJ kg^{-1}
h_{1c}	enthalpy of saturated liquid at condensing pressure	kJ kg^{-1}
h_{vc}	enthalpy of saturated vapour at condensing pressure	kJ kg^{-1}
h_{vd}	enthalpy of saturated vapour at discharge condition	kJ kg^{-1}
h_{ve}	enthalpy of saturated vapour at evaporating pressure	kJ kg^{-1}
m	mass of refrigerant	kg
$\dot{m}$	mass flow rate of refrigerant	kg s^{-1}
n	exponent in equation $pV^n = c$ (see section 9.8)	–
p	pressure	Pa or kPa
p_c	condensing pressure	Pa or kPa
p_d	discharge pressure	Pa or kPa
p_e	evaporating pressure	Pa or kPa
Q	rate of heat transfer to the system	W or kW
Q_c	rate of heat rejection at condenser	W or kW
Q_{ra}	actual refrigeration capacity	W or kW
q	specific heat energy	kJ kg^{-1}
q_c	heat rejected at condenser	J kg^{-1} or kJ kg^{-1}
q_j	heat transferred to cylinder jacket during compression	J kg^{-1} or kJ kg^{-1}

q_r	refrigerating effect	J kg^{-1} or kJ kg^{-1}
R	particular gas constant	kJ kg^{-1} K^{-1}
s	entropy	kJ kg^{-1} K^{-1}
s_1	entropy of saturated vapour at evaporating pressure	kJ kg^{-1} K^{-1}
s_2	entropy of saturated vapour at condensing pressure	kJ kg^{-1} K^{-1}
s_3	entropy of saturated liquid at condensing pressure	kJ kg^{-1} K^{-1}
T	Absolute temperature	K
T_c	absolute condensing temperature	K
T_d	absolute discharge temperature	K
T_e	absolute evaporating temperature	K
t_c	condensing temperature	°C
t_e	evaporating temperature	°C
V	volume	m^3
V_d	volume at discharge condition	m^3
V_e	volume of saturated vapour at evaporating pressure	m^3
v	velocity	m s^{-1}
v_s	specific volume	m^3 kg^{-1}
$\dot{v}$	volumetric flow rate	m^3 s^{-1}
W	rate of work done by the system	W or kW
W_r	power needed for compression	W or kW
w_r	work done in compression	J kg^{-1} or kJ kg^{-1}
z	elevation above a reference level	m
γ	c_p/c_v	–
η_v	volumetric efficiency	%

References

AAF Ltd. (1990): *The CFC Issue*, McQuay Snyder General, Cramlington, Northumberland.

ASHRAE Handbook (1997): Fundamentals 1.1.

Blackhurst, D.R. (1999): Recent Developments in Thermosyphon Cooling for Air Conditioning, *Proc. Inst. R.* 1998–99, 4.1–4.11.

British Refrigeration Association Specification (1996): *Guideline methods of calculating TEWI*, Issue 1, ISBN 1-870-623126, July.

BS 4434: (1989): Safety aspects in the design, construction and installation of refrigeration appliances and systems.

Butler, D. (1991): Refrigeration: Time to decide, *Building Services*, December, 19–21.

Butler, D. (1998): Testing alternative refrigerants for air conditioning systems, *Building Services*, 45–47.

Butler, D. and Perry, A. (1999): Novel forms of refrigeration, A BRE Investigation, *Building Services Journal*, February, 45–46.

Miro, C.R. and Cox, J.E. (1999): Ozone depletion and climate change, *ASHRAE Journal*, February, 22.

Pearson, S.F. (1990): Thermosyphon cooling, *Proc. Inst. R.*, 1989–90, 6–1 to 6–10.

Pearson, S.F. (1998): Enhanced Heat Transfer Surfaces: A Method of Improving Performance of Refrigerating Surfaces, *Proc. Inst. R.*, April.

Spalding, D.H. and Cole, E.H. (1961): *Engineering Thermodynamics*, Edward Arnold (Publishers) Ltd.

Trane Air Conditioning Manual (1961): Table 6–7, 343, The Trane Company, La Crosse, Wisconsin.

10

Air Cooler Coils

10.1 Distinction between cooler coils and air washers

The evaporative cooling of air by adiabatic saturation (see sections 3.5 and 3.6) may be improved by lowering the temperature of the feed water, but little improvement results if all the feed water is evaporated. The picture changes, however, if the rate of injection of feed water is increased so that not all of it is evaporated, but some falls into a sump and is recirculated. When the rate of water flow is speeded up its temperature has an increasingly important influence on the heat exchange with the airstream. For example, if an infinitely large quantity of water were circulated at a temperature t_w, then the air would leave the spray chamber also at a temperature t_w, in a saturated condition.

In air washers used for cooling and dehumidification processes (see section 3.4), the quantities of water circulated are very large compared with the quantity that could be totally evaporated. The temperature of the water thus plays a major part in determining the state of the air leaving the washer. Figure 10.1 illustrates the psychrometric considerations.

Direct cooling of this sort is not necessarily the most effective method. True contra-flow heat exchange, or even a reasonable approximation to it, cannot be easily obtained in an air washer. On the other hand, when chilled water is circulated through a cooler coil of more than three rows a good measure of contra-flow is realised. This permits a closer approach between the leaving air temperature and the entering-water temperature with cooler coils using chilled water or brine.

In comparison with the cooler coil the washer suffers from a number of other disadvantages:

(i)   It is more bulky.
(ii) Corrosion is a greater risk.
(iii) Maintenance is more expensive.
(iv) It uses an open chilled water circuit, thus causing the deposition of scale, rust, slime, etc. in the water chiller (evaporator), reducing its heat transfer efficiency and increasing the cost of the refrigeration plant.
(v)   Micro-organisms tend to establish colonies in the recirculated water tank. These may constitute a health risk if carried by the airstream into the conditioned space. Pickering and Jones (1986) discuss how an allergic response (humidifier fever) may be induced in some susceptible people who develop symptoms similar to those of mild influenza at the beginning of the working week. Immunity is acquired by the middle of the week but is lost over the weekend. The cycle can repeat.

Humidification by the injection of dry steam (see section 3.7) in the right place in a system is much better than the use of an air washer. It is sterile if not wet and perfectly safe.

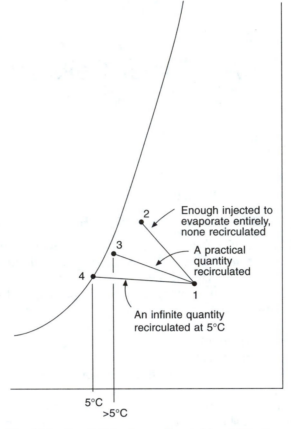

Fig. 10.1   Possibilities for cooling and humidification.

There is one point strongly in favour of air washers and sprayed cooler coils: the presence of the large mass of water in the sump tank and spray chamber gives a thermal inertia to the system, smoothing out fluctuations in the state of the air leaving the coil or washer, and adding stability to the operation of the automatic controls.

However, in spite of these favourable aspects, the air washer is very much out of fashion because the cooler coil is more efficient, and because of the five points mentioned.

10.2   Cooler coil construction

A cooler coil is not merely a heater battery fed with chilled water or into which cold, liquid refrigerant is pumped. There are two important points of difference: firstly, the temperature differences involved are very much less for a cooler coil than for a heater battery, and secondly, moisture is condensed from the air on to the cooler coil surface. With air heaters, water entering and leaving at 85°C and 65°C respectively may be used to raise the temperature of an airstream from 0° to 35°C, resulting in a log mean temperature difference of about 53°C for a contra-flow heat exchange. With a cooler coil, water may enter at 7°C and leave at 13°C in reducing the temperature of the airstream from 26°C to 11°C, a log mean temperature difference of only 7.6°C with contra-flow operation. The result is that much more heat transfer surface is required for cooler coils and, as will be seen in section 10.3, it is important that contra-flow heat exchange be obtained. The second point of difference, that dehumidification occurs, means that the heat transfer processes are more involved in cooler coils.

There are three forms of cooler coil: chilled water, direct expansion, and chilled brine. The first and third types make use of the heat absorbed by the chilled liquid as it is circulated inside the finned tubes of the coil to effect the necessary cooling and dehumidification of the airstream. The second form has liquid refrigerant boiling within the tubes, and so the heat absorbed from the airstream provides the latent heat of evaporation for the refrigerant.

Chilled water coils are usually constructed of externally finned, horizontal tubes, so arranged as to facilitate the drainage of condensed moisture from the fins. Tube diameters vary from 8 to 25 mm, and copper is the material commonly used, with copper or aluminium fins. Copper fins and copper tubes generally offer the best resistance to corrosion, particularly if the whole assembly is electro-tinned after manufacture. Fins are usually of the plate type, although spirally wound and circular fins are also used. Cross-flow heat exchange between the air and cooling fluid occurs for a particular row but, from row to row, contra- or parallel-flow of heat may take place, depending on the way in which the piping has been arranged. Figure 10.2(*a*) illustrates the case of a single serpentine tubing but a double serpentine, and other arrangements, are also used. In the double serpentine form shown in Figure 10.2(*b*) two pipes from the flow riser feed the first and third rows with two tubes from the second and fourth rows leading to the return header. Contra-flow connection is essential for chilled water coils in all cases. In direct expansion coils, since the refrigerant is boiling at a constant temperature the surface temperature is more uniform and there is no distinction between the parallel and contra-flow, the logarithmic mean temperature difference being the same. However, with direct-expansion cooler coils a good deal more trouble has to be taken with the piping in order to ensure that a uniform distribution of liquid refrigerant takes place across the face of the coil. This is achieved by having a 'distributor' after the expansion valve, the function of which is to divide the flow of liquid refrigerant into a number of equal streams. Pipes of equal resistance join the downstream side of the distributor to the coil so that the liquid is fed uniformly over the depth and height of the coil. It may be necessary to feed the coil from both sides if it is very wide. The limitations imposed by the need to secure effective distribution of the liquid refrigerant throughout the coil tend to discourage the use of very large direct-expansion cooler coils. Control problems exist.

All cooler coils should be divided into sections by horizontal, independently drained, condensate collection trays running across their full width and depth. Opinions seem to differ among manufacturers as to the maximum permissible vertical spacing between condensate drip trays. Clearly it depends on the sensible-total heat ratio (the smaller this is the greater the condensation rate), the spacing between the fins (the narrower the spacing the more difficult it is for the condensate to drain freely) and the face velocity (the faster the airflow the more probable the carryover of condensate). Fin spacings in common use are 316, 394 and 476 per metre (8, 10 and 12 per inch) and the thicknesses used lie between 0.42 and 0.15 mm. (Thinner fins, incidentally, tend to grip the tube less tightly at their roots and perhaps give poorer heat transfer.) Fins may be corrugated or smooth, the former reducing the risk of carryover while improving the heat transfer by a small increase in the surface area of the fins. An analysis of manufacturers' data suggests that for cooling coils having sensible-total heat ratios of not less than 0.65 the face velocities listed in Table 10.1 should not be exceeded without the provision of moisture eliminators.

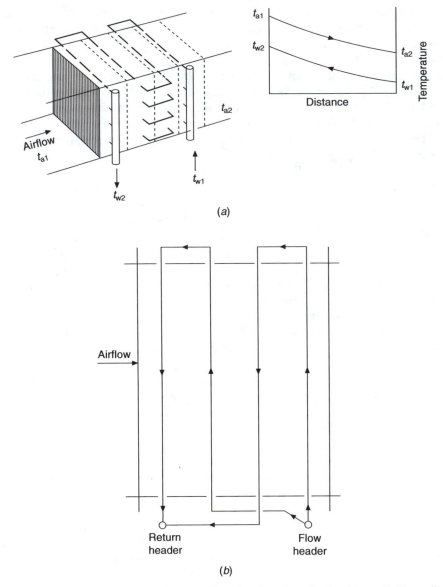

Fig. 10.2 (*a*) This arrangement shows contra-flow for the rows of a four row coil. Cross-flow heat exchange occurs in individual rows. (*b*) Double serpentine piping arrangement for a four row coil. Other arrangements are also used.

Table 10.1 Fin spacing and air velocity

Fin spacing per metre	Max. velocity m s^{-1}
316	2.5
394	2.2
476	2.1

The maximum vertical distance between intermediate drain trays should desirably not exceed 900 mm and more than 394 fins per metre should not be used with coils having large latent loads when the sensible-total heat ratios are less than 0.80. For sensible-total ratios less than 0.65 and for sprayed coils more than 316 fins per metre should not be used. When the sensible-total ratio is between 0.8 and 0.95 it is possibly safe to have drain trays 1200 mm apart, provided the face velocity and finning conforms with the suggestions in Table 10.1. Coils with sensible-total ratios exceeding 0.98 are virtually doing sensible cooling only and the risk of condensate carryover is slight. Water velocities in use are between 0.6 and 2.4 m s^{-1}, in which range the coils are self-purging of air. Water pressure drops are usually between 15 and 150 kPa and air pressure drops are dependent on the number of rows and the piping and finning arrangements. A coil that is doing no latent cooling offers about one-third less resistance to airflow. Typical air-side pressure drops for a four row coil with 2.25 m s^{-1} face velocity are 60 to 190 Pa, when wet with condensate.

Condensate trays should slope towards the drainage point and there must be adequate access for regular cleaning. It is important that the piping connection from the drainage point should be provided with a trap outside the coil. This must be deep enough to provide a water seal of condensate that will prevent air being sucked into the drainage tray in the case of draw-through coils, or blown out in the case of blow-through coils. Condensate will not drain freely away unless there is no airflow through the outlet point. The trap should feed condensate through an air gap into a tundish, prior to the condensate being piped to the drains. The air gap is necessary so that the presence of condensate drainage can be verified by observation and, for hygienic reasons, to ensure there is no direct connection between the main drains and the air conditioning system. See Figure 10.5.

After installation the aluminium fins on cooler coils do not give uniform condensation over their entire surface area until they have aged over about a year of use. ASHRAE (1996) mentions the development of a hydrophylic surface coating for aluminium fins that reduces the surface tension of the condensate and gives a more uniform distribution on the fins from the start.

Careless handling in manufacture, delivery to site and erection often causes damage to the coil faces, forming large areas of turned-back fin edges that disturb the airflow, collect dirt from the airstream and increase the air pressure drop. The fins in such damaged areas must be combed out after installation before the system is set to work.

Other materials are sometimes used for air cooler coils but ordinary steel coils should never be used because of the rapid corrosion likely. Stainless steel is sometimes used but it is expensive and, because its thermal conductivity is less than that of copper, more heat transfer surface is required.

Air cooler coils tend to be wide and short, rather than narrow and tall. This is because it is cheaper to make coils with this shape, there being fewer return bend connections to make (where tubes emerge from the coil casing). It is also because a short coil drains condensate away more easily: with a tall coil there is the likelihood of condensate building up between the fins at the bottom of the coil, inhibiting airflow and heat transfer and increasing the risk of condensate carry-over into the duct system.

A consequence of the wide shape of cooler coil faces is that airflow over them is likely to be uneven, the airstream tending to flow over the middle of the coil face. This is usually dealt with in air handling units by using multiple fans in parallel.

Blow-through coils are sometimes used but may give unsatisfactory results because the airflow discharged from a fan is very turbulent and, even if multiple fans in parallel are used, air distribution over the coil face will be uneven. The coil should be as far as possible from the fan outlet in order to give the turbulent airstream a chance to become smoother.

With draw-through coils, smooth holes having belled edges are normally provided in the casings where the tubes emerge to join the headers or return bends. The holes are slightly larger than the outer diameters of the tubes in order to provide a clearance space for thermal movement. It follows that, with blow-through coils, condensate will be blown out of such clearance spaces. This cannot be allowed and coils used for blow-through applications must have the clearance spaces sealed by the manufacturer.

Galvanised steel casings are often used for coils with copper tubes and copper or aluminium fins. This is a poor combination since copper and zinc in conjunction with slightly acidic condensate favour electrolytic corrosion. If possible, other materials should be used for cooler coil casings.

Drain cocks and air vents should always be provided for cooler coils using chilled water or brine.

10.3   Parallel and contra-flow

Consider a four-row coil in stylised form, as shown in Figure 10.3(*a*). Air enters the first row at state 0. Water enters at a temperature t_{wa} and, its temperature rising as it absorbs heat from the airstream, leaves at a higher value. Denote the mean surface temperature of the first row by t_{w1}. If the first row had a contact factor of unity, the air would be conditioned to saturation at a state 1′, as shown in Figure 10.3(*b*). In fact, the contact factor β is less than unity and the air leaves at a state denoted by 1. This is now the entry state for the second row, the mean surface temperature of which is t_{w2}. In a similar way, the exit state from the second row is 2, rather than 2′. This reasoning is applied to the succeeding rows, from which the leaving states of the airstream are denoted by 3 and 4. A line joining the state points 0, 1, 2, 3 and 4 is a concave curve and represents the change of state of the air as it flows past the rows under parallel-flow conditions. A straight line joining the points 0 and 4 indicates the actual overall performance of the coil. This condition line, replacing the condition curve, cuts the saturation curve at a point A when produced. By the reasoning adopted when flow through a single row was considered, one may regard the temperature of A as the mean coil surface temperature for the whole coil. It is denoted by t_{sm}.

Similar considerations apply when contra-flow is dealt with, but the result is different. A convex condition curve is obtained by joining the points 0, 1, 2, 3 and 4. Referring to Figure 10.4 it can be seen that, this time, air entering the first row encounters a mean coil surface temperature for the row of t_{w1} and that this is higher than the mean surface temperatures of any of the succeeding rows. As air flow through the four rows occurs, progressively lower surface temperatures are met. The tendency of the state of the air is to approach the apparatus dew points, 1′, 2′, 3′ and 4′, each at a lower value than the point preceding it, as it passes through the rows. The condition curve thus follows a downward trend as flow proceeds.

The result of this is that, by comparison with the case of a parallel flow, a lower leaving air temperature is achieved, greater heat transfer occurs, and the coil is more efficient, if it is piped for contra-flow operation.

The above analysis refers to what happens along the centre-line of the cooler coil. For any given row, the heat transfer is cross-flow and a different picture would emerge if any other section of the four-row coil, in the direction of airflow, were considered. The same argument could be adopted but the curve of the process line would not have the smooth

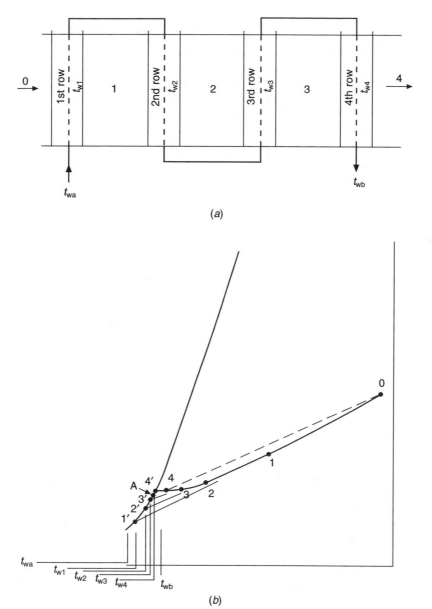

(a)

(b)

Fig. 10.3 Parallel flow. The points 0, 4 and A are in a straight line and A lies on the 100 per cent saturation curve. A is the apparatus dew point and its temperature, t_{sm}, is the mean coil surface temperature for the whole four rows of the coil. The temperatures t_{w1}, t_{w2}, t_{w3} and t_{w4} in figure (b) are the mean surface temperatures of the 1st, 2nd, 3rd and 4th rows in figure (a).

shape shown in Figures 10.3(b) and 10.4(b). There would also be some lateral variation in the psychrometric state of the air after each row. Since, for air conditioning purposes, one is only interested in the mean state of the air entering and leaving the coil it is common sense to show the condition line as a straight line joining the entering and leaving states, which must cut the saturation curve at the apparatus dew point when extended. The temperature of the apparatus dew point can still be regarded as the mean coil surface temperature.

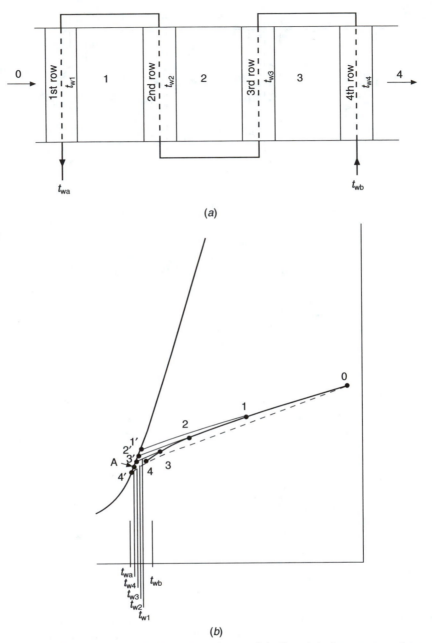

(a)

(b)

Fig. 10.4 Contra-flow. The points 0, 4 and A are in a straight line. A is the apparatus dew point and its temperature is the mean coil surface temperature, t_{sm}.

10.4   Contact factor

A psychrometric definition of this was given in section 3.4. Such a definition is not always useful—for example, in a cooler coil for sensible cooling only—and so it is worth considering another approach, in terms of the heat transfer involved, that is in some respects more informative though not precise.

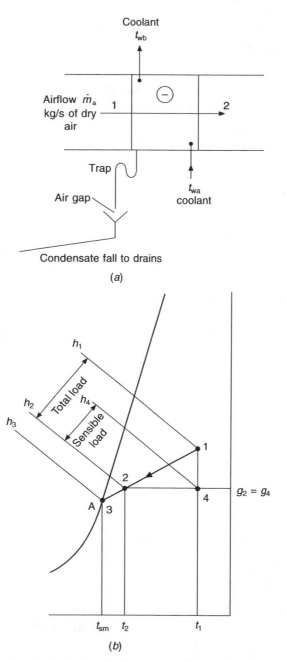

Fig. 10.5   The psychrometric relationship between sensible and total cooling load.

Coils used for dehumidification as well as for cooling remove latent heat as well as sensible heat from the airstream. This introduces the idea of the ratio S, defined by the expression:

$$S = \frac{\text{sensible heat removed by the coil}}{\text{total heat removed by the coil}}$$

In terms of Figure 10.5 this becomes

$$S = \frac{h_4 - h_2}{h_1 - h_2} \tag{10.1}$$

If the total rate of heat removal is Q_t when a mass of dry air $\dot{m}_a$ in kg s^{-1} is flowing over the cooler coil, then the sensible heat ratio S, can also be written as

$$S = \frac{\dot{m}_a c(t_1 - t_2)}{Q_t} \tag{10.2}$$

where c is the humid specific heat of the airstream.

The approximate expression for sensible heat exchange given by equation (6.7) allows equation (10.2) to be re-written as

$$Q_t = 1.25 \times \dot{v} \times (t_1 - t_2)/S \tag{10.3}$$

Sensible heat transfer, Q_s, can also be considered in terms of the outside surface air film resistance of the coil, R_a:

$$Q_s = (LMTD)_{as} A_t / R_a \tag{10.4}$$

where A_t is the total external surface of the coil and $LMTD_{as}$ is the logarithmic mean temperature difference between the airstream and the mean coil surface temperature, t_{sm}.

$$LMTD_{as} = [(t_1 - t_{sm}) - (t_2 - t_{sm})]/\ln[(t_1 - t_{sm})/(t_2 - t_{sm})] \tag{10.5}$$

Equation (10.4) then becomes

$$Q_s = \frac{(t_1 - t_2)}{\ln[(t_1 - t_{sm})/(t_2 - t_{sm})]} \cdot \frac{A_t}{R_a} \tag{10.6}$$

and therefore

$$1.25 \times \dot{v} \times (t_1 - t_2) = \frac{(t_1 - t_2)}{\ln[(t_1 - t_{sm})/(t_2 - t_{sm})]} \cdot \frac{A_t}{R_a}$$

whence

$$\ln[(t_1 - t_{sm})/(t_2 - t_{sm})] = (A_t/R_a)/(1.25 \times \dot{v}) \tag{10.7}$$

Since $\dot{v}$ equals $A_f \times v_f$, where A_f and v_f are the face area and face velocity, respectively, equation (10.7) becomes

$$\ln[(t_1 - t_{sm})/(t_2 - t_{sm})] = (A_t/A_f)/(R_a \times 1.25 \times v_f) = k$$

k is a constant for a given coil and face velocity but it takes no account of heat transfer through the inside of the tubes. We can now write

$$(t_2 - t_{sm})/(t_1 - t_{sm}) = \exp(-k)$$

Reference to Figure 10.5 shows that this is the approximate definition of the by-pass factor (see equation (3.4)) and hence equals $(1 - \beta)$ where β is the contact factor. If r is the number of rows and A_r is the total external surface area per row, $A_t = A_r \times r$ and k becomes $(A_r/A_f)(r)/(R_a 1.25 v_f)$. An approximate expression for the contact factor now emerges:

$$\beta = 1 - \exp\left(-\frac{A_r}{A_f} \cdot \frac{r}{1.25 R_a v_f}\right) \tag{10.8}$$

Note that the contact factor is independent of the psychrometric state and the coolant temperature, provided that the mass flow ratio of air to water (usually about unity) remains fairly constant.

EXAMPLE 10.1

A four-row coil with a face velocity of 2.5 m s^{-1} has a contact factor of 0.85. Calculate the contact factor for the following cases: (*a*) 3.0 m s^{-1} and four rows, (*b*) 2.0 m s^{-1} and four rows, (*c*) 2.5 m s^{-1} and six rows, (*d*) 2.5 m s^{-1} and two rows, (*e*) 2.5 m s^{-1} and eight rows. Assume that changes in the face velocity do not significantly alter R_a.

Answer

By equation (10.8) we can calculate that

$$0.85 = 1 - \exp\left(-\frac{A_r}{A_f} \cdot \frac{4}{1.25 \times R_a \times 2.5}\right)$$

$$0.15 = \exp\left(-1.28 \times \frac{A_r}{A_f R_a}\right)$$

$$\ln 0.15 = \left(-1.28 \times \frac{A_r}{A_f R_a}\right)$$

$$\frac{A_r}{A_f R_a} = \frac{-1.8971}{-1.28} = 1.482$$

This constant can now be used to answer the questions:

(*a*) β = 1 − exp[−1.482 × 4/(1.25 × 3.0)] = 0.79
(*b*) β = 1 − exp[−1.482 × 4/(1.25 × 2.0)] = 0.91
(*c*) β = 1 − exp[−1.482 × 6/(1.25 × 2.5)] = 0.94
(*d*) β = 1 − exp[−1.482 × 2/(1.25 × 2.5)] = 0.61
(*e*) β = 1 − exp[−1.482 × 8/(1.25 × 2.5)] = 0.98

10.5 Heat and mass transfer to cooler coils

Heat transfer to a cooler coil involves three stages: heat flows from the airstream to the outer surface of the fins and pipes, it is then transferred through the metal of the fins and the wall of the piping and, finally, it passes from the inner walls of the tubes through the surface film of the cooling fluid to the main stream of the coolant.

In general, dehumidification occurs as well as cooling so that the behaviour of cooler coils cannot be described in simple terms. A very approximate approach, adopted by some manufacturers, is to establish a *U*-value for the coil that is given a bias to account for the extra heat flow by virtue of condensation. The duty is then described by

$$Q_t = U_t A_t (LMTD)_{aw} \tag{10.9}$$

where U_t is the biased *U*-value, A_t the total external surface area and $(LMTD)_{aw}$ is the logarithmic temperature difference between the airstream and the water flowing, defined by

$$(LMTD)_{aw} = \frac{(t_1 - t_{wb}) - (t_2 - t_{wa})}{\ln[(t_1 - t_{wb})/(t_2 - t_{wa})]} \qquad (10.10)$$

using the notation of Figure 10.5. The biased thermal transmission coefficient is expressed by

$$1/U_t = R = SR_a + R_m + R_w \qquad (10.11)$$

where R_a is the thermal resistance of the air film when the external surface is dry, S is the sensible-total heat transfer ratio (see equation (10.1)), R_m is the thermal resistance of the tubes (R_t) plus the fins (R_f), and R_w is the resistance of the water film within the tubes. R_w must be multiplied by the ratio A_t/A_i, where A_t is the total external surface area and A_i is the internal surface area, so that it refers to the total external surface area. R is the total thermal resistance, air to water.

R_a is the reciprocal of the sensible heat transfer coefficient on the air side, h_a, which depends principally on the mass flow rate of the airstream. For standard air, a staggered arrangement of tubes, and 316 fins per metre (8 per inch), h_a is given by McAdams (1954) as:

$$h_a = 27.42v_f^{0.8} \qquad (10.12)$$

wherein v_f is the face velocity entering the cooler coil in ms^{-1}.

If the coil is only partially wet then there are two U-values: one using SR_a in equation (10.11) and referring to the wetted part of the external surface, and the other using R_a in the equation and referring to the dry part of the surface. Determining the boundary between the wet and dry areas is not straightforward but methods are given in by ARI (1991) and ASHRAE (1996).

The resistance of the fins when dry is described by ARI (1991) as:

$$R_f = \frac{1 - \eta}{\eta} R_a \qquad (10.13)$$

If the fins are wet then SR_a replaces R_a. The term η is the total surface effectiveness and is defined by

$$\eta = (\phi A_{fi} + A_p)/A_t \qquad (10.14)$$

where A_{fi} is the surface area of the fins, A_p the external area of the tubes, A_t the total external surface area and ϕ is the fin efficiency. The latter is an involved function depending on the practical features of coil construction. For plate fins (at least) its value seems to be virtually independent of face velocity or fin spacing. On the other hand, the number of rows, fin thickness and tube spacing are all significant, as Table 10.2 shows.

Using copper as a material instead of aluminium improves efficiency but the use of larger diameter tubing reduces it.

The resistance of the metal of the tubes, R_t, referred to the total external surface area, is given by McAdams (1954) as

$$R_t = \frac{A_t}{A_i} \frac{d_o}{2\lambda_t} \ln \frac{d_o}{d_i} \qquad (10.15)$$

wherein A_t/A_i is the ratio of the total external surface to the total internal surface area, λ_t is the thermal conductivity of the metal of the tubes, d_o is the outer tube diameter and d_i its inner diameter, both in metres.

Table 10.2   Approximate fin efficiencies, based on ARI (1991), for flat plate fins, 2.5 m s^{-1} face velocity and tubes of 15 mm nominal outside diameter. It is to be noted that McQuiston (1975) claims a reduction in fin efficiency occurs with increased latent load

Tube spacing mm	Fin thicknesses mm	Fin material	Fin efficiencies for various numbers of coil rows			
			2	4	6	8
37.5	0.15	Al	0.56	0.83	0.95	0.99
50.0	0.15	Al	0.42	0.73	0.90	0.95
37.5	0.42	Al	0.77	0.92	0.98	0.99
50.0	0.42	Al	0.66	0.87	0.95	0.97
37.5	0.15	Cu	—	0.92	0.97	—
50.0	0.15	Cu	—	0.84	0.94	—
37.5	0.42	Cu	—	0.96	0.99	—
50.0	0.42	Cu	—	0.94	0.98	—

For water at temperatures between 4.4°C and 100°C the thermal resistance of the water film inside the tubes, R_w, referred to their inner surface area, is given by the following simplified equation from ARI (1991):

$$R_w = d_i^{0.2}/[(1429 + 20.9t_{wm})v^{0.8}] \tag{10.16}$$

in which t_{wm} is the mean water temperature, v is the mean water velocity and d_i is the inside diameter of the tube in metres.

We can now calculate a U-value for a cooler coil and make an approximate estimate of a chilled water flow temperature to achieve a given duty.

EXAMPLE 10.2

A cooler coil is capable of cooling and dehumidifying air from 28°C dry-bulb, 19.5°C wet-bulb (sling), 55.36 kJ kg^{-1}, 0.8674 m^3 kg^{-1} and 10.65 g per kg to 12°C dry-bulb, 11.3°C wet-bulb (sling), 32.41 kJ kg^{-1}, 0.8178 m^3 kg^{-1} and 8.062 g per kg. 4.75 m^3 s^{-1} of air enters the coil and chilled water is used with a temperature rise of 5.5°C. The coil has the following construction: 6 rows, 2.64 m s^{-1} face velocity, 316 fins per metre, aluminium flat plate fins of thickness 0.42 mm, tubes spaced at 37.5 mm, face dimensions of 1.5 m width and 1.2 m height, tubes of 15 mm outer diameter and 13.6 mm inner diameter. Determine the following: (*a*) total external surface area, (*b*) total internal surface area, (*c*) ratio of external to internal areas, (*d*) U-value, assuming the coil is entirely wet with condensate, (*e*) logarithmic mean temperature difference and (*f*) chilled water flow temperature. Take the thermal conductivity of copper as 390 W m^{-1} K^{-1}.

Answer

(*a*) Number of fins = 1.5 × 316 = 474 over a width of 1.5 m.
Length of one fin in direction of airflow = 6 × 0.0375 = 0.225 m
Gross surface area of one fin

$$= (1.2 \times 0.000\,42 + 0.225 \times 0.000\,42 + 1.2 \times 0.225)2 = 0.5412 \text{ m}^2$$

Number of tubes per row = 1.2/0.0375 = 32.

Total number of tubes passing through fin plates = 6 × 32 = 192.

External cross-sectional area of one tube = $\pi \times 0.015^2/4 = 0.000\,177$ m^2

Since each fin is perforated on each of its sides by 192 tubes the net surface area of one fin is

$$0.5412 - 0.000\,177 \times 192 \times 2 = 0.4732 \text{ m}^2$$

Net surface area of all fins = 474 × 0.4732 = 224.3 m^2.

External perimeter of one tube = $\pi \times 0.015 = 0.0471$ m.

Gross external surface area of all tubes = 192 × 0.0471 × 1.5 = 13.56 m^2

Area of one tube obscured by one fin edge

$$= 0.000\,42 \times 0.0471 = 0.000\,019\,78 \text{ m}^2$$

Area of all tubes obscured by all fin edges

$$= 0.000\,019\,78 \times 192 \times 474 = 1.8 \text{ m}^2$$

Net area of all tubes = 13.56 − 1.8 = 11.76 m^2.

Total external surface area = A_t = 224.3 + 11.76 = 236.1 m^2.

(b) Internal perimeter of one tube = $\pi \times 0.0136 = 0.0427$ m.

$$\text{Total internal surface area of all tubes} = A_i = 192 \times 1.5 \times 0.0427$$
$$= 12.3 \text{ m}^2$$

(c) Ratio of external to internal surface areas

$$= A_t/A_i = 236.1/12.3 = 19.2$$

(d) Plotting the entering and leaving air states on a psychrometric chart, as a check that the coil performance is possible, establishes a contact factor of 0.906 and a mean coil surface temperature of 10.35°C. The total cooling load is then calculated as

$$Q_t = 4.75(55.36 - 32.41)/0.8674 = 125.7 \text{ kW}$$

By equation (6.6) the sensible component of the load is

$$Q_s = 4.75(28 - 12)358/(273 + 28) = 90.4 \text{ kW}$$

Hence $S = Q_s/Q_t = 0.72$.

By equation (10.12)

$$R_a = 1/h_a = 1/(27.42 \times 2.64^{0.8}) = 0.016\,77 \text{ m}^2 \text{ K W}^{-1} \text{ when dry}$$
$$R_a = 0.016\,77 \times 0.72 = 0.012\,07 \text{ m}^2 \text{ K W}^{-1} \text{ when wet}$$

From Table 10.2 the fin efficiency is 0.98 and by equations (10.14) the effectiveness of the surface is

$$\eta = (0.98 \times 224.3 + 11.76)/236.1$$
$$= 0.98$$

By equation (10.13) the resistance of the wet fins is

$$R_f = \frac{(1 - 0.98)}{0.98} \times 0.012\,07$$
$$= 0.000\,246 \text{ m}^2 \text{ K W}^{-1}$$

By equation (10.15) the resistance of the metal of the tubes is

$$R_t = 19.2 \times \frac{0.015}{2 \times 390} \ln \frac{0.0150}{0.0136}$$

$$= 0.000\ 036\ 2 \text{ m}^2 \text{ K W}^{-1}$$

Hence the resistance of the metal fins and tubes is

$$R_m = R_f + R_t$$

$$= 0.000\ 246 + 0.000\ 036\ 2$$

$$= 0.000\ 282 \text{ m}^2 \text{ K W}^{-1} \text{ when wet}$$

By equation (10.16) we can calculate R_w if the mean velocity of waterflow through the 32 tubes passing in series through the six rows is known.

$$\text{Mass flow of water} = 125.7/(4.19 \times 5.5)$$

$$= 5.454 \text{ kg s}^{-1}$$

Assuming a mean water temperature of 10°C the density is 999.7 kg m^{-3} and the total volumetric flow rate of water is $(5.454/999.7)1000 = 5.456$ litres s^{-1}.
The total cross-sectional area of the tubes through which this flows is

$$32 \times \pi \times 0.0136^2/4 = 0.004\ 649 \text{ m}^2$$

Hence the mean velocity of water flow is

$$5.456/(0.004\ 649 \times 1000) = 1.174 \text{ m s}^{-1}$$

By equation (10.16)

$$R_w = 0.0136^{0.2}/[(1429 + 20.9 \times 10) \times 1.174^{0.8}] = 0.000\ 227 \text{ m}^2 \text{ K W}^{-1}$$

referred to the internal tube surface. Hence, referred to the external surface area

$$R_w = 0.000\ 227 \times 19.2 = 0.004\ 358 \text{ m}^2 \text{ K W}^{-1}$$

Thus

$$U_t = 1/(0.004\ 358 + 0.000\ 282 + 0.012\ 07) = 1/0.0167 = 59.8 \text{ W m}^{-2} \text{ K}^{-1}$$

(e) $LMTD = Q_t/(U_t \times A_t) = 125\ 700/(59.8 \times 236.1) = 8.9°C$
(f) Using the notation of Figure 10.5 and equation (10.10) we have

$$8.9 = \frac{(28 - t_{wb}) - (12 - t_{wa})}{\ln[(28 - t_{wb})/(12 - t_{wa})]}$$

Since $t_{wb} = t_{wa} + 5.5$

$$8.9 = \frac{(22.5 - t_{wb}) - (12 - t_{wa})}{\ln[(22.5 - t_{wb})/(12 - t_{wa})]}$$

$$\ln[(22.5 - t_{wa})/(12 - t_{wa})] = 10.5/8.9 = 1.18$$

$$(22.5 - t_{wa})/(12 - t_{wa}) = \exp(1.18) = 3.25$$

$$t_{wa} = (3.25 \times 12 - 22.5)/2.25 = 7.3°C$$

The method of enhancing the U-value of a cooler coil in order to account for the dehumidification occurring, outlined above and used in example 10.2, cannot be regarded as anything other than very approximate. Some of the reasons are:

(a)   The heat transfer is cross-flow for any particular row, even though it is contra-flow from row to row. Hence the logarithmic mean temperature difference used in the foregoing should be modified to take account of this.

(b)   The airflow distribution over the face of the coil is not uniform, being concentrated in the centre with a tendency to stagnation at the corners. Hence the value of the heat transfer coefficient through the air film, h_a, calculated by equation (10.12), is not uniform over the coil face.

(c)   The use of the sensible-total heat transfer ratio, S, must be regarded as an approximation to account for dehumidification. Well-verified techniques for predicting the value of h_a are not available according to ASHRAE (1996) and it is best to use experimentally determined data, obtained from manufacturers.

(d)   If the bond between the root of the fin and the tube is not good, there is likely to be a significant loss of heat transfer. This may be true of some plate fins if the plates are too thin: there is then insufficient plate material to flow properly when punched and form a good, cylindrical collar (to separate the fins) in the process of manufacture. Furthermore, the collars may be cracked.

(e)   It is probable that not all the external surface of the fins and tubes is wet with condensate. In this case the method outlined will give an optimistic assessment of the performance.

(f)   The U-value is not constant throughout the cooler coil.

(g)   The psychrometric state of the air leaving the coil is not uniform across the face and hence calculations based on a single state will not be precise.

Note that the face velocity of 2.64 m s^{-1}, used in the example, would result in the carry-over of condensate from the fins, making the use of downstream eliminator plates essential.

Most manufacturers determine the performance of their cooler coils themselves but ASHRAE Standard 33 (1978) gives details of acceptable laboratory methods. ARI Standard 410 (1991) then offers procedures for predicting the results of such laboratory tests for other working conditions and coil sizes.

A more involved method of determining the performance of cooling and dehumidifying coils is given by ASHRAE (1991) and ASHRAE (1996) see also BS 5141: (1983).

It is worth noting that there is evidence from Rich (1973) that the heat transfer coefficient through the air film is essentially independent of the fin spacing. An unusual relationship between heat transfer coefficient and the number of rows has also been shown by Tuve (1936) and Rich (1975) when Reynolds numbers for the airflow are low (<10 000), the coefficient for deep coils being much less than that for shallow coils. A tentative explanation offered is that vortices shed by the tubes give non-uniform temperature distribution in the airstream for downstream rows.

10.6   Sensible cooling

Not any of the external surface of the cooler coil may be at a temperature less than the dew point of the entering air if dehumidification is to be avoided and sensible cooling achieved. In equation (10.11), for the U-value of a cooler coil, the component that is known with the greatest certainty is the value of the resistance through the water film, R_w, supposing that the flow is turbulent within the tubes (normally the case under design load conditions). The internal surface temperature of the tubes can be calculated using equations (10.11) and (10.16) if the sensible heat flow rate is known. This may then be regarded as the same as the external surface temperature of the tubes, for all practical purposes, the thermal resistance of the tube wall being insignificant in this respect.

EXAMPLE 10.3

If the coil in example 10.2 is used to cool 4.75 m^3 s^{-1} from 28°C dry-bulb, 19.5°C wet-bulb (sling) to 21°C dry-bulb, without dehumidification taking place, what is the minimum permissible chilled water flow temperature, if the chilled water flow rate is unaltered?

Answer

Although the coil is the same, the *U*-value is different because the external surface is not wet with condensate. The new *U*-value must be determined.

The sensible cooling duty is given by equation (6.6):

$$Q_s = 4.75 \times (28 - 21) \times 358/(273 + 28) = 39.547 \text{ kW}$$

As in example 10.2, by equation (10.12):

$$1/h_a = R_a = 0.016\ 77 \text{ m}^2 \text{ KW}^{-1} \text{ when dry}$$

As before, the fin efficiency is 0.98 and by equation (10.14) the effectiveness of the surface, η, is 0.98.

By equation (10.13) the resistance of the dry fins is:

$$R_f = \frac{(1 - 0.98)}{0.98} \times 0.016\ 77 = 0.000\ 342\ 2 \text{ m}^2 \text{ KW}^{-1}$$

The resistance of the metal of the tubes' walls is

$$R_t = 0.000\ 036\ 2 \text{ m}^2 \text{ KW}^{-1}$$

Thus

$$R_m = 0.000\ 342\ 2 + 0.000\ 036\ 3 = 0.000\ 378\ 5 \text{ m}^2 \text{ KW}^{-1}$$

Since the mean velocity of waterflow through the tubes is unchanged, the resistance of the water film, is

$$R_w = 0.004\ 358 \text{ m}^2 \text{ KW}^{-1}$$

Hence, by equation (10.11):

$$U_t = 1/(0.016\ 77 + 0.000\ 378\ 5 + 0.004\ 358)$$
$$= 1/0.021\ 51 = 46.5 \text{ Wm}^{-2} \text{K}^{-1}$$

If the inner surface of the tubes must not be at a temperature less than the dew point (t_d) of the entering airstream, then the minimum chilled water temperature, t_{wmin}, can be determined. Denoting the overall thermal resistance of the cooler coil by *R* and the leaving air temperature by t_2, we have

$$\frac{t_d - t_{wmin}}{t_2 - t_{wmin}} = \frac{R_w}{R}$$

whence

$$t_{wmin} = \frac{(Rt_d - R_w t_2)}{(R - R_w)} \tag{10.17}$$

See Figure 10.6.

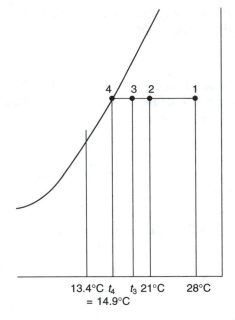

Fig. 10.6 The psychrometry for example 10.3.

The dew point of the airstream is 14.9°C and hence, for a six-row coil,

$$t_{wmin} = \frac{0.021\,51 \times 14.9 - 0.004\,358\,9 \times 21}{0.021\,51 - 0.004\,358} = 13.4°C$$

This is when the coil has six rows. However, the value of the minimum chilled water flow temperature will alter if the number of rows is changed, because so doing influences the effectiveness of the fins and hence changes the overall U-value, which is the reciprocal of R.

It does not necessarily follow that this chilled water flow temperature will give the sensible cooling duty. The duty obtained depends on the number of rows of the coil and, in turn, this has an effect on the fin efficiency and the effectiveness of the fins, which influence the thermal resistance of the fins, R_f. Since the ratio of the external to internal surface areas is unchanged the resistance of the tubes is not affected (see equation (10.15)).

EXAMPLE 10.4

Determine the number of rows required to achieve the sensible cooling duty for the cooler air forming the subject of example 10.3.

Answer

Successive assumptions are made for the number of rows and the corresponding U-values, minimum permissible chilled water flow temperatures and sensible cooling duties are determined.

The duty is 39.547 kW and, from example 10.2, the chilled water flow rate is 5.454 kg s^{-1}, hence the chilled water temperature rise is 39.457/(4.19 × 5.454) = 1.73 K, for the specified duty. The leaving chilled water temperature is thus t_{wmin} + 1.73. Hence air-to-water logarithmic mean temperature differences can be calculated for the various numbers of rows assumed, using equation (10.10). (There is a small error in doing this because, for the various duties determined, the chilled water temperature rise will not necessarily be 1.73 K.)

The calculations lead to the following tabulations:

Rows	A_{fi}	ϕ	ϕA_{fi}	A_p	A_t	η
6	224.3	0.98	219.8	11.82	236.1	0.98
5	186.9	0.96	179.4	9.85	197.8	0.96
4	149.5	0.92	137.5	7.88	157.4	0.92
3	112.15	0.86	96.4	5.91	118.1	0.87
2	74.8	0.77	57.6	3.94	78.74	0.78

Interpolations have been made in Table 10.2 to obtain values of ϕ, for odd numbers of rows.

Equation (10.14) is then used to determine η.

The resistance of the air-side film, when the coil is dry, has already been established as $0.016\ 77\ \text{m}^2\ \text{KW}^{-1}$, the resistance of the tubes has been determined as $0.000\ 036\ 2\ \text{m}^2\ \text{KW}^{-1}$ and the resistance of the water film has been calculated as $0.004\ 358\ \text{m}^2\ \text{KW}^{-1}$. Hence the sum of these, $R_a + R_t + R_w$, is $0.021\ 164\ 2\ \text{m}^2\ \text{KW}^{-1}$. The resistance of the fins is found from equation (10.13) and the total resistance and U-value calculated. The following tabulation results.

Rows	6	5	4	3	2
R_f	0.000 342 2	0.000 698 8	0.001 458 3	0.002 505 9	0.004 730 0
R	0.021 506 4	0.021 863 0	0.022 622 5	0.023 670 1	0.025 894 2
U_t	46.5	45.74	44.20	42.25	38.62

The dew point is 14.9°C, the leaving air temperature is 21°C and the values of R_w and R are known. Hence equation (10.17) can be used to determine the minimum permissible entering chilled water temperature, t_{wmin}. The following tabulation summarises the calculations.

Rows	6	5	4	3	2
R	0.021 506	0.021 863	0.022 622	0.023 670	0.025 894
Rt_d	0.3204	0.3258	0.3371	0.3527	0.3858
R_w	0.004 358	0.004 358	0.004 358	0.004 358	0.004 358
$R_w t_2$	0.091 518	0.091 518	0.091 518	0.091 518	0.091 518
$Rt_d - R_w t_2$	0.2289	0.2343	0.2456	0.2612	0.2943
$R - R_w$	0.017 15	0.017 50	0.018 26	0.019 31	0.021 54
t_{wmin}	13.35	13.39	13.45	13.53	13.66

Although the temperature rise of the chilled water is only 1.73 K for the design duty, it is reasonable to assume this is also the rise for all the duties since the calculations are aimed at deciding the number of rows to achieve the design duty. The leaving chilled water temperature, t_{wb} is then $t_{wmin} + 1.73°C$. The total external surface area, A_t, is proportional to the number of rows and the $(LMTD)_{aw}$ can be established by equation (10.10). The sensible cooling duty, Q_s, is then calculated using equation (10.9), as the following tabulation shows.

Rows	6	5	4	3	2
t_{wmin}	13.35	13.39	13.45	13.53	13.66
t_{wb}	15.08	15.12	15.18	15.26	15.39
$(28 - t_{wb})$	12.92	12.88	12.82	12.74	12.61
$t_2 - t_{wmin}$	7.65	7.61	7.55	7.47	7.34
$(LMTD)_{aw}$	10.06	10.01	9.95	9.87	9.74
A_t	236.1	197.8	157.4	118.1	78.7
U_t	46.50	45.74	44.20	42.25	38.62
Q_s (kW)	110.4	90.6	69.2	49.2	29.6

The required duty is 39.457 kW. Hence three rows of tubes are required.

The uncertainties, mentioned earlier, in the analysis of cooler coil performance imply that a small margin is prudent in the value of t_{wmin}. It is suggested that half a degree be added to the values calculated.

Alternatively, more involved procedures from ASHRAE (1996) and Kays and London (1985) using transfer functions are possible.

10.7 Partial load operation

The vast majority of cooler coils used in air conditioning perform under conditions of partial load for the greater part of their life. Operation under design conditions of full load is confined to a few hours per year in many cases, at least so far as the UK is concerned. In hot climates, full load conditions occur more often.

There are two ways in which the load on a cooler coil may reduce:

(i)   by reduction in the enthalpy of the moist air entering the coil, and
(ii) by a reduced demand on the part of the air conditioning system, necessitating an increase in the enthalpy of the air leaving the coil.

Lines of constant wet-bulb temperature are almost parallel to lines of constant enthalpy, on the CIBSE psychrometric chart. A reduction in the wet-bulb temperature of entering air is, therefore, one way in which a partial load condition can arise. Figure 10.7(*a*) illustrates this case. With a fall in load a reduction must occur in the temperature rise suffered by the chilled water passing through the coil. This, in turn, must mean a drop in the value of the mean coil surface temperature. Thus, the position of A falls to A′, along the saturation curve, as the entering wet-bulb reduces. For a given cooler coil the contact factor is constant if the ratio of air flow to water flow remains constant. Equation (10.8) implies this for the air side of the coil, and equation (10.6) suggests that, because the water-side resistance depends largely on the velocity of flow of the water through the tubes, the same is true for the water side. The constancy of the contact factor is an important point which allows geometrical methods to be used on the psychrometric chart in assessing the performance of coils under partial load conditions. Since changes in the state of the air entering the coil and variations in the temperature of the coolant have virtually no effect on the contact factor, the position of W, a point representing the state of the air leaving the coil, may be easily calculated from equation (3.1), defining the contact factor:

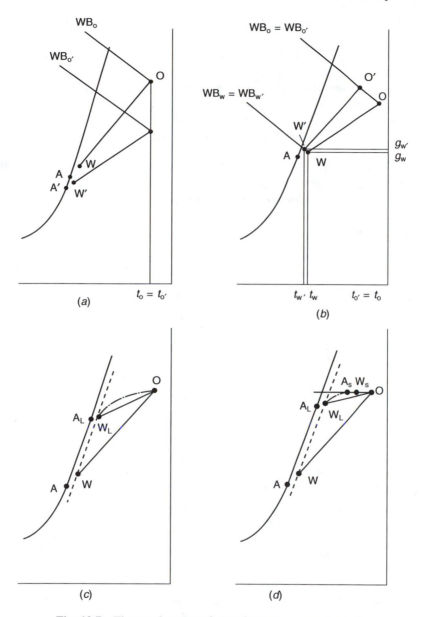

Fig. 10.7   The psychrometry for partial load on a cooler coil.

$$\beta = \frac{h_o - h_w}{h_o - h_a} = \frac{h_{o'} - h_{w'}}{h_{o'} - h_{a'}} \qquad (3.1)$$

It is possible that although the wet-bulb does not alter, the dry-bulb of the entering air may reduce. While this has practically no impact on the load (because the enthalpy is virtually constant), it does have an effect on the state of the air leaving the cooler coil. Figure 10.7(b) shows such a case. The comparative absence of a load change means that the mean coil surface temperature is unaltered, and so the position of A is fixed. The position of W is again easily found, using the definition of the contact factor given by equation (3.1). It can be seen that although the temperature of W' is less than that of W, its moisture content is higher. This is an aspect of the performance of a cooler coil that ought to be considered during selection, as well as operation with the design entering dry-bulb. The load may be unaltered but the performance may be quite unsatisfactory under such reduced dry-bulb conditions if this has not been taken account of.

As was discussed in chapter 8, a reduction in the sensible or latent heat gains in the conditioned space may require the supply of air at a state different from its design value. For example, if the latent heat gains fall off (the sensible heat gains remaining unaltered), the moisture content of the supply air must be elevated if the relative humidity in the room is to be kept constant. (This would be coupled with the supply of air at the same dry-bulb temperature, achieved by means of a reheater, as described in section 8.4.) There are three ways in which the state of air leaving a cooler coil using chilled water may be altered: (i) by varying the rate of water flow, (ii) by changing the temperature of the chilled water flowing on to the coil and by the use of face and by-pass dampers, dealt with in section 8.6.

If the state of the air entering a coil remains fixed but the rate of flow of the chilled water is reduced, the temperature rise suffered by the water increases. The mean coil surface temperature therefore goes up and the position of A rises up the saturation curve. At first, when this happens, the value of the contact factor stays virtually unaltered, little variation in the value of R_w taking place. So it is possible to draw a set of condition lines O–W, all having the same contact factor, β. In due course a limiting point, A_L, is reached. Any further reduction in the flow rate of the water causes such an increase in the value of R_w that the value of β can no longer be regarded as constant. The states of the air leaving the coil can be joined by a broken line from W to W_L, as shown in Figure 10.7(c). Above the limiting state, W_L, the value of β falls away and the locus of the air-leaving state is represented in the figure by the chain-dotted line which runs from W_L to O. This last part of the locus will not be parallel to a line of constant moisture content (except perhaps at its very end) since, the entering water temperature being assumed constant, some small amount of dehumidification always takes place.

A variation in the temperature of the chilled water flowing on to a coil with the flow rate of water constant, produces a condition rather similar to the one just considered. A progressive increase in the value of the temperature of the chilled water causes the mean coil surface temperature to rise and A moves up the saturation curve to a limiting position at A_L, as in Figure 10.7(d). Before this point the contact factor may be determined by geometrical methods on the psychrometric chart but afterwards it may not, although the value of β remains constant. As was mentioned in section 10.6, a chilled water temperature is eventually reached which results in the coil executing sensible cooling only. Under these conditions, air leaves the coil at state W_s and the locus from this point to O is along a line of constant moisture content.

10.8   The performance of a wild coil

When no control is exercised over either the temperature or the flow rate of the coolant, a coil is termed 'wild'. Such an operation may be quite satisfactory, and the method may have economic advantages, since the expense of three-way valves and thermostats is avoided. It is not quite true to assume that no control exists over the temperature of the coolant since, for chilled water coils, the water chiller produces water at a nominally constant temperature. Figure 10.7(a) gives a picture of what happens: as the entering wet-bulb drops, the load on the coil falls, and the mean coil surface temperature reduces, the position of A moving down the saturation curve. However, this will not continue indefinitely. The value of the temperature of the water produced by the chiller is a limiting factor. Under no-load conditions the temperature of A equals this value. So, provided the dew point of the air supplied to the conditioned space may be permitted to drop to such a low value, and that there is no objection to the refrigeration plant which produces the chilled water operating, the wild coil is acceptable. The economic question that must be answered is: will the saving in capital cost, by not using control valves and thermostats, be offset by the extra running cost of the refrigeration plant?

10.9 Sprayed cooler coils

Although they have fallen out of use for hygienic reasons, as Pickering and Jones (1986) explain, and would not be used today in commercial applications, sprayed cooler coils are worth considering as an academic illustration of a psychrometric process.

A sprayed cooler coil is a coil, fed with chilled water or liquid refrigerant in the usual way, positioned over a recirculation tank, with a bank of stand pipes and spray nozzles located a short distance from its upstream face. Mains water is fed to the tank through a ball-valve to make good any evaporative losses that may occur. Water is drawn from the tank by a pump and is delivered through the nozzles on to the face of the coil. This water then falls down the fins into the tank and is largely recirculated. Figure 10.8 illustrates this.

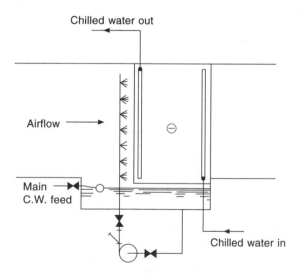

Fig. 10.8 A simple diagram of the plant arrangement for a sprayed cooler coil.

The prime function of the sprayed cooler coil is to provide humidification for operation of the air conditioning plant in winter. A cooler coil offers a very large surface area to the airstream passing over it, and it is by wetting this very large area that the spray nozzles achieve humidification. The nozzles, therefore, should not have a very large pressure drop across them, since this is associated with an atomisation effect, but should be placed close enough to the face of the coil to ensure that all the water delivered by them goes on to the coil. Only one bank of nozzles is necessary and, since this faces downstream, the humidifying efficiency associated with an atomised spray is small: of the order of 50 per cent. On the other hand, if atomisation is not relied on, but the surface of the coil is properly wetted instead, a better humidifying efficiency can be obtained. Only tests carried out by the manufacturer in the works or by the user on site, can provide reasonably accurate data as to the efficiencies likely. But, the contact factor of the coil itself provides some clue. If the coil were completely wetted throughout its depth, then it would perhaps be reasonable to take the humidifying efficiency as approaching the value of the contact factor. Since the coil is not completely wetted, the actual value will be less than this; a practical value is 80 per cent.

An effective distribution of water is about 0.82 litres per second on each square metre of face area of the coil. The pressure drop across the nozzles should be about 42 kPa. If more pressure than this is lost little is gained in effectiveness, but a penalty is paid in extra running costs for the pump.

When chilled water flows through the inside of the coil and spray water is circulated over the outside of the coil, accurate control of dew point is possible for a wide range of entering air conditions. Figures 10.9 and 10.10 illustrate how this may be achieved. If the coil is not sprayed, and is chosen to cool and dehumidify air from state 1 to 2 under design conditions in summer, it will produce a state 2′ if the entering air state is 1′, during autumnal or spring weather. This has a moisture content which is too low if control over dew point is necessary. If a sprayed coil is used, however, summer operation will, for all practical purposes, be the same as before but when the entering air state is 1′ it is now possible to produce a leaving state 2. For purposes of illustration the process can be regarded as one of adiabatic humidification from 1′ to 1″, followed by cooling and dehumidification to 2. This is shown in Figure 10.10. So, as with an air washer, the sprayed cooler coil, when properly controlled, can execute a process of cooling and humidification, as typified by the change 1′ to 2.

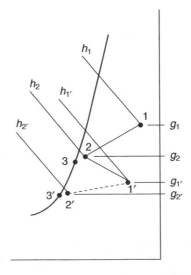

Fig. 10.9   Using a sprayed coil it is possible to achieve close control over the state of the air leaving the coil. Thus state 2 can be obtained for a variety of different entering states, such as 1 and 1′. If the coil were unsprayed the leaving state would be 2′ when the entry state was 1′. Sprayed cooler coils are rarely used, for hygienic reasons.

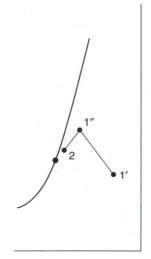

Fig. 10.10 One way of regarding the process of change across a sprayed cooler coil is to consider adiabatic saturation (state 1' to state 1") followed by cooling (state 1" to state 2).

There is one further advantage that the sprayed coil shares with the air washer. The presence of the large mass of water in the tank, at a temperature which is virtually the same as the leaving air temperature, provides a thermal reservoir which gives inertia to the control of the cooler coil. This is of value if a constant air leaving state is required, particularly in direct-expansion coils, which are sprayed. Inadequacies in the control over the coil output or fluctuations in the state of the air leaving the cooler coil are smoothed out. Stability is added.

During winter, of course, the wetted surface acts as an adiabatic humidifier, just as does an air washer. Control may be effected by means of a pre-heater or variable mixing dampers for fresh and recirculated air, as was discussed in sections 3.9 and 3.10. Under such circumstances the refrigeration plant would not be working, although, as is remarked in the next section, it may be desirable to run the pump used to circulate chilled water from the chiller to the coil.

10.10 Free cooling

Air–water systems (e.g. fan coil) require a supply of chilled water at a temperature of about 10°C throughout the year. The methods used for this purpose are:

(i)   Thermo-syphonic cooling. This was dealt with in section 9.10. The refrigeration plant is operated in winter, with its compressor shut down, to produce chilled water at a suitable temperature by a natural cooling process. An air-cooled condenser or a cooling tower is used.

(ii) A cooling tower can give water at a suitable temperature in winter, when the outside air wet-bulb is sufficiently low. The chilled water from the tower by-passes the condenser and is fed directly into the secondary circuit to the fan coil units. The chilled water produced is dirty because the circuit is open and subject to pollution from the atmosphere, and to corrosion. The larger particles of dirt are removed by a good quality water filter and the fan coil or other units fed with the water must be able to function with the level of contamination remaining. The pH of the water is monitored as a check on corrosion. Motorised, automatic valves direct the water to the condenser in summer (when the refrigeration plant must run) and to the units in winter (when the refrigeration plant is off). When using this method it is important to check the hydraulic balance between the chilled water and cooling water circuits: these are hydraulically separate in summer but interconnected in winter.

(iii)   If the water fed to the units must be clean, as is often the case, then a plate heat exchanger is inserted in the cooling water circuit. This transfers heat indirectly between the cooling water and secondary chilled water circuits for the fan coil units. The secondary chilled water flow temperature will be a degree or two warmer in winter, because of the temperature difference required at the heat exchanger. The advantages are that the secondary chilled water circuit is clean and the cooling and chilled water circuits are hydraulically independent.

(iv)   A dry cooler may be used. In simple terms this is a free-standing cooler coil over which outside air is drawn or blown, by a fan. A mixture of water and glycol is circulated through the coil, naturally chilled, and fed to the units. Owing to the viscosity and other physical properties of the aqueous glycol mixture (used as an anti-freeze measure) the thermal resistance of the surface film within the tubes is increased. The pressure drop through the system is also affected and appropriate allowances must be made for these effects when selecting the dry cooler, the fan coil or other units, and the pumps. Since only sensible heat transfer occurs, the surface areas needed for heat transfer in dry coolers tend to be rather large. The method is really only suitable for smaller installations.

(v)   Cooling at the primary air cooler coil. If the primary chilled water pump is kept running but the refrigeration plant is switched off, water will be circulated through the primary air cooler coil. When the outside air dry-bulb is less than about 5°C the flow of cold outside air over the cooler coil can chill the water pumped through its tubes to about 11°C or 12°C, which the secondary pump then circulates to the units. The freezing risk, when using the primary cooler coil in this way, must not be ignored.

(vi)   In the past, sprayed cooler coils were used in a manner similar to that outlined in (v), above. These were more effective in producing chilled water for secondary circuits because of the evaporative cooling effect of the spray water. The method has fallen into disuse because of the associated hygienic risks.

10.11   Direct-expansion coils

Direct-expansion coils differ from chilled water cooler coils in one important respect: performance depends not only on the considerations of air and coolant flow mentioned, but also on the point of balance achieved between the characteristic of the coil's performance and the characteristic of the performance of the condensing set to which the coil is connected. This aspect of the matter is dealt with in section 12.5.

10.12   Air washers

A diagrammatic picture of an air washer is given in Figure 10.11. Essentially, a washer consists of a spray chamber in which a dense cloud of finely divided spray water is produced by pumping water through nozzles. On its path through the chamber the air first

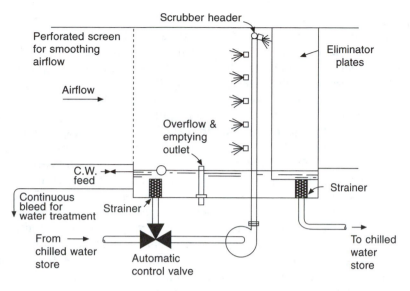

Fig. 10.11   Air washer: single bank of nozzles blowing upstream. Note that it is highly desirable for hygienic reasons to empty the spray water tank completely once in each twenty-four hours and to refill with fresh water. The air washer is rarely used today, for hygienic reasons.

passes an array of deflector plates or a perforated metal screen, the purpose of which is to secure a uniform distribution of air flow over the cross-section of the washer and to prevent any moisture accidentally being blown back up the duct. Within the spray chamber itself, the nozzles which atomise the water are arranged in banks, usually one or two in number, but occasionally three. The efficiency of the washer depends on how many banks are used and which way they blow the spray water, upstream or downstream. The banks consist of stand pipes with nozzles mounted so as to give an adequate cover of the section of the chamber. The arrangement is usually staggered, as shown in Figure 10.12. Finally, the air leaves the washer by passing through a bank of eliminator plates, the purpose of which is primarily to prevent the carry-over of unevaporated moisture beyond the washer into the ducting system. A secondary purpose is to improve the cleaning ability of the washer by

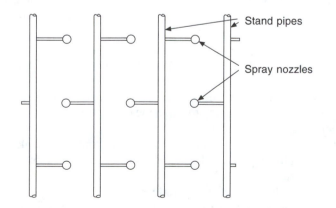

Fig. 10.12   Stand piping and nozzle arrangement for atomising water in an air washer.

offering a large wetted surface on which the dirt can impinge and be wasthed away. In general, washers are poor filters; they should not be relied on alone to clean the air.

Provided the temperature of the water circulated is less than the dew point of the entering air, a washer can dehumidify, and its cooling and dehumidifying capacity can be modulated by varying the temperature of the water handled by the pump. The usual way of doing this is by means of a 3-port mixing valve, as shown in Figure 10.11. However, in general, air washers are less useful than cooler coils in dehumidifying and cooling air. The reason for this is that contra-flow does not occur. As a droplet of water leaves a spary nozzle it travels for a short distance either parallel or counter to the direction of airflow, after that, it starts falling to the collection tank, and the heat transfer is according to cross-flow conditions. Near the surface of the water in the tank, even at the downstream end of the washer the air encounters water which is at its warmest. The consequence of all this is that a leaving air temperature which is less than the leaving water temperature is unobtainable.

To secure effective atomisation, the pressure drop across the nozzles used in the spray-chamber type of washer must be fairly large. If an attempt is made to modulate the capacity of the washer by reducing the water flow through the nozzles, it will be unsuccessful, because proper atomisation ceases below a certain minimum pressure. Typical nozzle duties vary from 0.025 litres s^{-1} with a pressure drop of 70 kPa through a 2.5 mm orifice to 0.3 litres s^{-1} with a drop of 280 kPa through a 6.5 mm orifice.

Efficiencies vary with the arrangement of the banks, and typical percentage values are:

1 bank downstream	50	
1 bank upstream	65–75	
2 banks downstream	85–90	For face velocities of airflow
2 banks upstream	92–97	over the section of 2.5 m s^{-1}
2 banks in opposition	90–95	
3 banks upstream	>95	

These efficiencies are based on the assumption that the section of the spray chamber is adequately covered by nozzles and that these are producing a properly atomised spray. A typical nozzle arrangement is 45 per m^2 yielding a water flow rate of about 2 litres s^{-1} over each m^2 of section of each bank.

As with cooler coils, the velocity of airflow is also significant: a value of 2.5 m s^{-1} over the face area is usually chosen. Little variation from this is desirable, either way, if the eliminators are to be successful. A range of 1.75 m s^{-1} to 3.75 m s^{-1} has been suggested.

Good practice in the design and use of air washers for air conditioning requires a water quantity of 0.11 litres s^{-1} per nozzle and 22 nozzle per m^2 of cross-sectional area in each bank, with a mean air velocity of 2.5 m s^{-1}.

Two banks, blowing upstream, are adequate, so the total circulated water quantity should be 0.44 litres s^{-1} m^{-2} of cross-section.

Washers must be long enough to allow adequate contact between the spray water and the air and to avoid the penetration of spray through the downstream eliminators or the upstream smoothing screen. There is therefore commonly 0.8 to 1.5 m between the banks with spaces of about 0.3 and 0.5 m before and after the initial and final banks, respectively. The tank is usually 450 mm deep with about 100 mm of freeboard. Washers are generally designed for air velocities of 2.0 to 3.0 m s^{-1} over their cross-section. For higher velocities the eliminators may not be fully effective unless specially designed for the purpose.

Washers have fallen out of favour for commercial applications and are now very seldom used.

Exercises

1. (*a*) Define the term 'contact factor' as applied to cooler coils and state the conditions under which it remains constant.

(*b*) Explain how the performance of a chilled water cooler coil varies under conditions of (i) varying water flow rate through the tubes, and (ii) varying water flow temperature.

(*c*) How is apparatus dew point, as applied to chilled water cooler coils, related to the flow and return temperatures of the chilled water passing along the tubes?

2. A cooler coil is chosen to operate under the following conditions:

Air on: 28°C dry-bulb, 20.6°C wet-bulb (sling), 28.3 m^3 s^{-1}.

Air off: 12°C dry-bulb, 8.062 g per kg dry air.

Chilled water on: 5.5°C.

Chilled water off: 12.0°C.

You are asked to calculate (*a*) the contact factor, (*b*) the design cooling load and (*c*) the water flow rate through the coil.

Answers

(*a*) 0.88; (*b*) 868 kW; (*c*) 31.8 kg s^{-1}.

3. An air-conditioning system comprises a chilled water cooler coil, reheater and supply air fan, together with a system of distribution ductwork. Winter humidification is achieved by steam injection directly into the conditioned space. It is considered economic, from an owning and operating cost point of view, to arrange that a constant minimum quantity of outside air is used throughout the year. The fixed quantity of outside air is 25 per cent by weight of the total amount of air supplied to the room. Using a psychrometric chart and the data listed below, calculate

(*a*)   the contact factor of the cooler coil,
(*b*)   the design load on the cooler coil in summer,
(*c*)   the load on the cooler coil under winter conditions, and
(*d*)   the winter reheat load.

Design data

Sensible heat gain in summer	25 kW
Latent heat gain in summer	0.3 kW
Sensible heat loss in winter	4.4 kW
Outside state in summer	28°C dry-bulb, 19.5°C wet-bulb
Outside state in winter	−1°C saturated
Inside state in summer and winter	21°C dry-bulb, 50 per cent saturation
Supply temperature in summer design conditions	12.5°C dry-bulb
Temperature rise due to fan power and duct heat gains	0.5°C

Assume that the ratio of latent to total heat removal by the cooler coil is halved under winter design conditions. Take the specific heat of dry air as 1.012 kJ kg^{-1} K^{-1}.

Answers

(a) 0.83; (b) 36.85 kW; (c) 20.05 kW; (d) 37.2 kW.

4.   (a) Air at a constant rate and at a state of 20.5°C dry-bulb, 17.9°C wet-bulb (sling), flows on to a cooler coil and leaves it at 11°C dry-bulb, 10.7°C wet-bulb (sling), when the coil is supplied with a constant flow rate of chilled water at an adequately low temperature. Calculate the state of the air leaving the coil if the load is halved when the state of the air entering the coil changes to 20°C dry-bulb with 14.2°C wet-bulb (sling). Make use of a psychrometric chart.

(b) State briefly how the contact factor of a cooler alters when (i) the number of rows is reduced and (ii) the face velocity is reduced, assuming that the ratio of air mass flow to water mass flow is constant.

Answers

(a) 10.7°C dry-bulb, 10.2°C wet-bulb, 30.02 kJ kg^{-1}, (b) (i) diminishes and (ii) increases.

5.   Moist air at 28°C dry-bulb and 20.6°C wet-bulb flow over a 4-row cooler coil, leaving it at 9.5°C dry-bulb and 7.107 g per kg. Using a psychrometric chart, answer the following:

(a) What is the contact factor?
(b) What is the apparatus dew point?
(c) If the air is required to offset a sensible heat gain of 2.4 kW and a latent heat gain of 0.3 kW, in the space being conditioned, calculate the weight of dry air which must be supplied to the room in order to maintain 21°C dry-bulb therein.
(d) What is the percentage saturation in the room?
(e) If the number of rows of the coil is decreased to 2, the rate of waterflow, the rate of airflow and the apparatus dew point remaining constant, what temperature and humidity will be maintained in the room?
Ignore any temperature rise due to duct friction or heat gain.

Answers

(a) 0.925, (b) 8.1°C, (c) 0.204 kg s^{-1}, (d) 49 per cent, (e) 26°C dry-bulb and 40 per cent saturation.

Notation

Symbol	Description	Unit
A_f	face area	m^2
A_{fi}	total surface area of the fins	m^2
A_i	total internal surface area	m^2
A_p	total surface area of the tubes	m^2
A_r	total surface area per row	m^2
A_t	total external surface area	m^2
c	specific heat capacity of humid air	kJ kg^{-1} K^{-1}

c_w	specific heat capacity of water	kJ kg^{-1} K^{-1}
d_i	inside tube diameter	m
d_o	outside tube diameter	m
g	moisture content of humid air	g kg^{-1} dry air
h	enthalpy of humid air	kJ kg^{-1} dry air
h_a	heat transfer coefficient for the surface film on the air side of a cooler coil	W m^{-2} K^{-1}
k	constant	–
$(LMTD)_{as}$	logarithmic mean temperature difference between the airstream and the mean coil surface temperature	K
$(LMTD)_{aw}$	logarithmic mean temperature difference between the airstream and the water stream	K
$\dot{m}_a$	mass flow rate of dry air	kg s^{-1}
$\dot{m}_w$	mass flow rate of water	kg s^{-1}
Q_s	rate of sensible heat transfer	W or kW
Q_t	rate of total heat transfer	W or kW
R	total thermal resistance	m^2 K W^{-1}
R_a	thermal resistance of the air film	m^2 K W^{-1}
R_f	thermal resistance of the fins	m^2 K W^{-1}
R_m	thermal resistance of the metal of the tubes and fins	m^2 K W^{-1}
R_t	thermal resistance of the tubes	m^2 K W^{-1}
R_w	thermal resistance of the water film	m^2 K W^{-1}
r	number of rows of a cooler coil	–
S	ratio of sensible to total heat transfer	–
t, t_a	dry-bulb temperature of an airstream	°C
t_d	dew-point temperature of an airstream	°C
t_{sm}	mean coil surface temperature	°C
t_w	chilled water or saturated air temperature	°C
t_{wa}, t_{wb}	entering, leaving chilled water temperature	°C
t_{wm}	mean chilled water temperature	°C
t_{wmin}	minimum chilled water temperature	°C
t_{w1}, t_{w2}	mean coil surface temperature of the first, second, etc. rows of a coil	°C
t'	wet-bulb temperature of an airstream	°C
U	thermal transmittance coefficient	W m^{-2} K^{-1}
U_t	biased U-value for a cooler coil to account for total heat flow, air-to-air	W m^{-2} K^{-1}
v	mean chilled water velocity	m s^{-1}
v_f	face velocity over a cooler coil	m s^{-1}
$\dot{v}$	rate of volumetric airflow over the face of a coil	m^3 s^{-1}
β	contact factor	–
η	total surface effectiveness	–
λ_t	thermal conductivity of the metal of the tubes	W m^{-1} K^{-1}
ϕ	fin efficiency	–

References

ARI 410 (1991): Forced-circulation air-cooling and air-heating coils. *Standard 410-91*, air-conditioning and Refrigeration Institute, Arlington, VA.

ASHRAE (1978): Methods of testing forced circulation air-cooling coils and air-heating coils, *ASHRAE Standard 33–1978*.

ASHRAE Handbook (1996): *Heating, Ventilating and Air Conditioning Systems and Equipment*, chapter 21.

BS5141: 1983: Specification for heating and cooling coils, Part 1, 1975 (1983) Methods of testing for rating of cooling coils.

Kays, W.M. and London, A.L. (1985): *Compact Heat Exchangers*, McGraw-Hill Book Company, New York.

McAdams, W.H. (1954): *Heat Transmission*, 3rd edition, McGraw-Hill Book Company, New York.

McQuiston, F.C. (1975): Fin Efficiency with Combined Heat and Mass Transfer, *ASHRAE Trans.* **81**, Part 1, 350–355.

Pickering, C.A.C. and Jones, W.P. (1986): Humidifier Fever, Health and Hygienic Humidification, Technical Note TN 13/86, BSRIA.

Rich, D.G. (1973): The Effect of Fin Spacing on the Heat Transfer and Friction Performance of Multi-row, Smooth, Plate Fin-and-tube Heat Exchangers, *ASHRAE Trans.* **79**, Part 2, 137–145.

Rich, D.G. (1975): The Effect of the Number of Tube Rows on Heat Transfer Performance of Smooth, Plate Fin-and-tube Heat Exchangers, *ASHRAE Trans.* **81**, Part 1, 307–319.

Tuve, G.L. (1936): Performance of Fin-tube Units for Air Heating, Cooling and Dehumidifying, *ASHVE Trans.* **42**, 99.

11

The Rejection of Heat from Condensers and Cooling Towers

11.1 Methods of rejecting heat

All heat gains dealt with by an airconditioning system must be rejected at the condenser. To accomplish this, the practical condenser for air conditioning must be water-cooled, evaporative-cooled or air-cooled.

Air-cooled, dry heat exchangers have also been used. These are capable of producing cooling water at a temperature within 11 K of the ambient dry-bulb. Thus cooling water might be produced at 40°C if the ambient air was at 29°C dry-bulb. Such temperatures are too high to be of practical value in air conditioning. The heat exchangers are also bulky and expensive in both capital and running costs.

Except in the comparatively rare cases where a supply of lake or river water may be drawn on, or the even rarer cases where mains water is available, the water used by a water-cooled condenser must be continuously recirculated through a cooling tower. Figure 11.1 illustrates what occurs. Water is pumped through the condenser and suffers a temperature rise of $(t_{w1} - t_{w2})$ as it removes the heat rejected by the refrigerant during the process of condensation. The cooling water then flows to the top of a cooling tower (an induced-draught type is illustrated) whence, in falling to a catchment tank at the bottom of the tower, it encounters airflow. Contra-flow heat exchange occurs, the water cooling by evaporation from t_{w1} to t_{w2} and the air becoming humid in the process. The heat lost by the water constitutes a gain of enthalpy to the air. The water is then ready for recirculation to the condenser at the desired temperature. Some water is necessary from the mains, to make good the evaporative losses from the system but, since 1 kg of water liberates about 2400 kJ during a process of evaporative cooling (whereas 1 kg rising through 5 degrees absorbs only 21 kJ), the amount of make up required is quite small, being of the order of 1 per cent of the circulation rate in this respect. There is some loss also due to the carryover of droplets by the emergent airstream but, even allowing for this, the loss is still little more than 1 per cent. This loss is termed drift and it must be reduced to virtually zero, by the provision of drift eliminators, to minimise the risk of contaminating the local surroundings. Drift should not be confused with pluming: this consists of steam which has condensed to water droplets and appears as a cloud coming out of some cooling towers under certain conditions of operation (see section 3.12).

Evaporative condensers operate in a similar fashion but more directly. In Figure 11.2 it can be seen that the pipe coils of the condenser are directly in the path of the air and

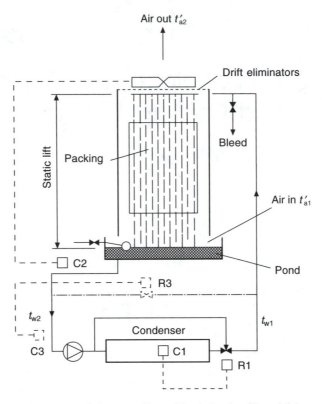

Fig. 11.1 Induced draught cooling tower. Control is shown by C1 and R1 over the condensing pressure associated with a reciprocating machine. Immersion thermostat C2 cycles the fan. An alternative arrangement for centrifugal machines is shown by a chain dotted line. Immersion thermostat C3 controls the flow water temperature on to the condenser by means of a motorised butterfly valve R3.

water streams and that evaporative cooling takes place directly on the outer surface of the tubes.

With air-cooled condensers, no water is used at all. The refrigerant condenser consists of a coil of finned tube, and over this air is drawn or blown by means of a propeller, axial flow or centrifugal fan. As with cooling towers and evaporative condensers, several fans may be used.

The evaporative condenser is probably the most efficient way of rejecting heat to the atmosphere, if it is properly designed. This is reflected in the typical air quantities required:

$\quad$ 30–60 $\ $ litres s^{-1} kW^{-1} for evaporative condensers,

$\quad$ 40–80 $\ $ litres s^{-1} kW^{-1} for cooling towers and

$\quad$ 140–200 litres s^{-1} kW^{-1} for air-cooled condensers.

ASHRAE (1996) suggest that 80–160 litres s^{-1} kW^{-1} are likely with face velocities of 2–4 ms^{-1}.

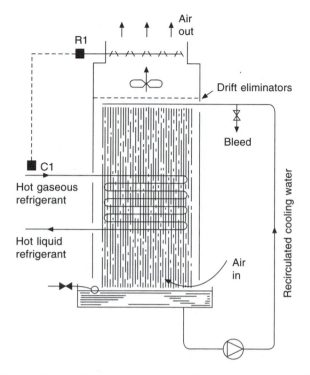

Fig. 11.2 Evaporative condenser. C1 is a pressure controller which keeps the refrigerant condensing
at a constant pressure through the agency of the motorised modulating dampers R1.

11.2 Types of cooling tower

Cooling towers may be broadly classified as (i) induced draught, (ii) forced draught and
(iii) cross flow.

Induced-draught towers consist of a tall casing with a fan at the top, rather similar to that
shown diagrammatically in Figure 11.1. The casing of the tower is packed with a material,
the purpose of which is to provide a large wetted surface for evaporative cooling. Many
different materials are in use, ranging from wooden slats to plastics. The performance of
a cooling tower does not depend on the type of filling used but rather on its arrangement
and on the uniformity and effectiveness of the water distribution. A constant air velocity
through the filling is also important in securing good performance.

The induced-draught tower has the advantage that any leakage will be of ambient air into
the tower, rather than vice versa. This means that there is less risk of the nuisance caused
by water and humid air leaking outwards to the vicinity than arises in other towers. An
important advantage is that humid air and drift (if any) leaving the tower is discharged at
a fairly high velocity, with some directional effect. There is thus much less risk of short-
circuiting, and the influence of wind pressure is reduced. It is true, however, that the fan
and water, being in the stream of humid air leaving the tower, are more liable to corrosion,
but adequate anti-corrosive protection should minimise this disadvantage.

Forced-draught towers, as their name implies, push air through the tower. Figure 11.3
illustrates this. Leakage is out of the tower but the fans and motors are not in the humid
airstream. Short-circuiting is something of a problem, and forced-draught towers should,
therefore, not be used in a restricted physical environment.

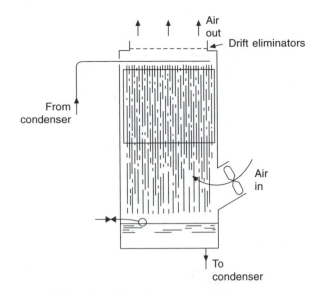

Fig. 11.3 Forced draught cooling tower.

Although the height of a tower is of importance in producing a long path for contra-flow evaporative heat exchange to take place, the cross-section of the tower is also important because the larger its value, for a given height, the greater will be the wetted surface area available. It follows that the volume occupied by a tower dictates its cooling capacity, rather than height alone. A choice is thus offered in choosing a tower: a tall tower or a short tower may be adopted, but in either case the volume will be about the same. Because of their unsightly obtrusion on skylines, cooling towers are sometimes criticised by architects and town planners. There is, then, sometimes considerable pressure to reduce their height, and low silhouette towers are chosen.

The cross-draught tower provides a low silhouette, as Figure 11.4 shows. Such towers are really a combination of induced-draught and cross-draught, but are designated cross-draught, nevertheless.

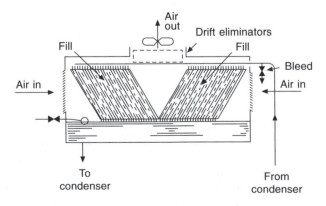

Fig. 11.4 Cross-draught cooling tower.

The casings of towers may be of metal, fibreglass, wood or concrete, or a combination of these. It is best to use the least corrodable material possible, both for the casing and the fill. For this reason, and also because of the opportunity it presents of giving the tower an architecturally harmonious anonymity, cooling-tower casings and ponds are sometimes erected by the builders; only the fill, the fan and the motor, and the water piping, being installed by the mechanical services' sub-contractors.

The design of towers is continually undergoing change, much of it effective. As a result, towers of great compactness are sometimes possible. One such type of tower uses a rotating water header coaxial with the induced-draught fan. This, coupled with the use of a cellulose fill, rather like corrugated cardboard, which presents a very large surface area, produces a uniform distribution of water and constancy of airflow.

An indirect cooling tower is sometimes used. This is similar to the arrangement in Figure 11.2 but the tubes shown crossing the middle of the tower convey cooling water for the condenser, not refrigerant. The cooling water provided is then not polluted by the spray water and the airstream in the cooling tower. On the other hand, heat transfer between the water in the piping and the stream of air and spray water outside the piping is less efficient.

There are also forms of tower that do not use mechanical draught. One of these uses the density difference of the warm humid air within the tower and the cooler air outside. Its application is for power stations and it is not used for air conditioning because of its bulk. Another type of tower that sometimes does find application in air conditioning uses water jets to induce airflow in parallel with the water stream. Such towers are either downflow or crossflow. Although they are relatively quiet and free from vibration, air velocities into and out of the towers are low and they are consequently susceptible to changes in wind speed and direction. They are best suited to cases where some variation in the cooling water flow temperature is tolerable.

To provide a small margin, that covers operation beyond design conditions, it is good practice to select cooling towers to operate in a wet-bulb that is one half to one degree above the design value adopted for the rest of the system.

11.3   Theoretical considerations

A full theoretical treatment yielding a design complete in every detail is not possible in this text. A theoretical approach takes the design only so far; beyond this, the design can be completed by assigning empirical values to the unsolved variables.

A heat balance must be struck between the water and the air:

$$\dot{m}_w c_w (t_{w1} - t_{w2}) = \dot{m}_a (h_{a2} - h_{a1}) \tag{11.1}$$

where $\dot{m}_w$ = mass flow rate of water in kg s^{-1}

c_w = specific heat capacity of water in kJ kg^{-1} K^{-1}

t_{w1} = inlet water temperature in °C

t_{w2} = outlet water temperature in °C

$\dot{m}_a$ = mass flow rate of air in kg of dry air per s

h_{a1} = enthalpy of air at inlet in kJ kg^{-1} dry air

h_{a2} = enthalpy of air at outlet in kJ kg^{-1} dry air

Although $\dot{m}_w$ decreases by evaporation as it flows through the tower, it is regarded here as a constant for simplicity. It can he shown that the enthalpy difference between the water and the air at any point in the tower is the force promoting heat and mass transfer, and Jackson (1951) shows that the following expression can be derived:

$$Z = \frac{2.61\dot{m}_w(t_{w1} - t_{w2})}{ksa\Delta h_m} \tag{11.2}$$

where k = the coefficient of vapour diffusion in kg water per s m^2 for a unit value of Δh_m,

   Δh_m = the mean driving force, in kJ kg^{-1},

   s = the wetted surface area per unit volume of packing, in m^{-1},

   a = the cross-sectional area of the tower in m^2,

   Z = the height of the tower in m.

Using equation (11.1), a complementary form of equation (11.2) exists:

$$Z = \frac{2.61\dot{m}_a(h_{a2} - h_{a1})}{ksa\Delta h_m} \tag{11.3}$$

The 'volume transfer coefficient', ks, is independent of atmospheric conditions and can be expressed by re-arranging equation (11.2).

In the vicinity of the water in the tower there is a thin film of saturated air, at a temperature t_w, the same as the temperature of the water. This air has an enthalpy h_w which is greater than that of the ambient air, h_a. Thus, for any given value of t_w, the water temperature, there exists an enthalpy difference, Δh, between the enthalpy of the film, h_w, and the enthalpy of the ambient air:

$$\Delta h = h_w - h_a \tag{11.4}$$

The mean value of this is Δh_m and is termed the mean driving force. Figure 11.5 shows this diagrammatically. The equilibrium line represents the variation in the value of h_w with

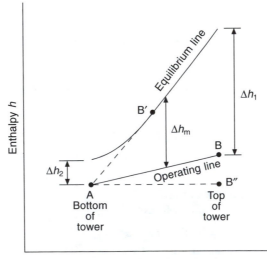

Fig. 11.5   An enthalpy-water temperature diagram showing the mean driving force for a cooling tower (Δh_m).

respect to the water temperature, t_w. The operating line shows the variation of the enthalpy of the air, with respect to t_w.

The driving force, promoting heat transfer at any value of t_w, is the difference given by equation (11.4). Since heat exchange occurs through the height of the tower, the value of t_w reduces as it falls through the tower. Thus, Δh_2 represents the driving force at the bottom of the tower and Δh_1 at the top. The mean value occurs at some intermediate position.

The operating line is straight, being determined by equation (11.1). Hence, the slope of the line depends on the ratio of water flow to air flow, $\dot{m}_w/\dot{m}_a$. When $\dot{m}_w/\dot{m}_a$ is zero, the driving force is zero and the tower is infinitely tall. Such a condition would arise if the operating line were AB′, tangential to the equilibrium curve, because at the point of tangential touching, Δh is zero. The converse case is AB″ parallel to the abscissa. Here the tower is not tall but, to secure the wetted surface area, it must be of large cross-section. It is evident that increasing the value of $\dot{m}_a$ with respect to $\dot{m}_w$, for a given tower, increases the mean driving force and produces a bigger tower capacity. Advantage is taken of this to secure optimal control of a tower. See section 11.6.

A typical value of $\dot{m}_w/\dot{m}_a$ is 1.0.

EXAMPLE 11.1

Assuming a water-to-air mass flow ratio of 1.0 and an ambient wet-bulb of 20°C (sling), calculate the air quantity likely to be handled by a cooling tower used to cool water from 32° to 27°C, for a refrigeration plant having a coefficient of performance of 4.

Answer

Heat rejected at the condenser, per kW of refrigeration
>   = 1.25 kW (because the COP is 4)

> Water flow rate through the condenser
>   = 1.25/(4.2 × 5)
>   = 0.0596 kg s^{-1}

Hence the air flow rate through the cooling tower is also 0.0596 kg s^{-1}.

Assuming an induced-draught tower, the fan handling saturated air at 20°C wet-bulb, then the humid volume is 0.8497 m^3 kg^{-1} and the air flow rate is

$$0.0596 \times 0.8497 = 0.0506 \text{ m}^3\text{s}^{-1} \text{ for each kW of refrigeration}$$

11.4   Evaporative condensers

Whereas a condenser/cooling tower arrangement requires a system of water distribution piping, this is almost entirely absent in the evaporative condenser. Only enough water need be circulated to ensure that the outside surface of the condenser coils is completely wet. The heat exchange is solely latent, and less water is required in circulation than is necessary with a condenser/cooling tower, where a sensible heat exchange occurs in the shell-and-tube condenser.

The evaporative condenser is thus more compact and is cheaper. It suffers from lack of flexibility, and oil return and other problems demand that the condenser should not be too far from the compressor. A cooling tower, on the other hand, may be a considerable distance away, the condenser then being adjacent to the compressor.

Scaling on the tubes of an evaporative condenser may be something of a problem, particularly if a high condensing temperature is used. Cleaning difficulties, caused by the presence of scale, dirt and corrosion, and by the close nesting necessary for the tubes (the use of finning being impossible), have considerably reduced the popularity of the evaporative condenser in recent years.

11.5   Air-cooled condensers

Unlike evaporative condensers, air-cooled condensers have a capacity which is related to the dry-bulb temperature of the ambient air, rather than to its wet-bulb temperature. If working condenser pressures are not to become excessively high, making the plant expensive to run, large condenser surface areas must be used. This has set a limit on the practical upper size of air-cooled condensers. Their use in air conditioning has been commonly confined to plants having a capacity of less than 70 kW of refrigeration, although they have been used for duties as high as 2000 kW, in temperate climates.

The hot gas discharged from the compressor is desuperheated over approximately the first 5 per cent of the heat transfer surface, followed by condensation over the succeeding 85 per cent, with a small drop in the condensing temperature, related to the frictional pressure loss. A certain amount of sub-cooling of the liquid can then occur. Additional heat transfer surface may be provided to assist the sub-cooling, achieving an increase of about 0.9 per cent in the cooling capacity for each degree of liquid temperature drop, according to ASHRAE (1996).

Propeller fans, direct-coupled to split-capacitor driving motors, are most commonly used to promote airflow, although axial flow or centrifugal fans are also sometimes adopted. Fan powers are about 20 to 40 W for each kW of refrigeration capacity. Noise is often a problem and this is made worse if there are obstructions in the inlet or outlet airflow paths. Although vertical fan arrangements are possible, with horizontal cooling airflow paths, these are susceptible to wind pressures and it is recommended that the condenser coils should be horizontal with vertical cooling airflow paths. Propeller fans will not deliver airflow against any significant external resistance. It follows that ducting connections are then not possible if these fans are used.

A 20-degree difference between the entering dry-bulb temperature and the condensing temperature is often consistent with the avoidance of excessively large condenser surface areas. Air-cooled condensers are increasing in popularity because of the absence of water piping, the consequent simplicity of operation and the freedom from any health risk associated with the use of spray water.

One objection to their use is that the capacity of the refrigeration plant does not gradually reduce as the ambient dry-bulb rises but ceases suddenly when the high pressure cut-out operates. A partial solution is to arrange for some of the compressor to be unloaded when the condensing pressure rises, before it reaches the cut-out point. Continued operation at a reduced capacity is then possible beyond the design ambient dry-bulb. It is a good plan to select air-cooled condensers to operate in an ambient temperature two or three degrees higher than the design value chosen for the rest of the air conditioning system.

EXAMPLE 11.2

Assuming an airflow rate of 0.15 m^3 s^{-1} kW^{-1}, and a difference of 20 degrees between the ambient dry-bulb and the condensing temperature of R134a, determine (*a*) the condensing pressure and (*b*) the leaving air temperature for an installation using an air-cooled condenser in London if the refrigeration duty is 1 kW.

Answer

(*a*) The design outside dry-bulb temperature for London is about 28°C (see section 5.11).

At 48°C, the absolute condensing pressure of R134a is 1252.7 kPa (12.36 bar) (see Table 9.1). This is a manageable pressure.

(*b*) To cool at a rate of 1 kW in the evaporator, about 1.25 kW must be rejected at the condenser, assuming a coefficient of performance of 4. All this heat is rejected into the airstream and so, using equation (6.6), the leaving air temperature will be

$$t = 28° + \frac{1.25}{0.15} \times \frac{273 + 28)}{358}$$

$$= 35°C$$

11.6 Automatic control

It is always desirable to keep a stable condensing pressure with reciprocating compressors, particularly when the plant runs throughout the year in a non-tropical climate. The temperate winter experienced in the UK can result in condensing pressures which are too low for the plant to operate properly, unless steps are taken to limit the rate at which heat is rejected in the condenser.

Figure 11.1 shows the method adopted for a cooling tower associated with a reciprocating compressor. A direct control is exercised over condensing pressure: as the pressure tends to rise, pressure sensor C1 partially closes the by-pass port in the three-way valve R1, passing more cooling water through the condenser. Sometimes the cooling water flow temperature on to the condenser is controlled (for centrifugal compressors) but this is not preferred for reciprocating machines because of the more sluggish response to changes in condensing pressure and the need for this to be kept fairly constant—flow temperatures of less than 19°C are very undesirable with this mode of control. Occasionally a two-port throttling valve is used instead of a three-port valve but it must always be controlled directly from condensing pressure, never by water temperature. Centrifugal machines prefer to operate with the cooling water temperature on to their condensers controlled at a value that is allowed to fall from about 27°C in summer to a minimum of about 18°C, in order to improve low-load performance, conserve energy and mitigate the risk of surge (see section 12.13). With centrifugal compressors the method shown as a broken line in Figure 11.1 is suitable. A motorised. two-port, butterfly valve, R3, is located in a by-pass across the condenser. The valve is sized so that the water pressure drop through it when fully open equals, or is just less than, the static lift of the tower. Upon fall in flow temperature, sensed by C3, the valve R3 is progressively opened. When the valve is fully open all the water by-passes the tower. It is essential that the valve is beneath the static water level and that it is always flooded when fully open, otherwise control will be lost at partial load.

Induced draught towers are taller than cross-draught and cooling by natural convection, with the fan off is possible to some extent. To save the running cost of the fan it may be switched on–off by an immersion thermostat, C2 (Figure 11.1) with a set-point of, say, 24 ± 2°C. When the ambient wet-bulb is low enough for the tower to produce water at

22°C by natural means the fan cycles and the flow temperature onto the condenser varies between 22° and 26°C. Meantime, the condensing pressure is stabilised by C1 and R1.

De Saules and Pearson (1998), using the work of Braun and Diderrich (1990) and Whillier (1976), describe a way of reducing the energy consumption of cooling tower fans and their related water chillers. It is argued that set points of 26°C to 28°C, for the control of cooling water flow temperature, are unnecessarily high for much of the year and result in excessive energy consumption since compressor power is sensitive to cooling water flow temperature. Savings greater than those achieved by lowering the set-point of the water leaving the cooling tower can be achieved by increasing the fan speed in proportion to the ratio of the partial load to the design maximum load. It is claimed that scheduling the fan apeed in this way prevents the increased cooling tower fan power from exceeding the saving in compressor power. However, it is acknowledged that cooling water flow temperatures less than 19°C could adversely affect lubrication and, by implication, give other problems. While saving energy is a laudable aim it must be remembered that doing so is not the prime purpose of the installation, which is to provide satisfactory air conditioning at partial load as well as full load.

Evaporative condensers have their capacity controlled by means of a condenser pressure-sensing element, C1. This modulates the flow of air passing over the wetted coils by means of motorised dampers, as shown diagrammatically in Figure 11.2. It is customary to have auxiliary contacts on the damper motor control so that, when the dampers are fully closed, the fan is switched off. A cruder form of control, not acceptable in temperate climates when the refrigeration plant has to run throughout the winter, is just to cycle the fan. A better method is to vary fan speed in response to changes in condensing pressure. Some economy in running costs can be effected, if thought necessary, by arranging to switch off the pump when the outside dry-bulb temperature is low enough to permit the condenser to reject its heat without the aid of evaporation.

In the past air-cooled condensers have been used for smaller duties but in recent years, because of the health risks associated with the use of cooling towers and evaporative condensers, there has been a tendency to use air-cooled condensers for much larger duties. When used with small cooling loads (for example, some computer installations) it is not uncommon for the refrigeration plant to have to run in winter as well as in summer, the reason being that only a small, constant quantity of fresh air is handled by the air conditioning system. Controlling condensing pressure by cycling the condenser fan or varying its speed may not prevent excessive cooling by the wind, in very cold weather. Using motorised dampers can then also be unsatisfactory: they do not provide a hermetic seal when closed and will admit cold outside air, even if the fan is off.

An alternative control (Figure 11.6) is sometimes used. There are several variations of this, some probably patented, but in principle the intention is to reduce the capacity of the condensers by reducing the surface area available for heat transfer. The liquid refrigerant is backed-up inside the condenser by restricting its flow to the liquid receiver. This reduces the area of the condenser through which heat is transferred from the condensing vapour to the environment and thus varies the rate of heat rejection in a way that can be automatically controlled from condensing pressure. As the pressure falls (sensed by P1), the motorised valve R1 reduces the outflow of liquid to the receiver. The size of the liquid receiver needs to be increased somewhat so that the refrigerant demands of the system can be met, even though R1 is fully closed for a short while. It is also necessary to have a gas-pressure equalising line between the condenser and the receiver so that an adequate pressure is maintained on the upstream side of the expansion valve.

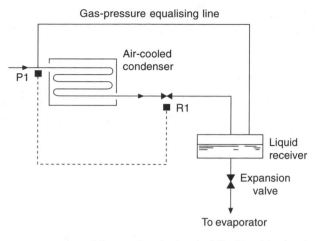

Gas-pressure equalising line

P1

Air-cooled condenser

R1

Liquid receiver

Expansion valve

To evaporator

Fig. 11.6 Condenser pressure control by varying the level of liquid refrigerant inside the condenser.

11.7 Practical considerations

Five major problems must be faced with cooling tower installations:

(i)   Corrosion
(ii) Scale formation
(iii) Protection against freeze up
(iv) Cavitation at the pump
(v)   Health risks

The answer to the corrosion problem is two-fold: first, a tower constructed from materials that are likely to be adequately resistant to corrosion for the expected life of the tower should be selected; secondly, due thought should be given to the use of water treatment.

Continuous evaporation means that the concentration of dissolved solids in the tank of the cooling tower will increase. Unless steps are taken to nullify the effects of this, in due course some scale or sludge formation is inevitable. This will reduce the performance of the cooling water system by being deposited on the heat transfer surfaces in the condenser and on the fill of the tower. Remedial action takes the form of allowing an adequate continuous bleed from the tower (usually arranged in the discharge pipe, as shown in Figure 11.1, so that water is not wasted when the pump is off) coupled with some form of water treatment. If the bleed rate is high enough, it may, under certain circumstances, be sufficient to minimise scale formation without the use of water treatment.

Bleed rates are often quite high. The rate desirable may be related to the hardness of the water, expressed in parts per million $CaCO_3$. For example, water which has a hardness of 100 ppm might need a continuous bleed equal to 100 per cent of the evaporation rate. At 200 ppm, it might require to be 200 per cent of the evaporation rate, and at 300 ppm, to be ten times the evaporation rate. This can sometimes mean a formidable consumption of water.

Water companies in the UK may sometimes demand that a break tank be interposed between the cooling tower (or other device, using mains water for cooling) and their mains. The break tank must be large enough to provide 4 hours' or half a day's supply of cooling water, in the event of a failure of the mains supply. They have three reasons for insisting on this:

(i)   A degree of stand-by is offered to the user.
(ii)   A break is necessary between the mains in order to prevent back-siphonage and contamination of the Company's water supply.
(iii) Fluctuations in demand are averaged; thus, the Company's mains can be sized for average, rather than peak demands.

There is a fourth possible reason why such a tank might be required. In a large office block (for instance), if the cooling tower supply is taken from the cold water storage tank used for supplying water to lavatories, there is reduced stand-by water available for lavatory flushing purposes. If the mains supply failed, the cooling tower would use this water, to the detriment of the hygiene in the building.

EXAMPLE 11.3

Estimate the capacity of the cooling-tower break tank necessary for a 720 kW air conditioning installation. The hardness of the mains water is 100 ppm and the plant runs for an 8-hour working day.

Answer

Assuming a coefficient of performance of 4, the heat rejected in the cooling tower is therefore

$$720 \times 1.25 = 900 \text{ kW}$$

Taking an approximate value of 2450 kJ kg^{-1} as the latent heat of water, then the evaporation rate will be 0.367 kg s^{-1}.

For the hardness quoted, a continuous bleed of 100 per cent is probably required. Thus, the break tank must have a capacity of 11 000 litres.

To prevent freeze-up, two precautions are necessary: first, the provision of an immersion heater with an in-built thermostat in the tank, and secondly the protection of all exposed pipework which contains water with weather-proofed lagging. Below the static water line (when the pump is off), such pipework should be traced with electric cable, thermostatically controlled to keep the water above 0°C. As an alternative, the tower may be drained in winter, if the refrigeration plant is off. But such drainage is a tedious maintenance chore and may not be acceptable to a user. Operation of a cooling tower in winter requires considerable forethought.

As much of the piping as possible must be kept beneath the static water level, otherwise when the pump stops, water above the level will drain from its piping into the tank and flood through the overflow on to the surrounding roof. In any case, the tank should have enough freeboard to accommodate the flow-back from the small amount of piping that must be above the water level, when the pump stops. Non-return valves are seldom an answer to this sort of problem because the available head is usually insufficient to give a tight shut off.

Vortices of air bubbles entrained in the water may form at the outflow from the cooling tower tank. UP to 10 per cent of the volume of the outflow may be air, in bad cases. The formation of vortices is thus very undesirable. It can be prevented, according to Denny and Young (1957), by having a large enough outflow diameter, or a greater depth of water above the outflow, or by the suitable positioning of baffles (see Figure 11.7).

A screen should be provided around the outflow pipe to filter out the larger pieces of debris which collect in the tank of any cooling tower. The size of the holes should be just

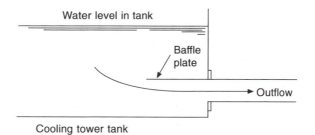

Water level in tank

Baffle
plate

Outflow

Cooling tower tank

Fig. 11.7 A vortex is possible at the outlet from the tank of a cooling tower with a shallow water depth.

smaller than the internal diameter of the water tubes in the condenser. Holes that are too small may be undesirable because the screen may clog up and frequently stop water flow. Birds seem to favour cooling towers in some districts and their droppings and feathers may prove a hazard. Frequent maintenance inspections should resolve this problem.

It is extremely bad practice to put a strainer at the suction side of the pump. The purpose of the strainer is to collect dirt and debris. Consequently, if the strainer is doing the job for which it was installed, the pressure drop across it will increase as time passes. After a while the pressure at pump suction will be so low that, for the particular water temperature, the water within the pump will boil. This is cavitation (see Jones (1997) and Pearsall (1972)): the performance of the pump diminishes significantly and, eventually, the impeller of the pump is destroyed. The purpose of the strainer is to protect the tubes of the condenser and the strainer must therefore always be put on the discharge side of the pump. The screen in the pond of the tower protects the pump.

To ensure that cavitation does not occur the pump should be located so that it is sufficiently below the static water level in the tank and always blows through the condenser. The pump manufacturer will quote the value of net positive suction head (abbreviated NPSH) required but the NPSH provided should always be in excess of this by a factor of at least three according to Grist (1974).

There is a health risk with the use of cooling towers and evaporative condensers, if contaminated spray water droplets of the right size are inhaled. (Five or 6 micron droplets may be inhaled into the deepest part of the lungs and 10 micron droplets can reduce to 3 microns in less than a second by evaporation.)

The risk that has received most attention is Legionnaires' Disease, described by the CIBSE (1991). This is an uncommon infection in the UK, there being about 100 to 300 cases reported annually, of which most seem to be associated with hot water service systems (particularly showers), rather than with cooling towers. It is a form of pneumonia and people most at risk are: smokers, alcoholics and those suffering from a respiratory disease, diabetes or cancer. The fatality rate is about the same (12 per cent) as with other forms of pneumonia and the disease is effectively treatable with the appropriate antibiotics.

The bacterium (*Legionella pneumophila*) exists in water naturally, throughout the UK and abroad, and has done so for many thousands of years. At temperatures below 20°C the bacterium is dormant and the multiplication rate is insignificant. Above this temperature multiplication increases and is greatest at 37°C, in laboratory specimens. At higher temperatures it reduces and ceases at 46°C. When the water temperature reaches 70°C the bacterium is killed instantaneously. Since the typical range of cooling water temperatures provided by cooling towers in the UK is from 32°C to 27°C, it is evident that steps must be taken to prevent contaminated spray from becoming a risk.

Of importance is the provision of a drift eliminator, as Wigley (1986) explains, and which, if properly specified and selected, can reduce the emission of liquid droplets from the outlet of a tower to 0.001 per cent of the cooling water circulation rate. Using a splash water distribution system for the packing, rather than nozzles, helps to reduce the drift. Correct commissioning is essential and first class operation and maintenance are vital. The Report of the Expert Advisory Committee on Biocides (1989) provides an exhaustive list of recommendations for good maintenance and generally offers excellent advice on water treatment, except for the matter of the location of strainers. Strainers must always be positioned on the discharge side of pumps.

It must not be forgotten that the location of cooling towers, evaporative condensers and air-cooled condensers should be right. The position chosen must not compromise in any way the free airflow into and out of the equipment, otherwise heat will not be rejected properly from the refrigeration plant and failure of its high pressure cut-out will be continual. The location chosen for cooling towers and evaporative condensers is also particularly critical in relation to the prevailing wind direction and the configuration of adjoining buildings; air that might convey small droplets of contaminated water (in spite of the use of drift eliminators) must not be discharged from towers or condensers in such a way that it could flow into fresh air intakes or open windows.

Exercises

1. Compare the merits of a forced convection evaporative condenser with those of a shell-and-tube water cooled type.

2. Compare briefly the relative merits of a shell-and-tube condenser with a cooling tower, and an evaporative condenser. Make a sketch of an evaporative condenser, indicating the main components.

3. (a) With the aid of neat sketches, distinguish between cooling towers and evaporative condensers. Name two other methods of rejecting the heat from the condenser of a refrigeration plant used for air conditioning.

(b) Discuss the practical limitations of the four methods mentioned in part (a) of this question. Give emphasis to application, running cost, maintenance and noise.

Notation

Symbol	Description	Unit
a	cross-sectional area of the tower	m^2
c_w	specific heat capacity of water	$kJ\ kg^{-1}\ K^{-1}$
h_a	enthalpy of humid air	$kJ\ kg^{-1}$ of dry air
h_w	enthalpy of a thin film of saturated air at a temperature t_w	$kJ\ kg^{-1}$ of dry air
k	coefficient of vapour diffusion for a unit value of Δh_m	$kg\ s^{-1}\ m^{-2}$
ks	volume transfer coefficient	
$\dot{m}_a$	rate of mass flow of air	$kg\ s^{-1}$
$\dot{m}_w$	rate of mass flow of water	$kg\ s^{-1}$

s	wetted surface area per unit volume of packing	m^{-1}
t	air temperature	°C
t_w	cooling water temperature	°C
Z	height of a cooling tower	m
Δh_m	mean driving force	$kJ\ kg^{-1}$

References

ASHRAE Handbook (1996): Heating, Ventilating and Air-Conditioning Equipment, SI Edition, 35.8.

Braun, J.E. and Diderrich, G.T. (1990): Near optimal control of cooling towers for chilled water systems, *ASHRAE Trans.* **96**, Part 2, 806–813.

Denny, D.F. and Young, G.A.J. (1957): *The Prevention of Vortices and Swirl at Intakes*, British Hydrodynamics Research Association, Publication SP583, July.

De Saules, T. and Pearson, C. (1998): Set-point and near optimum control for cooling towers, *Building Services Journal*, June, 47–48.

Grist, E. (1974): Net Positive Suction Head Requirements for Avoidance of Unacceptable Cavitation Erosion in Centrifugal Pumps, Cavitation, a Conference arranged by the Fluids Dynamics Groups of the I. Mech. E., Heriot-Watt University, Edinburgh, 153–162, September.

Jackson, J. (1951): *Cooling Towers*, Butterworth Scientific Publications.

Jones, W.P. (1997): *Air Conditioning Applications and Design,* 2nd edition, Edward Arnold Publishers.

Minimising the Risk of Legionnaires' Disease (1991): Technical Memorandum TM13: 1991, CIBSE.

Pearsall, I.S. (1972): *Cavitation*, M. & B. Monograph ME/10, Mills & Boon, London.

Report of the Expert Advisory Committee on Biocides (1989): Department of Health, HMSO.

Whillier, A. (1976): A fresh look at the calculation of performance of cooling towers, *ASHRAE Trans.* **82**, Part 1, 269–288.

Wigley, A.F. (1986): Cooling tower drift eliminators, *Hospital Engineering*, March (International Federation Issue No. 57).

Bibliography

1. W.H. McAdams, *Heat Transmission*, pp. 285–292. McGraw-Hill, 1942.
2. BS 4485: Water Cooling Towers (Milton Keynes: British Standards Institution) (1969–88).

12

Refrigeration Plant

12.1 The expansion valve

A pressure-reducing device must be placed in the liquid line before the evaporator. In air conditioning, three such devices are in use: the thermostatic expansion valve, the electronic expansion valve and the capillary tube. Expansion valves have to be protected from dirt and moisture by properly sized strainers, filters and driers on their upstream side.

(a) The thermostatic expansion valve

The primary function of a thermostatic expansion valve is to reduce the pressure from the high value prevailing in the condenser to the low one in the evaporator. Its secondary function is to meter the flow of the refrigerant so that the mass flow pumped by the compressor equals that fed through the evaporator. It also meters the flow of refrigerant so that the vapour leaving the evaporator is somewhat superheated, the purpose being to ensure that only gas is pumped and that no liquid enters the compressor (where it would cause damage) under any circumstances.

Figure 12.1 is a diagram of a thermostatic expansion valve. The downward force exerted above the diaphragm, D, corresponds to the saturation vapour pressure exerted by the fluid refrigerant in the partly filled bulb, B, located at the suction line leaving the evaporator. The upward force acting beneath the diaphragm is the sum of that exerted by the adjustable spring, S, and that corresponding to the evaporator pressure. If the vapour leaving the evaporator is superheated the temperature at the bulb will exceed that in the evaporator. Hence the saturated vapour pressure above D will be greater than the evaporator pressure beneath it. The imbalance is met by S, which can be adjusted manually to give any amount of superheat required. If the load on the evaporator falls (say if the enthalpy of the air on to the cooler coil reduces) insufficient heat is absorbed to vaporise all the refrigerant and provide the superheat wanted. The latter therefore diminishes, the temperature sensed by B falls and the saturation vapour pressure above D, exerted by B, also reduces. An imbalance then exists across D which is rectified by the spring, S, partly closing the valve and lessening the refrigerant flow. A new balance is established with the upward force at D, exerted by the sum of the evaporating pressure and the spring, matched by the downward force produced by the saturated vapour pressure from the bulb B. The value is usually set in the factory so that it begins opening when the superheat is about 5° (see point 1 in Figure 12.1(*d*)). The valve is arranged to deliver 100 per cent rated capacity when the superheat is increased by above another 3°, corresponding to its proportional band (see section 13.2), giving a total superheat of approximately 8°C (points 0 to 2 in Figure 12.1(*d*)). Expansion

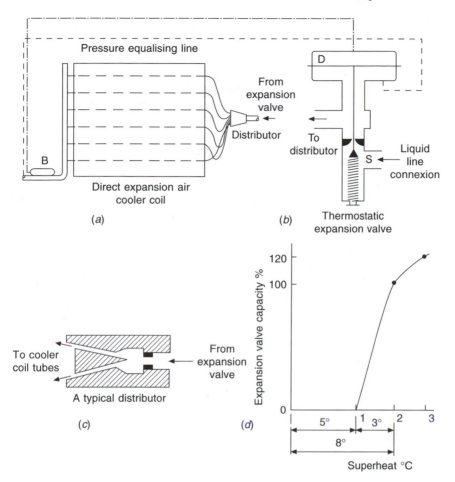

Fig. 12.1   The thermostatic expansion valve, its connection and performance.

valves are commonly rated at an evaporating temperature of about 5°C, this being necessary because capacity for a given superheat setting decreases with pressure, owing to the relationship between saturation temperature and pressure. It is not worthwhile reducing the superheat setting in an attempt to increase the capacity of an evaporator: the gain is insignificant. The valve is often big enough to give a further 20 per cent or so of capacity but this should not be used when sizing it. If the valve selected is too large proportional control will tend to be lost at partial load and the valve will hunt with the risk of allowing liquid into the compressor. Hunting is not only the result of over-sizing the expansion valve: it may also be because of overshoot and undershoot caused by the time lag of the control loop (see section 13.3). If the valve is too small not enough liquid passes through when fully open to meet the load, system capacity is reduced and the suction pressure will fall. It is important that condensing pressure be controlled at a high enough value to give a sufficient pressure difference across the valve for the correct flow of refrigerant to meet the duty. If condensing pressure is allowed to drop too low the suction pressure will also fall and reduced capacity will ensue; there will be the added risk of hermetic motor burn-out because of the lowered suction density.

The bulb should be clamped on top of a horizontal part of the suction line, between the two o'clock and ten o'clock positions, as close as possible to the evaporator outlet and before the equalising line connection. Although some expansion valves have internal pressure equalisation between their downstream side and the space beneath the diaphragm it is essential to use an external equalising line (see Figure 12.1) when there is an appreciable pressure drop between the expansion valve and the place in the suction line where the bulb is located. If an external equalising connection is not used with an evaporator circuit having a large pressure drop then, for the same evaporating pressure as with external equalisation, the closing pressure under the diaphragm will be that much larger. To achieve a balance, the pressure above the diaphragm exerted by the bulb will have to increase. This will be achieved by the expansion valve throttling the refrigerant flow until a higher superheat is developed at the evaporator outlet. The reduced refrigerant flow associated with this abnormally high superheat will give a serious fall in cooling capacity.

It is not possible to predict with any accuracy the performance of an expansion valve at partial duty because the valve is then over-sized and because of the influence of other factors, such as the design of the suction line, the location of the bulb, etc.

When sizing an expansion valve allowances must be made for the frictional losses through the condenser, the liquid line and its accessories (fittings, filters, driers, solenoid valves, isolating valves, distributor), the evaporator, the suction line and its accessories. Static head in the liquid line must also be taken into account. It is to be noted that pressure drop through the expansion valve is approximately proportional to the square of the mass flowrate of refrigerant and hence an increase in the condensing pressure will not give a proportional increase in flowrate and capacity.

A thermostatic expansion valve does not operate smoothly at less than about 50 per cent of its rated capacity.

Thermostatic expansion valves should not be used with flooded evaporators because a significant fall in capacity results from a few degrees of superheat.

(b)   The electronic expansion valve

The electronically controlled, motorised expansion valve is a most efficient alternative to the thermostatic expansion valve. Four methods of providing power for actuating the valve movement are in use:

(i)   *Heat operated.* An electrically heated coil, wound round a bimetallic element, receives an electronic signal from a controller, via a processor, and changes the temperature of the element, modifying its flexure. The element is connected mechanically to the valve stem and hence the valve movement is related to the electronic signal received. An alternative is to use a volatile material, encased in an electrically heated chamber, as the driving force. The expansion of the material is translated into a linear movement to operate the valve in response to the electronic signal.

(ii)   *Magnetically modulated.* Valve movement responds to the pressure of a spring, actuated by the position of an armature that is smoothly modulated by a direct current electromagnet. The movement of the armature is related to the current flowing in the electromagnet which, in turn, depends on the electronic signal received from the controller.

(iii)   *Modulated pulse width.* The expansion valve is a two-position solenoid valve with an exceedingly long cyclic life of millions of operations. The short durations of the solenoid valve in the open or closed positions are dictated by the load, through the electronic signal received.

(iv) *Multiphase step motor.* The rotary motion of the motor is translated into a linear movement to operate the valve, which can be very accurately positioned to open or close, in small increments, according to the electronic signal received. The quality of the valve is better than that of conventional expansion valves and its stroke is significantly longer, giving closer control.

Since it is easier to modify the electronics than to try and adapt the valve, electronic expansion valves readily accept the new refrigerants and seem able to get the best performance from a refrigerant that exhibits glide. The quality of control depends on the electronics, rather than the design of the valve.

The fact that a valve is motorised permits the addition of integral and derivative actions (see section 13.11) to the simple proportional control offered by the self-acting, thermostatic expansion valve. When coupled with electronics and small, dedicated computers, this opens the way for much more effective control over reciprocating refrigeration plant, used to chill water.

A self-acting expansion valve operates with about 8 degrees of superheat, when fully open, measured at the outlet from the evaporator. With the motorised valve, the amount of superheat can be measured by the temperature difference between two thermistors (see section 13.4), one located in the evaporator to sense the saturated evaporating temperature, and the other positioned in a passage within the compressor, just before the suction valve. Since, with the hermetic compressors normally used (see section 12.6), the suction gas rises in temperature by about a further 11 degrees as it flows through the motor driving the compressor, it follows that the gas entering the compressor could be as much as 19 degrees above the evaporating temperature. There is no virtue in having superheat, other than as a safety measure to prevent liquid entering the compressor under reduced load conditions. Hence, with a motorised, electronically controlled, expansion valve, its movement can be continuously monitored and its position adjusted so as to give zero superheat at outlet from the evaporator, whilst still maintaining the necessary, safe amount of superheat at entry to the compressor. The benefit of this is that all the evaporator surface can be used for evaporation, none being required for superheating, with a consequent significant increase in cooling capacity.

Tracking the valve position from an electronic processor that receives information regarding temperature from several thermistors has several other advantages, some of which are:

(i) With air-cooled machines the condensing pressure can be reduced under partial load by allowing some or all of the fans to continue running. This lowers the compression ratio, improves the volumetric efficiency and reduces the power absorbed by the compressor. (Water-cooled condensers, using conventional methods of pressure control, cannot achieve this in the same way.)

(ii) The evaporating pressure can be increased for design load conditions, as a direct consequence of being able to reduce the amount of superheat. This also helps to lower the compression ratio.

(iii) Chilled water return temperature can be measured and, since it is representative of the load on the system, used to raise the set-point of the temperature sensor controlling the flow temperature of the chilled water from the evaporator, under partial load conditions. This also raises the evaporating pressure and helps to minimise energy consumption. The re-set value of the chilled water flow temperature can be related to other control variables but it must be remembered, however, that even though the cooling load for the whole building may have reduced, there might still be components

in the total load that need chilled water at the lower design temperature. This would impose a limitation.
(iv)   A consequence of the above and the quick response of the electronic control system is that refrigeration capacity can be reduced by cylinder unloading (see section 12.6) in a manner that gives closer control over the chilled water flow temperature than is possible with conventional, electro-mechanical methods using step-controllers actuated from return chilled water temperature alone.
(v)   Measurement of the rate of change of chilled water return temperature allows the rate at which cylinders are loaded or unloaded to be varied, in order to match the rate at which load changes occur during start-up and shut-down.

(c)   The capillary tube

This is a small bore tube, principally used in domestic refrigerators and freezers. The diameter and length of the tube relate to the pressure difference between the condenser and the evaporator and restrict the flow of refrigerant while the compressor is running. Cutting the tube to a particular length provides the required pressure drop. When the compressor stops the refrigerant continues to flow until the pressures equalise. The capillary is additionally used as a restrictor to meter refrigerant flow in small, air-cooled, room air conditioning units. It has also been effectively adopted for water-loop, air conditioning/heat pump units. The flow of refrigerant is reversed when the units change their mode of operation, the roles of the evaporator and condenser being exchanged. This is acceptable to a capillary tube but would be impossible for an expansion valve.

EXAMPLE 12.1

Refrigerant 134a evaporates at 2°C in a direct expansion (abbreviated DX) cooler coil. What surplus pressure must the spring in the thermostatic expansion valve exert to maintain 8°C of superheat?

Answer

From Table 9.1 for the properties of saturated Refrigerant 134a we find that the saturation vapour pressure at 2°C is 314.5 kPa absolute and at 10°C (= 2° + 8 K) it is 414.99 kPa. The spring must therefore be adjusted so that it exerts a force corresponding to a pressure surplus of 414.99 − 314.5 = 100.49 kPa.

12.2   The distributor

Figure 12.1(c) illustrates a simple distributor. Its purpose is to give a uniform distribution of refrigerant throughout the evaporator. Thorough mixing of the liquid and vapour leaving the expansion valve is achieved by the turbulence created when the mixture passes at high velocity through the nozzle in the distributor. A homogeneous mixture then flows past the conical deflector and through multiple, small-diameter pipes, of equal friction, that connect the outlet of the distributor to the inlet tubes on the evaporator. The distributor may be located horizontally or vertically but the latter gives better distribution.
   The significant pressure drop across a distributor causes some of the liquid passing to flash to vapour, increasing turbulence and helping to improve mixing. The total pressure drop across an expansion valve and distributor is the same as would occur across an

expansion valve used alone, for an equal refrigerating effect. Pressure drop varies with the load on the evaporator and this cannot be compensated for by adjusting the setting of the thermostatic expansion valve.

Distributors should always be close to their own expansion valves and external pressure–equalising lines must be used. When a hot gas valve is used (see section 12.9) the gas is sometimes injected at the distributor through a side connection between the nozzle and the cone. Turbulence and mixing is improved when hot gas enters under partial load conditions. Distributors are not always adopted for very small installations.

12.3 Float valves

These are also used to reduce the pressure from the high to the low sides and to meter the refrigerant flow. Two forms are available:

(i) *High-side float valve*. When vapour condenses in the shell of the condenser the liquid level therein tends to rise. The float in the valve rises and the latter opens, liquid flows through the port, some flashing to gas as its pressure reduces with a fall in temperature. The high-side float valve works rather like a steam trap and liquid is fed into the evaporator at the same rate as gas is pumped out of it by the compressor. Most of the charge in the system resides in the evaporator and its quantity is critical: too much will cause liquid to flow from the flooded evaporator into the suction line and too little will starve the evaporator and give insufficient cooling capacity. Usually used with single evaporators, condensers and compressors, distribution difficulties may arise if used with multiple evaporators.

(ii) *Low-side float valve*. This is similar in function to the high-side valve but it opens when the level in the evaporator falls, admitting liquid after it has expanded through the valve port with a drop in pressure and temperature. Figure 12.2 illustrates a low-side float valve applied to a submerged evaporator.

12.4 Evaporators for liquid chilling

Liquid refrigerant inside the evaporator absorbs heat from the air being cooled or the water chilled and boils, to efect the refrigeration duty. Two types of evaporator are in use: a flooded version used mostly for water chilling and a dry expansion form used for both chilling water and cooling air.

(i) *Flooded shell-and-tube evaporator*. These are best used when there must be a small temperature difference between the fluid being chilled (such as water) and the refrigerant. Water passes through the inside of the tubes and is chilled by heat transfer to the boiling refrigerant on the outside of the tubes, which often have an extended surface. Sufficient space is left in the shell above the refrigerant level for liquid droplets to separate from the vapour removed through the suction outlet. Liquid eliminators may be provided for this purpose. Refrigerant level may be controlled by a low- or a high-side float valve. The water velocity within the tubes is in the range 1 to 3 m s^{-1} but it is probably best to select for velocities less than 2.5 m s^{-1} to minimise the risk of tube erosion.

New evaporator tubes quickly acquire a film of oxides and other impurities, even before use, and this provides an extra element of thermal resistance, termed the fouling factor, in the overall resistance between the water and the refrigerant. Chillers should be sized using a fouling factor of 0.000 088 m^2 K W^{-1} (0.0005 ft^2 h °F/Btu) if clean water is to be chilled

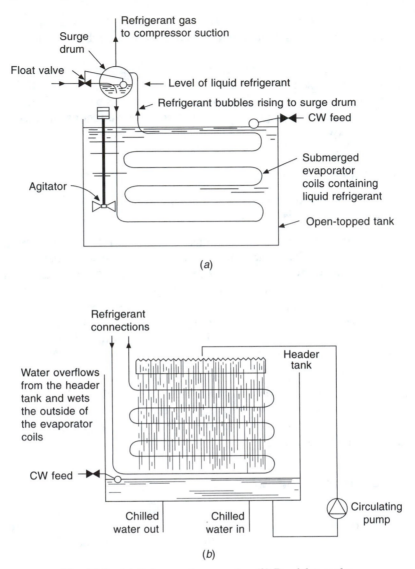

Fig. 12.2   (*a*) Submerged evaporator, (*b*) Baudelot cooler.

in a nominally closed circuit. If the water being chilled is in an open circuit (as with an air washer) the fouling factor adopted must be twice this, namely, 0.000 176 m^2 K W^{-1} (0.001 ft^2 h °F/Btu). The difference between the two fouling factors corresponds to between 5 per cent and 10 per cent of the overall thermal resistance, water-to-refrigerant and is significantly large. See also the discussion on fouling factors in section 12.14.

   Shell-and-tube chillers are commercially available as part of air or water-cooled packages, complete with condensers and reciprocating or centrifugal compressors. It is important that the chilled water flow rate within the tubes is kept at a fairly constant value in order to avoid the risk of freezing the water, bursting the tubes and causing expensive damage to the refrigeration plant. It is also important that the plant is properly controlled: packages with reciprocating compressors must only be controlled from return water temperature and never from flow water temperature, because of the risk of freeze-up at partial load (see section 12.10). Moreover, for similar reasons, the minimum practical design chilled flow temperature is about 6.5°C for a plant with a reciprocating compressor, although as low as 5°C is possible with safety when using a centrifugal or screw machine with sufficiently refined capacity control.

The connection from the feed and expansion tank (or from the pressurisation unit) must be made at the suction side of the pump, which should blow through the evaporator because of its relatively high pressure drop. The whole of the chilled water system is then at a pressure above the datum imposed at pump suction by the feed-and-expansion connection, with less risk of cavitation, or dissolved gases coming out of solution within the pump impeller.

(ii) *Submerged evaporators.* If coils containing boiling refrigerant are submerged beneath the water or brine being chilled in an open tank the water can be reduced in temperature to as low as 0°C, with safety: as ice forms it pushes against the surface and does not crush the evaporator coils. An agitator is necessary to improve heat transfer (see Figure 12.2(*a*)) and baffles may also be fitted in the tank. Such evaporators are commonly used for special industrial, rather than commercial, applications. In these cases it is vital that the supplier of the refrigeration plant should establish at the design stage the fluid flow characteristics over the tubes: the presence of a glycol in the water, to permit sub-zero temperatures to be obtained, can change the value of the Reynolds Number enough to give transitional or even fully laminar flow, with a dramatic fall in the heat transfer coefficient on the water side.

(iii) *Baudelot coolers.* As Figure 12.2(*b*) shows, heat transfer is promoted by arranging for water to flow downwards over the outside of the evaporator tubes. Water can safely be produced at temperatures approaching 0°C. Baudelot coolers are not used in commercial applications.

For both the submerged evaporator and the Baudelot cooler the open chilled water vessel is its own feed-and-expansion tank and the large mass of chilled water present confers stability on the system performance by virtue of its thermal inertia. See section 12.10 for the relevance of storage in commercial applications.

(iv) *Dry expansion evaporators.* Water flows over the outside of the tubes, through the shell and past baffles. Liquid refrigerant is metered by a thermostatic expansion valve to produce about 8 degrees of superheat at the outlet to the suction line. The tubes can have extended surfaces and high rates of heat transfer are possible. Evaporators of this type are common in commercial liquid chilling packages with reciprocating compressors. The strictures on water flow, control and chilled water temperature, mentioned above, are especially important here. Dry expansion chillers are often used for cooling duties below about 140 kW and flooded evaporators are common above this. A liquid-suction heat exchanger is desirable for dry evaporators.

(v) *Plate heat exchangers.* These comprise parallel plates of stainless steel, separated by gaskets and bolted together. Liquid refrigerant and water to be chilled flow between alternate plates with a high rate of heat exchange. Some flexing in the steel sheets is possible and it is claimed that a measure of freezing can be tolerated, although this is highly undesirable because of the potentially dire consequences to the refrigeration system.

12.5 Direct-expansion air cooler coils

We have already seen in chapter 10 how the capacity of an air cooler coil using chilled water may be expressed and these considerations also apply when a refrigerant is evaporated directly inside the tubes of the coil, except that the logarithmic mean temperature difference is between the airstream and the evaporating temperature. If it is assumed that the coil is entirely wet with condensate then wet bulbs may be used as an indication of performance since they are almost parallel to lines of constant enthalpy.

Catalogues often rate DX coils for a range of evaporating temperatures and offer a variety of dry- and wet-bulb temperatures for the entering state of the air, against which the condition of the air leaving the coil is quoted. Interpolation is possible to a limited extent and Table 12.1 gives a small part of a typical catalogue, modified to suit the working of examples which follow.

Table 12.1   Typical performance of a direct expansion air cooler coil with aluminium plate fins spaced at 317 per metre and for an entering air state of 23.9°C dry-bulb, 16.7°C wet-bulb (sling)

Evap. temp. °C	4 Rows leaving state		6 Rows leaving state	
	Dry-bulb °C	Wet-bulb °C	Dry-bulb °C	Wet-bulb °C
Face velocity: 2.0 m s⁻¹				
4.4	10.7	9.8	8.2	7.9
7.2	12.2	11.3	10.2	9.9
Face velocity: 2.5 m s⁻¹				
4.4	11.8	10.7	9.2	8.7
7.2	13.0	11.9	10.9	10.5

EXAMPLE 12.2

The mixture of fresh and recirculated air on to a DX cooler coil in an air handling unit is at 23.9°C dry-bulb, 16.7°C wet-bulb (sling). Use Table 12.1 to determine the state of the air leaving a 4-row coil with a face velocity of 2.5 m s⁻¹ when the evaporating temperature is 5.6°C. Check the psychrometry.

Answer

By interpolation we have a leaving state of 12.3°C dry-bulb and 11.2°C wet-bulb (sling). Figure 12.3 shows the psychrometry and the practicality of the coil performance: the points MWA lie in a straight line and the saturation curve is cut at A. The contact factor is $(23.9 - 12.3)/(23.9 - 9.8) = 0.82$, which is appropriate for a four-row coil with a face velocity of 2.5 m s⁻¹.

EXAMPLE 12.3

Assuming the results of example 12.2 and a rise of 1.2 K to cover fan power and duct gain, calculate the supply air quantity and cooling load to maintain a room at 22°C dry-bulb, 8.071 g kg⁻¹ (about 48 per cent saturation) in the presence of sensible heat gains of 31.13 kW, when it is 28°C dry-bulb, 19.5°C wet-bulb (sling), outside.

Answer

$$\text{Supply air temperature} = 12.3° + 1.2° = 13.5°C$$

$$\text{By equation (6.6) supply air quantity} = \frac{31.13}{(22 - 13.5)} \times \frac{(273 + 13.5)}{358}$$

$$= 2.931 \text{ m}^3 \text{ s}^{-1} \text{ at } 13.5°C$$

Plotting the relevant states on a psychrometric chart (Figure 12.3) and reading off enthalpies across the coil and the specific volume at the supply state we have

$$\text{Cooling load} = \frac{2.931}{0.822} \times (46.66 - 32.02) = 52.2 \text{ kW of refrigeration}$$

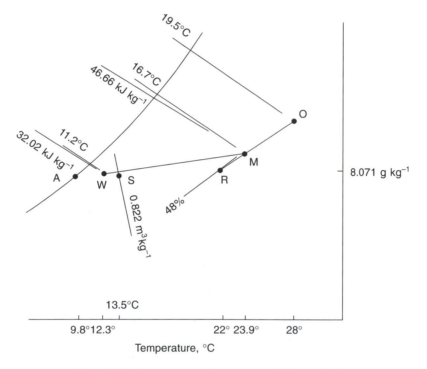

Fig. 12.3 Psychrometry for examples 12.2 and 12.3.

The performance of the air cooler coil can be expressed graphically and this shows its behaviour at partial load and what happens when it is piped up to a compressor, condenser and expansion valve. Two characteristic curves may be plotted (see Figure 12.4): one for the air-side perfromance and the other for the refrigeration-side behaviour, with the assumption that the coil surface is entirely wet with condensate. Referring to Figure 12.4 we see that the abscissa can be interpreted as wet-bulb temperature or evaporating temperature, as convenient. The point P can then be identified by the design cooling load (52.2 kW) and the wet-bulb temperature leaving the coil under design conditions (11.2°C). If the wet-bulb leaving the coil were the same as that entering, namely, 16.7°C, this would imply no change of enthalpy either and would correspond to zero cooling load. We can therefore plot

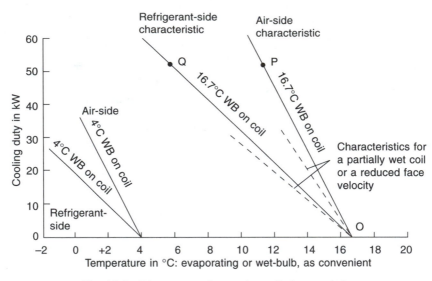

Fig. 12.4   Direct-expansion cooler coil characteristics.

a zero load point, O, at 16.7°C and 0 kW duty. The points O and P lie on the air-side characteristic and may be joined by a straight line for all practical purposes and for simplicity. Reference to Table 12.1 and example 12.2 shows that we can achieve a leaving wet-bulb of 11.2°C when the entering wet-bulb is 16.7°C and we have an evaporating temperature of 5.6°C. Another design duty point, Q, is plotted at a cooling load of 52.2 kW and an evaporating temperature of 5.6°C in Figure 12.4. The zero load point, O, also corresponds to an evaporating temperature of 16.7°C, the same as the on-coil and off-coil wet-bulb. The points O and Q are then joined by a straight line to give the refrigerant-side characteristic for the coil under the same conditions as the air-side characteristic.

If the entering wet-bulb temperature reduces, the pair of lines (air and refrigerant) moves to the left, remaining parallel to the design pair but having a different zero load point, at a value equal to the lower entering wet-bulb. For example, Figure 12.4 shows such a pair starting from a zero load point at 4°C on the abscissa. If the coil is only partly wet or if the face velocity falls, capacity is reduced for a given evaporating temperature, as shown by the pair of broken lines in Figure 12.4.

12.6   The reciprocating compressor

It is customary when considering the performance of reciprocating compressors, to do so for a particular condensing temperature because it is the performance of the condensing set that is really being considered. As the condensing pressure against which the compressor has to deliver the gaseous refrigerant rises, more work is expended per kW of refrigeration, and the refrigeration capacity of the condensing set reduces. On the other hand, as suction pressure at inlet to the cylinders of the compressor rises, less work need be done to secure the same refrigerating effect. Furthermore, with an increase in suction pressure, the density of the gas entering the cylinders also increases and so, for a given swept volume, the mass of refrigerant handled by the compressor becomes greater. A picture is thus formed of the characteristic behaviour of a condensing set. Figure 12.5 illustrates this.

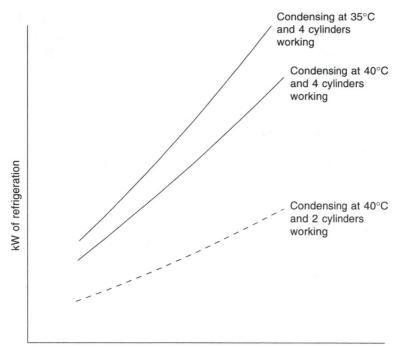

Fig. 12.5 Condensing set performance.

A compressor has a rising capacity characteristic; the higher the suction pressure the greater the capacity. The characteristic lines for different condensing temperatures are virtually straight, particularly if suction pressure is used as the abscissa. If the saturation temperature corresponding to suction pressure is used, superheat being ignored, some slight curvature results. Although straight lines can be used, the inaccuracy resulting being accepted, it is better to draw the lines as curves if the data are available for this purpose. They usually are, since compressor characteristics cannot be readily guessed. Reference must be made to the manufacturers for the information needed to plot the characteristic.

The capacity of a compressor, for a given condensing temperature, depends directly on the mass of refrigerant being pumped. Thus, if the speed of rotation or the number of cylinders used is altered, the refrigeration capacity of the compressor will change in direct proportion. This fact enables a family of characteristics to be plotted, provided, of course, that the initial information is to hand from the manufacturers.

Although the picture may eventually change, the majority of refrigeration compressors in use today are of the reciprocating type. These have from 1 to 16 cylinders, arranged in line, in *V* or *W* configuration, or sometimes in a radial disposition. The *V* or *W* arrangements are commonest. Most machines are driven by electric motors and are close-coupled or direct-coupled, although vee-belt drives are also used. Engines of various sorts have also been employed to drive compressors.

In common with most other types of compressor, reciprocating machines are available in open, hermetic or semi-hermetic form. The compressor of an open machine is driven by a separate motor, through a direct-coupling or a vee-belt. A seal is provided where the compressor shaft emerges from the casing but there is always the risk of some leakage of refrigerant gas out of the system. The driving motor is cooled in a conventional manner. With the hermetic machine the compressor and the driving motor share the same shaft (they are close-coupled) and both are contained in a welded casing. Comparatively cold, suction gas flows over the stator winding of the motor before entering the compressor. The advantages of the hermetic arrangement are that leakage at a shaft seal is eliminated, the shaft is shorter and more rigid, the bearing arrangement is simplified, the machine is quieter and the motor is gas-cooled. A potential disadvantage is that, if the control of the related air conditioning system is poorly designed or ineptly operated, or if the compressor is allowed to start too frequently, there is a greater risk of motor burn-out than with open machines. If a burn-out occurs, the consequent reaction with the refrigerant poisons the rest of the refrigeration system, necessitating thorough cleaning, or even replacement. Semi-hermetic machines are similar but have a removable cover bolted on to the end of the casing to facilitate occasional maintenance. Welded, hermetic machines are used for the smaller duties, up to about 70 kW of refrigeration and bolted, semi-hermetics are used for larger loads. Packaged, water-chilling units having multiple, semi-hermetic compressors are used with capacities up to about 700 kW of refrigeration.

The principal method adopted for regulating the capacity of a reciprocating compressor is cylinder unloading. Cylinders are unloaded by arranging to hold the suction valve open, the discharge valve staying closed because of the high pressure on its other side. As the piston moves up and down, gas is circulated through the suction valve, no compression is achieved and the cylinder makes no contribution to the refrigeration duty. The suction valve is held open mechanically, using the pressure developed by the oil lubrication system. A fall in the suction pressure is usually the signal that initiates cylinder unloading. A suitable differential gap separates the measured values of suction pressure at which loading and unloading occur. A less common alternative is to measure some external, physical property, such as the temperature of the water returning to a chiller. A fall in this is then the indication for unloading. An electrical signal from the temperature sensor may be used to open a solenoid valve in a pipeline feeding lubrication oil to the cylinder unloading mechanism. The use of suction pressure is neater because all the unloading mechanism is within the compressor whereas, with the other method, external connections must be made. Figure 12.5 illustrates what happens when a pair of cylinders in a four-cylinder machine is unloaded.

Proper lubrication is essential and must be ensured under all operational conditions. Halogen type refrigerants are miscible in oil and tend to collect in the crankcase of reciprocating machines. When the compressor starts, the low pressure developed in the crankcase causes the refrigerant to boil and the mixture foams. Foaming is bad for the lubrication of the system and should be prevented by the provision of a crankcase heater. This is electrically energised when the machine is off and is de-energised when the compressor runs. The crankcase heater is wrapped externally around the casing of hermetic machines but is fitted within the crankcase of semi-hermetic and open compressors. Other refrigerants that do not use mineral oil as a lubricant (see section 9.11) have their own characteristics, which must be established.

The inefficiency of the motor reduces the coefficient of performance of hermetic and semi-hermetic machines, in comparison with open-drive compressors. This is shown by equations (12.1) and (12.2):

$$COP \text{ (hermetic)} = \frac{\text{Refrigeration capacity (kW)}}{\text{Input power to motor kW}} \qquad (12.1)$$

$$COP \text{ (open)} = \frac{\text{Refrigeration capacity (kW)}}{\text{Input power to shaft (kW)}} \qquad (12.2)$$

12.7 The air-cooled condensing set

The heat-rejection capacity of an air-cooled condenser is almost directly proportional to the difference between the condensing temperature and the dry-bulb of the air entering, as Figure 12.6 shows. Enough extra surface is sometimes provided to sub-cool the liquid by 8° or 9°C but to see how a condenser actually performs we must look at its behaviour when piped to a compressor to form a condensing unit, or set.

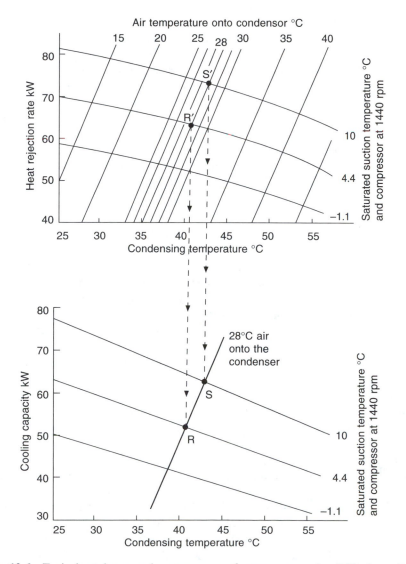

Fig. 12.6 Typical condenser and compressor performance, assuming 8.3° sub-cooling.

Compressor capacity can be described in terms of: cooling effect (refrigeration) against saturated suction temperature or pressure (Figure 12.5) for a given condensing temperature, cooling effect against condensing temperature for a given saturated suction temperature, or as heat rejected (cooling effect plus compressor power) against condensing temperature for a given saturated suction temperature. The upper part of Figure 12.6 shows compressor performance as heat rejection rate and the lower part as cooling effect, both against condensing temperature.

The intersections of the compressor and condenser curves in the upper part of Figure 12.6 indicate the condensing temperature obtaining for given suction and entering air temperatures and these points of intersection can then be transferred to the lower part of the figure to give the corresponding refrigeration effects. Thus the point R′ (28°C on to the condenser, 4.4°C suction temperature) represents a heat rejection rate of 63.4 kW at a condensing temperature of 40.6°C and transforms to the point R (28°C on to the condenser, 4.4°C suction temperature) and a refrigeration effect of 52.2 kW at the same condensing temperature. Similarly, S′ transforms to S, and so on. In this way it is possible to plot characteristic curves representing cooling capacity of a condensing set at a given dry-bulb on to the condenser against saturated suction temperature. From Figure 12.6 we can deduce that when air at 28°C enters the condenser the refrigeration duties are 62.9 kW, 52.2 kW and 43.1 kW for suction temperatures of 10°C, 4.4°C and −1.1°C, respectively. This is shown in Figure 12.7 by a curve for the condensing set passing through the point R, with air on at 28°C. An alternative approach is to express the performance of the condensing set for a given condensing temperature instead of a given air temperature on to it. For example, the refrigeration duties would then be quoted as 64.5 kW, 52.2 kW and 42 kW for respective suction temperatures of 10°C, 4.4°C and −1.1°C when the condensing temperature was 40.6°C.

12.8   Condensing set–evaporator match

The foregoing must be taken a step further by plotting the performance of the evaporator on the same co-ordinates as the condensing set. The intersection of the refrigeration-side characteristic of the evaporator and that of the condensing set will then tell us the actual performance of the combination. A pressure drop occurs in the suction line and it is conventional to regard this as the difference of two saturated pressures and to speak of a corresponding difference in saturated temperature. The suction line is usually sized for a pressure drop corresponding to a fall in saturated temperature of about 1°C. In Figure 12.7 this is assumed to be 1.2°C and thus a temperature of 5.6°C in the evaporator gives a suction temperature of 4.4°C. This allows the point Q to be established in Figure 12.7 at 52.2 kW and an evaporating temperature of 5.6°C. Referring to Figure 12.4 we see that this is also a point Q on the refrigerant-side characteristic, used in examples 12.2 and 12.3. From Figure 12.6 we can also plot curves for the condensing set performance at other air temperatures entering the condenser on Figure 12.7 and this will show how the match between the evaporator and the condensing unit changes as summer passes and winter approaches.

If the wet-bulb temperature on to the cooler coil falls the evaporator characteristic moves to the left (Figure 12.7) and the balance point with the condensing set shifts, with a reduction in evaporating and suction temperatures. Thus when the entering wet-bulb is 9.6°C the evaporating temperature is about +0.4°C (point W) and the suction temperature is about −0.8°C (point X), if the dry-bulb on to the condenser is 28°C but the balance drops

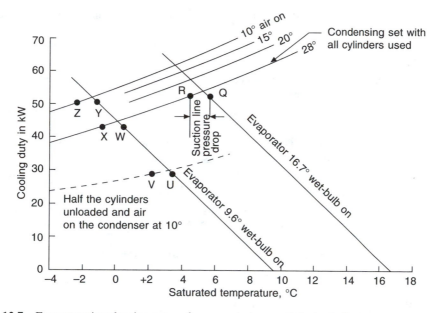

Fig. 12.7   Evaporator/condensing set performance in terms of air dry-bulb temperature on to the condenser.

to about −1.2°C evaporating (point Y) and −2.4°C suction (point Z) when the air on the condenser is as low as 10°C (Figure 12.7). Low evaporating temperatures will give low surface temperatures with the risk of frosting on the DX coil if below 0°C. Frost forms rapidly because the latent heat of fusion is not great (235 kJ kg^{-1}) and impedes the airflow, causing the evaporator characteristic to rotate to the left about the zero load point and so lowering the evaporating temperature and pressure even further. Hermetic compressors rely on the mass flow of suction gas to cool their stator windings and with a fall in suction pressure, density and mass flow rate also fall. Hence the risk of motor burn-outs at low load and the need to determine the risk of this by establishing where the evaporator–condensing set balance occurs at the lowest anticipated load.

12.9   The control of direct-expansion cooler coils and condensing sets

Remembering that the thermostatic expansion valve does not control cooler coil capacity suitable methods for doing so are as follows.

(*a*) *On–off switching.* Simple switching of a single evaporator–condensing set circuit, or of several, independent circuits, may be done satisfactorily from a room or return-air thermostat (together with a step controller if necessary). Typical air change rates in commercially air conditioned premises are 10 to 20 per hour, implying that 6 to 3 minutes is required for the effect of a change in supply air temperature to appear as a change in return air temperature. For a condition of 50 per cent sensible heat gain we would expect equal running and off times for the compressor, suggesting from 12 to 6 minutes between successive starts, which gives 5 to 10 starts per hour under the most adverse partial load condition. The allowable, maximum number of starts per hour should be checked with the supplier of the refrigeration plant. On–off switching *must never be attempted* from a thermostat located after the cooler because of the certainty of very rapid compressor cycling: when the machine is switched off only a second or two will elapse before the warm, uncooled air that is entering the coil passes over the thermostat and switches the compressor on again.

When several independent cooler coils (each with its own condensing set) are combined for cooling and dehumidification they may be arranged in three ways: in parallel across the face of the air handling unit (to give face control), in series in the direction of airflow (to give row control), or interlaced.

The commonest arrangement is probably when arranged for face control. A return air thermostat switches off the independent coils as a rise in temperature is sensed, allowing uncooled air to by-pass the active coil sections. When they are active, all the refrigeration circuits are equally loaded and this has the great advantage of simplicity and stability. If some of the coil sections are off, the state leaving the coil combination is a mixture of uncooled air that has by-passed the active sections with air that has been cooled and dehumidified.

If the coil sections are arranged in series there is no problem of by-passed air but there are difficulties with the refrigeration circuits, which do not handle equal loads when active. This arrangement is not recommended.

It is possible to arrange coils in series with a common fin block but to interlace their piping circuitry so that each coil section carries the same load when active. A four-row coil, for example, would have a common block of fins, extending over the whole of the cross-sectional area of the air handling unit, but might have two distributors and expansion valves (one for each separate condensing set). Liquid from the distributors would be fed symmetrically through the height, width and depth of the coil, so that each refrigeration piping circuit had the same cooling duty. When one section was switched off, the fins of that section would be available to help cooling and dehumidification throughout the whole of the coil, aiding the performance of the active section and mitigating the by-pass effects suffered by the simple parallel arrangement of face control. As a very simple indication, one circuit might feed rows one and three for the upper part of the coil, with the other circuit feeding rows two and four. For the lower half of the coil, the roles would be reversed, the first circuit feeding rows two and four while the second circuit fed rows one and three. The finning and piping arrangements would be identical for each circuit and the loads would be equalised. In most actual cases the interlacing arrangement is more involved than in the above simplified example.

(b) Cylinder unloading. Since capacity is directly proportional to refrigerant mass flow rate it is also proportional to volumetric flow rate for a given suction pressure. Hence it will be proportional to the number of cylinders in use. The broken line in Figure 12.7 shows the characteristic for the condensing set if half its cylinders are unloaded and air enters the condenser at 10°C. We see now that if the wet-bulb on to the evaporator is 9.6°C the evaporating and suction temperatures will rise to about 3.4°C (point U) and 2.2°C (point V) from their former values of −1.2°C (point Y) and −2.4°C (point Z) respectively. Cylinder unloading can be used to counter the risks of falling suction pressure with reduced wet-bulbs on to the evaporator. See section 12.6.

(c) Back-pressure valve (BPV). If an automatic valve is placed in the suction line (see Figure 12.8) it can be used to give a nominally constant evaporating pressure by increasing the pressure drop in the suction line. Figure 12.8 shows that the drop in the suction line is represented by the points 2 and 3 under the design load with the valve fully open. When

the wet-bulb temperature on the evaporator falls the characteristic shifts to the left, balancing at the points 2′ and 3′. By partly closing the valve the natural drop in the suction line (between points 2″ and 3″) is increased to give a new evaporator balance at the point P. The distance between the points 3″ and P represents the pressure drop across the partly closed valve. A pilot valve is sometimes used in conjunction with a thermostatic bulb in the airstream after the cooler coil to give proportional control over the leaving air temperature. The use of a back pressure valve increases the drop in the suction pressure and it should therefore never be used with hermetic compressors.

(*d*) *Hot gas valve (HGV)*. The function of this is to stabilise the evaporating pressure (if a self-acting valve is used) or to control the temperature of the air leaving the coil (if a motorised valve is used), by injecting gas from the hot gas line. In the latter case it is possible to achieve tight temperature control if the right quality of valve is chosen. Suction pressure does not fall in either case and hot gas valves are very suitable for use with hermetic compressors. *There is only one satisfactory way of using a hot gas valve* and that is by injecting the hot gas into the evaporator to impose a false load. Figure 12.8 shows that the evaporator balance can be kept at the point 3 by feeding in enough hot gas to add the false load represented by the vertical distance between the points 3 and P, the suction condition remaining at point 2. The best place to inject the hot gas is into a hot gas header, as Figure 12.8 shows. This then gives the best possible control. An alternative which is cheaper but not so controllable is to put gas into the side of the distributor.

EXAMPLE 12.4

Calculate the mass flow rate of hot gas necessary to control the evaporating pressure at 303.6 kPa for the case considered in Figure 12.8 if the entering wet-bulb is 13°C and a hot gas valve is used instead of a back-pressure valve.

Answer

The false load to be imposed on the evaporator is 100 kW – 75 kW = 25 kW. From Table 9.1, liquid Refrigerant 134a at 1°C and 303.6 kPa has a latent heat of 197.9 kJ kg^{-1} by interpolation. Hence the required mass flow rate is 25/197.9 = 0.126 kg s^{-1}.

12.10   The performance of water chillers

The capacity of reciprocating compressors is most commonly controlled in steps by cylinder-unloading, with corresponding steps in the chilled water flow temperature, when using a conventional thermostatic expansion valve. Control from a temperature sensor, located in the flowline from a chiller, is not a safe proposition and it should always be done from a thermostat in the chilled water return. In the simpler case, variations in suction pressure may be used to unload cylinders, some variation in the chilled water flow temperature being tolerated. It is vital that the chilled water flow temperature is not allowed to approach too closely to freezing point.

EXAMPLE 12.5

A plant has a design duty of 235 kW when chilling water from 13°C to 7°C with an evaporating temperature of 3°C. The compressor has six cylinders which are unloaded in

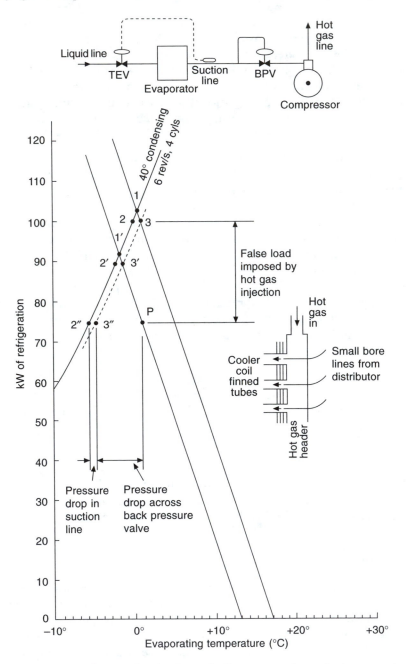

Fig. 12.8 Performance diagram showing how a back pressure valve and a hot gas valve may be used to control evaporating temperature. Back pressure valves should not be used with hermetic and semi-hermetic compressors.

pairs through a step-controller from a thermostat immersed in the return chilled water line. Establish the temperature settings for the step-controller to unload and load the cylinders, within the following restraints: maximum allowable chilled water flow temperature, 7°C;

minimum allowable evaporating temperature, 1°C; set-point of the low water temperature cut-out, 3.5° ± 1°C.

The compressor has the following performance when running at constant speed on six cylinders and at a constant condensing pressure:

Evaporating temperature (°C):	0	3	6	9	12
Refrigeration duty (kW):	210	235	260	285	310

Answer

Since the machine runs at a constant speed the volumetric flow rate of refrigerant is directly proportional to the number of loaded cylinders. Hence, because the suction density is constant for a given evaporating temperature, we can say the mass flow rate, and so the refrigeration capacity, is also directly proportional to the number of loaded cylinders at any given evaporating temperature. The following table can be compiled:

No. of cylinders	Evaporating temperature				
	0°C	3°C	6°C	9°C	12°C
	Duty in kW				
6	210	235	260	285	310
4	140	156.7	173.3	190	206.7
2	70	78.3	86.7	95	103.3

Three compressor characteristics can now be drawn (Figures 12.9(a) and (b)). As with air cooler coils the water chilling evaporator has a characteristic performance that can be expressed in two ways: as kW of refrigeration against evaporating temperature or chilled water flow temperature. Thus in Figure 12.9(a) we can plot a pair of points, A and B, at 3°C and 7°C, respectively, representing the evaporating and chilled water flow temperatures at a design duty of 235 kW. These points lie on the refrigerant-side and water-side characteristics of the evaporator for a return water temperature of 13°C and they share a common zero load point at C, when the flow and return water temperatures and the evaporating temperature are each 13°C. In a fashion similar to that adopted earlier for air cooler coils, additional pairs of characteristics may be drawn, parallel to the design pair, for other return water temperatures.

As the load on the chiller diminishes the return water temperature drops and the pair of evaporator characteristics shifts to the left (Figure 12.9(a)). When the evaporator refrigerant characteristic cuts the compressor characteristic for six cylinders at 1°C (point D) the load is 219.3 kW and water is being chilled to 4.6°C (point E) from a return water temperature of 10.2°C (point F). We have now reached one of the restraints imposed for safe operation and the machine must be unloaded from six to four cylinders. The return water temperature to effect this is 10.2°C. After unloading, the balance moves from D to G, with evaporation at 3.5°C (point G) and a chilled water flow temperature of 6.1°C (point H). Upon further fall in load the return water temperature reduces to 7.1°C (point K) with an evaporating temperature of 1°C (point I) and a chilled water flow temperature of 3.4°C (point J). Two more cylinders are unloaded when the thermostat in the return chilled water line senses 7.1°C (point K) and a new balance is struck at an evaporating temperature of 3.8°C (point

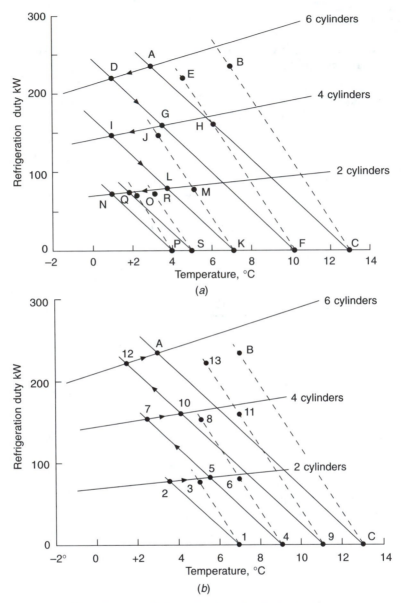

Fig. 12.9 (a) Chiller unloading. (b) Chiller loading.

L) and a chilled water flow temperature of 5.1°C (point M). As the load continues to decline the return water temperature continues to fall and we see that if it dropped to 4.1°C (point P) the evaporating temperature would be 1°C (point N) but the chilled water flow temperature would be 2.3°C (point O), at a value less than the low water temperature cut-out point of 2.5°C (= 3.5° − 1°). We may observe here that if we adopt a return water temperature of 5°C (point S) for unloading the last two cylinders and switching off the compressor, evaporation will be at 1.8°C (point Q) and the chilled water flow temperature will be 3.1°C (point R), giving us a margin of 0.6°C above the low water temperature cut-out value.

When the system load starts to increase and the return water temperature rises to 7°C (point 1 in Figure 12.9(*b*)) the machine is started with two cylinders operating and balances at an evaporating temperature of 3.6°C (point 2) and a water flow temperature of 5.0°C (point 3). As the load continues to rise the pair of evaporator characteristics shifts to the right (Figure 12.9(*b*)) until the chilled water flow temperature reaches 7°C (point 6) when the evaporating temperature is 5.6°C (point 5) and the return water temperature is 9.1°C (point 4). Two more cylinders are then loaded (four in all now operating) and the balance moves to evaporation at 2.5°C (point 7) with a water flow temperature of 5.15°C (point 8). With a progressive load increase the return chilled water temperature rises to 11.1°C (point 9) when the system is evaporating at 4.1°C (point 10) and the water is being produced at 7°C (point 11). The compressor then loads fully, working with six cylinders, balancing anew at evaporating and flow temperatures of 1.5°C (point 12) and 5.4°C (point 13), respectively. If the load goes on rising to its design duty the balance will eventually return to points A and B, chilled water being produced at 7°C when the evaporating temperature is 3°C, from a return water temperature of 13°C (point C).

The loading and unloading set-points for the step-controller can now be summarised:

Table 12.2 Loading and unloading set-points for example 12.5

Ch.w. ret. temp. °C	Evap. temp. °C	Ch.w. flow temp. °C	Duty kW	Cylinder change: From —	To —	New evap. temp. °C	New ch.w.flow temp. °C	New duty kW
Unloading								
10.2	1	4.6	219.3	6	4	3.5	6.1	160.6
(F)	(D)	(E)				(G)	(H)	
7.1	1	3.4	144.9	4	2	3.8	5.1	78.3
(K)	(I)	(J)				(L)	(M)	
5.0	1.8	3.1	74.4	2	0	—	5.0	0
(S)	(Q)	(R)						
Loading								
7.0	—	7.0	0	0	2	3.6	5.0	78.3
(1)						(2)	(3)	
9.1	5.6	7.0	82.25	2	4	2.5	5.15	154.7
(4)	(5)	(6)				(7)	(8)	
11.1	4.1	7.0	160.6	4	6	1.5	5.4	223.2
(9)	(10)	(11)				(12)	(13)	

The letters and numbers in brackets in the table refer to Figures 12.9(*a*) and (*b*).

Within reason, it does not matter how frequently cylinders are loaded or unloaded except for the last step, which cycles the compressor motor on–off. It is most important that the motor is not switched on too often because the inrush of electric current on starting causes the windings to overheat and repeated starts at short intervals will burn them out. Typical allowable starts per hour are: 4 to 6 for hermetic, reciprocating compressors, 8 to 10 for open, reciprocating machines and 2 for hermetic centrifugals.

When the cooling load is a little less than the maximum refrigerating capacity the plant runs for a lot of the time and when it is almost zero the compressor will be mostly off. If the cooling load is half the average cooling capacity of the last step of refrigeration, the machine will start and stop most frequently and the following analysis may be considered.

Average cooling capacity of last step of refrigeration:	q_r	kW
Average cooling load on system:	$0.5q_r$	kW
Surplus capacity available for cooling down the water in the system:	$0.5q_r$	kW
Mass of water in the system:	m	kg
Specific heat capacity of water:	4.187	kJ kg^{-1} K^{-1}
Allowable variation in return chilled water temperature:	Δt	K
Time taken to cool down the water in the system:	θ	s

Hence

$$0.5q_r = (4.187m\Delta t)/\theta$$

and the cooling down time is therefore

$$\theta = (2 \times 4.187m\Delta t)/q_r \qquad (12.3)$$

If the cooling load is assumed as a constant and equals $0.5q_r$, the time taken to warm up the water in the system by Δt after the compressor has shut down will equal that taken to cool it down when the compressor was running, for all practical purposes. The time between successive starts will then be 2θ and if the allowable frequency of starting is f times per hour we can write

$$f = 3600/2\theta. \text{ By substitution in equation (12.1) we have}$$
$$f = (3600q_r)/(2 \times 2 \times 4.187m\Delta t) = (215q_r)/(m\Delta t) \qquad (12.4)$$
$$m = (215q_r)/(f\Delta t) \qquad (12.5)$$

EXAMPLE 12.6

For the case of example 12.5 and Figures 12.9(*a*) and (*b*), determine the minimum amount of water allowable in the system if the number of starts per hour for the compressor is limited to six.

Answer

When the return chilled water temperature falls to 5°C the last step of the chiller switches off, immediately prior to which the refrigeration capacity was 74.4 kW (point Q). The return water temperature then starts to rise and when it reaches 7°C (point 1) the compressor is started on its first step and the capacity is initially 78.3 kW (point 3). The return chilled water then starts to fall again towards 5°C. Thus

$$q_r = (74.4 + 78.3)/2 = 76.35 \text{ kW}$$

When the load on the cooler coil and the heat gain to the chilled water pipes in the system is half this value the greatest number of starts per hour will prevail and since this must be limited to six we can use equation (12.5) to establish the necessary mass of water in the system, m, to ensure this:

$$m = (215 \times 76.35)/(6 \times (7 - 5)) = 1368 \text{ kg}$$

12.11 The screw compressor

This is a positive displacement machine used with refrigerants at pressures above atmospheric and available in single or twin rotor form. It is applied in air conditioning for chilling water and is usually, but not always, water-cooled.

(a) The single screw compressor (see Figure 12.10)

The machine comprises a single screw (or main rotor) having six, carefully machined, helical grooves mounted in a cast iron, cylindrical casing with suction pressure at one end and discharge pressure at the other. Two planet wheels (otherwise termed gaterotors), each having eleven teeth and made from a special plastic material, engage in the helical grooves on each side of the main rotor through slots on opposite sides of the rotor casing. Refrigerant gas in the suction chamber is able to enter all six grooves in the screw as far as the tooth of the planet wheel engaging with a groove, on opposite sides of the rotor. Gas cannot pass along the grooves that are not fully engaged with the teeth of the planet wheel because a plate at the discharge end of the casing prevents this. There are, however, two ports in the plate that will permit gas to be discharged from two of the grooves as the rotation of the screw brings them opposite the port openings.

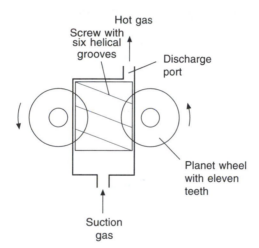

Fig. 12.10 Simplified diagram of a single screw compressor.

In the first, suction stage of the operation, the screw rotates and gas from the suction chamber fills part of one of the grooves. As rotation proceeds this volume of suction gas moves along the groove until the engagement of the next tooth on the planet wheel closes off the groove from the suction chamber. This volume of gas in the groove is still at the suction pressure and is enclosed by: the engaging tooth from the planet wheel, the surfaces of the helical groove itself, the cylindrical casing of the screw and the plate at the discharge end of the screw. In the next phase of the operation, further rotation of the screw causes the tooth to move along the helical groove, reducing the volume of the trapped gas and compressing it. Finally, as the other end of the groove comes opposite the opening in the plate at the discharge end of the casing, the trapped gas is delivered at high pressure into a discharge manifold. This process takes place in the helical grooves on both sides of the screw and the manifold receives hot, high pressure gas from both discharge ports. From the above it is seen that the planet wheels separate the high and low pressure regions of the casing as they mesh with the screw.

The screw is driven by an electric motor in a hermetic or semi-hermetic arrangement. The planet wheels idle and virtually no power is lost to them from the screw, except a very small amount by friction. Frictional losses are small because of the natural lubricating property of the plastic material of the planet wheel teeth, their compliance, and the close machining tolerances adopted. A further advantage of the design is that bearing constraints are small and operating lives can be as much as 200 000 hours.

The screw can tolerate some liquid feedback.

A version having only one planet wheel is also used for some applications. Operation is similar.

Lubrication can be by oil injection, its function being to lubricate, cool and seal between high and low pressures, and actuate the capacity unloading mechanism. An oil separator is then needed in the high pressure refrigerant and an oil cooling system is also required.

A favoured alternative is to use liquid refrigerant that is virtually free from oil. (There may be some small amount of oil present because of the need for oil in the bearings.) This is injected into the compressor casing. The advantages are that no oil separator is required and no external oil cooler is needed. Since the frictional losses are very small where the planet wheel teeth engage the screw, the only functions of the injected refrigerant are to seal and to cool.

Economisers are often fitted and volumetric and isentropic efficiencies are high (see Figure 12.11).

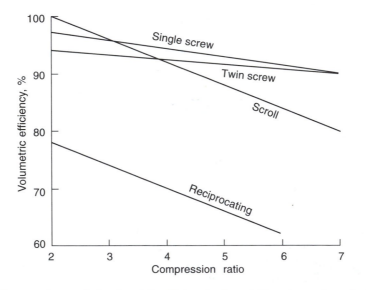

Fig. 12.11   Comparative, typical volumetric efficiencies for scroll, reciprocating, single screw and twin screw compressors.

Capacity control is by means of a slide valve that changes the place where compression begins, resulting in variable compressor displacement. Actuation is from a piston driven by oil pressure or, if an oil-free system is used, by an electric motor. Unloading may be proportional by continuous modulation of the slide valve position, or in steps.

Both water-cooled and air-cooled machines are available and the range of cooling capacities is from about 15 kW to 4500 kW, depending on the operating conditions, whether or not oil is used for lubrication, and the use of water or air for condenser cooling. Low noise and vibration levels are likely because of the absence of valves and the small variations in torque.

(b)   The twin screw compressor

This is similar to the single screw but comprises a pair of intermeshing rotors or screws: the male screw has four or five helical lobes and the female screw six or seven helical grooves, in 4 + 6, 5 + 6 or 5 + 7, male–female combinations. As with the single screw machine the process is in three stages: suction, compression and discharge.

The two screws are contained in a compression chamber having a suction opening at one end and a discharge opening at the other. As the screws rotate the space between one pair of male lobes and female grooves is presented to the suction opening. Suction gas flows into the lobe/groove space until the screws have rotated enough for the exposed end of this space to pass beyond the suction opening. The entire length of the lobe/groove space is then full of gas at the suction pressure.

As the rotation of the screws continues the male lobe reduces the volume of the trapped gas in the female groove which, being contained by the plate at the discharge end of the compression chamber, is compressed.

The third phase of the process occurs as rotation continues and the end of the lobe/groove space starts to pass the discharge port, allowing the hot gas to flow into the discharge manifold. Multiple lobe/groove spaces pass across the suction and discharge openings in rapid sequence and the flow of compressed gas is smooth and continuous, for all practical purposes.

The most common arrangement is for the male rotor to be driven from a two-pole electric motor. The female screw is then driven by the male rotor, being separated from it by a film of oil. Alternatively, the female rotor is driven instead of the male rotor. A less popular method is to drive the female screw through synchronised timing gears from the male screw but this has proved very noisy in early installations.

Rolling contact between the meshing screws minimises frictional wear and volumetric efficiencies are high (see Figure 12.11).

If both rotors are driven, using synchronised timing gears, no oil is needed since there is no contact between the two screws. However, with other arrangements oil is needed for lubrication, sealing and cooling. The oil takes up much of the heat of compression and discharge temperatures are less than 88°C. An oil separator is necessary in the discharge hot gas line and cooling is provided for the oil, either by an external heat exchanger or by the injection of liquid refrigerant into the compression process.

Economisers are sometimes used to improve performance, a secondary suction port introducing gas between the main suction and discharge ports.

Capacity control is by a sliding valve which, as it opens, allows suction gas to escape to the suction side of the screws, without being compressed. This can give smooth proportional control over capacity down to 10 per cent of design duty but some machines are controlled in steps, 100 per cent, 75 per cent, 50 per cent, 25 per cent and 0 per cent being typical. The electrical power absorbed is usually proportional to the refrigeration duty down to about 30 per cent load but thereafter efficiency is poorer, about 20 per cent power being absorbed at 10 per cent duty.

The machines are mostly water-cooled but air-cooling is possible and the range of refrigeration capacities available is from about 120 kW up to more than 2000 kW.

Like the single screw, the twin screw can tolerate some liquid floodback.

Machines are hermetic, semi-hermetic or open and the motors in hermetic or semi-hermetic compressors are best cooled by suction gas. However, one make of machine uses discharge gas for this purpose, thus avoiding the need for an oil separator at discharge. A hermetic machine should not then be used with high condensing temperatures to give heat pumping or heat reclaim. Instead, an open compressor should be used, where the driving motor is air-cooled in a conventional manner.

12.12 The scroll compressor

This is a positive displacement machine that achieves compression in a way that is different from either the reciprocating or screw compressors. The process of compression is accomplished by a pair of scrolls as illustrated in Figure 12.12. Referring to the figure, the scroll having an open end on the left-hand side of sub-figure b is fixed and cannot move. The other scroll has its right-hand end fixed but the remainder is driven by an electric motor to orbit about the centre-line of the fixed scroll. Suction gas enters the two open slots formed by the fixed ends of each of the scrolls (sub-figure b) and, as orbiting commences, a pair of trapped gas volumes is formed (sub-figure c). As the orbiting proceeds these two trapped volumes move radially towards the centre of the scrolls and are compressed. Meanwhile, the open ends of the scroll are again able to admit another pair of volumes of suction gas. For clarity of illustration, Figure 12.12 only shows the radial progress of one pair of trapped gas volumes from suction at the outer slots to compression at the middle and discharge at one end of the centre of the scrolls. It can be seen in the figure that, at the middle of the scrolls, the end of the fixed scroll does not move but the end of the orbiting scroll does, to effect compression. As orbiting continues, a series of pulses of hot gas is discharged at the centre of the scrolls, virtually as a continuous stream of high pressure gas.

There are no valves and fewer moving parts than in the reciprocating compressor. Frictional and other losses are reduced because of the close manufacturing tolerances and the isentropic efficiency (equation (9.19)) is improved. The compressor has a higher volumetric efficiency (see Figure 12.11) and a better coefficient of performance. It is quieter and vibration is less.

Scroll compressors are available in the range of 3 to 50 kW of refrigeration, with air-cooled condensers. Application has been world-wide in residential and commercial air conditioning. Capacity control is usually through an inverter giving variable compressor speeds from 15 to 50 Hz with corresponding variable flowrates of refrigerant. A benefit of this is a soft start for the electric driving motors.

In comparison with air-cooled reciprocating machines, the scroll condensing set has a flatter characteristic performance curve of cooling capacity versus outside air temperature. Most are hermetic and there is virtually no maintenance.

12.13 Centrifugal compressors

Reciprocating compressors running at about 24 rev s^{-1} have capacities of the order of 35 kW per cylinder when used for conventional air conditioning applications. The maximum number of cylinders on one machine is about 16, so this puts the maximum capacity of a piston machine at about 550 kW of refrigeration, excluding freak machines with large strokes and bores. Although one reciprocating compressor of this size is likely to be cheaper than a centrifugal machine of like capacity, the complication of controlling its output (cylinder unloading) may put it in an unfavourable light compared with a centrifugal

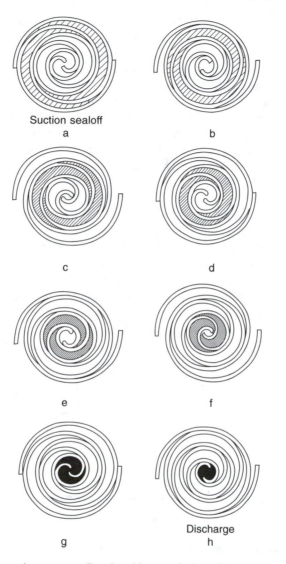

Fig. 12.12 Scroll compression process. Reprinted by permission of the American Society of Heating, Refrigeration and Air Conditioning Engineers, Atlanta, Georgia, from the 1992 ASHRAE Handbook— *Systems and Equipment SI.*

compressor, which can have a modulating control exercised over its capacity and chilled water flow temperature. Centrifugal systems are available for capacities as low as 280 kW of refrigeration, but they come into their own, economically speaking, at about 500 kW with a maximum at about 20 000 kW.

Whereas a reciprocating compressor is a positive displacement device, a centrifugal compressor is not. If the flow of gas into a piston machine is throttled it will continue to pump, even though the amount delivered is small, provided that its speed is kept up by an adequate input of power to its crankshaft. There is no 'stalled' condition. This is not so with a centrifugal compressor. The rotating impeller of a centrifugal compressor increases the pressure of the gas flowing through its channels by virtue of the centrifugal force resulting from its angular velocity. The velocity of the impeller is constant in a radial direction but the linear velocity in a direction at right angles to the radius of the impeller increases as the radius gets greater. The energy input to the gas, which is rotated within the impeller, thus increases towards the periphery of the wheel. This input of energy is what causes the gas to flow outwards through the impeller against the pressure gradient, that is, from the low pressure prevailing at the inlet eye to the high pressure existing at the periphery. The function of the impeller casing, or the volute, is to convert the velocity pressure of the gas leaving the wheel to static pressure, with as much efficiency as possible.

In addition to the radial movement imposed on the gas by the impeller, the gas stream tends to rotate relative to the impeller. This is illustrated in Figure 12.13(*a*). On an absolute basis, any particular particle of gas will tend not to rotate but, since the wheel is rotating, the particle will appear to rotate relative to the wheel. The point P_1 is facing the convex side of an impeller blade initially but, later in the process of rotation, denoted by P_4, it is facing the concave side of the preceding impeller blade. The effect of this is to produce a circulatory movement of the gas within the impeller, as shown in Figure 12.13(*b*). It can be seen that this circulatory movement assists the flow to the periphery of the wheel, produced by centrifugal force, on the concave side of a blade but retards it on the convex side. The effect introduces losses which can be minimised by using a wheel with narrow channels between the impeller blades.

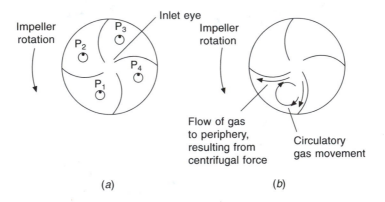

Fig. 12.13   Circulatory losses in the impeller of a centrifugal compressor.

For a given compressor, running at a given speed, the pressure–volume characteristic curve is virtually a straight line, as shown in Figure 12.14, if no losses occur. Losses do occur, however. They are the circulatory losses just described, losses due to friction, and losses caused by the fact that the gas entering the impeller has to change direction by 90 degrees, as well as having a rotation imposed on it. These entry losses may be modified by adjusting the swirl of the gas before it enters the inlet eye of the impeller. There is a proper angle of swirl for each rate of gas flow, that is, for each load. Variable-inlet guide vanes are fitted on all modern centrifugal compressors. Their position is adjusted to suit variations in the load, which permits a modulating control of output with little alteration in efficiency. The intention is that the machine should operate at a design point which involves the minimum loss, at the maximum efficiency.

A centrifugal impeller is designed to pump gas between the low suction pressure and the high condensing pressure. If the condensing pressure rises, the difference between these

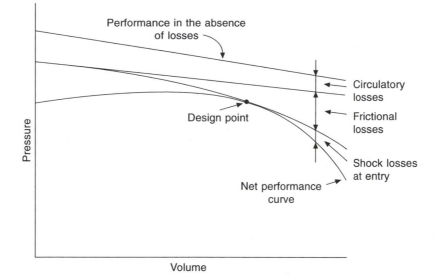

Fig. 12.14   A pressure-volume diagram for a centrifugal compressor showing how the various losses give the characteristic curved shape.

two pressures exceeds the design value and the compressor quite soon finds the task of pumping beyond its ability. Thus, whereas, the reciprocating machine will continue to pump, but at a steadily reducing rate as the condensing pressure rises, the pumping capacity of the centrifugal compressor falls rapidly away. This is illustrated in Figure 12.15(*a*). A similar behaviour is seen if the suction pressure is reduced, the condensing pressure being held at a constant value, as Figure 12.15(*b*) shows.

This feature of the performance of the centrifugal machine gives rise to a phenomenon termed 'surging'. When the pressure difference exceeds the design pumping ability of the impeller, flow ceases and then reverses, because the high condensing pressure drives the gas backwards to the lower suction pressure. Pressure in the evaporator then builds up, and the difference between the high and low sides of the system diminishes until it is again within the ability of the impeller to pump. The flow of gas then resumes its normal direction, the pressure difference rises again, and the process repeats itself.

This oscillation of gas flow and rapid variation in pressure difference which produces it is *surging*. Apart from the alarming noise which surging produces, the stresses imposed on the bearings and other components of the impeller and driving motor may result in damage to them. Surging continuously is most undesirable, but some surge is quite likely to occur from time to time, unless a very careful watch is kept on the plant. This is particularly possible with plants which operate automatically and are left for long periods unattended. Surge is likely to occur under conditions of light load (when the suction pressure is low) coupled with a high condensing temperature.

The proper use of inlet guide vanes can give a modulating control of capacity down to 15 per cent or even, it is claimed, to 10 per cent of design full load.

The high heads necessary for air conditioning applications may be developed in two ways: either by running the impeller fast enough to give the high tip speed wanted or by using a multi-stage compressor. High tip speeds can be obtained by using large-diameter impellers, but if diameters are excessively large, structural and other difficulties arise.

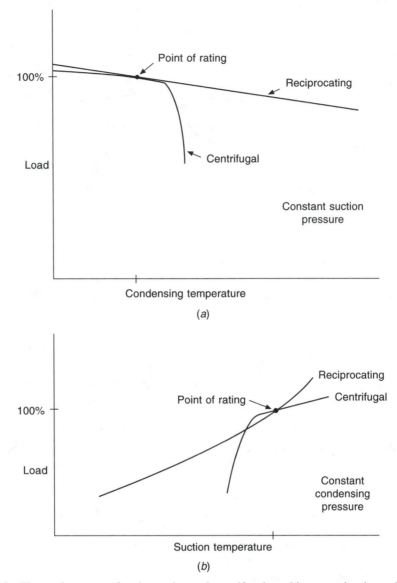

Fig. 12.15   The performance of reciprocating and centrifugal machines may be shown in terms of both condensing and suction pressures.

High speeds of rotation are usually produced by using speed-increasing gears which multiply the 48 rev s^{-1} normally obtainable from a two-pole induction motor. Using an inverter to increase the frequency of the electrical supply is an alternative possibility. Full load running speeds lie in the range from 30 Hz to 1500 Hz with compression ratios from 2 to 30.

Although speed-increasing gear is noisy it has largely displaced two-stage compression, one reason being that gear ratios can be selected to suit the application, with maximum efficiency. Figure 12.16(a) shows a diagrammatic arrangement of two-stage compression with an 'intercooler' added. The purpose of this is to improve the performance of the cycle

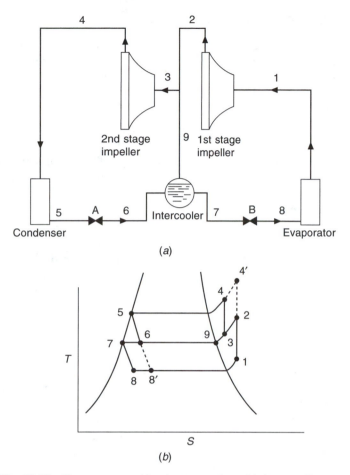

Fig. 12.16   Two-stage centrifugal compression with inter-cooling.

by two expansion devices, A and B, and feeding some of the low pressure gas at state 9, obtained after valve A, to the suction side of the second stage compressor. Figure 12.16(*b*) describes the cycle in terms of the temperature–entropy changes involved.

Hot, high-pressure liquid leaves the condenser at a state denoted by 5. The pressure and temperature of the fluid is dropped by passing it through the first expansion device, A. Some liquid flashes to gas in the process, the flash gas being fed from the intercooler (merely a collection vessel) to the intermediate stage of compression. The remaining liquid passes from the intercooler through the second expansion device, B, and thence to the evaporator. Gas leaves the evaporator at state 1 and enters the first stage impeller, being compressed to state 2. When gas at state 2 mixes with gas at state 9, from the intercooler, it forms state 3 and enters the second stage impeller, leaving this at state 4. This then enters the condenser, and the process is repeated.

If the intercooler is not used, the cycle will follow the line 1–4′–5–8′–1. The refrigerating effect, represented by the area beneath 8′–1, is clearly less than the effect produced when an intercooler is used, represented by the area beneath 8–1.

Most condensers used with centrifugal machines are water-cooled but it is possible to get packaged air-cooled equipment in the range from about 450 kW to 1200 kW of refrigeration.

12.14 The water-cooled condenser

The rejection of heat from condensers was discussed in chapter 11. The treatment here is confined to water-cooled condensers. There are four types of condenser in use: shell-and-tube, shell-and-coil, tube-in-tube, and plate heat exchanger.

(i) Shell-and-tube

Hot, superheated gas usually enters at the top of the shell and is desuperheated and condensed to a liquid at high pressure and temperature by coming in contact with horizontal tubes which convey the cooling water. Some sub-cooling of the liquid below the saturation temperature corresponding to the pressure prevailing inside the shell also occurs and is desirable. The condensed liquid is usually only enough to fill about the lower fifth of the shell, but the shell must be large enough to contain the full charge of refrigerant when repair or maintenance work is carried out on other components of the system. There is thus no need for a separate liquid receiver.

The capacity of a condenser to reject heat depends on the difference of temperature between the condensing refrigerant and the cooling water. Thus, its capacity may be increased by raising the condensing temperature or by increasing the rate of water-flow. Raising the condensing temperature has the side effect of lowering the capacity of the compressor, as has been observed earlier.

Some choice is open to the designer of a refrigeration system: a small condenser may be chosen, giving a high condensing temperature, or vice-versa. The first choice will result in a motor of larger horsepower to drive the compressor, with consequent increased running costs. The alternative will give lower running costs, but the size and capital cost of the condenser will be greater.

A typical choice of water flow rate is about 0.06 litres s^{-1} for each kW of refrigeration, rising in temperature from 27°C to 32°C, as it flows through the condenser.

Pressure settings for safety and other purposes are expressed in BS 4434 and are related to good practice. The values chosen are particularly important if the plant is to be used as a heat pump, the heat normally rejected at the condenser being put to good use. Recommended settings are:

High-pressure cut-out setting:	design working pressure + 10%
Maximum working pressure:	high-pressure cut-out setting/0.9
Test pressure:	maximum working pressure + 30% (if a steel shell) or + 50% (if a cast iron shell).

Condensers are generally used with dirty water from a cooling tower or other source and a proper allowance for the fouling factor must be made. A typical value is 0.000 175 6 m² K W^{-1} (0.001 ft² h °F/Btu) but, if tubes with an enhanced internal surface area (such as longitudinal or spiral grooves) are used, the allowance for fouling must be greatly increased.

Starner (1976) claims that making an allowance for a fouling factor does more than merely increase the thermal resistance. To achieve the same overall heat transfer it becomes necessary to increase the surface area of the evaporator and the presence of additional tubes lowers the water velocity therein, giving a further fall in the overall heat transfer coefficient. Suitor *et al.* (1976) imply that the build-up of fouling as time passes does not increase the thermal resistance in a linear fashion: it is claimed that the fouling resistance increases asymptotically with time.

Only about 1 degree of sub-cooling is normally achieved but greater sub-cooling can be obtained if it is arranged to submerge the tubes in the refrigerant condensate.

It is important, when considering plant layout, that at least the full length of the condenser shell is allowed for tube withdrawal or mechanical cleaning.

EXAMPLE 12.7

If the design working pressure for a refrigeration plant is 15 bar determine the setting of the high-pressure cut-out switch, the maximum working pressure and the test pressure.

Answer

HP cut-out setting:	$15 \times 1.1 = 16.5$ bar
Max working pressure:	$16.5/0.9 = 18.3$ bar
Test pressure:	$18.3 \times 1.3 = 23.8$ bar if steel
or	$18.3 \times 1.5 = 27.4$ bar if cast-iron

(ii) Shell-and-coil

Cooling water flows through one or more coils within a shell and refrigerant condenses on the outer surface of the coil(s). Condensers of this sort are used for small, self-contained packaged plant in the range 2 to 50 kW of refrigeration. Mechanical cleaning is not possible for the inside of the tubes but sometimes chemical methods are adopted. Water pressure drops may be greater than for shell-and-tube condensers.

(iii) Tube-in-tube

These are used in packaged plant over the range 1 to 180 kW of refrigeration. They comprise a coil of tubes, one within the other. The water may flow through the central tube with the refrigerant condensing in the annular tube, or vice-versa. Mechanical cleaning of the water tube is not possible but chemical methods may be used. Sometimes the tubes are straight and then mechanical cleaning on the water side is possible, provided that the waterflow is through the central tube.

(iv) Plate heat exchangers

An array of parallel, stainless steel plates is used, hot gas condensing on one side of alternate plates with cooling water flowing on the other side. Gaskets of appropriate materials are used to separate the plates. Heat transfer rates are usually higher than with shell-in-tube condensers but mechanical cleaning is not possible. Heat exchangers of this type are also sometimes used as water chillers (see section 12.4).

12.15 Piping and accessories

From a thermodynamic point of view, the only essential substance in the system is the refrigerant, anything else is undesirable. Of the other substances which may find their way into a system, only a lubricant is necessary, and this is only because the compressor requires it for proper operation. Thus air, water and dirt are not wanted, and every possible step must be taken to ensure that they are absent. Their total elimination may be impossible but care in the proper choice of materials, and in assembly and charging, can minimise their presence.

The piping used must be of a material which will not be attacked by the refrigerant (for example, copper is acceptable with the fluorinated hydrocarbons but not with ammonia) and, apart from having the necessary structural properties, it must be clean and dehydrated, prior to use. The same holds true for all fittings, valves, and other components used.

Piping must be large enough to permit the flow of the refrigerant without undue pressure drop and, most important, it must permit the lubricant in the system to be carried to the place where it is wanted, that is, to the crankcase. Inevitably, some is discharged with the high-pressure refrigerant from the compressor, into the condenser. When the refrigerant is in the liquid state, it mixes well with mineral oil, if it is a fluorinated hydrocarbon, and both are easily carried along. When the refrigerant is a gas, however, mixing does not occur. If mineral oil is used it turns into a mist and the pipes must be sized to give a refrigerant gas velocity high enough to convey the lubricant. It is clear from this that a problem may arise under conditions of partial load, when the system is not pumping the design quantities of gas. Mineral oil usually collects on the walls of the piping and drains down into the low points of the system. For this reason, proper oil return to the crankcase can be assisted by pitching the pipes. An oil separator between the compressor and the condenser is also useful, particularly in air-conditioning applications where large variations of load occur. Lubricants that are not mineral oils will have their own characteristics.

An oil pressure failure switch is essential. This is interlocked with the compressor motor starter so that if the pressure falls below a minimum value, the compressor is stopped and damage from lack of lubrication is prevented.

Water in a system can cause trouble in a number of ways and to minimise this a dryer is inserted in the piping. This should not be regarded as a valid substitute for proper dehydration in the assembly and setting to work, but should be considered in addition to it.

Strainers should be fitted upstream of the expansion valve and any solenoid valve which may accompany it.

Sight glasses are essential. They should be placed in the liquid line between the condenser or liquid receiver and the expansion valve. Their purpose is to verify the presence of an adequate charge in the system and to monitor the state of the liquid entering the expansion valve.

Gauges are also essential although they are not always fitted on some of the modern packaged plant. They permit a proper check to be kept on the running of the plant and they are of assistance in commissioning the plant and in locating faults.

Gauge glasses are necessary, although they may not always be present. Their purpose is to establish the levels of the liquid in the condensers and receivers, this being of particular use in charging the system.

If the condensing pressure is less than the ambient air pressure, air and possibly other non-condensible gases will leak into the condenser and an automatic purging system becomes necessary. (Note that some small leakage of refrigerant into the atmosphere will always occur when the automatic purge operates.) Non-condensible gases collect in the condenser and their presence raises the total pressure (in accordance with Dalton's law), reducing the refrigeration capacity and increasing the power needed for compression. Oxygen in the gases may cause mineral oil to oxidise in the presence of a high temperature and the non-condensible gases can form a film over the heat transfer surfaces within the condenser. Webb *et al.* (1980) showed how small percentages of non-condensible gases cause a large decrease in the film heat transfer coefficient with shell and tube condensers.

12.16 Charging the system

Before any charge of refrigerant is put into a system, it must be thoroughly pressure-tested for leaks on both the low and high side of the compressor. Anhydrous carbon dioxide should be used for this and all joints and connections carefully inspected, soapy water being used for a bubble test. A watch on the pressure gauges will indicate if a serious leak is present. Following this, a small amount of refrigerant should be added to the system, and all joints, pipework and connections gone over with a hallide torch. (This is a trace technique for detecting leaks when fluorinated hydrocarbons are used. Any escaping refrigerant reacts with the flame of the torch to produce a green colour.)

The system is regarded as free from leaks if, after having been left under pressure for 24 hours, no variation is observed in the gauge readings, due attention being paid to variations in the ambient temperature, which will alter the gauge readings.

When the system is free from leaks, it must be dehydrated. This is accomplished by using an auxiliary vacuum pump (never the refrigeration compressor itself) to pull a hard vacuum on the entire system. Dehydration can be accomplished in this way only if liquid water in the system is made to boil, the vapour then being drawn off. At 20°C, the saturation vapour pressure is 2.337 kPa absolute. That is, a vacuum of about 99 kPa. The application of external heat to the system may be of help in attaining the required amount of dehydration.

Following the above testing and dehydration procedure, the ancillary systems should be set to work. By this is meant the associated air handling and water handling plant, together with their automatic controls. It is not possible to charge and set to work a refrigeration system unless this is done. For example, unless the cooling-water pump is circulating an adequate amount of water through the condenser, the refrigerant plant will fail continually on its high-pressure cut-out.

The system should be charged from a cylinder or drum of refrigerant through a dehydrator. For any given system, if the design duty is to be obtained, only one weight of refrigerant is the correct amount for it to operate between the design condensing and evaporating pressures. For packaged plants of well-known size and performance, a measured weight of refrigerant should give the desired operation. However, for plants of a more bespoke character, sight glasses and gauge glasses may be used to advantage in assessing the necessary amount of the charge.

The location of a charging point varies according to the size and type of plant. Very small reciprocating machines may have no special point for charging; this then has to be done through the gauge connection at the suction valve. In this case it is important that only vapour (i.e. no liquid) refrigerant is charged. For larger installations the charging point is either between the expansion valve and the evaporator or in the liquid line or at the liquid receiver. In these cases liquid can be fed in.

Notation

Symbol	Description	Unit
f	frequency of starting a compressor motor	h^{-1}
m	mass of chilled water in a system	kg
q_r	average cooling capacity of the last step of refrigeration	kW
Δt	allowable variation in the return water temperature	K
θ	time taken to cool down or warm up the mass of water in a system	s

References

Starner, K.E. (1976): Effect of fouling factor on heat exchanger design, *ASHRAE Journal* **18** (May), 39.

Suitor, J.N., Marner, W.J. and Ritter, R.B. (1976): The history and status of research in fouling of heat exchangers in cooling water service, Paper No. 76-CSME/CS ChE-19, National Heat Transfer Conference, St Louis, MO.

Webb, R.L., Wanniarachchi, A.S. and Rudy, T.M. (1980): The effect of non-condensible gases on the performance of an R11 centrifugal water chiller condenser, *ASHRAE Trans.* **86**(2), 57.

Bibliography

1. W.B. Gosney, *Principles of Refrigeration,* Cambridge University Press, 1982.
2. W.H. Carrier, R.E. Cherne, W.A. Grant and W.H. Roberts, *Modern Air Conditioning, Heating and Ventilating,* Pitman.
3. W.F. Stoecker, *Refrigeration and Air Conditioning*, McGraw-Hill, 1958.
4. *Trane Refrigeration Manual,* The Trane Company, La Cross, Wisconsin.
5. *ASHRAE Handbook* 1996, Heating, Ventilating and Air Conditioning Equipment, SI.
6. J.L. Threlkeld, *Thermal Environmental Engineering*, Prentice-Hall, 1970.
7. ARI Standard 750-87 Thermostatic Refrigerant Expansion Valves (ANSI/ARI 750-87).
8. BS 4454: 1989, Safety aspects in the design, construction and installation of refrigeration appliances and systems.
9. E. Purvis, *Scroll Compressor Technology,* 1987, ASHRAE Heat Pump Conference, New Orleans, USA.
10. *ASHRAE Handbook* 1998, Refrigeration, SI.

13
Automatic Controls

13.1 The principle of automatic control

The steady-state condition is rare and, in general, the design capacity of a system will exceed the load because loads change continually. The state maintained in the conditioned space will not stay constant if plant capacity is uncontrolled.

As an example, consider the plenum ventilation system illustrated in Figure 13.1. When the outside temperature rises above its design winter value fabric heat losses will diminish and the heater battery capacity will be too great, with a consequent rise in room temperature. To keep this at a nominally constant value four things must occur. Firstly, it must be measured. Secondly, the change in temperature must be used to send a signal to the heater

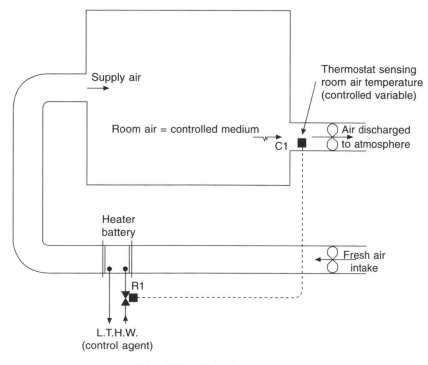

Fig. 13.1 Closed loop system.

battery. Thirdly, the strength of the signal must produce a matching change in battery output. Finally, the time taken for all this to happen must not be so long that further load changes occur in the meantime and the battery output becomes significantly out of phase with the load.

Note that a temperature change must be measured if the battery capacity is to be altered, implying that temperature cannot be kept constant. Such a deviation is inherent in control systems but a sustained deviation (offset) can be prevented, with a lapse of time, in certain refined systems, described later. Several methods are adopted for the signal transmission but its strength must be related to the measured deviation so that the capacity of the regulating device at the battery can gauge its response to this.

The principle of using an observed deviation in the load to give a corrective response in capacity is termed 'negative feedback' and a control system using this principle is called a 'closed loop'. Another form of control system is an 'open loop', which does not use negative feedback but, instead, regulates capacity in a pre-arranged manner. For example, since fabric heat loss is proportional to the difference between room and outside temperature, and since heater battery capacity is related to the flow temperature of the hot water, it is possible to reduce the value of the latter as the outside air temperature rises, according to a calculated schedule. Room temperature could then be kept constant without actually having been measured. Such a system (termed 'compensated control') works as an open loop. An open loop system usually contains sub-systems which are themselves closed loops, in order to work properly.

13.2   Definitions

To establish a basis for discussion agreement must be reached on the meaning of the terms used. A glossary of terms is provided by BS 1523: Part 1: 1967 which augments those given below.

1. Controlled variable. The quantity or physical property measured and controlled, e.g. room air temperature.

2. Desired value. The value of the controlled variable which it is desired the control system should maintain.

3. Set point. The value of the controlled variable set on the scale of the controller, e.g. 21°C set on a thermostat scale.

4. Control point. The value of the controlled variable that the controller is trying to maintain. This is a function of the mode of control, e.g. with proportional control and a set point of $21° \pm 2°C$, the control point will be 23°C at full cooling load, 21°C at 50 per cent load and 19°C at zero load.

5. Deviation. The difference between the set point and the measured value of the controlled variable at any instant, e.g. with a set point of 21°C and an instantaneous measured value of 22°C the deviation is +1°C.

6. Offset. A sustained deviation caused by an inherent characteristic of the control system, e.g. with a set point of $21° \pm 2°C$ the offset is + 2°C at a continuous full cooling load, 23°C being maintained.

7. Primary element (also termed a sensor). That part of a controller which responds to the

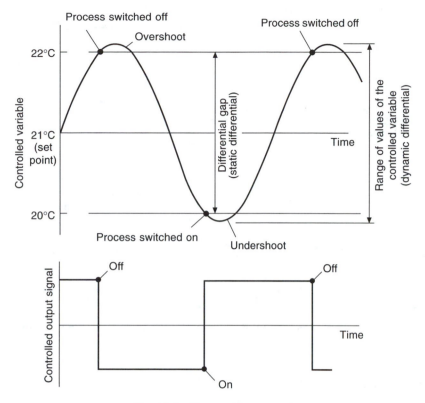

Fig. 13.2 Two-position control.

value of the controlled variable in order to give a measurement, e.g. a bimetallic strip in a thermostat.

8. Final control element. The mechanism altering the plant capacity in response to a signal initiated at the primary element, e.g. a motorised valve.

9. Automatic controller. A device which compares a signal from the primary element with the set point and initiates corrective action to counter the deviation, e.g. a room thermostat.

10. Differential (also Differential gap). This refers to two-position control and is the smallest range of values through which the controlled variable must pass for the final control element to move between its two possible extreme positions, e.g. if a two-position controller has a set point of 21° ± 2°C the differential gap is 4°C. See Figure 13.2.

11. Proportional band. Also known as throttling range, this refers to proportional control and is the range of values of the controlled variable corresponding to the movement of the final control element between its extreme positions, e.g. a proportional controller with a set point of 21° ± 2°C has a proportional band of 4°C. See Figure 13.3.

12. Cycling. Also known as hunting this is a persistent, self-induced, periodic change in the value of the controlled variable.

13. Open loop system. A control system lacking feedback.

14. Closed loop system. A control system with feedback so that the deviation is used to

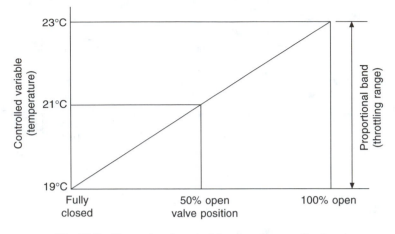

Fig. 13.3 Proportional control for a cooling application.

control the action of the final control element in such a way as to tend to reduce the deviation.

15. Deadtime. The time between a signal change and the initiation of perceptible response to the change.

13.3 Measurement and lag

The quality of control an automatic system gives depends initially on the accuracy of measurement of the controlled variable: tight control cannot be expected from coarse measurements. Accuracy also depends on the response of one part of the system to a signal from another and this is typified by lag—the sum of the lags of individual components in the system being its total time lag.

Consider the system in Figure 13.1. A change in heat loss from the room causes its temperature to depart from the desired value and, because of the deviation, heat is transferred between the thermostat, Cl, and the air around it. Enough heat must flow into the primary element for an adequate response to occur and this is not instantaneous, since it depends on the mass of the element, its specific heat capacity and its thermal conductivity. For example, a bimetallic strip must absorb sufficient energy to produce the thermal expansion required to make or break an electrical contact. Suppose the thermostat, C1, in Figure 13.1 controls a heater battery by means of a two-position valve, R1. Some time is occupied for the passage of the signal between C1 and R1, an instance of the distance velocity lag which, for the case of a pneumatic control system, would be measurable. On arrival at the valve the signal must be translated into corrective action, with any amplification necessary. If C1 had sensed a fall in temperature and R1 were pneumatically actuated air would have to be admitted to the space above the diaphragm (see Figure 13.4(*a*)) in order to open the valve, time being needed for this to take place. Further time elapses because of the mechanical inertia of the valve and because the water must be accelerated to a higher flow rate. The increased flow rate of hot water gives a greater heat transfer through the tube walls and fins of the battery into the airstream, occupying more time. There is then the distance–velocity lag as the air flows along the duct from the heater battery to the supply grille. Finally, there is the air diffusion lag in the room itself, probably the largest of all the lags: the thermostat in the extract duct (or perhaps on the wall of the room) would not know that corrective action had been taken until the air delivered from the supply grille had spread throughout the room and been extracted. For example, this would occupy six minutes if the supply air change rate were ten per hour.

The total time lag is the sum of all these (and other) individual time lags and is significant, setting a limit on the speed of response of the system as the load changes. There is the possibility that the response of the system being controlled (as distinct from the automatic control system) may be out of phase with the corrective information fed from the measuring element, with implications of instability. It is also evident that the change in the value of the controlled variable will always exceed the change in the feedback, because of the time lags. Thus a thermostat may have a differential gap of only two degrees but the room temperature may fluctuate by three or more degrees because an instantaneous response is impossible.

We can also see that there are features of the system being controlled, as well as those of the automatic control system itself, that influence performance when load changes occur. The quality of the controlled system must match that of the controlling system.

13.4 Measurement elements

(a) General characteristics

Certain terms, relating specifically to measurement, are in common use and are described briefly in the following:

(i) *Accuracy*. This is expressed as a percentage of the full scale of the instrument, e.g. if a maximum error is 4° for a thermometer having a scale from 0°C to 100°C its accuracy is 4 per cent. Where an instrument includes several components (e.g. a detector and a transmitter) their individual errors combine to give a compound error, expressed by the root of the sum of the squares of the component errors. Thus two components with individual errors of 4° and 1° would give a compound error of $\sqrt{17}$ or 4.12°. Accuracy is the ability of an instrument to indicate the true value of the measured variable.

(ii) *Dead band*. This is the range of values of the measured variable to which the instrument does not respond.

(iii) *Deviation*. This is the instantaneous difference between the value of the measured variable and the set point.

(iv) *Drift*. A gradual change between the set point and successive measurements of the controlled variable, not related to the load.

(v) *Hysteresis*. This is when the change of response to an increasing signal from a measuring element, with respect to time, is different from the response to a decreasing signal, in a repeating pattern.

(vi) *Repeatability (precision)*. This is the change of the deviation about a mean value. It is an indication of reliability and is the ability of the instrument to reproduce successive measurements in agreement with one another. It is not the same as accuracy.

(vii) *Sensitivity*. Instruments do not respond instantaneously and the largest variation in measured variable that occurs before an instrument starts to respond is an indication of its sensitivity.

(viii) *Stability*. This is the independence of the measured property from changes in other properties. Drift is absent.

(ix)   *Time constant.* This is the time taken for an exponential expression to change by a fraction 0.632 (= $1 - e^{-1}$) of the time needed to reach a steady-state value.

(x)   *Transducer.* A device that translates changes in the value of one variable to corresponding changes in another variable.

(b)   Temperature measurement

The commonest elements in use are as follows:

(i)   *Bi-metallic strips.* A pair of dissimilar metals having different coefficients of linear thermal expansion are joined firmly together and hence bend on change in temperature, making or breaking contacts or moving a flapper over a compressed air nozzle. To increase mechanical movement bi-metallic strips may be wound in helical or spiral form. Accuracies are about 1 per cent and the range of suitability is –180°C to 540°C. This is not really a very accurate technique and is unsuitable for use at a distance.

(ii)   *Liquid expansion thermometer.* A bulb filled with a liquid having a suitable coefficient of volumetric expansion (e.g. alcohol or toluene) is connected by a capillary to bellows or a Bourdon tube. The former produces a linear motion to make contacts or move a flapper, in response to thermal expansion, whereas the latter, comprising a spirally wound tube, unwinds on rise of temperature to give a rotational movement at its centre to operate an indicator or a controller. Accuracy is about 1 per cent and the range of application is –45°C to 650°C. Long capillary tubes through ambient temperatures other than that at the bulb introduce error that requires compensation. The practical limiting distance between measurement and indication or other use is about 80 m with the use of repeater transmitters or 30 m without. The fluid content of the bulb is critical and depends on the temperature range of the application: insufficient liquid may evaporate at higher temperatures to exert a saturation vapour pressure. In other cases, excessive hydraulic pressure may damage the controller if the instrument is used beyond the upper limit of its range.

(iii)   *Vapour pressure thermometer.* A liquid–vapour interface exists in the bulb and the saturation vapour pressure transmitted along the capillary depends only on the temperature at the bulb. No capillary compensation is therefore needed. Accuracy is about 0.5 per cent and the range of suitability is from –35°C to 300°C. The volatile liquids used to fill the bulb are usually toluene, alcohol and methyl chloride. These thermometers are generally only for use either above or below a given ambient temperature: if this is crossed there is the risk of the vapour condensing within the capillary and significant error occurring. By using a combination of immiscible, volatile and non-volatile liquids in the fill it is possible to deal with this difficulty.

(iv)   *Gas-filled thermometer.* If the fill is above its critical temperature over the whole range of temperatures in the application it is a gas and its volumetric thermal expansion follows Charles' Law, since it operates at a virtually constant volume. Absolute pressure within the system is then directly proportional to absolute temperature. The range of use is –85°C to 540°C.

(v)   *Thermocouples.* An electric current flows through a pair of dissimilar metals joined to form a loop if one junction is hotter than the other. The current generated is used as the feedback variable and, depending on the metals adopted, the application is from –260°C to 2600°C. In practical applications the reference (cold) junction is kept at a constant temperature by automatic compensation using a thermistor in a bridge circuit.

(vi) *Resistance thermometer.* The variation in electrical resistance of a conductor with temperature change is used to provide feedback over a range of applications from $-265°C$ to $1000°C$. Accuracy is about 0.25 per cent. A Wheatstone bridge circuit is commonly used, one resistance being the temperature sensing element. Remote set-point adjustment and other refinements are readily possible. This instrument is simple, accurate, stable and has a virtually linear response.

(vii) *Thermistors.* The measuring elements are oxides of metals with an inverse, exponential relation between temperature and electrical resistance. Their response is rapid and although it is non-linear it is possible to reduce the non-linearity by combining them with resistors. Application is limited to a maximum temperature of $200°C$.

(viii) *Infra-red thermometers.* These respond either to thermal radiation or light falling upon a detector and usually give a visual display. For thermal radiation the detector may be a thermopile, composed of suitable metallic oxides, a thermocouple junction, or a semiconductor. The response is delayed according to the time lag of the thermal capacity of the detector. With light-responsive detectors a change in the electrical property of the detector is caused by the absorption of light. The field of thermal radiation caused by adjoining surfaces must be considered and the nature of the surface (emissivity) of the object emitting the radiation affects the readings.

(c) Humidity measurement

(i) *Mechanical.* Human hair, parchment, cotton, moisture-sensitive nylon or polymer and other hygroscopic materials vary in length as their moisture contents change according to fluctuations in the ambient relative humidity. The mechanical movement that occurs is used to provide the feedback. Response is mostly non-linear, except in the middle range of humidities. Other hygroscopic materials cannot match the performance of moisture-sensitive nylon or polymer. Range 5 per cent –100 per cent, ±5 per cent.

(ii) *Electrical resistance.* A double winding of gold or platinum wire over a cylinder of suitable material is coated with a hygroscopic film of lithium chloride. Changes in the relative humidity alter the electrical resistance of the lithium chloride between the wires and this is used to provide the feedback to the rest of the control system. Accuracy is about ± 1.5 per cent but deteriorates with age. Elements of this type only cover a band of approximately 20 per cent change in relative humidity so multiple elements must be used for control over a wider range. Other versions employing graphite and other materials are available but with poorer accuracies, of the order of 2 to 3 per cent.

(iii) *Electrical capacitive.* A thin film of non-conductive polymer material is coated on both sides of metal electrodes which are mounted inside a perforated plastic capsule. The relationship between humidity and capacity is non-linear with rising humidity. The non-linearity is dealt with electronically and the sensor is compensated against temperature in an amplifier circuit, to give an output signal in the range 0 per cent to 100 per cent relative humidity with an accuracy of ±2 per cent. Many other humidity sensing techniques are available but capacitive sensors appear to be most popular.

(d) Dew point measurement

(i) *Direct reading.* A sample is passed through an observation chamber, one side of which is of polished metal. A thermocouple measures the temperature of this surface. A second chamber, on the other side of the polished metal, contains evaporating ether which cools the surface (under manual control). Dew is observed to form on the polished surface and the thermocouple then indicates the dew point temperature of the air being sampled and passed through the observation chamber.

(ii) *Indirect type*. A pair of gold wire electrodes is wound over a cloth, impregnated with lithium chloride and covering a tube or bobbin. The bulb of a filled system or the element of a resistance thermometer measures the temperature of the air within the bobbin, the dew point of which is to be determined. When a low voltage is applied across the electrodes current flows between them through the lithium chloride on the cloth, generating heat and evaporating water from the cloth. The concentration of lithium chloride increases and its electrical resistance alters. When the cloth is dry the salt is in a solid crystalline state and becomes non-conducting. Moisture is re-absorbed from the air in the bobbin because of the hygroscopic nature of lithium chloride, and a current flows again. A stable temperature within the bobbin is reached when the thermal and electrical fluxes are in balance. At this condition it is a property of lithium chloride that the air inside the bobbin is at about 11 per cent relative humidity. This is then related with the measured temperature to give an indication of the dew point. The range of suitability is from –45°C to 70°C with a minimum humidity of 12 per cent. The accuracy is about 2 per cent. Performance is critically dependent on the cleanliness of the air. The best accuracy is in the range –23°C to +34°C and above 40°C, dew point. Contamination spoils accuracy and the response is slow.

(iii) *Calculation based on measurement of temperature and vapour pressure*. A resistance element measures temperature and, simultaneously, a capacitive element in the form of a thin film of polymer absorbs water vapour molecules. The rate at which the molecules are absorbed is directly proportional to the partial pressure of the water vapour. Dew point is calculated electronically. The output is dry-bulb temperature, vapour pressure, dew point and moisture content. Accuracy of ±2 per cent is claimed in the range –60°C to + 80°C dew point. Stability is also claimed and the output can be related to a digital control system through a microprocessor.

(e) Enthalpy measurement

Measurements of dry-bulb temperature and relative humidity, made simultaneously, are interpreted as enthalpy. Enthalpy can also be calculated from measurements of temperature and humidity, and displayed or used for control purposes.

(f) Flow rate measurement

Several methods have been used, principally as a safety cut-out to protect the compressors of water chillers if the water flow rate decreases.

(i) *Orifice plates*. A standard orifice to a BS specification with pressure tappings located in the correct places is fitted in the pipeline and the pressure drop provides a flow rate indicator. A reduced flow rate gives a fall in pressure drop which is used to effect the cut-out. The method is not satisfactory when used with chilled water because scale and dirt, always present in a system, collect at the orifice and gradually increase the pressure drop, implying that the flow rate is increasing when, in fact, the reverse is the case.

(ii) *Spring-load blade*. A flexible metal blade is immersed in the pipeline and set during commissioning to spring back and make an electrical contact that stops the compressor if the water flow rate diminishes. Difficulties in setting the blade during commissioning have arisen.

(iii) *Vortex shedding flowmeter.* For the case of turbulent flow the presence of a blunt obstruction in a pipeline causes the formation of vortices which grow larger and are shed as flow occurs past the obstruction. The frequency of vortex shedding is proportional to water velocity and hence to the flow rate, if the relevant cross-sectional area is known. The Reynolds number must exceed 3000. Problems with the viscosity of the water have occurred when used with water chillers.

(iv) *Venturi meter.* A properly made venturi meter, with the correct pressure tappings, is a preferred solution. No orifice or blade is present and the entry and exit sections of the venturi tube taper smoothly to the throat, with little opportunity for dirt or scale to collect. A reduction in the flow rate through the throat increases the static pressure therein. Obstructions upstream and downstream cause errors and the downstrem section should expand gradually to recover velocity pressure and reduce the overall loss (see section 15.4). The meter should be fitted in long straight lengths of piping. There are other more complicated methods of flow rate measurement.

(g) CO_2 measurement

This has become an accurate process and can be adopted as a method of indicating probable indoor air quality and controlling the rate of fresh air supplied. The most commonly used technique for this purpose is by Non-Dispersive Infra Red (NDIR) sensors. Carbon dioxide strongly absorbs radiation at a wavelength of 4200 nm when illuminated by an infra-red light. Air is sampled into a chamber formed by a pair of parallel, gas-permeable membranes and the chamber is flooded with infra-red radiation. An infra-red detector at the end of the chamber measures the amount of radiation absorbed at 4200 nm and transmits a signal to the control system. Accuracies from 150 to 50 ppm are claimed and the response time can be quite short, but it depends on the instrument used. Calibration may be held for months but dirt, condensation and mechanical shock can spoil this, according to ASHRAE (1997). Other techniques are also used.

13.5   Types of system

There are five types in common use:

1. Self-acting. The pressure, force or displacement produced as a signal by the measuring element is used directly as the power source at the final control element. A good example is the thermostatic expansion valve (Figure 12.1) in which the vapour pressure generated in the bulb by the temperature measured is transmitted along a capillary and acts above the diaphragm of a valve, causing it to move and so regulate the flow of refrigerant. This form of system is simple (no external power is needed) and proportional in nature, but has a fairly wide proportional band because of the need to produce enough power to move the final control element between its extreme positions.

2. Pneumatic. Compressed air is piped to each controller which, by bleeding some to waste reduces the air pressure to a value related to the measured value of the controlled variable. This reduced pressure is then transmitted to the final control element. Economy in the use of compressed air may be achieved by modifying the system shown in Figure 13.4(*a*) so that air is only bled to waste when the final control element is actually moving. Sizing the pressure-reducing orifice shown in the inlet pressure line in the figure is critical: if it is too big an excessive amount of air will be bled to waste and if it is too small airflow into the space above the valve diaphragm will be slow and the response of the final control element sluggish. It is therefore important that dirt, moisture or oil are absent from the compressed air supply. It is usual to have two air compressors (one stand-by) and to store the compressed air in a receiver at a suitable pressure up to a maximum of 8.6 bar (gauge). As the air is used by the controls the stored pressure falls through a fairly wide differential, at the lower end of which the duty compressor is started. The receiver is sized to keep compressor starts to less than 12 per hour. Industrial quality compressed air, used for driving tools and machines is not at all suitable for automatic controls, which must have instrument quality compressed air.

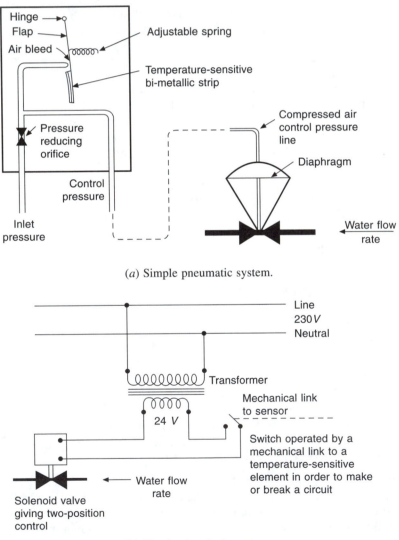

(*a*) Simple pneumatic system.

(*b*) Simple electrical system.

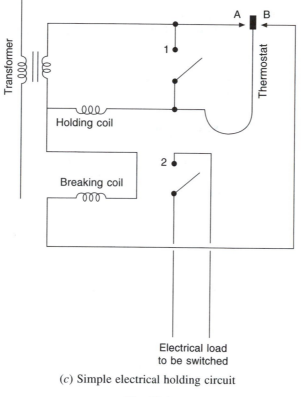

(*c*) Simple electrical holding circuit

Fig. 13.4

This must be clean, free of oil and dried to a dew point low enough to satisfy the application: driers should be in duplicate and may be by mechanical refrigeration or chemical means, with an automatic changeover.

3. Hydraulic. Oil or water is used to transmit a signal in a way similar to that for compressed air systems but the application is for much higher power transmission than pneumatic systems can cope with.

4. Electrical. There are two basic functions: switching and resistance variation. If large thermostats of slow response are employed the electric current switched may flow directly through the contacts of the thermostat but continuous switching generates high temperatures locally and the contacts will burn and even weld together. Indirect switching can be achieved (Figure 13.4(*b*)) by the mechanical displacement of a temperature-sensitive element (for example the cam on the spindle of a step-controller, rotated by a Bourdon tube and filled system). Another technique is to use a holding circuit, as shown in Figure 13.4(*c*). A rise in temperature makes the contact A, say, and current flows through the holding coil, which pulls in switches 1 and 2, magnetically, making the main power circuit. Only a small current flows through A. When the temperature drops the power is maintained after A is broken until the contact at B is made. This then energises the breaking coil via switch 1 and B, which opens switches 1 and 2 magnetically and turns off the power supply to the load. The other application is the variation of the resistance of one leg of a Wheatstone bridge. Figure 13.5 shows a classical circuit for giving proportional control over a final control element (or a 'modulating valve'). When in balance no current flows through W1 and W2, normal running and reverse running coils of an electric motor that drives the modulating valve, V. These are used to regulate the flow of hot water, say, to a heater battery. On change in temperature sensed by a thermostat, T, mechanical displacement of the linkage moves the wiper across the potentiometer windings represented by R1 and R2. The resistance on one side of the bridge (R1 + R3) is then different from that on the other (R2 + R4). Current flows through the coils C1 and C2 of the balancing relay which is consequently moved magnetically to make a contact at *a* or *b*, energising W1 or W2. The valve motor then runs in its forward or reverse direction, depending on which winding is energised, opening or closing the valve. To prevent the valve merely running to its extreme positions there is a mechanical connection between the valve position and a wiper on the balancing potentiometer of the bridge which is moved in a reverse direction to that of the first wiper, so tending to bring the bridge back into balance by making (R1 + R3) = (R2 + R4) again. The modulating valve thus moves in a series of small increments and provides stable

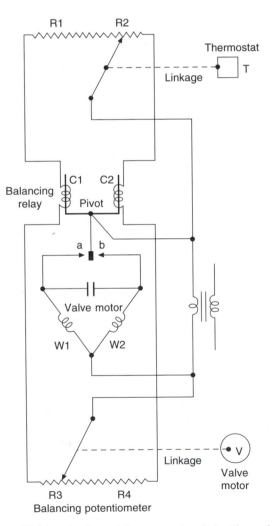

Fig. 13.5 Classical bridge circuit used for conventional electric modulating valves.

control. Technological advances have brought improvements in electrical modulating valves, with variable position, solenoid-type valves achieving proportional control over water flow in a simpler fashion than illustrated in Figure 13.5.

5. Electronic.   Electronic controls have almost entirely displaced pneumatic controls in commercial applications, since about 1985, because of the technical advances in the subject and the development of digital techniques and computers. Much smaller signal strengths are transmitted from the sensors to amplifiers, which measure phase changes and decide on the strength of output signals for control purposes. Digital controllers convert electronic signals from the sensors into numbers which can then be used for calculations by a microcomputer according to a program. Data can be stored, programs can be changed, set points altered, and many other operations carried out, locally or at a distance. See section 13.16.

13.6   Methods of control

There are three methods of control in common use in air conditioning: two-position, proportional and, to a lesser extent, floating. All are capable of sophistication, aimed at improvement, whereby offset is reduced or stability enhanced.

The form of control chosen must suit the application. For example, one essential feature of floating control is that it works best with a system having a small thermal storage and a predictable response. If it is misapplied, the system may hunt as a result of excessive lag and may be difficult to set up if the response is not well known. It is also essential that any method of control chosen should be as simple as is consistent with the results desired. Elaboration may result in some controls being redundant, the system becoming temperamental and wearisome to commission. There is more to commissioning an air conditioning installation than setting 21°C on a thermostat and switching the plant on.

13.7   Simple two-position control

There are only two values of the manipulated variable: maximum and zero. The sensing element switches on full capacity when the temperature falls to, say, the lower value of the differential and switches the capacity to zero when the upper value of the differential is reached. Figure 13.2 illustrates the way in which the controlled variable alters with respect to time when an air heater battery, say, is switched on and off. Overshoot and undershoot occur because of the total lag. It can be seen that the range of values of the controlled condition exceeds the differential gap and that if the lag is large, poor control will result. This is not to denigrate the method; it is an excellent, simple and cheap form of control for the correct application.

13.8   Timed two-position control

An exact match between load and capacity does not generally exist for two-position control; for example, if the load is 50 per cent then, on average, the plant will be on for half the time and off for the other half. This would be so if the plant were on for 30 minutes in each hour, or if it were on for every alternate minute in each hour. But it is clear that a 30-minute on–off cycle would produce a much wider swing in the controlled variable than would a one-minute cycle. The lower the value of the room air temperature, then the longer the period for which the heater battery, say, must be on, in a small plenum heating installation. Thus, for large loads the ratio of 'on' to 'off' time would not be to one but, say, 0.8 'on' to 0.2

'off'. Simple two-position control can be improved to obtain a timed variation of capacity, which gives a smaller differential gap, by locating a small heater element next to the temperature-sensitive element in the thermostat. If the ambient temperature is low, the heating element loses heat more rapidly and so takes a longer time for its temperature and the temperature of the adjacent sensitive element to reach the upper value of the differential gap which switches off the heater battery. Thus, the heater battery is on for longer periods when the ambient room air temperature is low.

13.9   Floating action

Floating control is so called because the final control element floats in a fixed position as long as the value of the controlled variables lies between two chosen limits. When the value of the controlled variable reaches the upper of these limits, the final control element is actuated to open, say, at a constant rate. Suppose that the value of the controlled variable then starts to fall in response to this movement of the final control element. When it falls back to the value of the upper limit, movement of the final control element is stopped and it stays in its new position, partly open. It remains in this position until the controlled condition again reaches a value equal to one of the limits. If the load alters and the controlled variable starts to climb again, then, when the upper limit is again reached, the final control element will again start to open and will continue to open until either it reaches its maximum position or until the controlled variable falls again to a value equal to the upper limit, when movement will cease. If the load change is such that the controlled variable drops in value to below the lower limit, say, then the final control element will start to close, and so on. Thus, the final control element is energised to move in a direction which depends on the deviation; a positive deviation gives movement of the element in one direction and a negative deviation causes movement in a reverse direction. There is a dead or floating band between the two limits which determines the sign of the deviation.

Figure 13.6 illustrates an example of floating control. A room is ventilated by a system as shown. The capacity of the LTHW heater battery is regulated by means of a motorised two-port valve, R1, controlled by floating action from a thermostat, C1, located in the supply air duct. There is negligible lag between C1 and R1. It is assumed that the variation in the state of the outside air is predictable and so the response rate of the controller can be properly chosen.

When the load is steady, the valve is in a fixed position, indicated by the line AB in Figure 13.6(*b*), provided that the supply air temperature lies between the upper and lower limits, as indicated by the line AA′, in Figure 13.6(*c*). Under this steady-state condition, the capacity of the heater battery matches the load. If the outside air temperature increases, but the valve position remains unchanged, the supply air temperature starts to rise, as shown by the line A′B in Figure 13.6(*c*). When the air temperature in the supply duct attains a value equal to the upper limit (point B), the valve starts to close (point B in Figure 13.6(*b*)). The rate of closure remains fixed, having been previously established during the initial process of setting up the controls, the battery output is continuously decreased, and the supply air temperature, after some overshoot, starts to fall. When the temperature reaches the upper limit (point C in Figure 13.6(*c*)), the movement of the valve stops (point C in Figure 13.6(*b*)). Suppose that this new valve position (line CD in Figure 13.6(*b*)) gives a heater battery output which is less than the load. The supply air temperature will continue to drop, until the lower limit is reached at the point D in Figure 13.6(*c*). This will then make the valve start to open (line DE, Figure 13.6(*b*)) and, after some undershoot, the

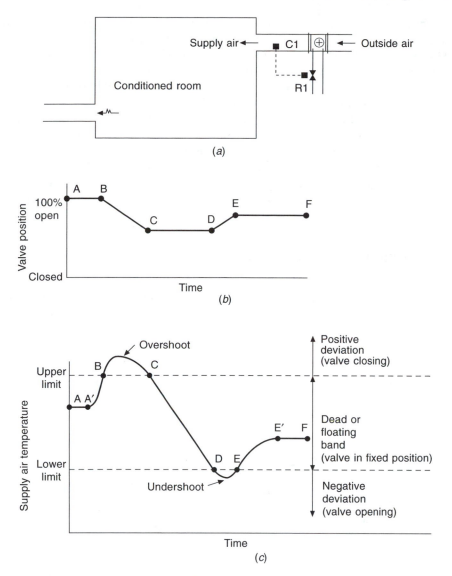

Fig. 13.6 Floating action.

supply air temperature will start to climb, passing the value equivalent to the point E, Figure 13.6(*c*), and stopping the opening movement of the valve. The battery may now have an output that will permit the supply air temperature to be maintained at some value between the upper and lower limits. If this is the case, then the air temperature will climb but will level off, along a curve EE′, the valve position remaining constant. The line EF in Figure 13.6(*b*) will then indicate the fixed valve position that gives a steady value of the supply air temperature, corresponding to the line E′F in Figure 13.6(*c*).

This example is intended only to describe floating action. It is not intended as a recommendation that floating control be applied to control ventilating systems. In fact, for plenum systems in particular, the form of control is unsuitable; apart from the question of lag, load changes are so variable that although the speed at which the valve is arranged to open may be quite satisfactory during commissioning, it may prove most unsatisfactory for the other rates of load change which will prevail at other times.

The slopes of the lines BC and DE are an indication of the rate at which the valve closes and opens. With single-speed floating action this rate is determined once and for all when the control system is set up. One method whereby a choice of speed is arranged is to permit the valve to open intermittently with short adjustable periods of alternate movement and non-movement. If the timing device is set to give short periods of movement and long periods of non-movement, the slope of the lines BC and DE will be small.

Sophistication may be introduced to floating control, the speed at which the valve opens and closes being varied according to the deviation or to the rate of change of deviation, in the same way that proportional control may be modified, as described in section 13.11. If the dead band is tightened, the control tends to become less stable, degenerating towards two-position.

13.10   Simple proportional control

If the output signal from the controller is directly proportional to the deviation, then the control action is termed simple proportional. If this output signal is used to vary the position of a modulating valve, then there is one, and only one position of the valve for each value of the controlled variable. Figure 13.7 shows this. Offset is thus an inherent feature of simple proportional control. Only when the valve is half open will the value of the controlled variable equal the set point. At all other times there will be a deviation: when the load is a maximum the deviation will be the greatest in one direction, and vice-versa.

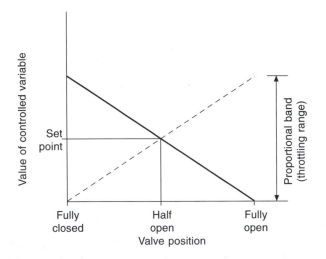

Fig. 13.7   The operation of a proportional controller depends on the temperature of the water handled by the valve: the full line is for the case of hot water and the broken one for chilled water.

The terms direct action and reverse action are used to denote the manner in which the final control element moves in response to the signals it receives from the primary sensor element. For example, suppose that a room suffers a heat gain and that this is offset by means of a fan-coil unit fed with chilled water, the output being regulated by means of an electrically actuated, motorised, two-port modulating valve. If, when the room temperature rises the controlling thermostat sends an increasing signal to the valve motor this is termed direct acting. Supposing that the signal drives the motor in a direction that closes the valve; it will reduce the flow of chilled water and is clearly wrong. A reversing relay must be added to the control circuit to drive the motor in the other direction, when chilled water is handled and open the valve, upon receiving a signal of increasing strength. The controller is now reverse acting and the broken line in Figure 13.7 shows this. If hot water flows through the fan coil unit in winter, to deal with a heat loss, the relay is kept out of the circuit and the controller acts directly so that a rise in room temperature sends a signal of increasing strength to the motor, causing the valve to move towards its closed position. This is shown by the full line in Figure 13.7.

An alternative view of simple proportional control is obtained by introducing the concept of 'potential value' of the controlled variable (due to Farrington (1957)). If there is not a match initially between load and capacity, the controlled variable will gradually change, approaching a steady value at which a match will prevail. This steady value, attainable exponentially only after an infinity of time, is the potential value of the controlled variable. Every time the capacity of the plant is altered there is a change in the potential value. Capacity variation can thus be spoken of in terms of potential correction, ϕ: if a deviation of θ occurs in the controlled variable then a potential correction of ϕ must be applied. Simple proportional control can be defined by an equation:

$$\phi = -k_p \theta + C_1 \tag{13.1}$$

where k_p is the proportional control factor and C_1 is a manual re-set constant that determines the set point of the controlled variable.

Thus, under a constant load condition, sustained deviation is the rule. This offset corresponds to a value of ϕ and when, after a load change, the controlled variable has responded to this correction, it settles down with some decaying overshoot and undershoot at a steady value. Such a steady value is the control point. Hence, even though the set point is unaltered (at the mid-point of the proportional band) the control point takes up a variety of values, depending on the load.

To minimise offset k_p should be large and increasing its value is known as increasing the control sensitivity, corresponding to narrowing the proportional band. For a given operating condition too much sensitivity (or too narrow a proportional band) can induce hunting, proportional control degenerating to two-position control.

13.11   Refined proportional control

The offset which occurs with simple proportional control may be removed to a large extent by introducing reset. The set point of a proportional controller is altered in a way which is related either to the deviation itself or to the rate at which the deviation is changing.

A controller the output signal of which varies at a rate which is proportional to the deviation is called a proportional controller with integral action.

When the output signal from a proportional controller is proportional to the rate of change of the deviation, the controller is said to have derivative action. Since derivative action is not proportional to deviation, it cannot be used alone but must be combined with another control action. Such a combination is then termed 'compound control action'.

Figure 13.8 illustrates some of the behaviours of different methods of control action. Deviation is plotted against time. Curve A, for proportional control plus derivative action,

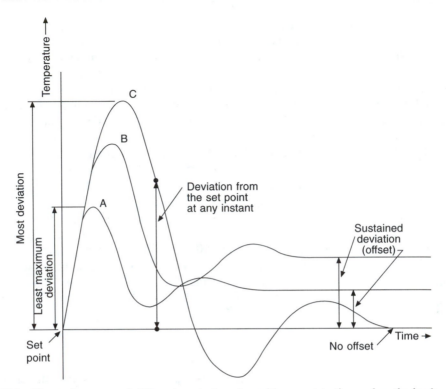

Fig. 13.8   The performance of different control modes, with respect to time, when the load suddenly changes: A proportional plus derivative, B simple proportional, C proportional plus integral. See Eckman (1958).

reaches a steady state quite rapidly but some offset is present: Curve B shows the case of simple proportional control; there is a larger maximum deviation than with curve A and the offset is greater. Curve C is for proportional plus integral control; the maximum deviation is greater than for the other two curves and the value of the controlled variable oscillates for some time before it settles down to a steady value without any offset at all.

Proportional plus derivative plus integral control is not shown. Its behaviour is similar to curves A and C but the controlled variable settles down more quickly. There is no offset.

In terms of potential correction, integral action may be defined by

$$\frac{d\phi}{dt} = -k_i \theta \tag{13.2}$$

whence

$$\phi = -k_i \int \theta \, dt + C_2 \tag{13.3}$$

where k_i is the integral control factor and C_2 is a constant of integration. This shows that the potential correction is proportional to the integral of the deviation over a given time.

(Incidentally, floating control can be regarded as a form of integral action that applies to a potential correction whenever any deviation occurs, the magnitude of the correction being independent of the size of the deviation, depending only on its sign, positive or negative.)

Proportional plus integral control (often abbreviated as P + I) is much used and can be expressed by the sum of equations (13.1) and (13.3):

$$\phi = -k_p \theta - k_i \int \theta \, dt + C \tag{13.4}$$

where C is the combined constant.

Derivative action may be similarly defined by

$$\phi = -k_d \frac{d\theta}{dt} \tag{13.5}$$

where k_d is the derivative control factor.

Proportional plus integral plus derivative (abbreviated P + I + D) control can be expressed by adding equations (13.4) and (13.5):

$$\phi = -k_p\theta - k_i \int \theta \, dt - k_d \, d\theta/dt + C \tag{13.6}$$

13.12 Automatic valves

It has been stated elsewhere in literature on the subject that a control valve is a variable restriction, and this description is apt. It follows that a study of the flow of fluid through an orifice will assist in understanding the behaviour of a control valve.

The loss of head associated with air or water flow through a duct or pipe is discussed in section 15.1, and the equation quoted is:

$$H = \frac{4flv^2}{2gd} \tag{15.2}$$

This implies that the rate of fluid flow is proportional to the square root of the pressure drop along the pipe and to the cross-sectional area of the pipe. We can therefore write a basic equation for the flow of fluid through any resistance, for example a valve:

$$q = Ka(2g(h_u - h_d))^{1/2} \tag{13.7}$$

where q = volumetric flow rate in $m^3 \, s^{-1}$,

a = cross-sectional area of the valve opening, in m^2,

h_u, h_d = upstream and downstream static heads, in m of the fluid,

K = a constant of proportionality.

If it is assumed that the position of the valve stem, z, is proportional to the area of the valve opening, then

$$q = K_1 z(2g(h_u - h_d))^{1/2} \tag{13.8}$$

where K_1 is a new constant of proportionality.

Unfortunately, the picture is not as simple as this and the flow rate is not directly proportional to the position of the valve stem; the constants of proportionality are not true constants (they depend on the Reynold's number, just as does the coefficient f in equation (15.2)) and the area of the port opened by lifting the valve stem is not always proportional to the lift. A more realistic picture of the behaviour is obtained if the flow of fluid is considered through a pipe and a valve in series, under the influence of a constant difference of head across them. Figure 13.9 illustrates the case.

The head loss in the pipe, plus that across the valve, must equal the driving force produced by the difference of head between the reservoir, H_1, and the sink, H_0.

The Fanning friction factor, f, is sometimes re-expressed as the Moody factor, $f_m (= 4f)$. Using the Moody factor the head lost in the pipe, h_1, is given by $(f_m l v^2)/(2gd)$. Also $q^2 = a^2 v^2 = (\pi d^2/4)^2 v^2$ and hence

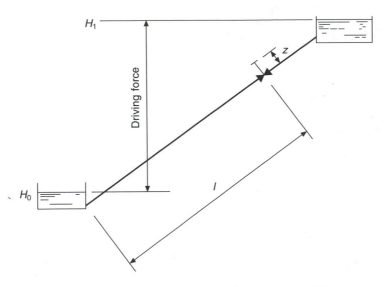

Fig. 13.9   A source and sink of water provide a constant head that is used as the driving force for a simple analysis of valve performance.

$$h_1 = (8f_{\mathrm{m}}lq^2)/(\pi^2 d^5) \tag{13.9}$$

The head lost across the valve is then $(H_1 - H_0 - h_1)$. Thus equation (13.8) for the flow through the valve can be rewritten:

$$q = K_1 z\{2g[H_1 - H_0 - (8f_{\mathrm{m}}lq^2/\pi^2 gd^5)]\}^{1/2}$$

$$q^2 = K_1^2 z^2 2g[H_1 - H_0 - (8f_{\mathrm{m}}lq^2/\pi^2 gd^5)]$$

Write $\beta = (16f_{\mathrm{m}}lK_1^2/\pi^2 d^5)$ and the equation simplifies to

$$q = K_1 z \left[\frac{2g(H_1 - H_0)}{(1 + \beta z^2)} \right]^{1/2} \tag{13.10}$$

So it is seen that even if K_1 is a constant, the flow rate is not directly proportional to the lift, z. In the equation, βz^2 is the ratio of the loss of head along the pipe (equation (13.9)) to that lost across the valve (from equation (13.8)):

$$\beta z^2 = \frac{8f_{\mathrm{m}}lq^2}{\pi^2 gd^5} \bigg/ \frac{q^2}{K_1^2 z^2 2g} = \frac{16f_{\mathrm{m}}lK_1^2}{\pi^2 d^5} \cdot z^2 \tag{13.11}$$

For the smaller pipe sizes the influence of d^5 increases and βz^2 becomes large. Thus, for a constant difference of head the flow rate falls off as the pipe size is reduced, as would be expected, the drop through the valve reducing as the drop through the pipe increases. Hence, the presence of a resistance in series with the valve alters the flow through the valve, under conditions of a constant overall difference of head. Whether this conclusion is valid also for the case of circulation by a centrifugal pump through a piping circuit depends on where the pressure–volumetric flow rate characteristic curve for the piped circuit intersects that of the pump: for intersections on or near the flat part of the pump curve and for comparatively small volumetric flow rate changes, elsewhere on the curve, it is probably valid.

If a valve is to exercise good control over the rate of flow of fluid passing through it, the ideal is that q/q_0 shall be directly proportional to z/z_0, where q_0 is the maximum flow and z_0 is the maximum valve lift (when the valve is fully closed, q and z are both zero). A direct proportionality between these two ratios is not attainable in practice.

The ratio of the pressure drop across the valve when fully open to the pressure drop through the valve and the controlled circuit, is termed the 'authority' of the valve. For example, if a valve has a loss of head of 5 metres when fully open and the rest of the piping circuit has a loss of 15 metres, the authority of the valve is 0.25.

The effect of valve authority on the flow-lift characteristic of a valve can be seen by means of an example.

EXAMPLE 13.1

Using the foregoing theory show how the authority of a valve affects the water flow rate for a given degree of valve opening.

Answer

To simplify the arthmetic divide equation (13.10) throughout by $K_1\sqrt{2g}$, choosing new units for q and z. We have then

$$q = z[(H_1 - H_0)/(1 + \beta z^2)]^{1/2} \tag{13.10a}$$

Denote the maximum flow rate by q_0, when the valve stem is in its position of maximum lift, z_0. For any given valve position other than fully open the valve lift ratio is then z/z_0 and the corresponding water flow ratio is q/q_0. It follows that z/z_0 can vary from 0 to 1.0, as also can q/q_0. Equation (13.11) shows that βz^2 equals h_1/h_v where h_v is the loss of head through the valve. The authority of the valve is then defined by:

$$\alpha = h_v/(h_1 + h_v) = 1/(1 + \beta z_0^2) \tag{13.12}$$

Because α is for full flow conditions when the valve is completely open we must use z_0 in equation (13.12).

Let us suppose that $H_1 - H_0$ is 100 units of head and that the valve lift, z, varies from 0 units (fully closed) to 1 unit (fully open). We may now consider various valve sizes, changing the size of the connected pipe circuit each time so that the head absorbed by both always equals the available driving head of 100 units.

(i) Suppose a small valve is used, with $h_v = 50$ units. Then, for the special case of $z = 1.0$, equation (13.12) yields $\alpha = 0.5$ and the constant β is 1.0. We may use this value of β and $(H_1 - H_0) = 100$ in equation (13.10a) and we have $q = 10z[1/(1 + z^2)]^{1/2}$, which may be applied for all values of z between 0 and 1.0 to give corresponding flow rates, tabulated as follows for the case of $\alpha = 0.5$.

z	0.1	0.2	0.3	0.5	0.7	0.9	1.0
z/z_0	0.1	0.2	0.3	0.5	0.7	0.9	1.0
q	0.995	1.961	2.874	4.472	5.735	6.690	7.071
q/q_0	0.141	0.277	0.406	0.632	0.811	0.946	1.0

Any particular value of q/q_0 is determined by dividing the value of q by the value of q_0, which is 7.071. Thus when $z = 0.2$, $q/q_0 = 1.961/7.071 = 0.277$.

(ii) Suppose a larger valve is used, with $h_v = 20$. Since $H_1 - H_0$ is a constant at 100 we must consider a larger piping system with a loss of $h_1 = 80$. Hence $\alpha = 20/(20 + 80) = 0.2$ and, by equation (13.12), the constant β is now 4. Equation (13.10a) then becomes $q = 10z[1/(1 + 4z^2)]^{1/2}$. We may proceed like this for bigger valves, such that $\alpha = 0.4, 0.2, 0.1, 0.05$, and so on. Similar tabulations can be done and the results are shown as curves in Figure 13.10. We see that the greater the authority the nearer the characteristic is to a straight line. For the control of fluid flow an authority of 0.5 is often chosen.

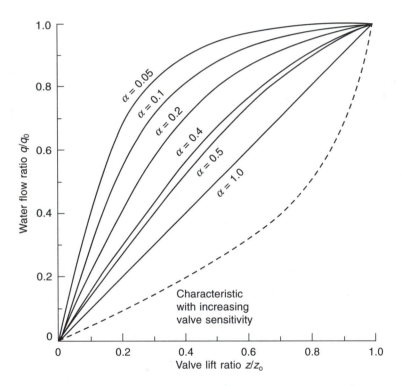

Fig. 13.10   Characteristic curves for a conventional valve (full lines) and a valve with an increasing sensitivity (broken line).

In air conditioning a near-linear relationship is desired between valve lift and heat transfer from a cooler coil or heater battery, rather than with fluid flow rate. This introduces a complication because the sensible heat transfer capacity of either is not directly proportional to fluid flow. This is typified by Figure 13.11 where we see that, for a water temperature change of about 10 degrees, there is a proportional reduction in heat transfer from 100 per cent to about 85 per cent or 90 per cent, as the flow drops to about 50 per cent but thereafter the capacity falls off rapidly. A valve with the behaviour shown in Figure 13.10 would not be suitable for controlling cooling or heating capacity. Instead, the ports and plugs of valves are modified, or 'characterised', to give them an increasing valve sensitivity, defined as $\Delta q/\Delta z$. An arbitrary example of this is shown as a broken line in Figure 13.10. As with uncharacterised valves, the position of the curve depends on the authority but the shape

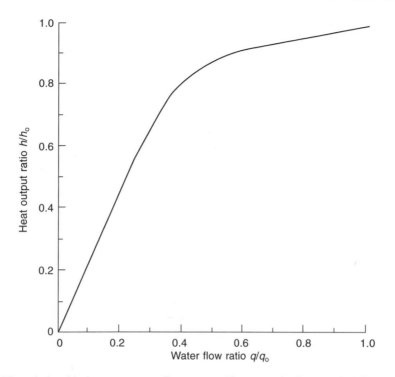

Fig. 13.11 The relationship between water flow rate and heat transfer for a typical finned tube cooler or heater coil. The curve is for a water temperature rise/drop of about 10 degrees. If the water temperature change is more than this the curve moves closer to the diagonal position.

depends on the design of the port or the plug. The valve seat may also be contoured to influence the control achieved. Different possibilities are illustrated in Figure 13.12.

If a characterised valve is used to regulate the sensible capacity of a coil we can see the sort of result obtained by using the broken line in Figure 13.10 with the curve in Figure 13.11 that relates q/q_o with h/h_o, the heat output ratio. Reading the data from the two curves and tabulating we have:

z/z_o	0.1	0.2	0.3	0.4	0.5	0.6	0.7	0.8	0.9
q/q_o	0.045	0.090	0.145	0.190	0.250	0.315	0.395	0.500	0.650
h/h_o	0.09	0.21	0.33	0.45	0.58	0.70	0.81	0.885	0.95

This is a simplification because the curves from Figures 13.10 and 13.11 were not for an actual characterised valve or an actual cooler coil. Nevertheless, the results plotted in Figure 13.13 show that an approximation to the ideal, linear relationship between valve lift ratio (z/z_o) and heat output ratio (h/h_o) can be achieved. Different forms of characterization are possible and each would have an equation different from the simplification of equation (13.10a). One popular form of characterisation is termed 'equal percentage', where, for each linear increment of the valve lift, the flow rate is increased by a percentage of the original flow rate. In practice it turns out that an authority of between 0.2 and 0.4 gives the best resemblance to the ideal performance. Figure 13.14 shows the curves usually obtained. It also illustrates the minimum possible controllable flow.

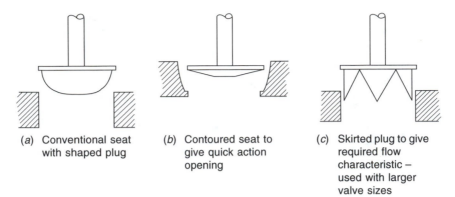

(a) Conventional seat with shaped plug

(b) Contoured seat to give quick action opening

(c) Skirted plug to give required flow characteristic – used with larger valve sizes

Fig. 13.12 Various ways of characterising valve performance.

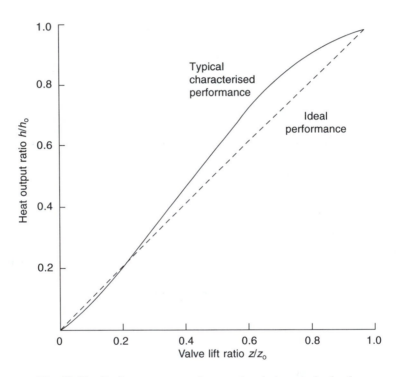

Fig. 13.13 Performance curve for a notional characterised valve.

Two-port valves may be single- or double-seated (see Figure 13.15), arranged to be normally open (upon failure of power or air pressure or upon plant shut-down) or normally closed, depending on system operational requirements or safety. Valves with single seats can give a tight shut-off but there is always an out-of-balance pressure across the plug which is a restricting factor in valve selection: a valve must be able to withstand the maximum likely pressure difference. Double seated valves (Figure 13.15) do not have the same out-of-balance forces across the pair of plugs, although there is still a pressure drop across the valve as a whole. Such valves will not give tight shut-off.

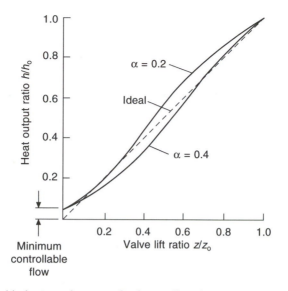

Fig. 13.14 Relationship between heat transfer from a finned tube heater battery or cooler coil and valve lift, for a typical equal percentage characterised valve with authorities of 0.2 and 0.4.

Three-port valves are intended to give constant flow rate but do not always achieve this. In Figure 13.16(*a*) we see a diagram of a mixing valve, defined as a valve in which two fluid streams, A and B, mix to give a common stream, AB. In Figure 13.16(*b*) it is seen that the constancy of the combined flow depends on how the individual plugs, A and B, are characterised. A diverting valve is much less used than a mixer and is defined as one which splits a single entering fluid stream into two divergent streams. A mixing valve, as shown in Figure 13.16(*a*), *must not, under any circumstances*, be piped up as a diverter. Figure 13.16(*c*) shows that this is an unstable arrangement: as the plug moves from its central position towards either seat the pressure drop across the plug increases and forces it on to the seat. The plug then bounces off the seat on to the opposite seat and control is impossible. Figure 13.16(*d*) shows a stable diverting valve arrangement with two plugs: as a plug moves towards a seat the pressure drop increases and tends to push the pair of plugs back to the neutral, central position. The actuator then has something to do and stable control is obtained.

Three-port valves can be used in mixing or diverting applications, as Figure 13.17 shows. The most common and cheapest arrangement is sub-figure (*a*) but there may be

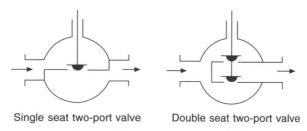

Single seat two-port valve   Double seat two-port valve

Fig. 13.15 Single and double seat two-port valves.

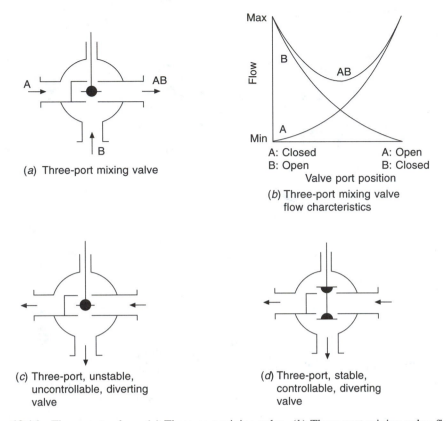

Fig. 13.16   Three-port valves. (*a*) Three-port mixing valve. (*b*) Three-port mixing valve flow characteristic. (*c*) Three-port, unstable, uncontrollable, diverting valve. (*d*) Three-port, stable, controllable, diverting valve.

occasions when sub-figure (*c*) is preferred. If the arrangement in Figure 13.17(*c*) is used then the considerations for valve characterisation are different from those considered hitherto. The water flow rate through the cooler coil (or heater battery) is constant and the cooling or heating capacity is proportional to the logarithmic mean temperature difference, air-to-water, as equation (10.9) shows, the *U*-value for the coil being unchanged. This implies that the control of water flow rate should be related to valve lift in a linear fashion, requiring a valve authority nearer to 0.5 (see Figure 13.10).

The desirable properties of characterisation for the plugs of a three-port mixing valve are asymmetrical: the plug regulating flow through the cooler coil should have an increasing sensitivity while that controlling flow through the port open to the by-pass should be characterised to give a constant, combined flow out of the valve. Such asymmetrical characterisation is sometimes available from manufacturers.

The resistance to flow through the by-pass should be the same as that through the cooler coil and its connections. This can be done by sizing down the by-pass pipe and/or providing a regulating valve in it (as shown in Figure 13.17).

In a throttling application the pressure drop across the two-port valve increases as the valve closes and eventually the force exerted by the valve actuator (through a pneumatic diaphragm or an electric motor) is insufficient to continue closing the valve smoothly. Furthermore, because of the necessary clearance between the plug or skirt and the seating,

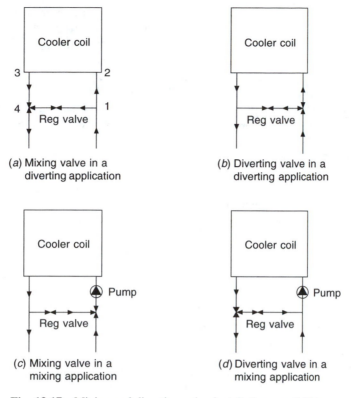

Fig. 13.17 Mixing and diverting valve installation possibilities.

the uncontrolled leakage becomes an increasing proportion of the flow as the valve closes. There is thus always a minimum flow rate for proportional control and beyond this the control degenerates to two-position. The curves shown in Figures 13.10, 13.13 and 13.16(*b*) are therefore not quite correct: flow does not modulate smoothly down to zero as the valve closes and there is a minimum controllable flow, as Figure 13.14 illustrates. This gives rise to the concept of rangeability, defined as the ratio of maximum to minimum controllable flow, and turn-down ratio, defined as the ratio of maximum to minimum usable flow. The difference arises from the possibility that the valve may be oversized and never required to be fully open in a particular application, the turn-down ratio then being less than the rangeability quoted by the valve manufacturer. Commercial valves can have rangeabilities of up to 30:1 but better-made industrial valves may have as much as 50:1. However, because of the non-linearity of heat exchanger capacity, with respect to flow rate, and because the valve may often be slightly oversized, proportional control is only possible down to 15 or 20 per cent of heat exchanger capacity. Oversizing the heat exchanger will make this worse.

It must be remembered, when assessing valve authority, that it is the pressure drop in the part of the piping circuit where variable flow occurs that is relevant. Thus for a two-port valve the authority is a fraction of the pump head for the index circuit. This is also the case for two-port valves in parallel sub-circuits because a regulating valve in each branch is adjusted to make the branch circuit resistance the same as that for the remainder of the index circuit. With three-port valves the authority is also related to the circuit in which the flow is variable. Figure 13.17(*a*) shows an example: the three-port valve at the point 4 should have a loss when fully open that is a suitable fraction (0.2 to 0.4) of the resistance (1 to 2) + (cooler coil) + (3 to 4). The regulating valve in the by-pass would be adjusted during commissioning so that the pressure drop from 1 to 4 through the by-pass equalled that from 1 to 4 through the cooler coil and its connections.

If valve friction, forces associated with the fluid flow through the valve, or fluctuating external hydraulic pressures prevent a pneumatically-actuated valve from taking up a position that is proportional to the control pressure, a valve stem positioner is needed. Such a positioner ensures there is only one valve position for any given actuating air pressure. In critical applications it may also be necessary to provide positioners for motorised modulating dampers.

The flow coefficient (or capacity index), A_v, is used when sizing valves, in the formula $q = A_v \sqrt{\Delta p}$, a version of equation (13.7) in which q is the flow rate in m^3 s^{-1} and Δp is the pressure drop in Pa. For a given valve, A_v represents the flow rate in m^3 s^{-1} for a pressure drop of 1 Pa.

13.13   Automatic dampers

A single butterfly damper disturbs the airflow considerably. It is therefore better, both for balancing during commissioning and for the automatic control of airflow, to use multi-leaf butterfly dampers which may be in parallel or opposed-blade configurations.

Every butterfly damper has an inherent characteristic relating airflow with blade position and Figure 13.18 shows two typical inherent curves. Such characteristics are for the case of a constant pressure drop across the damper, regardless of blade angle. It is evident that as the blades close in a real installation the pressure drop across the damper will not be constant and the characteristic will become different: it is then called an installed characteristic (see Figure 13.19) and its shape and position depend on the authority, α, defined in the way adopted for valves and related to the part of the duct system where variable flow occurs. It appears to be common practice to select automatic dampers in terms of another parameter, δ, defined as the ratio of the pressure drop through the system, excluding the damper, to that through the damper fully open, for the case of maximum airflow.

The pressure drop across a fully open damper depends on: damper construction, blade shape, damper dimensions, frame intrusion into the airstream, and the ratio of the cross-sectional area of the fully open damper to that of the duct in which it is fixed. (The inherent characteristic is also a function of the damper-duct area ratio, but most significantly so only when this is less than 0.5.) Typical pressure drops, fully open, are of the order of 10 to 15 Pa for parallel blade dampers used in a mixing application, as shown in Figure 13.20.

Seals of various sorts can be provided to minimise leakage when a damper is nominally shut. A simple automatic damper, not having any special seals, has a typical leakage rate that can give a face velocity over its section of about 0.25 m s^{-1} for a pressure drop across it of 375 Pa, according to one manufacturer. A square law can probably be used to give a rough relationship between velocity and pressure drop. With special construction and seals a low-leakage damper can give velocities as low as from 0.01 to 0.025 m s^{-1} for the same pressure drop, it is claimed.

One of the commonest applications of motorised dampers in air conditioning is to vary the mixing proportions of recirculated and fresh air (see section 3.10), as shown in Figure 13.20. A controller C1, positioned in the ducting to sense condition W, would be used to vary the mixing proportions of fresh and recirculated air in winter, so that the dry-bulb

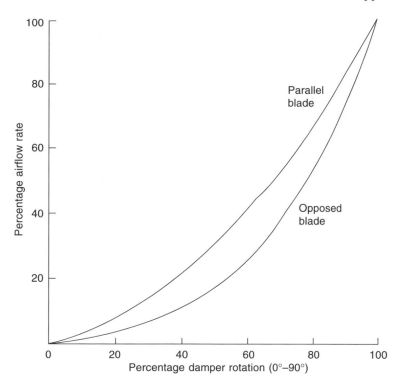

Fig. 13.18 Typical inherent damper characteristic.

temperature of state W was maintained. The dampers would be moved by means of the damper motor (or group of damper motors) R1. Consider the variations of pressure which may occur as the mixing proportions alter. If the discharge air and the variable fresh-air dampers have been chosen to have a pressure drop across them of 15 Pa, then, when 100 per cent of fresh air is handled the static pressure at state R is +15 Pa and the static pressure at state M is −15 Pa. The difference of pressure over the recirculation dampers is therefore 30 Pa. When the recirculation dampers are fully open, the discharge air and the variable fresh air dampers being fully closed, the pressure difference should still be 30 Pa if the volumes handled by the supply and extract fans are not to vary. Since virtually constant pressure operation is the rule, parallel-blades are a better choice than opposed-blades as they have a more nearly linear inherent characteristic. A flat S-shaped curve can be obtained if the dampers are properly selected. See Figure 13.19.

To get a good installed characteristic dampers are sometimes arranged so that they are never more than 45° open, because of the approximate linearity of the first half of characteristic curve (see Figure 13.19). To secure the correct balance of face velocity and pressure drop it is sometimes necessary to adopt a cross-sectional damper area less than that of the duct in which the damper is to be mounted. It is occasionally recommended that gradually expanding and contracting duct connections be made to the damper but this is scarcely ever a practical proposition and is probably not worth doing, in the majority of cases.

13.14 Application

Although no two designs are ever alike, there are basic similarities in all designs, particularly for comfort conditioning. To illustrate the application of automatic control consider the

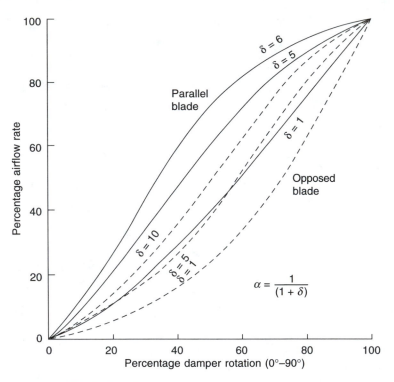

Fig. 13.19 Typical installed damper characteristic.

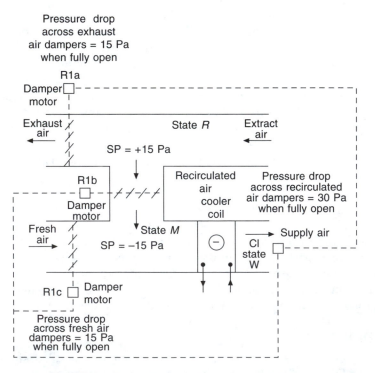

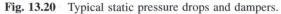

Fig. 13.20 Typical static pressure drops and dampers.

plant and psychrometry shown in Figures 13.21(a) and 13.21(b). The intention is to control the dry-bulb temperature and relative humidity in a room. For simplicity, the temperature rise through the extract fan is ignored. The mode of control is to be proportional plus integral over both temperature and relative humidity.

In the summer design case, minimum outside air at state O_s mixes with maximum recirculated air at state R, to form a mixture state, M_s. The temperature of the air leaving the cooler coil, t_w, is sensed by controller C1a, which regulates the flow of chilled water through the cooler coil by means of the three-port valve, R1a. Since the method of control is proportional plus integral there will be no offset in the value of t_w.

When the enthalpy of the outside air exceeds that of the air recirculated from the room it is economical to use as much recirculated air as possible and minimise the amount of fresh air handled. Accordingly, the enthalpy of the outside air is sensed by an enthalpy controller, C1b, located in the fresh air duct. The set point of C1b is the design enthalpy of the air in the room. When the outside air has an enthalpy exceeding the set point, the motorised dampers are moved to positions giving minimum fresh air with maximum recirculated and minimum discharged air. This is achieved by means of damper motors R1b (fresh air), R1c (recirculated air) and R1d (discharged air). If the enthalpy of the outside air has a value less than that of the recirculated air but more than the enthalpy of state W, the dampers move to the position that gives 100 per cent fresh air and the group of motors, R1b, R1c and R1d, is controlled in sequence with the three-port valve R1a.

At the same time, room temperature is controlled by a thermostat C2, located in the extract air duct, close to the extract grille, so as to sense a temperature that is representative of the whole of the room. Upon fall in temperature sensed by C2, motorised valve R2, in the *LTHW* line from the reheater battery, is opened. Since proportional plus integral control is specified for C2, the value of the room temperature sensed will return to the set point, after a short while. There will be no offset.

As the summer passes and winter approaches, the state of the outside air moves from O_s towards O_w and, in due course, the dampers move to give 100 per cent fresh air. Less and less cooling will be needed and C1a will progressively open the by-pass port of R1a, closing the port to the cooler coil at the same time. Eventually, all the chilled water is by-passing the cooler coil because the outside air temperature is the same as the set point of C1a, 100 per cent fresh air being handled. Any further reduction in the temperature of the outside air will make C1a operate the damper motors to vary the proportions of fresh and recirculated air in order to give a constant value of temperature, t_w ($= t_{mw}$), without offset, since control is proportional plus integral.

Humidity is controlled by a humidistat, C3, located in the extract air duct. Upon fall in the value of relative humidity motorised valve R3 is opened to inject dry steam into the supply airstream as near as possible to the supply air grille or diffuser. (Sometimes dry steam is injected in the conditioned space itself.) The steam injection process for design winter conditions is seen on the psychrometric chart (Figure 13.21(b)) by the line from B to S_w.

Upon rise in humidity sensed by C3, the control between C1a and R1a is overridden and C3 closes the by-pass port of the chilled water valve, R1a, opening the port to the cooler coil. An interlock is included to ensure that humidifying and dehumidifying cannot occur simultaneously. In this illustration, since C3 exercises proportional plus integral control, there will be no offset in the relative humidity.

In formulating a scheme for the automatic control of an air-conditioning system it is essential that a schematic or flow diagram be prepared, after the fashion of Figure 13.21(a)

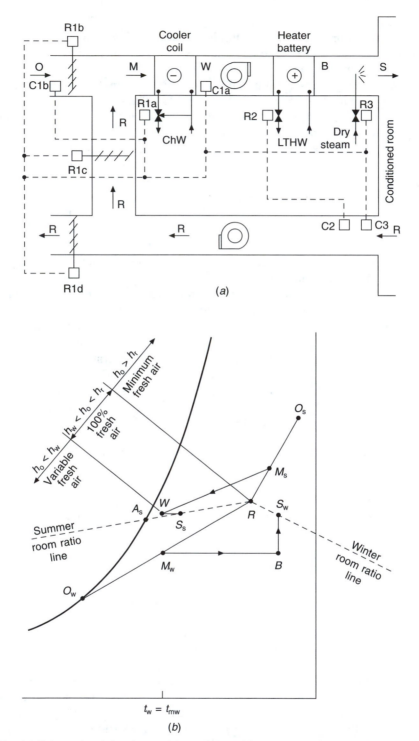

Fig. 13.21   (*a*) Schematic of the plant and controls to achieve control over temperature and humidity in a room. (*b*) Psychrometry for Figure 13.21(*a*). Subscript s refers to summer design. Subscript w refers to winter design. Proportional plus integral control gives no offset in dry-bulb temperature t_w (= t_{mw}). Temperature rise through the extract fan is ignored for simplicity of illustration.

and that all relevant data be shown on it. Such relevant data would be, for example, all airflow and waterflow rates and all psychrometric states. It is also essential that a schedule of operation of the automatic controls be drawn up. This should include such information as the set points, proportional bands (or differential gaps) and location of all controllers (thermostats, humidistats, etc.), and the flow rates, pressure drops when fully open related to an authority between 0.2 and 0.4, location, size, etc., of all regulators (motorised valves, dampers, etc.). A short description of the action of each regulator on, say, rise in temperature and also on plant shut-down, should be included in the schedule. All controllers and regulators should be denoted in a systematic fashion, such as C1, R1, C2, R2 and so on.

It is also necessary that the sequence of operation of the components of the air conditioning plant be stated, due care being paid to safety considerations, e.g. the chilled water pump must start before the refrigeration compressor does.

A system cannot operate unless it is well designed, properly installed and competently commissioned. Commissioning cannot be started unless regulating valves for balancing water flow have been provided in the right places in the pipework, together with appropriate points for pressure tappings and temperature measurement. The designer must supply the commissioning team with adequate details about the system. Hence the necessity of a good schematic, supported by comprehensive controller and regulator schedules, referred to earlier. As well as setting the controls correctly, commissioning involves balancing the airflow rates in the duct systems and the waterflow rates in all the main and subsidiary circuits, adopting the principles of proportional balancing given in the CIBSE Commissioning Codes for air and water (1986).

It follows that operating instructions must also be provided by the designer (particularly if Building Management Systems are to be used). These are best written throughout the development of the design and should always be available for the commissioning team as well as the future occupier of the building.

13.15   Fluidics

A frictional pressure drop occurs when air flows over a surface, pulling the airstream towards it. If the pressure at the surface is artificially increased slightly, by an external agency, the airstream can be persuaded to leave the surface and seek a more stable pattern of flow. The techniques of *fluidics* exploit this so-called 'Coanda' effect to provide a simple manner of switching an airstream between two stable configurations by using small pressure differences, no moving mechanical parts being involved. Fluidic switching has been used to a limited extent in air conditioning in the past but has not yet been extensively applied.

13.16   Control by microprocessors and Building Management Systems (BMS and BEMS)

Instead of adopting the analogue philosophy of the foregoing an alternative approach to automatic control is one of calculation, making use of microcomputer technology. A microprocessor is then the main feature of the control system. Data on temperature, flow rates, pressures, etc., as appropriate, are collected from sensors in the system and the treated spaces and stored in the memory of the processor. Provided that equations defining the performance of the control elements, the items of plant and the behavioural characteristics of the systems controlled have been developed and fed into the micro-processor as algorithms, deviations from the desired performance can be dealt with by calculation, the plant output being varied accordingly. Mathematical functions replace control modes, such as $P + I + D$. For example, if room temperature rose in a space conditioned by a constant volume reheat system, the correct position of the valve in the *LTHW* line feeding the heater battery could be calculated and corrected as necessary, to bring the room temperature back to the set point as rapidly as possible, without any offset. Data can be stored to establish trends and anticipation can be built into the program so that excessive swings in controlled conditions may be prevented. Furthermore, self-correction can be incorporated so that the control system learns from experience and the best possible system performance is obtained. While this implies that commissioning inadequacies and possibly even design faults can be corrected it is a mistake to rely on this: optimum results are really only obtainable, and the cost of the installation justified, from systems that have been properly designed, installed and commissioned. Under such circumstances it is then feasible to extend the scope of microprocessor control to include the management of all the building services with an economic use of its thermal and electrical energy needs.

The functions of a building management system (BMS) or building energy management system (BEMS) are monitoring and control of the services and functions of a building, in a way that is economical and efficient in the use of energy. Furthermore, it may be arranged that one system can control a group of buildings.

There are three types of system:

(i)   *Central systems.* These placed a heavy duty on communications, were unreliable and expensive. They are no longer popular.

(ii) *Distributed intelligence systems.* The outstations, where all data processing is done, have computing power, allowing local decisions and reprogramming to be made, with transmission to a central unit, if required.

(iii) *Open systems (OSI).* Different systems in different buildings (or even in the same building), including lifts, lights, security etc., may speak in different languages or use different procedures (protocols). This makes it difficult to harmonise monitoring and control. Translation devices, termed gateways, must then be used. The aim of these is to unify control of all the systems in a building and, as convenient, to unify the control and energy management of a group of buildings. See Scheidweiler (1992).

Building energy management systems have not always been a success in the past. A failure to understand the way a system worked has led to some of such systems being by-passed. Control systems and building energy management systems must be easier to understand; systems should be simple enough for the maintenance people to comprehend and be able to operate and override as necessary.

The communications network between computer terminals and related equipment within buildings is often based on the use of fibre optics. The dominant language used by such systems is called Transfer Control Protocol, or Internet Protocol (TCP/IP)—which is the basis of the Internet. This has been developed for use as a site communications backbone that connects information technology networks and the building services. The use of advanced digital signal processing can automatically adapt to a wide range of wire types and compensate for different methods of installation. The use of such open systems will allow users to control buildings through the Internet and will enable systems from different manufacturers to be dealt with.

Exercises

1. (*a*) Explain what is meant by (i) proportional control and (ii) offset. Give an example of how offset is produced as a result of proportional action.

(*b*) Show by means of a diagram with a brief explanation, the basic layout and operation of a simple electrical *or* a simple pneumatic proportional control unit, suitable for use with a modulating valve and a temperature-sensitive element producing mechanical movement.

2. An air conditioning plant treating two separate rooms comprises a common chilled water cooler coil, constant speed supply fan, two independent re-heaters (one for each room) and a single dry steam humidifier (for one of the rooms only). A constant speed extract fan is provided and motorised control dampers arrange for the relative proportions of fresh, recirculated and discharged air to be varied, as necessary. Draw a neat schematic sketch of a system that will maintain control over temperature in one room and control over temperature and humidity in the other room, throughout the year. Illustrate the performance in summer and winter by a sketch of the psychrometry involved.

Notation

Symbol	Description	Unit
A_v	flow coefficient (capacity index)	–
a	cross-sectional area of a valve opening	m^2
C	combined constant	–
C_1	Manual re-set constant	–
C_2	constant of integration	–
d	diameter of a pipe or duct	m or mm
f	dimensionless coefficient of friction (Fanning)	–
f_m	dimensionless coefficient of friction (Moody)	–
g	acceleration due to gravity	$m\ s^{-2}$
H_1, H_o	head in a reservoir or sink, in m of fluid	m
h	static head in m of fluid	m
h_d	downstream static head in m of fluid	m
h_1	head lost along a pipe in m of fluid	m
h_u	upstream static head in m of fluid	m
h_v	head lost across a valve in m of fluid	m
K, K_1	constants of proportionality	–
k_d	derivative control factor	–
k_i	integral control factor	–
k_p	proportional control factor	–
l	length of a pipe	m
p	pressure	Pa
q	volumetric flow rate	$m^3\ s^{-1}$
q_o	maximum volumetric flow rate	$m^3\ s^{-1}$
t	time	s
v	mean velocity of water flow or airflow	$m^3\ s^{-1}$
z	valve lift	m
z_o	maximum valve lift	m
α	valve or damper authority	–

β $16f_{\mathrm{m}}lK_1^2/\pi^2 d^5$ –

δ $\dfrac{\text{system pressure drop excluding damper}}{\text{fully open damper pressure drop}}$ –

ϕ potential correction in appropriate units   –

θ deviation in appropriate units   –

References

ASHRAE Handbook (1997): Fundamentals, 14.29–14.31.

CIBSE (1986): Commissioning Codes: Series A, Air Distribution Systems, Series W, Water Distribution Systems.

Eckman, D.P. (1958): *Automatic Process Control*, John Wiley.

Farrington, G.H. (1957): *Fundamentals of Automatic Control*, Chapman & Hall.

Scheidweiler, A. (1992): *Dangerous Liaisons*, Electrical Design.

Bibliography

1. Engineering Manual of Automatic Control, Minneapolis-Honeywell Regulator Company, Minneapolis 8, Minnesota.
2. J.E. Haines, *Automatic Control of Heating and Air Conditioning*, McGraw-Hill, 1961.
3. Damper Manual, Johnson Service Company, Milwaukee, Wisconsin.
4. J.T. Miller, *The Revised Course in Industrial Instrument Technology*, United Trade Press Ltd, 1978.
5. W.H. Wolsey, A theory of 3-way valves, *J. Inst. Heat. Vent. Engrs.*, 1971, **39**, 35–51.
6. BS 1523: Part I: 1967. Glossary of Terms used in Automatic Controlling and Regulating Systems, Part I, Process and kinetic control, British Standards Institution.
7. F. Evans, Capacity index and velocity pressure factors, Building Services Engineering and Technology, 1981, **2**, No. 4, 191–2.
8. K. Letherman, *Automatic Controls for Heating and Air Conditioning*, Pergamon Press, 1981.
9. T.J. Lechner, Fluidic circuits for measurements and control, *ASHRAE Journal*, Feb 1972, 40–45.
10. R.W. Haines, *Control Systems for Heating, Ventilating and Air Conditioning*, 3rd edition, Van Nostrand Reinhold, New York, 1983.
11. W.F. Stoecker and D.A. Stoecker, *Microprocessor Control of Thermal and Mechanical Systems*, Van Nostrand Reinhold, New York, 1989.
12. *ASHRAE Handbook*, 1999, Heating, Ventilating, and Air Conditioning Applications, Chapter 45.
13. *Control Dampers*, Holdfire Equipment Ltd, Stonehouse, Gloucestershire, 1976.

14

Vapour Absorption Refrigeration

14.1 Basic concepts

Following the procedure adopted in section 9.1, let a liquid be introduced into a vessel in which there is initially a vacuum, and let the walls of the container be maintained at a constant temperature. The liquid at once evaporates, and in the process its latent heat of evaporation is abstracted from the sides of the vessel. The resultant cooling effect is the starting point of the refrigeration cycle to be examined in this chapter, just as it was in the beginning of the vapour compression cycle considered in chapter 9.

As the liquid evaporates the pressure inside the vessel rises until it eventually reaches the saturation vapour pressure for the temperature under consideration. Thereafter, evaporation ceases and the cooling effect at the walls of the vessel is not maintained by the continued introduction of refrigerant. The latter merely remains in a liquid state and accumulates in the bottom of the container. To render the cooling process continuous it is necessary, as we have already seen earlier, to provide a means of removing the refrigerant vapour as fast as it forms. In the vapour compression cycle this removal is accomplished by connecting the evaporator to the suction side of a pump. A similar result may be obtained by connecting the evaporator to another vessel containing a substance capable of absorbing the vapour. Thus, if the refrigerant were water, a hygroscopic material such as lithium bromide could be used in the absorber. The substance used for this purpose is termed the 'absorbent'.

In order to obtain closed cycles for both refrigerant and absorbent the next stage in the process must be the release of the absorbed refrigerant at a convenient pressure for its subsequent liquefaction in a condenser. This is accomplished in the 'generator', where heat is applied to the absorbent–refrigerant solution and the refrigerant is driven off as a vapour.

The absorber and generator together take the place of the compressor in the vapour-compression cycle. So far as the refrigerant is concerned, the rest of the absorption cycle is similar to the compression cycle, i.e. the vapour is liquefied in the condenser and brought into the evaporator through an expansion valve or an orifice. As for the absorbent, on leaving the generator it is, of course, returned to the absorber for another cycle.

In an absorption refrigeration system cooling water is required for both the condenser and the absorber.

The principal advantages of the absorption cycle over other refrigeration systems are that it can operate with low-grade energy in the form of heat, indirectly as steam or high temperature hot water, or directly as gas, oil, hot exhaust gases, or solar heat. Furthermore, it has few moving parts. Theoretically, only a single pump is required, that needed for conveying the absorbent-refrigerant solution from the low-pressure absorber to the comparatively high-pressure generator. In practice, two more pumps are frequently used, one to recirculate solution over cooling coils in the absorber and another to recirculate the refrigerant over chilled water coils in the evaporator. The basic single effect absorption-refrigeration cycle without these refinements is illustrated in Figure 14.1.

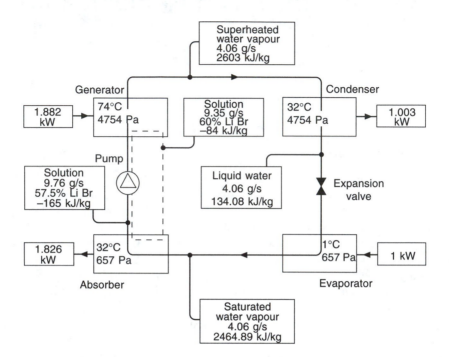

Fig. 14.1   A single effect absorption-refrigeration cycle using lithium bromide as the absorbent and water as the refrigerant.

14.2   Temperatures, pressures, heat quantities and flow rates for the lithium bromide–water cycle

If it is assumed that the temperature of the liquid refrigerant (water) leaving the condenser is t_c and that the temperature of evaporation is t_e, then it is easy to calculate the mass flow rate of refrigerant, $\dot{m}_r$, which has to be circulated per kW of refrigeration. This calculation depends on the further assumption

$$h_{ve} = 2463 + 1.89t_e \text{ kJ kg}^{-1} \tag{14.1}$$

where h_{ve} is the enthalpy of the refrigerant vapour leaving the evaporator.

This is approximately true for the pressures and temperatures in common use in air conditioning.

Similarly, for the enthalpy of the liquid leaving the condenser, h_{1c}, we can write (in the case of water),

$$h_{1c} = t_c c_1$$

$$= 4.19t_c \text{ kJ kg}^{-1} \tag{14.2}$$

Then,

$$\dot{m}_r(h_{ve} - h_{lc}) = \text{kW of refrigeration}$$

and

$$\dot{m}_r = 1/(2463 + 1.89t_e - 4.19t_c) \tag{14.3}$$

EXAMPLE 14.1

If a vapour absorption system using water as a refrigerant evaporates at 1°C and condenses at 32°C, determine the mass flow rate of refrigerant circulated per kW of refrigeration.

Answer

By equation (14.3)

$$\dot{m}_r = 1/(2463 + 1.89 \times 1 - 4.19 \times 32)$$
$$= 0.000\ 429\ 0\ \text{kg s}^{-1}\ \text{kW}^{-1}$$

From psychrometric tables the condensing and evaporating pressures are 4.754 kPa and 0.657 kPa, respectively.

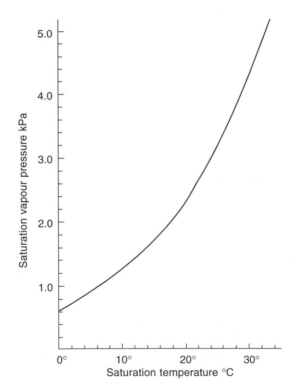

Fig. 14.2 Saturation vapour pressure of water vapour.

Considering now the absorber-generator part of the system, it is assumed that the temperature in the absorber is t_a and that the generator is operating at a temperature t_g. Knowing these temperatures and pressures p_c and p_e, the following data are obtained from tables or from a chart such as Figure 14.3, which gives the properties of lithium bromide–water solutions:

Solution leaving absorber (at p_e and t_a)	concentration enthalpy	$C_a\%$ h_a kJ kg^{-1}
Solution leaving generator (at p_c and t_g)	concentration enthalpy	$C_g\%$ h_g kJ kg^{-1}

EXAMPLE 14.2

Determine the concentrations and enthalpies of the solution in the absorber and the generator, for example 14.1.

Answer

Referring to Figure 14.3, we read off the following:

At 0.657 kPa and 32°C, $C_a = 57.5\%$ and $h_a = -165$ kJ kg^{-1}
At 4.754 kPa and 74°C, $C_g = 60\%$ and $h_g = -84$ kJ kg^{-1}

It is now possible to calculate the mass flow rate of solution that must be circulated to meet the needs of the cycle. If $\dot{m}_{sa}$ is the mass flow rate of solution leaving the absorber

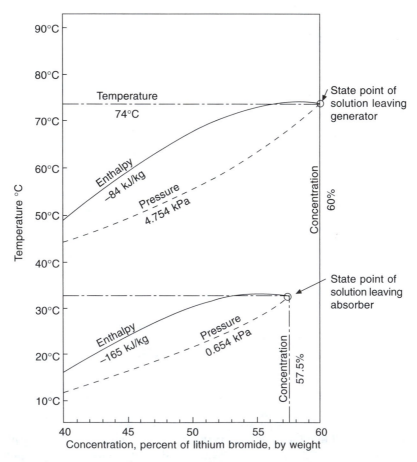

Fig. 14.3 Use of a chart giving properties of lithium bromide–water.

and $\dot{m}_{sg}$ the flow rate leaving the generator, both in kg s^{-1} kW^{-1}, then for a mass balance the following must hold—

$$C_a \dot{m}_{sa} = C_g \dot{m}_{sg} \tag{14.4}$$
$$\dot{m}_{sa} = \dot{m}_{sg} + \dot{m}_r \tag{14.5}$$

Substituting for $\dot{m}_{sa}$ from equation (14.5) in equation (14.4), we obtain

$$\dot{m}_{sg} = \frac{C_a \dot{m}_r}{C_g - C_a} \tag{14.6}$$

EXAMPLE 14.3

Using the data of the previous examples, determine the mass flow rate of solution that must be circulated in the absorber and in the generator.

Answer

From equation (14.6),

$$\dot{m}_{sg} = \frac{57.5 \times 0.000\,429}{60 - 57.5} = 0.009\,87 \text{ kg s}^{-1} \text{ kW}^{-1}$$

and using equation (14.5),

$$\dot{m}_{sa} = 0.009\,87 + 0.000\,429 = 0.010\,30 \text{ kg s}^{-1} \text{ kW}^{-1}$$

An equation similar to (14.1) can be used to determine h_{vg}, the enthalpy of the superheated water vapour leaving the generator,

$$h_{vg} = 2463 + 1.89 t_g \tag{14.7}$$

EXAMPLE 14.4

Using the earlier data and equation (14.7), determine the enthalpy of the water vapour leaving the generator.

Answer

$$h_{vg} = 2463 + 1.89 \times 74 = 2603 \text{ kJ kg}^{-1}$$

With the information now available it is possible to calculate the heat balance for the whole cycle.

In the absorber

$$\begin{bmatrix} \text{heat of} \\ \text{entering} \\ \text{water} \\ \text{vapour} \end{bmatrix} + \begin{bmatrix} \text{heat of} \\ \text{entering} \\ \text{solution} \end{bmatrix} - \begin{bmatrix} \text{heat of} \\ \text{leaving} \\ \text{solution} \end{bmatrix} = \begin{bmatrix} \text{heat to be} \\ \text{removed at} \\ \text{absorber} \end{bmatrix}$$

$$\dot{m}_r h_{ve} + \dot{m}_{sg} h_g - \dot{m}_{sa} h_a = H_a \text{ kW per kW of refrigeration} \tag{14.8}$$

In the generator

$$\begin{bmatrix} \text{heat of} \\ \text{leaving} \\ \text{water} \\ \text{vapour} \end{bmatrix} + \begin{bmatrix} \text{heat of} \\ \text{leaving} \\ \text{solution} \end{bmatrix} - \begin{bmatrix} \text{heat of} \\ \text{entering} \\ \text{solution} \end{bmatrix} = \begin{bmatrix} \text{heat to be} \\ \text{supplied to} \\ \text{generator} \end{bmatrix}$$

$$\dot{m}_r h_{vg} + \dot{m}_{sg} h_g - \dot{m}_{sa} h_a = H_g \text{ kW per kW of refrigeration} \tag{14.9}$$

In the condenser

$$\begin{bmatrix} \text{heat of} \\ \text{entering} \\ \text{water} \\ \text{vapour} \end{bmatrix} - \begin{bmatrix} \text{heat of} \\ \text{leaving} \\ \text{liquid} \end{bmatrix} = \begin{bmatrix} \text{heat to be} \\ \text{removed at} \\ \text{condenser} \end{bmatrix}$$

$$\dot{m}_r h_{vg} - \dot{m}_r h_{1c} = H_c \text{ kW per kW of refrigeration} \tag{14.10}$$

EXAMPLE 14.5

Using the data of preceding examples, calculate H_a, H_g and H_c. Check the heat balance.

Answer

From equations (14.8) and (14.1) and referring to Figure 14.1,

$$H_a = 0.000\,429(2463 + 1.89 \times 1) + 0.009\,87 \times (-84) - 0.010\,30 \times (-165)$$
$$= 1.0574 - 0.8291 + 1.6995$$
$$= 1.9278 \text{ kW per kW of refrigeration}$$

From equation (14.9),

$$H_g = 0.000\,429 \times 2603 + 0.009\,87 \times (-84) - 0.010\,30 \times (-165)$$
$$= 1.1167 - 0.8291 + 1.6995$$
$$= 1.987 \text{ kW per kW of refrigeration}$$

From equation (14.10) and (14.2),

$$H_c = 0.000\,429(2603 - 4.19 \times 32)$$
$$= 1.0592 \text{ kW per kW of refrigeration}$$

Summarising these results,

Heat removed		Heat added	
From absorber	1.928	To generator	1.987
From condenser	1.059	To evaporator	1.000
Total	2.987		2.987
	kW/kW		kW/kW

14.3 Coefficient of performance and cycle efficiency

In chapter 9 the coefficient of performance was defined as the ratio of the energy removed at the evaporator to that supplied at the compressor. In the absorption refrigeration cycle, the energy for operating the system is applied through the generator, hence the coefficient of performance may be defined as the ratio of the refrigerating effect to the rate of energy supplied to the generator:

$$COP = 1/H_g \tag{14.11}$$

$$= \frac{\dot{m}_r(h_{ve} - h_{lc})}{\dot{m}_r h_{vg} + \dot{m}_{sg}h_g - \dot{m}_{sa}h_a}$$

The highest possible coefficient of performance would be obtained by using reversible cycles. Thus, the rate of heat supplied per kilogram of refrigerant ($Q_g = H_g/\dot{m}_r$) to the generator at temperature T_g might be used in a Carnot engine rejecting its heat to a sink at temperature T_c. The efficiency of this engine would be

$$(T_g - T_c)/T_g = W/Q_g$$

or

$$Q_g = (T_g W)/(T_g - T_c) \tag{14.12}$$

where W is the rate of work done, i.e. the area CDEF in Figure 14.4.

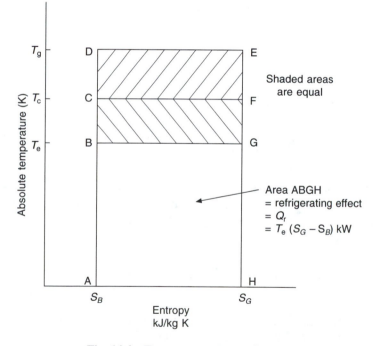

Fig. 14.4   Temperature-entropy diagram.

If this work is used to drive a Carnot refrigerating machine then the rate of work input to this, the area BCFG, equals W above, and the ratio of areas BCFG and ABGH is

$$W/Q_r = (T_c - T_e)/T_e$$

or

$$Q_r = (T_e W)/(T_c - T_e)$$ (14.13)

where Q_r is the refrigerating effect in kW.

Using equations (14.12) and (14.13) an expression for the maximum possible coefficient of performance is obtained:

$$COP_{max} = Q_r/Q_g$$
$$= \frac{(T_g - T_c)T_e}{(T_c - T_e)T_g}$$ (14.14)

An improved performance can be achieved by using two generators, the first working at a higher steam pressure and temperature than the second. Vapour from the first generator passes to the second where, by condensation, it provides the source of thermal energy. Coefficients of performance of such double effect machines better than those of machines with single generators are possible and typical values are 0.92 to 1.0, according to ASHRAE (1998).

EXAMPLE 14.6

Calculate the coefficient of performance for the cycle in example 14.5 and compare this with the maximum obtainable.

Answer

From equation (14.11),

$$COP = \frac{1}{1.987} = 0.50$$

From equation (14.14),

$$COP_{max} = \frac{(347 - 305)274}{(305 - 274)347} = 1.07$$

In practice, the criterion of performance is more commonly the amount of steam required to produce one kW of refrigeration. For the lithium bromide–water cycle the figure is around $0.72 \text{ g s}^{-1} \text{ kW}^{-1}$ and about $0.38 \text{ g s}^{-1} \text{ kW}^{-1}$ for double effect machines, according to ASHRAE (1998).

14.4   Practical considerations

Figure 14.5 is a more practical flow diagram for a single effect lithium bromide–water absorption refrigeration system. It represents a water chiller producing water at 6.7°C. The evaporator works at 10.13 kPa, or 4.4°C, and is equipped with a recirculating pump which sprays the liquid refrigerant (water) over the bundle of tubes carrying the water to be chilled. The absorber operates at 40.6°C and is also provided with a recirculating pump which sprays concentrated solution over the bundle of tubes carrying water from the cooling tower.

Weak solution from the absorber is pumped to the generator via a heat exchanger, where its temperature is raised from 40.6°C to 71.1°C. The generator is supplied with steam at 1.841 bar and in it the solution temperature is 104.4°C and the pressure 10.13 kPa. The heat exchanger effects an economy in operation since the cold solution is warmed from

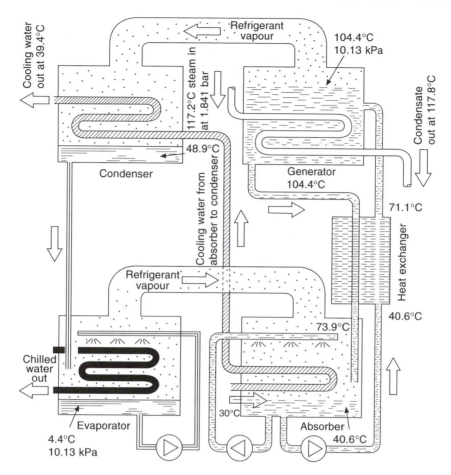

Fig. 14.5 Practical flow diagram for a single effect lithium bromide–water system.

40.6°C to 71.1°C before it enters the generator for further heating to 104.4°C, and warm solution is cooled from 104.4° to 73.9°C before it enters the absorber for further cooling to 40.6°C.

In the example illustrated, cooling water from the tower or spray pond at 30°C first passes through the absorber and then the condenser, which it leaves at 39.4°C. The machine shown in the diagram consumes, at full load, from 0.70 to 0.72 g s^{-1} of steam per kW of refrigeration and the cooling water requirement is about 0.07 litre s^{-1} kW^{-1}. If control is by varying the evaporation rate, steam consumption increases from about 0.72 to about 1.08 g s^{-1} kW^{-1}, as the load falls from 50 to 10 per cent of full capacity. With solution control, on the other hand, there is a reduction in steam consumption from 0.72 to about 0.61 g s^{-1} kW^{-1} as the load falls from 100 to 30 per cent and a small subsequent rise of 0.02 g s^{-1} kW^{-1} to 0.63 g s^{-1} kW^{-1}, when the load falls further through the range 30 to 10 per cent.

The output of the machine is controlled in two ways: (i) by varying the rate at which the refrigerant boils off in the generator, or (ii) by allowing some of the solution leaving the heat exchanger to by-pass the generator. The first method involves either modulating the heat supply to the generator or varying the flow of cooling water through the condenser.

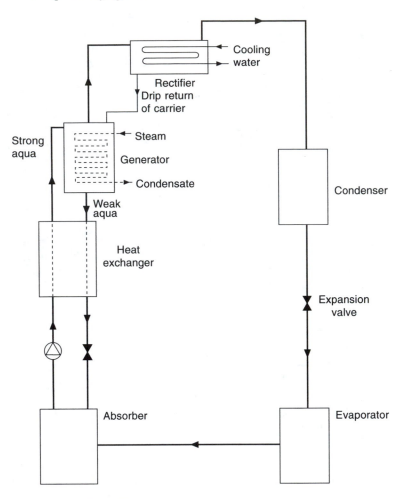

Fig. 14.6   Flow diagram for an aqua–ammonia system.

Electronic controls are used with modern machines and give better results than pneumatic or electric controls. Chilled water flow temperatures from 4°C to 15°C are possible.

High temperature hot water is best not used as a heat source in the generator as an alternative to steam. This is because the temperature drop that accompanies the water flow as it gives up heat introduces thermal expansions and stresses for which the heat exchanger was probably not designed. With steam this does not occur because it condenses at a constant temperature as it gives up its heat.

Lithium bromide machines are used with steam at pressures from 60 kPa to 80 kPa for single effect operation or 550 kPa to 990 kPa for double effect, according to ASHRAE (1998) and are available with cooling capacities from about 350 kW to about 6000 kW of refrigeration. Water-cooled systems are favoured, one reason being that there is less risk of crystallisation than with air-cooled machines. Crystallisation occurs with lithium bromide machines when the concentration of the absorbent in the solution becomes too high and the solution solidifies or turns into a slush, reducing the flow rate and causing the refrigeration process to fail. It can occur when the cooling water flow temperature falls rapidly during operation and causes liquid to be carried over from the generator to the condenser. Most machines automatically limit the heat input at the generator, according to changes in the cooling water temperature. The ability to use lower cooling water temperatures, without crystallisation occurring, is desirable, because it improves operational efficiency.

Air can leak into the system and hydrogen form therein, as a result of corrosion. The presence of these condensible gases reduces refrigeration capacity and increases the risk of crystallisation. A purge system is therefore needed. Corrosion inhibitors of various sorts are also used as are performance enhancers; these are necessary because of the low values of the heat and mass transfer coefficients of lithium bromide machines.

14.5 Other systems

The lithium bromide–water system is now almost the only one chosen for air conditioning applications. An alternative that has been used extensively in industrial work and, to a small extent, in air conditioning, is the system in which ammonia is the refrigerant and water the absorbent. Lower temperatures can be produced with this arrangement than are possible when water is used as a refrigerant, but there are difficulties. These arise because the absorbent vaporises in the generator as well as the refrigerant, and the system therefore has to be provided with a rectifier to condense and return as much of the absorbent as possible. A diagram of the system is shown in Figure 14.6.

Steam consumptions for water–ammonia systems generally exceed those attainable with lithium bromide–water systems and are in the region of 1.1 to 1.5 g s^{-1} kW^{-1}.

Srivastava and Weames (1997) also report the results of work on vapour adsorption machines.

Exercises

1. Show on a line diagram the essential components of a steam heated aqua–ammonia absorption refrigerating plant. Briefly outline the cycle of operation and describe the function of each component of the plant.

2. Make a neat sketch showing the layout of a lithium bromide–water absorption refrigeration plant using high-pressure hot water for the supply of energy to the generator. How may its capacity be most economically controlled? Show a suitable control system for this purpose in your diagram.

Notation

Symbol	Description	Unit
C_a	concentration of LiBr, per cent by weight, in the absorber	
C_g	concentration of LiBr, per cent by weight, in the generator	
COP	coefficient of performance	
COP_{max}	maximum possible coefficient of performance	
c_1	specific heat of the liquid refrigerant	kJ kg^{-1} K^{-1}
H_a	rate of heat removal at the absorber	kW per kW of refrigeration

H_c	rate of heat removal at the condenser	kW per kW of refrigeration
H_g	rate of heat addition at the generator	kW per kW of refrigeration
h_a	enthalpy of the solution leaving the absorber	kJ kg^{-1}
h_g	enthalpy of the solution leaving the generator	kJ kg^{-1}
h_{1c}	enthalpy of the refrigerant liquid leaving the condenser	kJ kg^{-1}
h_{ve}	enthalpy of the refrigerant vapour leaving the evaporator	kJ kg^{-1}
h_{vg}	enthalpy of the vapour leaving the generator	kJ kg^{-1}
$\dot{m}_r$	mass flow rate of refrigerant circulated	k gs^{-1} per kW of refrigeration
$\dot{m}_{sa}$	mass flow rate of solution leaving the absorber	k gs^{-1} per kW of refrigeration
$\dot{m}_{sg}$	mass flow rate of solution leaving the generator	k gs^{-1} per kW of refrigeration
p_c	condensing pressure of the refrigerant	kPa
p_e	evaporating pressure of the refrigerant	kPa
Q_g	heat supplied at the generator, in kJ kg^{-1} of refrigerant	
Q_r	refrigerating effect, in kJ kg^{-1} of refrigerant	
T_c	$273 + t_c$	K
T_e	$273 + t_e$	K
T_g	$273 + t_g$	K
t_c	temperature of the refrigerant leaving the condenser	°C
t_e	temperature of the refrigerant leaving the evaporator	°C
t_a	temperature of the refrigerant in the absorber	°C
t_g	temperature of the refrigerant in the generator	°C
W	rate of work done	kW per kW of refrigeration

References

ASHRAE Handbook (1998): Refrigeration, SI Edition, 41.1–41.12.

Srivastava, N.C. and Weames, I. (1997): A review of developments in vapour-adsorption refrigeration and heat pump systems *J. Inst. Energy* **LXX,** No. 485, December, 116–127.

Bibliography

1. W.F. Stoecker, *Refrigeration and Air Conditioning,* McGraw-Hill Book Company.
2. W.H. Carrier, R.E. Cherne, W.A. Grant and W.H. Roberts, *Modern Air Conditioning, Heating and Ventilating,* Pitman.
3. W.B. Gosney, *Principles of Refrigeration,* Cambridge University Press, 1982.
4. *ASHRAE Handbook* 1997 Fundamentals.
5. *ASHRAE Handbook* 1996 Heating, Ventilating and Air Conditioning Systems and Equipment.

15

Airflow in Ducts and Fan Performance

15.1 Viscous and turbulent flow

When a cylinder of air flows through a duct of circular section its core moves more rapidly than its outer annular shells, these being retarded by the viscous shear stresses set up between them and the rough surface of the duct wall. As flow continues, the energy level of the moving airstream diminishes, the gas expanding as its pressure falls with frictional loss. The energy content of the moving airstream is in the kinetic and potential forms corresponding to the velocity and static pressures. If the section of the duct remains constant then so does the mean velocity and, hence, the energy transfer is at the expense of the static pressure of the air. The magnitude of the loss depends on the mean velocity of airflow, V, the duct diameter d and the kinematic viscosity of the air itself, v. It is expressed as a function of the Reynolds number (Re), which is given by

$$(Re) = \frac{Vd}{v} \tag{15.1}$$

By means of first principles it is possible to derive the Fanning equation and to state the energy loss more explicitly:

$$H = \frac{4flV^2}{2gd} \tag{15.2}$$

where H = the head lost, in m of fluid flowing (air),

f = a dimensionless coefficient of friction,

g = the acceleration due to gravity in m s^{-2}

V and d have the same meaning as in equation (15.1).

If pressure loss through a length of duct, l, is required we can write

$$\Delta p = \frac{2fV^2 \rho l}{d} \tag{15.3}$$

because pressure, p, equals $\rho g H$, ρ being the density of the fluid.

The Fanning equation provides a simple picture but further examination shows that f assumes different values as the Reynolds number changes with alterations in duct size and mean air velocity, as Figure 15.1 shows. The curves shown for small and large ducts are lines of constant relative roughness (0.01 and 0.000 01, respectively), defined by the ratio

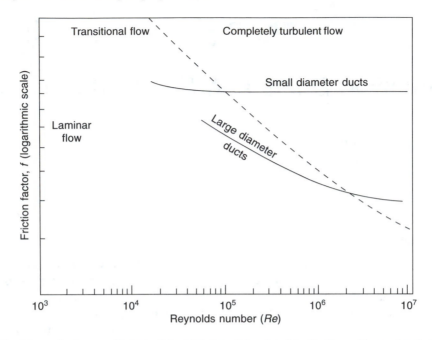

Fig. 15.1   Dimensionless coefficient of duct friction, f, is related to the Reynolds number but, without exception, turbulent flow always occurs in duct systems.

k_s/D_h, where k_s is the absolute roughness and D_h is the mean hydraulic inside diameter of the duct, in the same units, see equation (15.14).

A further appeal to first principles yields the equation

$$f = 2C(Re)^n \tag{15.4}$$

Again the simplicity is misleading; the determination of C and n requires considerable research effort, as Figure 15.1 implies. C and n are not true constants, but f can be expressed approximately by the Poiseuille formula:

$$f = \frac{16}{(Re)} \tag{15.5}$$

This applies only to streamline flow where (Re) is less than 2000. For turbulent flow, Re being greater than 3000, equation (15.4) no longer holds good and instead the Colebrook White function is used

$$\frac{1}{\sqrt{f}} = -4 \log_{10} \left(\frac{k_s}{3.7d} + \frac{1.255}{(Re)\sqrt{f}} \right) \tag{15.6}$$

Attempts have been made to rearrange the Fanning equation by making use of equation (15.6) and an experimental constant but the CIBSE uses a more sophisticated equation, due to Colebrook and White (1937 and 1939):

$$Q = -4(N_3 \Delta p d^5)^{1/2} \log_{10} \left(\frac{k_s}{3.7d} + \frac{N_4 d}{(N_3 \Delta p d^5)^{1/2}} \right) \tag{15.7}$$

where Q = the rate of airflow in $m^3 \; s^{-1}$

Δp = the rate of pressure drop in Pa per metre of duct run

d = the internal diameter of the duct in metres

k_s = the absolute roughness of the duct wall in metres

$N_3 = \pi^2/32\rho = 0.308\ 42\rho^{-1}$

$N_4 = 1.255\pi\mu/4\rho = 0.985\ 67\mu\rho^{-1}$

ρ = the density of the air in $kg \; m^{-3}$

μ = the absolute viscosity of the air in $kg \; m^{-1} \; s^{-1}$

Equation (15.7) is not solvable in a straightforward manner but ASHRAE (1997a) gives a simplified equation, for an approximate determination of the friction factor, due to Altshul and Kiselev (1975) and Tsal (1989), is as follows

$$f' = 0.11(k_s/D_h + 68/(Re))^{0.25} \qquad (15.8)$$

If f', determined from the above, equals or exceeds 0.018, then f is to be taken as the same as f'. If f' is less than 0.018 then the value of f is given by

$$f = 0.85f' + 0.0028 \qquad (15.9)$$

The CIBSE (1986a) has published a chart (see Figure 15.2), that relates volumetric airflow rate, duct diameter, mean air velocity and pressure drop rate. It refers to the following conditions:

clean galvanised sheet steel ductwork, having joints and seams made in accordance with good commercial practice,

standard air at 20°C dry-bulb, 43% relative humidity and 101.325 kPa barometric pressure

air density 1.2 $kg \; m^{-3}$

absolute viscosity 1.8×10^{-5} $Ns \; m^{-2}$ (or $kg \; m^{-1} \; s^{-1}$)

absolute roughness 0.15 mm (as for typical galvanised steel)

Good, approximate corrections to the pressure drop rate (Δp) for changes in air density, arising from variations in barometric pressure (p_{at}) and dry-bulb temperature (t), can be applied by using the following equation:

$$\Delta p_2 = \Delta p_1 \frac{p_{at2}(273 + 20)}{101.325(273 + t_2)} \qquad (15.10)$$

The CIBSE (1986a) quotes more refined corrections, taking into account additional factors and relating barometric pressure to changes in altitude.

The influence on the pressure drop rate can be considerable when ducts are made from materials other than galvanised sheet steel. Extensive correction factors are published by CIBSE (1986a) but some typical correction factors are given in Table 15.1. Where a range of values is given in Table 15.1 the value of the factor depends on the equivalent duct diameter.

Using equations (15.7) to (15.9), as appropriate, the pressure drop rate can be determined for a duct of any material, given the necessary details of absolute roughness, by means of the following equation:

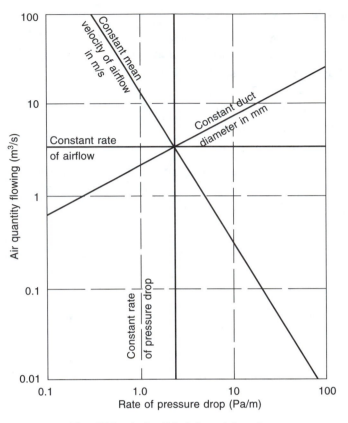

Fig. 15.2   A simplified duct-sizing chart.

Table 15.1   Pressure drop correction factors for ducts of various materials

Material	Absolute roughness, k_s (mm)	Correction factors for various pressure drop rates (Pa m^{-1})			
		0.5	1.0	2.0	5.0
Galvanised sheet steel	0.15	1.0	1.0	1.0	1.0
Galvanised steel spirally wound	0.075	0.95	0.94	0.93	0.92
Aluminium sheet	0.05	0.93	0.91	0.90	0.88
Cement render or plaster	0.25	1.07	1.08	1.08	1.09
Fair faced brick	1.3	1.42 to 1.41	1.50 to 1.45	1.54 to 1.48	1.63 to 1.54
Rough brick	5.0	2.18 to 1.97	2.46 to 2.04	2.62 to 2.12	2.76 to 2.23

$$\Delta p = \frac{4 f \rho Q^2}{D_h A^2} \qquad (15.11)$$

Flexible ducts are usually made in a spiral form and the pressure drop depends on the material of manufacture and the extent to which the spiral is tightened. Pressure drops can be very much greater than those in equivalent, spirally-wound, steel ducting. As an example, a 200 mm diameter, flexible, straight duct made from a three-ply laminate of aluminium and polyester wound on to a helical steel wire fully extended, has a pressure drop rate of about 1.3 Pa m^{-1} when conveying 100 litres s^{-1} of standard air. A similar, straight, spirally-wound steel duct has a pressure drop rate of about 0.75 Pa m^{-1}. Manufacturers' published data should be used. Manufacturers also point out that pressure drop rates vary significantly from published data if flexible ducting is not fully extended. It is claimed that the pressure drop rate could double in a flexible duct that is only 75 per cent extended.

Flexible ducting lined with 25 mm of glass fibre (or the equivalent) is available. Significant attenuation of noise is claimed but there is the risk of noise break-out through the wall of the flexible duct into the surrounding space.

Circular section, permeable cloth ducting is also used, principally for air distribution in industrial applications. Being permeable, the air is diffused into the room uniformly over the entire length of the duct and low velocity air distribution achieved. The recommended materials are polypropylene, polyester and nylon. Cotton should not be used because of its hygroscopic nature and the risk of promoting the growth of microorganisms. About 100 Pa is needed inside the ducting to achieve airflow although with special construction this can be reduced to about 20 Pa. The cloth duct provides a measure of terminal filtration and hence periodic laundering is necessary.

15.2   Basic sizing

Equation (15.12) shows the basis of sizing: since the air flow rate (Q) is known, the cross-sectional area of the duct (A) can be established if a suitable mean velocity (V) is chosen.

$$Q = AV \tag{15.12}$$

There is no general agreement on duct system classification but low velocity systems are often regarded as those in which the maximum mean velocity is less than 10 m s^{-1}, medium velocity as having maximum mean velocities between 10 and 15 m s^{-1} and high velocity systems as those with maximum mean velocities not exceeding 20 m s^{-1}. As a general principle, velocities should be kept as low as is reasonably possible and 20 m s^{-1} should never be exceeded. Duct systems are classified in HVCA (1998) by pressure as well as velocity, low pressure being up to +500 Pa or down to −500 Pa, medium pressure up to +1000 Pa or down to −750 Pa, and high pressure up to +2500 Pa or down to −750 Pa. Large negative pressures are undesirable in comfort air conditioning systems because extract ducts with large sub-atmospheric pressures are not stable, tending to collapse if deformed. On the other hand, large sub-atmospheric pressures are essential for industrial exhaust systems and systems used for pneumatic conveying, but such ducts are constructed with this in mind.

Three methods of sizing are used: velocity, equal pressure drop and static regain.

(a)   Velocity method

A suitable mean velocity is chosen (CIBSE 1986a) for a section of the system that is considered to be critical, usually in terms of noise. This is often the main duct, after the fan discharge. Equation (15.12) is then used for sizing this section.

The volumetric airflow rate handled by subsequent sections of the supply main reduces as air is fed through the branches and reference to Figure 15.3 shows that if the velocity is kept constant the rate of pressure drop increases as the volumetric flowrate falls. This is

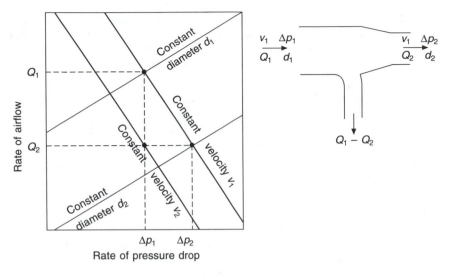

Fig. 15.3   The change of pressure drop rate at constant velocity.

undesirable because the total pressure loss through the system becomes too large and the risk of regenerated noise increases. Hence it is necessary to reduce the mean air velocity in the main as air is fed through branches. Here lies the difficulty of using the method: it is not always easy to obtain recommended mean air velocities for the downstream main duct when it is handling less airflow. One approach is to consider the ducting at the end of the system that delivers air into the last air distribution terminal (supply grille, supply diffuser, variable air volume supply device etc.). Manufacturers invariably quote recommended velocities for good air distribution without the generation of undesirable noise. A conservative interpretation of these recommendations should be adopted. Common sense and engineering prudence may then be used to proportion velocities between the fan outlet and the air distribution terminals. Manufacturers' maximum velocities should never be exceeded and supply air terminals should never be selected to operate at the extreme bounds of their quoted volumetric ranges.

EXAMPLE 15.1

Size the ducting shown in Figure 15.4, given that the velocity in the main after the fan is to be 7.5 m s^{-1} and the velocity in any branch duct is to be 3.5 m s^{-1}.

Answer

Refer to Figure 15.4. Common sense suggests that the velocity in section *BC* should be the mean of 7.5 and 3.5, namely, 5.5 m s^{-1}. Hence the following table is compiled, using

Section	Q m^3 s^{-1}	V m s^{-1}	A m^2	d mm	Δp Pa m^{-1}
AB	1.5	7.5	0.2	505	1.15
BC	1.0	5.5	0.182	481	0.67
CD'	0.5	3.5	0.143	427	0.34

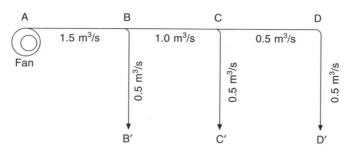

Fig. 15.4 Duct layout diagram for example 15.1.

equation (15.12) and making reference to the CIBSE duct sizing chart for the relevant pressure drop rates.

The duct run having the greatest total pressure drop is termed the index run. This is usually, but not always, also the longest duct run. It is evident that the pressure at C is enough to deliver the design airflow rate to the index terminal at D' but is more than enough to deliver the design rate through the branch to C'. A similar consideration applies at B. This is dealt with during commissioning when branch dampers at B and C are adjusted to absorb the surplus pressure and ensure the correct branch airflow. Sometimes it is possible to reduce the sizes of the branch ducts so that they absorb the surplus pressure without the need for dampering.

The correct and well-established method to be adopted for adjusting branch dampers is termed proportional balancing, developed by Harrison *et al.* (1965). A summary of the method is as follows. The main system damper is partly closed, the setting of the safety overload cut-outs on the supply fan motor starter are checked and the supply fan is switched on. After establishing the location of the index supply air terminal its balancing damper is fully opened and the airflow rate measured. In general this will be less than the design intention. Working backwards to the fan, successive air terminals are balanced to deliver the same percentage of the design airflow rate as that measured at the index terminal. After completing this and carrying out various system checks, the main system damper is opened, the total airflow rate measured and the fan speed adjusted to give the design duty. A similar technique is applied to extract systems. The detailed procedure given in the CIBSE Commissioning Code (1986b) must be followed.

(b) Equal pressure drop method

This method is commonly adopted for sizing low velocity systems. The following alternative approaches which, in principle amount to the same method, may be used:

(i)   Pick a maximum mean velocity for a critical section of duct, size the duct using equation (15.12), note the pressure drop rate by means of a duct sizing chart and size the rest of the system on this rate.
(ii) Pick a pressure drop rate and limiting maximum mean velocity that experience has proved to be suitable and size the whole system on this pressure drop rate, subject to the limiting velocity, and using a duct sizing chart.

The second method has been commonly used with a pressure drop rate of 0.8 Pa m^{-1} and a limiting velocity of about 8.5 or 9.0 m s^{-1}. ASHRAE (1997a) quotes a range of limiting velocities from 9 to 20 m s^{-1}, leaving the exact choice to the designer.

EXAMPLE 15.2

Size the duct system shown in Figure 15.4, using a constant pressure drop rate of 0.8 Pa m^{-1}, subject to a velocity limit of 6.0 m s^{-1}.

Answer

Refer to Figure 15.4 and to a duct sizing chart (CIBSE (1986a)) or better. Compile the following table:

Section	AB	BC	CD′	BB′	CC′
Q m^3 s^{-1}	1.5	1.0	0.5	0.5	0.5
Δp Pa m^{-1}	0.65	0.8	0.8	0.8	0.8
d mm	565	465	360	360	360
V m s^{-1}	6.0	5.9	5.1	5.1	5.1

(c)   Static regain method

When air passes through an expanding duct section its velocity reduces. The kinetic energy of the airstream, represented by its velocity pressure (see section 15.5), also reduces and, in the absence of losses by friction or turbulence, is transferred to the static pressure of the airstream, which rises accordingly. This increased static pressure is then available to offset friction and other losses in the downstream duct. The method is best applied to medium and high velocity systems where ample kinetic energy is initially available but, as the velocity is progressively reduced, less energy is available for transfer and the ducts tend to become too large. For these reasons it is usually only possible to size parts of systems. Refer to section 15.10(b), where static regain is dealt with.

15.3   Conversion from circular to rectangular section

An airstream having a circular section is the most efficient way of containing the airflow because the section of the jet then has the minimum ratio of perimeter to area, P/A. For this reason, the best way of ducting airflow is in ducts of circular section. If rectangular sections are used the corners of the ducts contain turbulence and represent a loss of energy. This is made worse if the ducts are of very large aspect ratio. The CIBSE (1986a) and HVCA (1998) recommend that the aspect ratio should not exceed 4 but engineering prudence suggests that the maximum should be 3. Similar considerations apply to the use of flat oval duct. Bearing in mind the above restriction on aspect ratio, it is not always good practice to regard a rectangular duct section as the best way of using building space. It may be better to use multiple spirally wound ducts of circular section.

Ducts should be sized initially to give circular sections with diameters read from a duct sizing chart to the best accuracy possible. After this, the sizes may be converted to the standard sizes of circular duct commercially available using CIBSE (1986a) data, or to rectangular or flat oval dimensions, if necessary.

The conversion from circular to rectangular section should be done so that the rectangular duct has the same surface roughness and conveys the same volumetric airflow rate with the same rate of pressure drop, as does the circular duct. An alternative approach, of little value in commercial air conditioning, is to convert so that the rectangular duct has the same surface roughness, mean velocity and pressure drop rate but carries a different volumetric airflow rate. The starting basis is the Fanning formula (equation (15.2)) and the derivation for the useful conversion, giving equal volumetric airflow rate and pressure drop is as follows:

The rectangular equivalent having the same volumetric airflow rate and the same rate of pressure drop:

$$H = \frac{4flV^2}{2gd} \tag{15.2}$$

Multiplying by ρg to convert to pressure

$$\Delta p_t = \frac{2\rho flV^2}{d} \tag{15.13}$$

The mean hydraulic diameter, D_h, is defined by

$$D_h = A/P \tag{15.14}$$

where A is the internal cross-sectional area and P is the internal perimeter and, for the case of circular ducts this is $(\pi d^2/4)/(\pi d)$ which is $d/4$. Thus d equals $4A/P$ and a substitution can be made in equation (15.13):

$$\Delta p_t = \frac{\rho flV^2}{2} \cdot \frac{P}{A} \text{ and, since } V = Q/A, \text{ we have}$$

$$Q = \sqrt{\frac{2\Delta p_t}{fl}} \sqrt{\frac{A^3}{P}}$$

For each duct, circular and rectangular, it is stipulated that the volumetric airflow rate, Q, must be the same and, since the pressure drop rates and surface roughnesses are also required to be equal Δp_t, f, ρ and l are also the same. Hence conversion is achieved by equating $\sqrt{A^3/P}$ for the circular and rectangular sections:

$$\sqrt{\frac{\pi^3 d^6}{4^3 \pi d}} = \sqrt{\frac{(ab)^3}{2(a+b)}}$$

where a and b are the sides of the rectangular duct. Re-arranging the equation yields the solution required:

$$d = 1.265 \sqrt[5]{\frac{(ab)^3}{(a+b)}} \tag{15.15}$$

If the surface roughnesses of the circular and rectangular ducts are not the same their friction factors are different. Denoting these by f_c and f_r, respectively, they may be incorporated in equation (15.15) to yield

$$d = 1.265 \sqrt[5]{\frac{f_c(ab)^3}{f_r(a+b)}} \tag{15.16}$$

A similar approach yields the following equation for flat oval ducts having overall dimensions of $a \times b$, the same surface roughness and conveying the same volumetric airflow rate with the same rate of pressure drop:

$$d = 1.453 \sqrt[5]{\frac{[ab + a^3(\pi/4 - 1]^3 f_{\mathrm{c}}}{[\pi a + 2(b - a)]f_{\mathrm{r}}}} \qquad (15.17)$$

Equations are seldom used for conversion. Tables are published by CIBSE (1986a) which are commonly used.

In terms of the smoothest and quietest likely airflow, other factors being equal, the preferred duct section shapes are as follows, in order of preference:

(1) spirally-wound circular
(2) circular duct rolled from flat sheet
(3) square
(4) rectangular or flat oval with aspect ratios not exceeding 3.

Rectangular ductwork should never be used with high velocity systems; it is expensive and noisy.

15.4 Energy changes in a duct system

When air flows through a system of duct and plant, the prime source of energy to make good the losses incurred by friction and turbulence is the total pressure of the air stream, defined by a simplified form of Bernoulli's theorem:

$$p_{\mathrm{t}} = p_{\mathrm{s}} + p_{\mathrm{v}} \qquad (15.18)$$

where

p_{t} = total pressure in Pa
p_{s} = static pressure in Pa
p_{v} = velocity (or dynamic) pressure in Pa

The velocity pressure corresponds to the kinetic energy of the airstream and the static pressure corresponds to its potential energy.

For the airstream to flow through the system of plant and ductwork, in spite of the losses from friction and turbulence, the total energy on its upstream side must exceed the total energy on its downstream side. If a pressure, p, propels a small quantity of air, δq, through a duct in a short time, δt, against a frictional resistance equal and opposite to p, then the rate at which energy must be fed into the system to continue the flow is $p(\delta q/\delta t)$. This is termed the air power, w_{a}, and is delivered to the airstream by the fan impeller.

air power = force × distance per unit time
 = pressure in N m^{-2} × area in m^2 × velocity in m s^{-1}
 = fan total pressure in N m^{-2} × volumetric airflow rate in m^3 s^{-1}

$$w_{\mathrm{a}} = p_{\mathrm{tF}} Q \qquad (15.19)$$

where

p_{tF} = fan total pressure in Pa or kPa (see equation (15.21))
Q = volumetric airflow rate in m^3 s^{-1}

In its passage through a fan the airstream suffers various losses. These are similar to those occurring in a centrifugal compressor and are illustrated in Figure 12.13. In addition, there are bearing losses. It follows that the power input to the fan shaft must exceed the output from the impeller to the airstream. The ratio of impeller output to shaft input is termed the total fan efficiency, η_t and this provides a definition of fan power:

$$w_f = w_a/\eta_t \tag{15.20}$$

The air power delivered to the airstream provides for the sum of the following:

The acceleration of outside air from rest to the velocity in the air intake, the frictional resistance of the air inlet louvres, the energy losses incurred by turbulence formed in the vena-contracta at entry, the frictional resistance by each item of plant, frictional resistance in the ducts and duct fittings, the frictional resistance of the index grille, losses incurred by the presence of turbulence anywhere in the system and the kinetic energy loss from the system (represented by the mass of moving air delivered from the index grille).

The energy loss by friction and turbulence would cause a temperature rise in the airstream if it were not exactly offset by the fall in temperature resulting from the adiabatic expansion accompanying the pressure drop. The only temperature rise occurs at the fan, where adiabatic compression takes place.

The power absorbed by the electric driving motor must exceed the fan power, during steady-state operation, because of the loss in the drive between the fan and the motor and because the efficiency of the motor is less than 100 per cent. There is also the matter of margins, discussed later in section 15.18.

The size of the fan depends on the airflow rate and the type of fan depends on the application, but the speed at which the fan must run and the size of the motor needed to drive the fan depend on the total pressure loss in the system of duct and plant. Hence it is necessary to calculate energy losses in the system. The following principles and definitions relate to such calculations:

(a) $p_t = p_s + p_v$ $\qquad\qquad\qquad\qquad\qquad\qquad\qquad\qquad\qquad\qquad\qquad$ (15.18)

This is a simplification of Bernoulli's theorem, stating that, in an airstream, the total energy of the moving air mass is the sum of the potential and kinetic energies. Energy is the product of an applied force and the distance over which it is acting. Hence, since pressure is the intensity of force per unit area, total pressure may be regarded as energy per unit volume of air flowing. (This is seen if the unit for pressure, N/m^2, has its numerator and denominator multiplied by metres, yielding Nm/m^3, which equals J/m^3.) Similarly, static pressure can be considered as potential energy per unit volume and velocity pressure as kinetic energy per unit volume.

A conclusion drawn from the above is that energy loss through a system corresponds to fan total pressure.

(b)   Energy loss corresponds to a fall of total pressure

It is a corollary of Bernoulli's theorem that a fall in energy should correspond to a fall in total pressure and so, when assessing the energy loss through a system, it is the change in total pressure that must be calculated. On the suction side of a fan the total pressure upstream exceeds that at fan inlet and on the discharge side of the fan the total pressure at fan outlet exceeds that downstream. The fan impeller replaces the energy dissipated, by elevating the total pressure between the fan inlet and the fan outlet.

(c)   Velocity pressure is always positive in the direction of airflow

For a given volumetric airflow rate in a duct of constant cross section, the mean velocity of airflow and the corresponding kinetic energy must be constant. If a loss of energy occurs, because of friction or turbulence, the corresponding fall in total pressure can only appear as an equal fall in static pressure. Thus it is the static pressure of an airstream that is the source of energy for making good losses. If the kinetic energy of the airstream is to be drawn upon then it is first necessary to reduce the velocity by expanding the duct section, in order to convert kinetic energy to potential energy (in the form of static pressure), as described by Bernoulli's theorem.

(d)   Fan total pressure, p_{tF}

This is defined by

$$p_{tF} = p_{to} - p_{ti} \tag{15.21}$$

where p_{to} is the total pressure at fan outlet and p_{ti} the total pressure at fan inlet.

(e)   Fan static pressure, p_{sF}

This is defined by

$$p_{sF} = p_{so} - p_{ti} \tag{15.22}$$

where p_{so} is the static pressure at fan outlet.

(f)   Fan static pressure by virtue of equation (15.18)

Fan static pressure is also defined by

$$p_{sF} = p_{tF} - p_{vo} \tag{15.23}$$

where p_{vo} is a velocity pressure at fan outlet based on a notional mean velocity v_{fo} defined by

$$v_{fo} = Q/A_{fo} \tag{15.24}$$

where Q is the volumetric airflow rate and A_{fo} is the area across the flanges at fan outlet.
   The velocity distribution over the outlet area of a fan is very turbulent and not easy to measure with accuracy. Hence a notional mean velocity at fan outlet, defined by equation (15.24), is determined and the corresponding velocity pressure is added to fan static pressure to define fan total pressure indirectly, by means of equation (15.23). Static pressure may be above or below atmospheric pressure, acting as a bursting or collapsing influence on the system. Hence the zero chosen for the expression of pressure is atmospheric pressure, and static and total pressures are given positive or negative values.

15.5   Velocity (dynamic) pressure

If velocity pressure is regarded as kinetic energy per unit volume then, denoting the mean velocity of airflow by V, mass by m and density by ρ, the following can be derived:

$$p_v = \frac{(0.5\, m\, V^2)}{(\text{volume})} = 0.5\rho V^2 \tag{15.25}$$

If the density of air is taken as 1.20 kg m^{-3} the equation becomes

$$p_v = 0.6V^2 \qquad\qquad (15.26)$$

and the converse is

$$V = 1.291\sqrt{p_v} \qquad\qquad (15.27)$$

15.6 The flow of air into a suction opening

Consider a simple length of ducting attached to the inlet side of a fan and another length of ducting attached to its outlet. Air is accelerated as it approaches the suction opening and, in order to produce this increase in kinetic energy, a negative potential energy has to be set up within the opening. Figure 15.5(a) illustrates that the air, in negotiating the entry to the duct, is compelled to undergo a change of direction (unless it happens to be on the centre-line of the duct) and that this involves the setting up within the duct of a pocket of turbulence which reduces the area of entry available to the air. The reduced area is termed the 'vena-contracta'.

Three immediate conclusions can be drawn from this.

(1) The velocity of airflow through the vena-contracta must be higher than that prevailing in the succeeding downstream length of duct.
(2) The curved paths followed by the air in the eddies within the pocket of turbulence involve the expenditure of energy—in accordance with Newton's first law of motion.
(3) There is a drop in total pressure as the air flows through the open end of a suction duct, because of conclusion (2) above, regardless of the presence of any grille at the opening. If a grille is present then the loss of total pressure will be greater.

An application of Bernoulli's theorem (equation (15.18)) permits the changes of total, static and velocity pressure to be determined as air enters the system. Figure 15.5(a) shows a plot of such pressure changes for a suction opening at the end of a duct, and (b) shows airflow into a 'no-loss' entry. The end of the duct is constructed in such a way that solid material occupies the space normally filled with turbulence and prevents the formation of a vena-contracta. Virtually no losses occur, and the static suction set up just within the open end of the duct, where it has attained its proper diameter, is numerically equal to the velocity pressure, but opposite in sign.

(No-loss entry pieces of this kind provide a very reliable and accurate method of measuring the rate of airflow through a duct. It is only necessary to take a careful measurement of the static pressure on the section where the taper has just ceased and to convert this to velocity, by means of equation (15.27). Such a method is usually suitable for laboratory uses only.)

15.7 The coefficient of entry (CE)

For normal entry-pieces not all the static suction is used to accelerate the air to the velocity prevailing in the downstream duct. Some of the potential energy of the suction set up is wasted in offsetting losses due to turbulence and friction. It is customary to express these losses in terms of the steady velocity pressure in the downstream duct, after any vena-contracta. Thus, the loss due to eddies formed at the vena-contracta can be written as ζ multiplied by this velocity pressure, where ζ is obviously less than unity.

The fact that there is a reduced area. A', available for airflow at the vena-contracta gives rise to the concept of a coefficient of area, defined by

$$C_A = A'/A$$

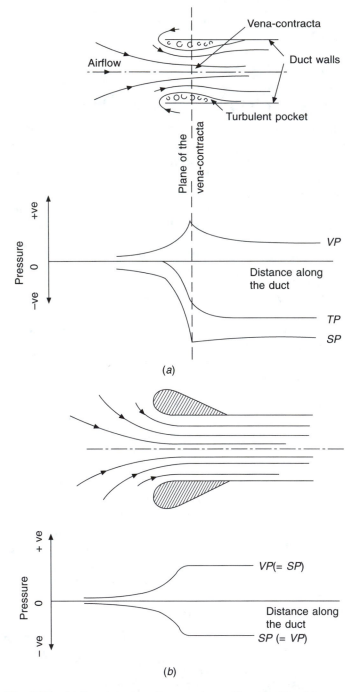

Fig. 15.5   Airflow into a suction opening, with and without loss.

where A is the cross sectional area of the duct.

There is also the concept of a coefficient of velocity. This arises from the fact that the velocity at the vena-contracta is less than that which would be attained if all the static pressure were to be converted into velocity. There is some friction between the air flowing through the vena-contracta and the annular pocket of turbulence which surrounds it. This is consequent on the energy transfer needed to maintain the eddies and whorls within the pocket. The coefficient of velocity is defined by

$$C_V = \frac{\text{actual velocity at the vena-contracta}}{\left(\begin{array}{c}\text{velocity which would be attained at the same}\\\text{section in the absence of losses}\end{array}\right)}$$

Since the quantity flowing is the product of velocity and area it can be seen that a flow or entry coefficient C_E, may be inferred as the product of the coefficients of velocity and area

$$C_E = C_V \times C_A$$

Thus, Q, the actual rate of flow of air entering the system, may be expressed by

$$Q = C_E \times Q'$$

where Q' is the theoretical rate which would flow if no loss occurred. It may also be written as

$$Q = C_E(A \times V')$$

But $V' = 1.291\sqrt{p_v}$, the theoretical maximum possible velocity, where p_v is equivalent to the static suction set up in the plane of the vena-contracta. Thus,

$$Q = C_E(A \times 1.291\sqrt{p_v})$$

15.8   The discharge of air from a duct system

Air in the plane of the open end of the discharge duct must be at virtually atmospheric pressure—since there is no longer any resisting force to prevent the equalisation of pressure. We can therefore say that the potential energy of the air leaving the system through an open end is zero. The kinetic energy is not zero however. There is an outflow of mass from the system. Applying Bernoulli's theorem, it is seen that the total energy of the airstream leaving the ducting must be equal to its kinetic energy or velocity pressure.

If a grille or diffuser is placed over the open end of the duct, the total pressure on its upstream side must be greater than that on its downstream side by an amount equal to the frictional loss incurred by the flow of air through the grille. If any change of velocity that may take place as the air flows between the bars of the grille is ignored, the velocity pressure upstream must equal the velocity pressure downstream. It follows that the static pressure of the upstream air is used to overcome the frictional loss through the grille. This is illustrated in Figure 15.6.

Denoting the upstream side of the grille by the subscript 1 and the downstream side by subscript 2, and assuming face velocities at the grille, we can write

$$p_{t1} = p_{t2} + \text{frictional loss past the grille}$$
$$p_{v1} + p_{s1} = p_{v2} + p_{s2} + \text{frictional loss past the grille}$$

but

$$p_{v1} = p_{v2}$$

hence

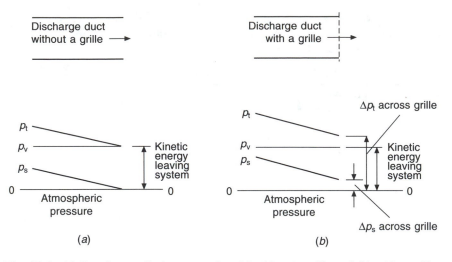

Fig. 15.6　Airflow from a discharge opening, (a) without a grille and (b) with a grille.

$$p_{s1} - p_{s2} = \text{frictional loss past the grille}$$

Since p_{s2} is virtually the same as atmospheric pressure—the zero datum—this becomes

$$p_{s1} = \text{frictional loss past the grille}$$

To sum up: the fan has to make good the energy loss at the end of the system incurred by the kinetic energy of the airstream leaving the system *and* the friction past the last grille.

15.9　Airflow through a simple duct system

Figure 15.7 shows a simple straight duct on the suction side of a fan, followed by a similar duct of smaller cross-section on its discharge side. There is a grille over both the inlet and outlet openings of the duct.

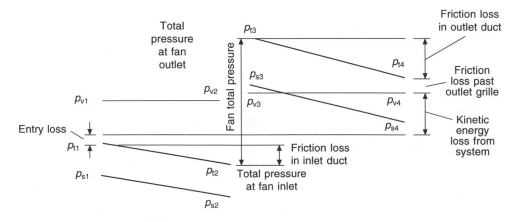

Fig. 15.7　Airflow through a simple, notional duct system.

Air entering the system is accelerated from rest outside the duct to a velocity V_1 prevailing within the duct, once the vena-contracta has been passed and the flow has settled down to normal. If, for convenience of illustration, the physical presence of the vena-contracta is ignored but the loss which it causes is included with the frictional loss past the inlet grille, the entry loss can be shown as occurring in the plane of entry. We can then write

$$\text{Entry loss} = 0 - p_{t1} = -p_{t1}$$
$$\text{static suction} = -p_{t1} - p_{v1} = -p_{s1}$$

The static depression immediately within the inlet grille is numerically greater than the velocity pressure by an amount equal to the energy loss past the inlet grille.

The loss due to friction along the duct between plane 1 and plane 2 equals $p_{t1} - p_{t2}$. The static depression is clearly $p_{t2} - p_{v2}$. It is also clear that the total pressure at fan inlet, p_{t2}, represents the loss of energy through the system up to that point, and that it comprises the frictional loss past the inlet grille, the loss past the vena-contracta and the loss due to friction in the duct itself.

On the discharge side of the system it is easier, for purposes of illustration, to start at the discharge grille. The change of static pressure past this grille equals $p_{s4} - 0$ but the change of total pressure across the grille equals $p_{t4} - 0$, or the friction plus the loss of kinetic energy. Between the plane 3 and 4, the loss of energy due to friction equals $p_{t3} - p_{t4}$, in the direction of airflow. The total energy upstream must exceed that downstream if airflow is to take place. Hence, p_{t3} must be greater than p_{t4} and so also must p_{t1} be greater than p_{t2}, in a similar way. But p_{t3} is the total energy at fan outlet and p_{t2} is the total energy at fan inlet. Consequently, the difference between these two quantities, p_{t3} and p_{t2}, must be the energy supplied to the system by the fan. This statement fits in with equation (15.21) and we can write

$$p_{tF} = p_{t3} - p_{t2} = |p_{t3}| + |p_{t2}|$$

numerically speaking, since p_{t2} has a negative sign. Thus, the fan total pressure can never be negative.

To sum up:

$$p_{tF} = [\text{friction past the inlet grille} + \text{energy loss at the vena-contracta}]$$
$$+ [\text{friction loss in the inlet duct}]$$
$$+ [\text{friction loss past the outlet grille} + \text{kinetic energy lost from the system}]$$
$$+ [\text{friction loss in the outlet duct}]$$
$$= (0 - p_{t1}) + (p_{t1} - p_{t2}) + (p_{t4} - 0) + (p_{t3} - p_{t4})$$
$$= (p_{t3} - p_{t2})$$

15.10 Airflow through transition pieces

It is customary to express the total pressure loss, Δp_t, through a duct fitting by the product of a coefficient, ζ, and a velocity pressure, p_v. The velocity used is the actual mean velocity, V, determined by the ratio Q/A, where Q is the volumetric airflow rate and A is the cross sectional area of the fitting. For the case of transition pieces in which the section is expanding or contracting, the smaller section, where the velocity is greater, is generally used for A. In the following examples, ζ values from the CIBSE (1986a) have been used but coefficients from other sources yield approximately similar pressure drops.

(a)   Expanders

It is possible to derive an expression from first principles according to Prandtl (1953) and Lewitt (1948) for the total pressure loss when an airstream flows through an abruptly expanding duct section (as in Figure 15.8):

$$\Delta p_t = {}^1\!/_2 \rho (V_1 - V_2)^2 \qquad\qquad (15.28)$$

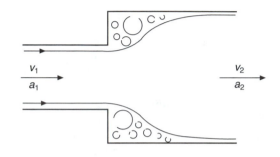

Fig. 15.8   Airflow through an abrupt expansion.

Since $V_2 = V_1(a_1/a_2)$ this can be re-written as

$$\Delta p_t = {}^1\!/_2 \rho V_1^2 (1 - a_1/a_2)^2$$

The term in the brackets is the coefficient, ζ, used to evaluate the total pressure loss, since ${}^1\!/_2 \rho V_1^2$ is the upstream velocity pressure. Thus, for abruptly expanding duct sections:

$$\zeta = (1 - a_1/a_2)^2 \qquad\qquad (15.29)$$

More recent work by Idelchik *et al.* (1986) and ASHRAE (1993) expresses the loss coefficient in general terms, referred to the included angle of the transition piece, θ, from 0° to 90° to cover the case of expanding sections and from 90° to 0° for reducing sections. It is to be noted that the coefficients for a 90° abrupt expander are somewhat greater than those predicted by equation (15.29). At the time of writing, the CIBSE (1986a) use coefficients for abrupt expanders in agreement with equation (15.29) but, for gradual expansion pieces, this is multiplied by a further coefficient to express the total pressure loss as:

$$\Delta p_t = \zeta_a \zeta_g p_v \qquad\qquad (15.30)$$

where ζ_a is the coefficient for an abrupt expander, ζ_g the coefficient for a gradual expander and p_v is the velocity pressure in the upstream, smaller duct section. ASHRAE (1997a) use a single coefficient.

If air flows through an expansion piece (Figure 15.9), without friction or turbulence, there is no loss and p_{t1} equals p_{t2}. Hence, by equation (15.18):

$$p_{s1} + p_{v1} = p_{s2} + p_{v2} \quad \text{and so}$$
$$p_{v1} - p_{v2} = p_{s2} - p_{s1}$$

This states that there is full conversion of kinetic energy (velocity pressure) to potential energy (static pressure). In other words, 100 per cent static regain occurs.

In the actual case (Figure 15.9) p_{t2} is less than p_{t1} and the loss is:

$$\Delta p_t = p_{t1} - p_{t2}$$

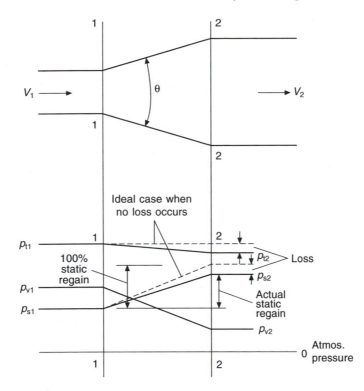

Fig. 15.9 Airflow through a gradual expander, with and without loss.

A full conversion of velocity pressure to static pressure does not take place and the static regain is given by the fall in velocity pressure minus the total pressure loss:

Static regain $= p_{s2} - p_{s1}$

$$p_{s2} - p_{s1} = p_{v1} - p_{v2} - \Delta p_t \qquad (15.31)$$

EXAMPLE 15.3

Calculate the total pressure loss and the static regain through a symmetrical, gradual expansion piece having an initial velocity of 20 m s^{-1} and a final velocity of 10 m s^{-1}, using (a) CIBSE (1986a) data and (b) ASHRAE (1993) data.

Answer

$$a_1/a_2 = V_2/V_1 = 10/20 = 0.5$$

(*a*) From CIBSE (1986a) data, $\zeta_a = 0.25$ and $\zeta_g = 0.8$.

$$p_{v1} = 0.6 \times 20^2 = 240 \text{ Pa by equation (15.26)}$$

Hence $\Delta p_t = 0.25 \times 0.8 \times 240 = 48$ Pa by equation (15.30) and, by equation (15.31), static regain is given by

$$p_{s2} - p_{s1} = 0.6 \times 20^2 - 0.6 \times 10^2 - 48 = 132 \text{ Pa}$$

One hundred per cent static regain would be $0.6 \times 20^2 - 0.6 \times 10^2$, or 180 Pa. Hence the percentage static regain is only $132 \times 100/180$, namely, 73 per cent.

(b) From ASHRAE (1993) data, $\zeta_g = 0.24$.
Hence

$$\Delta p_t = 0.24 \times 240 = 58 \text{ Pa}$$

and the static regain is

$$p_{s2} - p_{s1} = 0.6 \times 20^2 - 0.6 \times 10^2 - 58 = 122 \text{ Pa}$$

which is 68 per cent of the maximum possible.

Static regain can also occur in a main duct when its section remains constant as air is fed from it through a branch, or a supply grille. This is because the velocity in the main reduces when the volumetric airflow rate in it falls, after a branch or a grille.

(b) Static regain duct sizing

It is possible to use static regain methods according to Carrier (1960), Shataloff (1966) and Chun Lun (1983) for sizing but it is impossible to be precise about the loss incurred as air flows through an expander, because of variations in duct construction and the lack of exact data regarding loss coefficients. However the following offers an approximate solution.

Refer to Figure 15.10. The exact form of the expanding section, 0—1, is not defined except by assuming that the fraction of static regain occurring is denoted by r. The further

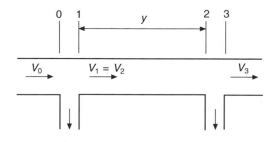

Fig. 15.10 Diagram of a duct system for the derivation of an approximate equation for duct sizing by static regain. Also refers to Example 15.4.

assumption is made that the rate of pressure drop in the duct section to be sized, 1—2, is proportional to the square of the mean velocity, $V_1(= V_2)$. The static regain across the section of duct, 0—1, is to be equated to the loss of total pressure in the downstream main duct from 1 to 2, in order to determine the mean velocity ($V_1 = V_2$) in the duct and so size it.

$$\text{Static regain from 0 to 1} = r(p_{v0} - p_{v1})$$

Total pressure loss in the main from 1 to 2 is given by $\Delta p_{t12} = y\Delta p_{t0}(V_1^2/V_0^2)$, where y is the duct length in m and Δp_{t0} is the pressure drop rate in the duct approaching section 0.

Equating the static regain to the downstream total pressure loss:

$$r(0.6V_0^2 - 0.6V_1^2) = y\Delta p_{t0}(V_1^2/V_0^2)$$
$$r(0.6V_0^4 - 0.6V_1^2V_0^2) = y\Delta p_{t0}V_1^2$$
$$V_1^2(y\Delta p_{t0} + 0.6rV_0^2) = 0.6rV_0^4$$

$$V_1^2 = (0.6rV_0^4)/(y\Delta p_{t0} + 0.6rV_0^2)$$

$$V_1 = V_0^2 \sqrt{\frac{0.6r}{(y\Delta p_{t0} + 0.6rV_0^2)}} \qquad (15.32)$$

EXAMPLE 15.4

Refer to Figure 15.10. The volumetric flow rate in the main duct before section 0 is 1.0 m^3 s^{-1} and 0.1 m^3 s^{-1} is fed through the first branch. The mean velocity in the main duct leading to the branch is 12 m s^{-1}, the diameter is 326 mm and the pressure drop rate CIBSE (1986a) is 5.5 Pa m^{-1}. Size the main duct, between sections 1 and 2, assuming that 70 per cent static regain occurs from 0 to 1. Consider two cases: (*a*) when the length of the duct 1–2 is 10 m and, (*b*), when it is 3 m.

Answer

(*a*) By equation (15.32)

$$V_1 = 12^2 \sqrt{\frac{0.6 \times 0.7}{10 \times 5.5 + 0.6 \times 0.7 \times 12^2}} = 8.68 \text{ m s}^{-1}$$

By equation (15.12)

$$A = 0.9/8.68 = 0.1037 \text{ m}^2$$

whence

$$d = 363 \text{ mm}$$

Check

$$\text{Static regain} = 0.7(0.6 \times 12^2 - 0.6 \times 8.68^2) = 29 \text{ Pa}$$

By CIBSE (1986a) the pressure drop rate for a duct of 363 mm diameter handling 0.9 m^3 s^{-1} is about 2.3 Pa m^{-1}. Hence the pressure drop is approximately 23 Pa over a length of 10 m.

(*b*) By equation (15.32)

$$V_1 = 12^2 \sqrt{\frac{0.6 \times 0.7}{(3 \times 5.5 + 0.6 \times 0.7 \times 12^2)}} = 10.64 \text{ m s}^{-1}$$

By equation (15.12)

$$A = 0.9/10.64 = 0.084\ 59$$

whence

$$d = 328 \text{ mm}$$

Check

$$\text{Static regain} = 0.7(0.6 \times 12^2 - 0.6 \times 10.64^2) = 13 \text{ Pa}$$

By CIBSE (1986a) the pressure drop rate for a 328 mm diameter duct handling 0.9 m^3 s^{-1} is about 3.3 Pa m^{-1}. Hence for a length of 3 m the total pressure drop is about 10 Pa.

Reading a pressure drop rate from the duct sizing chart in CIBSE (1986a) is not very accurate and therefore the above checks can be regarded as not unreasonable.

The difficulty when using equation (15.32) is in deciding on the value of the static regain factor *r*. It is suggested that the designer uses ASHRAE (1993) (which contains a large selection of duct transition pieces and branches) and chooses the type of branch that most closely resembles the actual fitting likely to be used. The coefficient given will then enable the loss to be calculated and a value for the static regain fraction, *r*, determined.

(c) Reducers

Airflow through an abrupt reducer is illustrated in Figure 15.11. The major source of loss is the pocket of turbulence that forms downstream is the narrower duct section, where the higher velocity prevails. Equation (15.28) is fundamental and can be used here to refer to the losses incurred by expansion from the vena-contracta at section *c—c*, to section 2—2, in Figure 15.11:

$$\Delta p_t = \tfrac{1}{2}\rho(V_c - V_2)^2$$

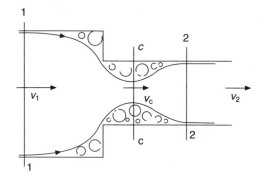

Fig. 15.11   Airflow through an abrupt reducer.

Since $V_c = V_2/C_A$ this can be rewritten as

$$\Delta p_t = {}^1\!/_2 \, \rho V_2^2 [(1/C_A) - 1]^2 \tag{15.33}$$

There is an insignificant error because the upstream loss has been ignored.

According to Lewitt (1942), experiment shows that, for waterflow, C_A is about 0.62 which yields 0.376 for the value of $[(1/C_A) - 1]^2$, the loss coefficient that multiplies the velocity pressure to give the total pressure loss. The CIBSE (1986a) quotes loss factors in terms of the ratio of the downstream to upstream areas, A_2/A_1, and these are used to multiply the downstream velocity pressure, p_{v2}. The same reference gives loss factors in terms of the included angle only, for the case of gradually reducing fittings and, unlike the procedure for expansion pieces, this is not multiplied by the coefficient for the corresponding abrupt fitting. Thus

$$\Delta p_t = \zeta_a p_{v2} \tag{15.34}$$

and

$$\Delta p_t = \zeta_g p_{v2} \tag{15.35}$$

where p_{v2} is the velocity in the downstream section of the reducer.

EXAMPLE 15.5

Calculate the total pressure loss through a symmetrical, gradual reducer, of circular section, having an included angle of 30° if the initial velocity is 10 m s^{-1} and the final velocity is 20 m s^{-1} using CIBSE (1986a) data.

Answer

For an included angle of 30° and a symmetrical section

$\zeta_g = 0.02$. Hence, by equation (15.35)
$\Delta p_t = 0.02 \times 0.6 \times 20^2 = 5$ Pa

Figure 15.12 illustrates the case of a gradual reducer. If there is no loss, the fall in static pressure is used to accelerate the air from V_1 to V_2 and exactly equals the increase in velocity pressure. In the real case, when there is a loss, the static pressure must provide the energy to do two things: to accelerate the air and to make good the loss.

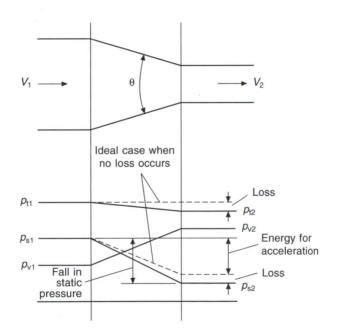

Fig. 15.12 Airflow through a gradual reducer, with and without loss.

15.11 Airflow around bends

As with other pieces of ducting, the loss incurred when air flows around a bend is expressed as a fraction of the mean velocity pressure in the bend, provided that the section is constant. If the section varies, precise calculations are not possible but, with a bit of common sense, a good approximation can frequently be obtained.

For a normal bend of constant cross-sectional area, the loss depends on three structural properties:

(i)   The curvature of the throat.
(ii)   The shape of the section.
(iii)   The angle through which the airstream is turned.

It is customary to express the curvature of the throat either in terms of the ratio of the throat radius R_t to the dimension W, parallel to the radius, or, in terms of the ratio of the centre-line radius R_c to the dimension W. Figure 15.13(*a*) illustrates these two methods. The mode using the centre-line radius is the more common and the one adopted by the CIBSE. A very large value of R_c/W implies that the air is only very gradually turned and that, turbulence having little opportunity to form, the loss is small. It is evident, however, that not only will skin friction play an increasing part if the bend is excessively gradual, but that the bend will be expensive to make and unsightly in appearance, occupying as it does a very large amount of space. A good practical value for R_c/W is 1.0.

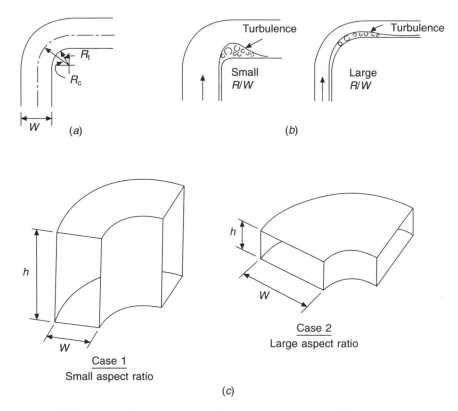

Fig. 15.13   (a) and (b) Airflow around a bend. (c) Bends of different aspect ratios.

The loss through a bend is approximately proportional to the angle turned through, with bends of 45° having ζ-values of 0.6 times those of similar 90° bends. When a pair of 90° bends with smoothly curved throats forms an off-set (Figure 15.14) the ζ-value for the combination is about 1.8 times that of a single 90° bend. Tsal (1989) and CIBSE (1986a) provide comprehensive information.

The shape of the section, or the aspect ratio, has an effect which is shown by Figure 15.13(*c*). Case 1 illustrates a duct where the aspect ratio, W/h, is small and the curvature

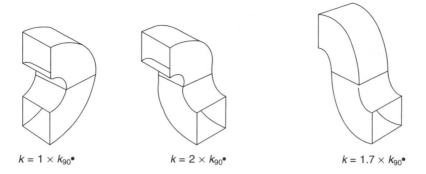

$k = 1 \times k_{90^\circ}$ $k = 2 \times k_{90^\circ}$ $k = 1.7 \times k_{90^\circ}$

Fig. 15.14 Duct bends in series.

of the throat is also small. Case 2 shows the reverse situation: W/h is large, the curvature of the throat is great, and the loss is comparatively large.

There are two methods of minimizing the energy loss round a bend: the use of splitters or the use of turning vanes.

Splitters are only of value when the aspect ratio (W/h) is large and R_t/W is small. They divide the duct into several sub-sections, reducing the turbulence, the noise and the pressure drop. Extensive information on the loss through bends fitted with splitters is provided in ASHRAE (1993) but none is given in CIBSE (1986a). It is usual to arrange the positioning of the splitters so that they cluster near to the throat of the bend. Experiment suggests that it is the curve ratio, defined as throat radius/heel radius, which determines the energy loss, rather than the aspect ratio directly. The consequence is that the splitters should be arranged to produce a number of sub-sections of the duct, each having the same curve ratio.

An expression for the curve ratio, C, in terms of the number of splitters, n, the throat radius, R_0, and the heel radius, R_{n+1}, can be obtained. The following statements hold good for the bend shown in Figure 15.15:

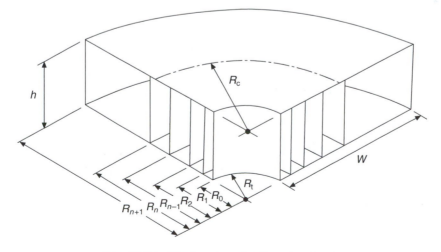

Fig. 15.15 The location of splitters in a bend.

$$\frac{R_0}{R_{n+1}} = K = \frac{\text{throat radius of the bend without splitters}}{\text{heel radius of the bend without splitters}}$$

$$C = \frac{R_0}{R_1} = \frac{R_1}{R_2} = \text{etc.} \dots = \frac{R_n}{R_{n+1}} \tag{15.36}$$

Thus $R_1 = R_0/C$, $R_2 = R_1/C = R_0/C^2$, etc. .... $R_{n+1} = R_0/C^{n+1}$

Therefore

$$C = \left(\frac{R_0}{R_{n+1}}\right)^{1/(n+1)} = K^{1/(n+1)} \tag{15.37}$$

Note also that

$$\frac{R_c}{W} = \frac{0.5(R_0 + R_{n+1})}{(R_{n+1} - R_0)}$$

Dividing above and below by R_{n+1}, this becomes

$$\frac{R_c}{W} = \frac{0.5(K + 1)}{(1 - K)} \tag{15.38}$$

Applying equations (15.36) to (15.38) shows that a single splitter would be closer to the throat than the centre-line. It is seldom necessary to use more than two splitters. The method is expensive in manufacturing cost.

One of the most effective bends, both from the point of view of accommodating the ductwork neatly and minimising the energy loss, is the mitred bend containing turning vanes. There are two types of turning vane, the simple kind and the so-called aerofoil sort. Of these, the former have an energy loss which is about double that of a conventional bend with R_c/W equal to unity, and the latter a loss somewhat less than the conventional bend. To achieve a fairly low pressure drop, the ratio of the height of the vanes to their spacing should be six, or greater.

Reference should be made to manufacturers' data for the loss through turning vanes.

15.12   Airflow through supply branches

Loss occurs along two paths when air flows through a brach-piece: there is a loss of energy for flow through the main and also for flow through the branch. The exact magnitude of the loss depends on the way in which the branch-piece is constructed but it is, nevertheless, possible to generalise. ASHRAE (1997a) gives extensive information on loss coefficients. The CIBSE (1986a) quotes multiplying factors, referred to the velocity pressure in the downstream main or the branch, as appropriate. Both of these are dependent on the ratio of the downstream to upstream velocities. Figure 15.16 illustrates this: there is a loss of total pressure between points 1 and 2, when airflow through the main is considered, and also a loss of total pressure (not necessarily the same) when air flows from point 1 to point 3, by way of the branch.

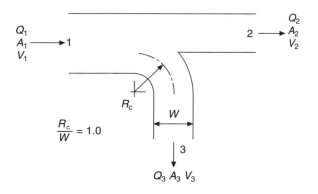

Fig. 15.16   Airflow through a supply branch piece.

EXAMPLE 15.6

Given that, in Figure 15.16, the velocities V_1, V_2 and V_3 are 10, 8 and 6 m s^{-1}, respectively, determine the total pressure loss and the static pressure change through the main (1–2) and the branch (1–3). Assume the section of the branch duct is square.

Answer

The CIBSE (1986a) quotes the following loss coefficients:
For the main,

$$V_2/V_1 = 0.8, \; \zeta_{1-2} = 0.18$$

For the branch,

$$V_3/V_1 = 0.6, \; \zeta_{1-3} = 3.5 \times \text{the factor for the equivalent bend}$$
$$= 3.5 \times 0.23. \text{ See CIBSE (1986a)}$$
$$= 0.805$$

Hence we have

For the main:

$$\Delta p_t = 0.18 \times 0.6 \times 8^2 = 6.9 \text{ Pa}$$
$$\Delta p_s = 0.6 \times 10^2 - 0.6 \times 8^2 - 6.9 = 14.7 \text{ Pa}$$

That is, a static regain has occurred in the main because the fall in velocity pressure exceeded the loss so there is a net increase in the potential energy or the static pressure.

For the branch:

$$\Delta p_t = 0.805 \times 0.6 \times 6^2 = 17.4 \text{ Pa}$$
$$\Delta p_s = 0.6 \times 10^2 - 0.6 \times 6^2 - 17.4 = 21.0 \text{ Pa}$$

and a static regain has taken place here also.

15.13   Flow through suction branches

The distribution of air in comfort conditioning installations is more important on the discharge side of the fan than on its suction side. The positioning and arrangement of extract grilles is not usually very critical, and for this reason supply branch-pieces receive more attention in the literature on the subject. However, in industrial exhaust systems the arrangement of the ducting and the configuration of the suction hoods and openings is of vital importance. For such applications it is thus necessary to know something of the losses incurred when air flows into a suction branch-piece. Loss coefficients for a large variety of suction branch configurations are available in American literature. See Idelchik (1986), ASHRAE (1993b) and ACGIH (1995).

15.14 Calculation of fan total and fan static pressure

The air volume handled determines the size of a fan and the purpose for which it is to be used dictates the type of fan to be chosen. The reason for calculating the fan total pressure is to establish the speed at which the fan should be run, most fans being belt-driven, and the power of the motor that ought to be selected to drive it.

EXAMPLE 15.7

For the system shown in Figure 15.17, size all ducts at a pressure drop rate of 1 Pa m^{-1}, subject to a velocity limit of 8.5 m s^{-1}, and calculate the fan total and fan static pressures. Use the CIBSE (1986a) duct-sizing chart and loss coefficients for fittings and the additional information provided below.

Air inlet at A	square-edged louvres with 80 per cent free area and a mean face velocity of 4 m s^{-1}
Filter at B	pressure drop 90 Pa, face velocity 1.5 m s^{-1}

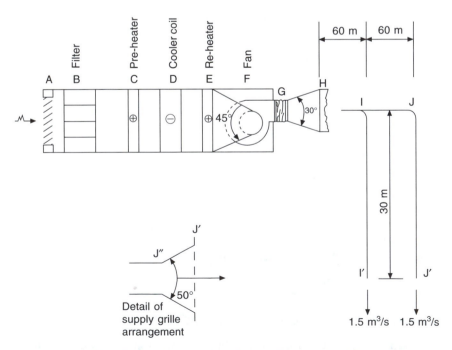

Fig. 15.17 Plant and duct diagram for example 15.7.

Pre-heater at C	pressure drop 50 Pa, face velocity 3 m s^{-1}
Cooler coil at D	pressure drop 150 Pa, face velocity 2 m s^{-1}
Re-heater at E	pressure drop 50 Pa, face velocity 3 m s^{-1}
Fan inlet at F	625 mm diameter
Fan discharge at G	600 mm × 500 mm
Supply grilles at I′ and J′	pressure drop 5 Pa, face velocity 2 m s^{-1}
Bend at J	centre-line radius 375 mm

Convert all circular duct sizes into equivalent rectangular sizes that have the same volumetric airflow rate and the same pressure drop.

Answer

The following information is determined from the CIBSE (1986a) or (1986c).

Duct section	Airflow rate m^3 s^{-1}	Pressure drop rate Pa m^{-1}	Circular		Rectangular	
			Diameter mm	Velocity m s^{-1}	Dimensions h × w mm mm	Velocity m s^{-1}
HI	3.0	1.0	675	8.4	500 × 750	8.0
IJ & II′	1.5	1.0	520	7.1	350 × 750	5.71

The following additional calculations must now be done:

Fan discharge area:	$0.6 \times 0.5 = 0.3$ m^2
Fan discharge velocity:	$3.0/0.3 = 10$ m s^{-1}
Fan inlet area:	$\pi \times 0.625^2/4 = 0.3068$ m^2
Fan inlet velocity:	$3/0.3068 = 9.778$ m s^{-1}

Establishing changes of total pressure is much easier than determining changes of static pressure. If the total pressure is known at a point in the duct system the static pressure at the same point can always be calculated by subtracting from it the velocity pressure, in accordance with Bernoulli's theorem (see equation (15.18)).

1. Outside Air Intake at A

Enough static suction must be developed immediately behind the louvres to provide the source of energy for accelerating the airstream from rest (at an infinite distance from the outside face of the louvres) to the face velocity and for offsetting the loss by friction and turbulence. The CIBSE (1986a) loss coefficient is 1.4 and hence

$$\Delta p_t = 1.4 \times 0.6 \times 4^2 = 13.44 \text{ Pa}$$

which is negative in the direction of airflow.
  Accumulated pressure downstream:

$$p_t = -13.44 \text{ Pa}, \ p_v = +9.6 \text{ Pa}, \ p_s = p_t - p_v = -23.04 \text{ Pa}$$

2.   Expander AB to Filter

The area ratio for this sudden expansion is the reciprocal of the velocity ratio and is 1.5/4.0 = 0.375 whence, the CIBSE loss coefficient is 0.39.

$$\Delta p_t = 0.39 \times 0.6 \times 4^2 = 3.7 \text{ Pa}$$

Accumulated pressures downstream:

$$p_t = -17.14 \text{ Pa}, \ p_v = +1.35 \text{ Pa}, \ p_s = -18.49 \text{ Pa}$$

Note that a static regain has occurred across the expander (from −23.04 Pa to −18.49 Pa) because the velocity has reduced, although this is irrelevant to the calculation of total pressure loss.

3.   Filter at B

$$\Delta p_t = 90 \text{ Pa}$$

Accumulated pressures downstream:

$$p_t = -107.14 \text{ Pa}, \ p_v = +1.35 \text{ Pa}, \ p_s = -108.49 \text{ Pa}$$

4.   Reducer BC to Pre-heater

This is a sudden contraction with an area ratio corresponding to 1.5/3.0, whence the CIBSE loss coefficient is 0.23.

$$\Delta p_t = 0.23 \times 0.6 \times 3^2 = 1.24 \text{ Pa}$$

Accumulated pressures downstream:

$$p_t = -108.38 \text{ Pa}, \ p_v = +5.4 \text{ Pa}, \ p_s = -113.78 \text{ Pa}$$

5.   Pre-heater at C

$$\Delta p_t = 50 \text{ Pa}$$

Accumulated pressures downstream:

$$p_t = -158.38 \text{ Pa}, \ p_v = +5.4 \text{ Pa}, \ p_s = -163.78 \text{ Pa}$$

6.   Expander CD to Cooler Coil

The sudden expansion from a velocity of 3 m s^{-1} to one of 2 m s^{-1} implies an area ratio of 0.67 and the CIBSE loss coefficient is 0.11.

$$\Delta p_t = 0.11 \times 0.6 \times 3^2 = 0.59 \text{ Pa}$$

Accumulated pressures downstream:

$$p_t = -158.97 \text{ Pa}, \ p_v = +2.4 \text{ Pa}, \ p_s = -161.37 \text{ Pa}$$

7.   Cooler Coil at D

$$\Delta p_t = 150 \text{ Pa}$$

Accumulated pressures downstream:

$p_t = -308.97$ Pa, $p_v = +2.4$ Pa, $p_s = -311.37$ Pa

8. *Reducer DE to Re-heater*

For this abrupt reducer the area ratio corresponds to 0.67 for which the loss coefficient is 0.11.

$$\Delta p_t = 0.11 \times 0.6 \times 3^2 = 0.59 \text{ Pa}$$

Accumulated pressures downstream:

$p_t = -309.56$ Pa, $p_v = +5.4$ Pa, $p_s = -314.96$ Pa

9. *Re-heater at E*

$$\Delta p_t = 50 \text{ Pa}$$

Accumulated pressures downstream:

$p_t = -359.56$ Pa, $p_v = +5.4$ Pa, $p_s = -364.96$ Pa

10. *Transformation EF to Fan Inlet*

This is a gradual, symmetrical reducer and if we take the included angle as 45° the CIBSE loss coefficient is 0.04.

$$\Delta p_t = 0.04 \times 0.6 \times 9.778^2 = 2.29 \text{ Pa}$$

Accumulated pressures downstream:

$p_t = -361.85$ Pa, $p_v = +57.37$ Pa, $p_s = -419.22$ Pa

These are the pressures at the fan inlet.

   It is now convenient to start at the index grille, J′, and work backwards to the fan discharge.

11. *Index Discharge Grille, J′*

The frictional loss is given by the manufacturers at 5 Pa and this is the change of static pressure across the grille. The total pressure loss is the sum of this friction and the kinetic energy loss from the system, corresponding to the velocity pressure at the grille face.

$$\Delta p_t = 5 + 0.6 \times 2^2 = 7.4 \text{ Pa}$$

Accumulated pressures upstream:

$p_t = +7.4$ Pa, $p_v = +2.4$ Pa, $p_s = +5$ Pa

12. *Expander J″J′ to Index Grille*

The air velocity falls from 5.71 m s^{-1} to 2 m s^{-1} so the area ratio is 2/5.71 = 0.35, whence the CIBSE loss coefficient is 0.42. Since it is a gradual expander with an included angle of 50° (Figure 15.17) there is a further multiplier which the CIBSE (1986c) gives as 1.0. In other words this included angle is the equivalent of a sudden expansion.

$$\Delta p_t = 1.0 \times 0.42 \times 0.6 \times 5.71^2 = 8.22 \text{ Pa}$$

Accumulated pressure upstream:

$$p_t = +15.62 \text{ Pa}, \ p_v = +19.56 \text{ Pa}, \ p_s = -3.94 \text{ Pa}$$

(Notice that there is actually a negative static pressure in the beginning of the expander, even though the ductwork is on the discharge side of the fan. This is quite possible and arises because the static regain is +8.94 Pa, i.e. from −3.94 to +5 Pa.)

13.   *Straight Duct JJ′*

$$\Delta p_t = 1 \text{ Pa m}^{-1} \times 30 \text{ m} = 30 \text{ Pa}$$

Accumulated pressures upstream:

$$p_t = +45.62 \text{ Pa}, \ p_v = +19.56 \text{ Pa}, \ p_s = +26.06 \text{ Pa}$$

14.   *Bend at J*

The value of h/W (the reciprocal of the aspect ratio) is 350/750 = 0.47 and the ratio of the centre-line radius to the width is 1.0. The CIBSE loss coefficient is therefore 0.31.

$$\Delta p_t = 0.31 \times 0.6 \times 5.71^2 = 6.06 \text{ Pa}$$

Accumulated pressures upstream:

$$p_t = +51.68 \text{ Pa}, \ p_v = +19.56 \text{ Pa}, \ p_s = +32.12 \text{ Pa}$$

15.   *Straight Duct IJ*

$$\Delta p_t = 1 \text{ Pa m}^{-1} \times 60 \text{ m} = 60 \text{ Pa}$$

Accumulated pressures upstream:

$$p_t = +111.68 \text{ Pa}, \ p_v = 19.56 \text{ Pa}, \ p_s = +92.12 \text{ Pa}$$

16.   *Supply Branch Piece I*

For flow through the main, the velocity ratio is 5.71/8 = 0.71. The CIBSE loss coefficient for this is 0.25, referred to the downstream main velocity pressure.

$$\Delta p_t = 0.25 \times 0.6 \times 5.71^2 = 4.89 \text{ Pa}$$

A static regain occurs as the air flows through the main, past the branch.
Accumulated pressures upstream:

$$p_t = +116.57 \text{ Pa}, \ p_v = +38.4 \text{ Pa}, \ p_s = +78.17 \text{ Pa}$$

17.   *Straight Duct HI*

$$\Delta p_t = 1 \text{ Pa m}^{-1} \times 60 \text{ m} = 60 \text{ Pa}$$

Accumulated pressures upstream:

$$p_t = +176.57 \text{ Pa}, \ p_v = +38.4 \text{ Pa}, \ p_s = +138.17 \text{ Pa}$$

18.   Expander GH at Fan Discharge

The area ratio corresponds to $8/10 = 0.8$, whence the CIBSE (1986a) loss coefficient is less than 0.09 so this value can be taken as a simplification. The CIBSE (1986a) also quote a further multiplying factor of 0.8 because it is a gradual expander with an included angle of $30°$.

$$\Delta p_t = 0.8 \times 0.09 \times 0.6 \times 10^2 = 4.32 \text{ Pa}$$

Accumulated pressures upstream:

$$p_t = +180.89 \text{ Pa}, \ p_v = +60 \text{ Pa}, \ p_s = +120.89 \text{ Pa}$$

These are the pressures at fan discharge.
    Using equation (15.21) for the fan total pressure we have

$$p_{tF} = +180.89 - (-361.85) = 542.74 \text{ Pa}$$

    Also, by means of equation (15.22) we have

$$p_{sF} = +120.89 - (-361.85) = 482.74 \text{ Pa}$$

Equation (15.23) verifies this:

$$p_{sF} = +542.74 - (+60) = 482.74 \text{ Pa}$$

Clearly, using two decimal places in the above calculations is academic. More realistic results would be $p_{tF} = 543$ Pa and $p_{sF} = 483$ Pa. These would be rounded off, upwards, after the application of margins (see section 15.18).
    As is seen in later sections (15.18 and 15.19), it does not follow that the design duty will necessarily be obtained: a great deal depends on how the fan is installed in the system and the extent to which the kinetic energy in the airstream leaving the fan is converted to static pressure by suitable reductions in velocity.

15.15   The interaction of fan and system characteristic curves

A system of ductwork and plant has a characteristic behaviour that can be shown by a curve relating the total pressure loss with the air quantity flowing through it. Equation (15.3) showed that pressure drop depends on the square of the air velocity and equation (15.12) allows us to extend this idea and say that the total pressure drop, Δp_t, is proportional to the square of the volumetric airflow rate, Q. A system characteristic curve is thus a parabola, passing through the origin, expressed by

$$\Delta p_{t2} = \Delta p_{t1}(Q_2/Q_1)^2 \qquad (15.39)$$

    The behaviour of a fan also be shown by a characteristic curve that relates fan total pressure with volumetric airflow rate but this can only be established by test. Figure 15.18 shows how the shape of the fan curve depends on the type of impeller and it can be seen that an impeller with many, shallow, forward-curved blades has a point of inflection but one with a few, deep, backward-curved blades does not.
    When a duct system is coupled with a fan running at a particular speed the duty obtained is seen by the intersection of their respective, characteristic pressure–volume curves. Such an intersection defines a point of rating on the fan characteristic, shown by P_1 or P_2 in Figure 15.19. Changes in the fan speed do not alter the position of the point of rating on the fan curve but cause it to take up different positions on the system curve, in accordance with the fan laws (see section 15.16). Thus, in Figure 15.19, P_1 and P_2 are the same point

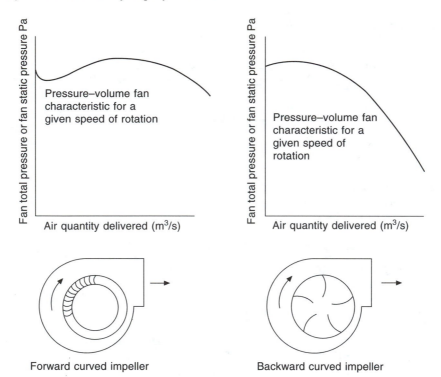

Fig. 15.18 Pressure–volume diagrams for fans with forward- and backward-curved impeller blades.

of rating on the fan curve but occupy different positions on the system curve, depending on the speed at which the fan runs.

EXAMPLE 15.8

Plot the pressure–volume characteristic for the system forming the basis of example 15.7.

Answer

From example 15.7 the total pressure loss in the system for a flow rate of 3 m³ s⁻¹ is 542.74 Pa, which is 543 Pa, for all practical purposes. Using equation (15.39) we can write $\Delta p_{t2} = 543(Q_2/3)^2$ and the following table is compiled:

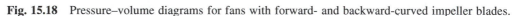

$Q\,(\text{m}^3\,\text{s}^{-1})$	0.5	1.0	1.5	2.0	2.5	2.75	3.0	3.25
$\Delta p_t\,(\text{Pa})$	15	60	136	241	377	456	543	637

This information is plotted in Figure 15.19.

A fan should be chosen that has a characteristic, for a given running speed, that passes through the system curve at the desired duty, namely, 3.0 m³ s⁻¹ and 543 Pa, in the cases of examples 15.7 and 15.8. If, for some reason, the fan curve does not intersect the system curve at the right point then the fan speed may be changed, in accordance with the fan laws (section 15.16), to give the correct duty.

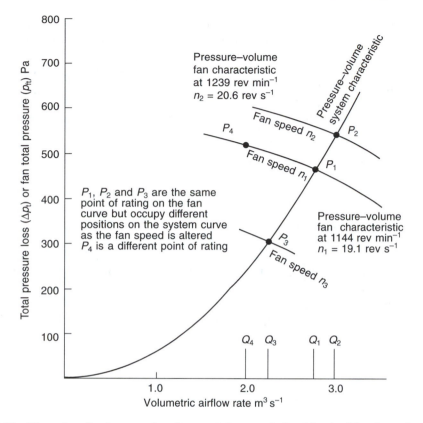

Fig. 15.19 The point of rating on a fan characteristic curve is fixed but it slides down the system characteristic curve as the fan speed is reduced.

EXAMPLE 15.9

A fan has a pressure–volume characteristic given by the following information, obtained from a test, when running at 1144 rev min^{-1} (19.1 rev s^{-1}):

Q (m^3 s^{-1})	1.5	2.0	2.5	3.0	3.5
p_{tF} (Pa)	542	520	488	443	388

Determine the quantity of air handled if it is installed in the system forming the subject of examples 15.7 and 15.8.

Answer

The data are plotted in Figure 15.19 and show an intersection with the system characteristic at P_1, yielding a duty of 2.77 m^3 s^{-1} at a fan total pressure of 463 Pa.

15.16 The fan laws

It is usual for a particular type of fan to be made as a set of different sizes, all exhibiting dynamical similarity because it is not a practical proposition to test every single fan made.

Their performances are then related by what are termed the fan laws, which are good enough for all practical purposes, even if not exact. The amount of testing that must be done by a manufacturer is then much reduced.

The laws apply to a given point of rating on the pressure–volume characteristic of the fan and those of most use in air conditioning are as follows:

A. *For a given fan size, duct system and air density*

1 The volume handled varies directly as the fan speed.
2 The pressure developed varies as the square of the fan speed.
3 The power absorbed varies as the cube of the fan speed.
4 The total efficiency is constant.

The duct system has a pressure loss depending on the square of the volume handled (see equation (15.39)); thus it follows that the first two laws correspond to the law for the system. The consequence of this is that the fan will always operate at the same point of rating, for a particular system, and that this point will move up or down the system curve as the fan speed is changed. We see this in Figure 15.19.

The third law is a consequence of the first two, in accordance with equations (15.19) and (15.20). Hence the fan efficiency must be constant.

B. *For a given fan size, duct system and speed*

1 The volume handled remains constant.
2 The fan total pressure, fan static pressure and the velocity pressure at fan discharge vary directly as the density.
3 The fan power absorbed varies directly as the density.
4 The total efficiency is constant.

EXAMPLE 15.10

(*a*) If the case of Example 15.9 is for operation at 0°C and 101.325 kPa barometric pressure with a total efficiency of 84 per cent determine the absorbed fan power.
(*b*) Determine the air quantity handled, the fan total pressure and the absorbed fan power if the temperature is 35°C and the barometric pressure is 85 kPa.

Answer

(*a*) By equations (15.19) and (15.20) we have

$$\text{Fan power} = \frac{2.77 \times 0.463}{0.84} = 1.527 \text{ kW}$$

(*b*) The air quantity handled is unchanged at 2.77 m^3 s^{-1} because the speed is unchanged, in accordance with law B1.
Using law B2 we have

$$\text{Fan total pressure} = 463 \times \frac{(273 + 0)}{(273 + 35)} \times \frac{85}{101.325} = 344 \text{ Pa}$$

Hence, since the fan total efficiency is unchanged because the point of rating is not altered, we have, by equations (15.19) and (15.20):

$$\text{Fan power} = \frac{2.77 \times 0.344}{0.84} = 1.134 \text{ kW}$$

Alternatively, using law B3, we could have calculated

$$\text{Fan power} = 1.527 \times \frac{(273 + 0)}{(273 + 35)} \times \frac{85}{101.325} = 1.135 \text{ kW}$$

EXAMPLE 15.11

For the conditions of example 15.9, at what speed must the fan run to deliver 3 m^3 s^{-1}?

Answer

The first fan law in group A applies and hence the new speed, n_2, is given by

$$n_2 = 1144 \times 3.0/2.77 = 1239 \text{ rev min}^{-1} = 20.6 \text{ rev s}^{-1}$$

The fan curve representing this speed is that shown in Figure 15.19, intersecting the system curve at the point P_2. Note that P_2 is the same point of rating on the fan curve but shifted up the system curve from position P_1, by the speed increase.

15.17 Maximum fan speed

When the impeller runs, stresses are developed in the shaft, impeller blades, back plate and other rotating parts. These stresses limit the speed at which the fan can safely operate. The critical speed of the fan shaft depends on the nature of the material, its mass, its diameter and the distance between the bearings. The critical speed must be much greater than the maximum allowable running speed, otherwise the bending and torsional stresses set up in the shaft material will approach dangerous limits. The fan manufacturer sets the maximum allowable, safe speed but, in general terms, maximum running speeds should not exceed 66 per cent of the critical speed according to ASHRAE (1993) and a limit of 55 per cent is better, when the effects of operational wear and temperature are considered.

All fans used should be dynamically balanced by the manufacturer, as well as statically balanced.

15.18 Margins

Excessively large margins on design performance are bad practice because they impose restrictions on fan selection and prevent operation at best efficency. They should not be set. Nevertheless, there is a gap between design and commissioning: there may be design errors, late changes in the required performance, unforeseen installation difficulties, mistakes in the interpretation of the design intention by the erector of the duct system, and so on. There is also the possibility of air leakage. The *HVCA* (1998) deals with air leakage, giving acceptable leakage limits and specifying testing procedures. The CIBSE (1986b) deals with allowable leakage rates in detail.

Bearing in mind the above, it is desirable to augment the design duty by a small amount, for the purpose of fan selection. It is suggested that the volumetric airflow rate should be increased by 5 per cent and, since pressure drop in a system is proportional to the square of the airflow rate, the fan total pressure should be increased by 10 per cent.

Although the fan manufacturer usually selects the driving motor it is often useful to be able to make an estimate of the probable motor power at an early stage in the design. As is seen later, backward-curve impellers have a non-overloading characteristic curve for fan power versus volumetric airflow rate, whereas forward-curved impellers do not. It is suggested that, for the purpose of making a preliminary estimate of the power of the motor needed to drive a fan, margins of 25 per cent be added for backward-curved impellers and 35 per cent for forward-curved impellers. These allowances would be applied to the fan powers determined from equation (15.20), including the margins of 5 per cent and 10 per cent for the airflow rate and fan total pressure, respectively. The power would then be rounded up to the next commercial motor size.

It is to be noted that when the fan starts, enough torque has to be available to accelerate the impeller to the running speed in a reasonably short period of time, depending on the characteristics of the starter. About 18 seconds is a reasonable period. See Keith Blackman (1980).

When commissioning a system it is often necessary to increase the fan speed to achieve the design duty, in accordance with fan law A1. The success of this depends on the driving motor having enough power.

15.19   Power–volume and efficiency–volume characteristics

Two further characteristic curves, power–volume and efficiency–volume, are also used to express the performance of a fan at a given speed and these are shown in Figure 15.20(*a*), (*b*) and (*c*), for three different types of fan. In Figure 15.20(*a*), for a centrifugal fan with backward-curved blades on its impeller, we see that peak efficiency corresponds to a point on the pressure–volume curve to the right of its maximum pressure where it is beginning to slope downwards fairly sharply. This is an advantage in that changes in fan total pressure (or system resistance) do not give large changes in the volume delivered. The curve for the power absorbed is seen to reach a maximum value just after the position of greatest efficency and to fall away thereafter.

The performance of a fan with forward-curved impeller blades is quite different (Figure 15.20(*b*)). Not only is peak efficiency to the left of the hump on the pressure–volume curve but the power absorbed goes on rising as the airflow increases. Moreover, if selection is to the right of the highest efficiency, small changes in fan total pressure give comparatively large changes in the air volume handled.

Figure 15.20(*c*) illustrates the behaviour of an axial flow fan and we see there is a non-overloading power characteristic, like that of a backward-curved fan. Maximum efficiency also corresponds to a steeply sloping position on the pressure–volume characteristic.

The point of rating chosen on the pressure–volume curve for a fan running at a certain speed identifies a particular total efficiency. When the fan is installed in a duct system, having a characteristic intersecting that of the fan at the chosen point, this point of rating stays in a fixed position on the fan curve, even if the speed is changed (see section 15.16). It follows that the fan efficiency will then remain constant at the original value as the speed is altered. The fan power absorbed can then be calculated according to the third law in group A when the speed is changed.

EXAMPLE 15.12

(*a*) For the case of examples 15.10 and 15.11 determine the fan power when the fan is running at 1239 rev min^{-1} (20.6 rev s^{-1}) handling air (i) at 0°C and 101.325 kPa and (ii) at 35° and 85 kPa.

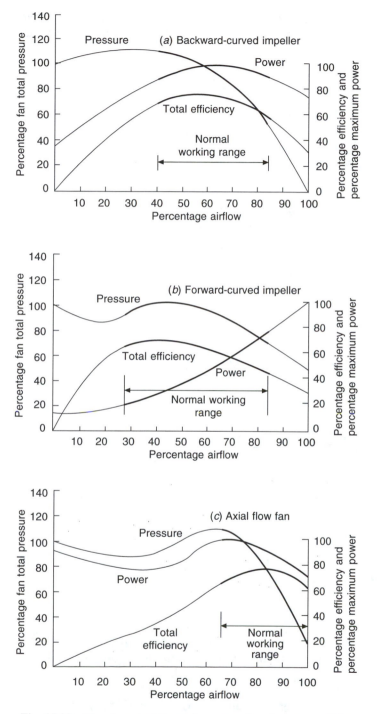

Fig. 15.20   The characteristic performance of various types of fan.

(*b*) Specify a suitable duty for the fan in (i), above, and estimate the probable power of the driving motor, assuming that the fan impeller has backward-curved blades.

Answer

(*a*)   (i) At 0°C and 101.325 kPa:

Using fan law 3 in group A we have

$$\text{Fan power} = 1.527 \times (1239/1144)^3 = 1.94 \text{ kW}$$

Alternatively, law 2 could be used to calculate the fan total pressure:

$$p_{tF} = 463 \times (1239/1144)^2 = 543 \text{ Pa}$$

Since the point of rating is unchanged (otherwise the fan laws could not have been applied) the efficiency remains at 84 per cent and, by equation (15.20), we then have

$$w_f = 3.0 \times 0.543/0.84 = 1.94 \text{ kW}$$

(ii)   At 35°C and 85 kPa:

Using fan law 3 in group B we have

$$\text{Fan power} = 1.94 \times \frac{(273 + 0)}{(273 + 35)} \times \frac{85}{101.325} = 1.44 \text{ kW}$$

Alternatively, law 2 in group B could be used to calculate the fan total pressure:

$$p_{tF} = 543 \times \frac{(273 + 0)}{(273 + 35)} \times \frac{85}{101.325} = 404 \text{ Pa}$$

Since the point of rating is unchanged and the volume handled remains at 3.0 m³ s⁻¹ (by fan law 1 in group B), the fan power is determined from equation (15.20):

$$w_f = 3.0 \times 0.404/0.84 = 1.44 \text{ kW}$$

(*b*) Applying the margins suggested earlier and considering the operating conditions at 0°C and 101.325 kPa the duty specified would be

$$1.05 \times 3.0 = 3.15 \text{ m}^3 \text{ s}^{-1} \text{ and } 1.1 \times 543 = 597 \text{ Pa}$$

It is reasonable to assume the same total fan efficiency as before and, applying a factor of 1.25 (because the impeller is backward-curved) the motor power can be estimated as

$$1.25 \times 3.15 \times 0.597/0.84 = 2.8 \text{ kW}$$

This would be rounded up to a motor size of 3 kW.

   An impeller with backward-curved blades must run faster than one with forward-curved blades to give the same volumetric airflow rate, as Figure 15.21 shows. In the vector diagram we see that the tip speed vector at the blade tip must be greater for the backward-curved impeller than for the forward-curved, if the resultant vectors are to have equal lengths, implying the former must run at a greater rotational speed. The consequences are that fans with forward-curved impeller blades are smaller, cheaper, slower running and quieter than those having backward-curved blades for the same airflow rate, although the latter are more efficient. The forward-curved impeller can only develop fan total pressures up to about 750 Pa but there is virtually no limit with backward-curved fans. The usual

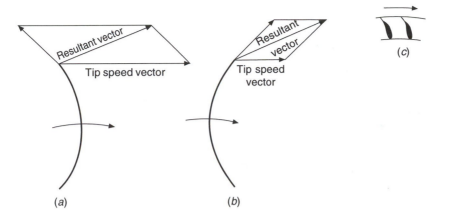

Fig. 15.21   Vector diagrams at the blades of impellers and a picture of the blades of a backward-curved impeller with aerofoil section blades.

choice for conventional low velocity systems is consequently the forward-curved fan. For medium and high velocity systems a backward-curved blade with an aerofoil section (Figure 15.21(*c*)) is commonly used.

To achieve the required performance from a centrifugal fan it is essential that the ductwork approach to the inlet eye is smooth and straight. The greater the departure from this ideal the more the reduction in performance. When single or double inlet fans are mounted in chambers, or air handling units (see section 15.21 and Figure 15.24), without ducted connections, at least three-quarters of an inlet diameter should be allowed between the inlet eye and the nearby wall of the chamber and at least 1.5 inlet eye diameters between the adjoining inlets of pairs of fans. The duct connection to the discharge side of the fan must also be straight and smoothly expanding. Ths is to give the airstream leaving the fan outlet a chance to expand and become less turbulent. The full effective length of straight duct (*L*) to achieve this, according to Keith Blackman (1980), depends on the mean velocity of airflow (*V*) in the duct and is given in equivalent diameters by

$$L/D_e = 2.5 + 0.2(V - 12.5) \qquad (15.40)$$

where D_e is the equivalent duct diameter of the fan outlet, **subject to a minimum length of 2.5 equivalent diameters.**

Figure 15.22 illustrates the pattern of airflow at the outlet from a centrifugal fan. The true picture is more complicated than the figure suggests: there are many rotational components in the airstream and a complete recovery of their kinetic energy and a conversion to useful static pressure is difficult, if not impossible (as much as eighteen equivalent diameters may be needed). It is only the static pressure of an airstream that is immediately available to offset frictional losses, and it follows that it is highly desirable to convert as much as possible of the velocity pressure into static pressure. A measure of recovery occurs as the air flows through the rest of the supply system because the air velocity is progressively reduced in the process of duct sizing and, if this is done in a well engineered manner, significant static regain will result.

However, steps for recovery should be taken as soon as the air leaves the fan, the aim being to smooth the airflow into a fully developed, symmetrical profile as soon as possible. Equation (15.40) shows that this can mean the provision of quite a long length of straight duct.

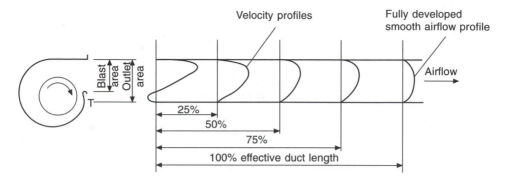

Fig. 15.22   Velocity profiles and effective duct lengths for the recovery of velocity pressure and conversion to useful static pressure at the outlet from a centrifugal fan.

The presence of a damper too close to the fan discharge greatly upsets the airflow and gives much higher pressure losses. Similarly, bends should not be too close to the fan outlet and, in the case of a branch it is especially important to have an adequately long section of straight duct between it and the fan.

The axial flow fan may appear to offer advantages in plant layout because of the straight through direction of airflow. However it is often convenient to use a centrifugal fan because the change of direction through 90°, suffered by the airflow, tends to reduce the length of plant room needed. Nevertheless axial flow fans can often be useful, provided that gradual reducers and expanders, of adequate straight length, are fitted at inlet (especially) and outlet. Axial flow fans invariably require silencers, particularly if the fan total pressure exceeds about 70 Pa.

15.20   Fan testing

In the UK performance testing is covered by BS 848: Part 1: 1980. The standard is comprehensive and internationally well accepted. It covers three types of installation:

  A: Free inlet and free outlet.
  B: Free inlet and ducted outlet.
  C: Ducted inlet and free outlet.
  D: Ducted inlet and ducted outlet.

The standard prescribes the parameters for the testing rigs that may be adopted and, in particular, ensures that adequate steps are taken to provide smooth airflow conditions at inlet and at outlet. An interpretation of a test arrangement for a type D installation, according to Keith Blackman (1980), is shown in Figure 15.23. A low loss entry piece (see sections 15.6 and 15.7) enables an accurate determination of airflow rate to be made by equating the measured static depression to velocity pressure. A measurement of the static pressure at fan inlet is comparatively easy in the large inlet chamber where smooth airflow conditions prevail. Establishing the velocity at fan outlet is more difficult because of the turbulent nature of the airflow. Static pressure is easier to measure and hence performance can be conveniently expressed as airflow rate against fan static pressure (see equation (15.22)).

The airflow rate is divided by the area across the flanged outlet from the fan (see Figure 15.22) to establish a notional mean air velocity at fan outlet. The corresponding velocity pressure is added to the measured fan static pressure to give the fan total pressure in accordance with equation (15.23).

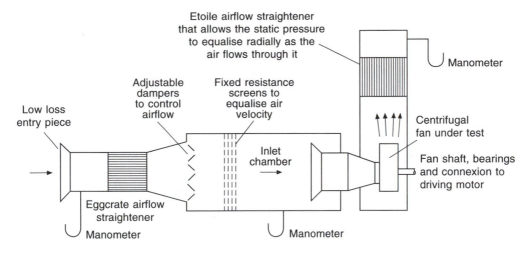

Fig. 15.23   An interpretation of a testing arrangement for a type D installation (ducted inlet and ducted outlet) according to BS 848: Part 1: 1980. Based on a method adopted by Woods of Colchester Ltd (incorporating Keith Blackman) and used with their kind permission.

It is to be noted that the fan static pressure represents the sum of the total pressure losses before the fan inlet and the static pressure available to overcome the losses after the fan outlet, ignoring any benefit that may be obtained by converting the velocity pressure at fan outlet into static pressure. Hence, to take full advantage of the fan performance, every effort should be made to recover as much as possible of the velocity pressure at fan discharge, to augment the static pressure available.

It is further to be noted that, since the way the fan is installed on the site is never the same as the way in which its performance was tested according to BS 848 (1980), it is to be expected that there will be some shortfall in performance, unless this has been foreseen at the design stage and allowed for in the calculation of system total pressure loss. The discrepancy is sometimes spoken of as system effect. In this respect it is most important that smooth inlet and outlet duct connections are made to the fan.

15.21   The performance of air handling units

The majority of fans used in building services today are fitted in air handling units. Under such circumstances the performance of a fan is quite different from that established by BS 848. Air handling units should be tested according to BS 6583 (1985).

If a centrifugal fan is installed in a cabinet, as is the case with the fan section of an air handling unit, the proximity of the cabinet sidewall to the fan inlet eye imparts a swirl to the air stream entering the impeller and affects its performance. This must be taken into account. ASHRAE (1993) and AMCA (1973) give a relationship between the distance from the fan inlet eye to the cabinet wall and the loss of total pressure incurred. See Figure 15.24. It is recommended that a cabinet wall should not be closer than 0.75 D to the fan inlet and that adjacent fans in a cabinet should not be closer to each other than 1.5 D, where D is the diameter of the fan inlet. Similar considerations apply to the roof and floor of the cabinet and the centre-line of the inlet eyes should be equidistantly located between them.

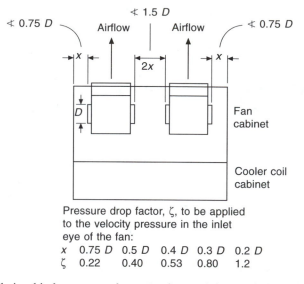

Pressure drop factor, ζ, to be applied
to the velocity pressure in the inlet
eye of the fan:

x	0.75 D	0.5 D	0.4 D	0.3 D	0.2 D
ζ	0.22	0.40	0.53	0.80	1.2

Fig. 15.24   The relationship between total pressure loss and the proximity of a cabinet wall or an adjacent fan. A cooler coil section is attached when testing a fan cabinet to BS 6583: 1985. Separate tests are done for the cases of draw-through units (as shown) and blow-through units. If, as is common, multiple fans are used, they are driven from a common impeller shaft.

In order to test and specify the performance of the fan section of an air handling unit to BS 6583 (1985) the fan cabinet is attached to a cooler coil section. This is to simulate actual operating conditions. The combination is then regarded as a fan and tested to BS 848 (1980). Thus the characteristics of the fan in the cabinet and the influence of a cooler coil section in close proximity are taken into account. Separate tests are carried out when the cooler coil section is located on the discharge side of a fan cabinet.

The difference between the total pressure at inlet to the cooler coil cabinet and the static pressure at the outlet from the fan cabinet, is measured against volumetric airflow rate for each fan speed considered, and the results plotted for the fan and cooler coil cabinet combination. The total pressure loss through the cooler coil cabinet is established according to BS 5141: 1975 for each airflow rate and is added to the curve for the fan and coil cabinet combination. This then gives a curve for the fan cabinet alone. By subtracting the pressure drop (determined according to the appropriate BS) for any other plant items to be included in the air handling unit, the manufacturer can establish a curve for the performance of the unit as a whole. Such a characteristic curve expresses the static pressure offered by the unit to offset external total pressure losses. Figure 15.25 illustrates this. Some of the velocity pressure at fan discharge can be profitably converted to static pressure if a smooth expansion piece, of adequate length, is fitted. This is then also available for offsetting part of the external total pressure loss. See section 15.10 and Figure 15.22.

The external total pressure loss is usually termed the external resistance and is the sum of the total pressure losses on both the suction and discharge sides of the air handling unit. It should include the extra pressure loss incurred by the presence of the dirt on the filter, just before the filter is to be replaced.

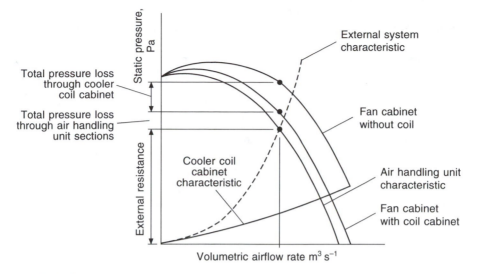

Fig. 15.25 The performance of an air handling unit according to BS 6583: 1985.

15.22 Methods of varying fan capacity in a duct system

There are four basic ways in which the capacity of a fan-duct system can be altered:

(1) Changing the fan speed
(2) Partly closing a damper in the duct system
(3) Partly closing variable position guide vanes at the inlet eye of a centrifugal fan or changing the blade pitch angle of an axial flow fan
(4) Varying the effective impeller width of a centrifugal fan.

1. The first method can be used to increase or decrease the air quantity handled, provided the driving motor is of adequate size. If the fan speed is increased from n_1 to n_2 rev s^{-1}, the pressure–volume curve for the fan shifts to a new position on the same system curve, the point of rating on the fan curve remaining unchanged, as indicated by the points P_1 and P_2 in Figure 15.19. It can be seen that the new point of intersection is P_2, and the fan delivers a large air quantity, Q_2, at a higher fan total pressure, p_{tF2}, through the same duct system. Conversely, a reduction in speed from n_1 to n_3 gives a fan-system intersection at point P_3, the fan curve having been lowered.

For most constant volume systems the fan speed is only altered during commissioning, in order to achieve the design duty, and this is conveniently done by changing the pulleys. However, variable air volume systems must have the fan speed modulated almost continuously as the required duty varies. The most efficient way of doing this is by a frequency converter. An inverter rectifies the alternating current to a direct current supply, and then uses this to produce a smooth, variable frequency, alternating current output. Controlling the frequency of the output provides the desired speed modulation for the induction motor driving the fan.

2. If a damper in the duct system is closed it changes the position of the system characteristic. The damper is part of the system and, in common with all other system components, there is a certain pressure drop across it for the flow of a certain quantity of air past it. If, by some unspecified means, the flow of air through the damper is kept fixed, regardless of the position of the damper, an experimental relationship can be found between the pressure drop across the damper and its position, for a fixed airflow. Section 13.13 discusses this.

The result of partly closing a damper which forms part of a system is shown in Figure 15.26(*a*). It can be seen that the system curve rotates upwards in an anti-clockwise direction about the zero volume point, as the damper is closed.

Figure 15.26(*b*) indicates what happens when a damper is used to reduce the capacity of a fan-duct system. If the fan speed remains constant at n rev s^{-1} and the damper in the duct is fully open, the duty is Q_1 at p_{tF1}. In order to obtain a reduced airflow Q_2, the damper is partly closed and the point of intersection moves from P_1 to P_2, which is now a different point of rating on the fan characteristic and so the fan has a different efficiency. The required capacity is obtained at a fan total pressure of p_{tF2}. The failing of the method is that it is wasteful of air power. It is however, convenient.

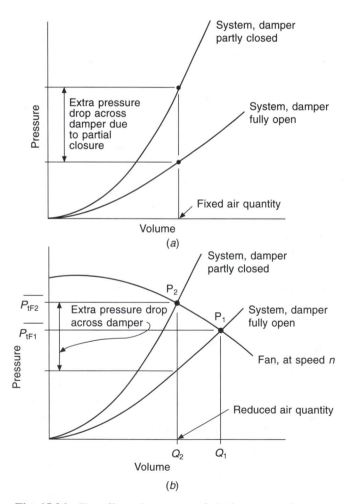

Fig. 15.26 The effect of opening and closing system dampers.

EXAMPLE 15.13

If the output of the fan-duct system used in example 15.9 is reduced to $2 \text{ m}^3 \text{ s}^{-1}$ by partly closing a main damper, calculate the air power wasted across the damper.

Answer

When the fan runs at 19.1 rev s^{-1} the duty of the system is $2.77 \text{ m}^3 \text{ s}^{-1}$ at 463 Pa. Reference to Figure 15.19 shows that if the air quantity is to be $2 \text{ m}^3 \text{ s}^{-1}$ the system curve must cut the fan characteristic at 520 Pa. The system loss at $2 \text{ m}^3 \text{ s}^{-1}$, on the other hand, is only 241 Pa (see example 15.8). The damper must therefore be closed enough to absorb the difference of $520 - 241 = 279$ Pa if $2 \text{ m}^3 \text{ s}^{-1}$ is to be delivered. The system curve rotates anti-clockwise about the origin as the main damper is closed until it intersects the fan characteristic at the point P_4 in Figure 15.19 at a duty of $2 \text{ m}^3 \text{ s}^{-1}$ and 520 Pa fan total pressure, hence

$$\text{Wasted air power} = 2 \times 0.279 = 0.56 \text{ kW}$$

If the fan efficiency were known for a speed of 19.1 rev s^{-1} and a flow rate of $2 \text{ m}^3 \text{ s}^{-1}$ the wasted fan power could be calculated by equation (15.20). In a similar way, as a filter gets dirty the pressure drop across it increases, the system curve rotates counter-clockwise, the airflow rate reduces and the fan power absorbed across the filter rises.

EXAMPLE 15.14

A system of plant and ductwork has a total pressure loss of 531 Pa, of which 90 Pa is across a clean filter, when handling $3 \text{ m}^3 \text{ s}^{-1}$. A fan having the following pressure–volume characteristic at 8 rev s^{-1} is connected to the system:

$\text{m}^3 \text{ s}^{-1}$	0	0.5	1.0	2.0	2.5	3.0	3.5
Pa	360	385	410	433	424	400	343

(*a*) Determine the duty achieved with a clean filter.
(*b*) Determine the speed at which the fan must run in order to obtain a duty of $3 \text{ m}^3 \text{ s}^{-1}$ with a clean filter.
(*c*) Calculate the duty at the revised fan speed when the filter is dirty and its resistance is 180 Pa.

Answer

(*a*) The system characteristic with a clean filter may be established by assuming a square law:

$\text{m}^3 \text{ s}^{-1}$	0.5	1.0	1.5	2.0	2.5	3.0	3.5
Pa	15	59	133	236	369	531	723

This, together with the fan data for running at 8 rev s^{-1} is plotted in Figure 15.27. The two curves intersect at P, yielding a duty of $2.66 \text{ m}^3 \text{ s}^{-1}$ at a fan total pressure of 418 Pa when the filter is clean.
  (*b*) Using the first fan law in group A the new speed, n, required to obtain $3 \text{ m}^3 \text{ s}^{-1}$ is:

$$n = 8.0 \times 3.0/2.66 = 9.02 \text{ rev s}^{-1}$$

The intersection of the fan curve at this speed with the characteristic of the system with a

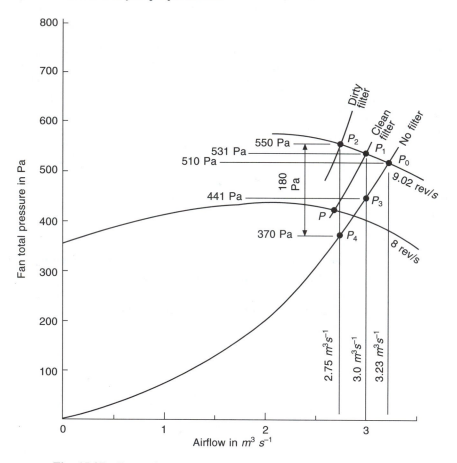

Fig. 15.27   Fan and system characteristic curves for example 15.14.

clean filter is shown by the point P_1 in Figure 15.27, with a fan total pressure of 531 Pa, of which the clean filter is absorbing 90 Pa.

(*c*) The loss through the system without a filter is thus $531 - 90 = 441$ Pa for an airflow of 3 m³ s⁻¹, shown by the point P_3 in Figure 15.27. It is seen that this system curve (without a filter) cuts the fan characteristic for 9.02 rev s⁻¹ at P_0, giving a duty of 3.23 m³ s⁻¹ at 510 Pa. When the clean filter is installed the duty is given by the point P_1 but as the system runs and the filter gets dirty the effect is to rotate the system curve in an anti-clockwise direction about the origin until it intersects the fan curve at the point P_2. Ths point is vertically above the point P_4 and the distance between them is $550 - 370 = 180$ Pa, as Figure 15.27 shows. The duty with a dirty filter is thus 2.75 m³ s⁻¹ at 550 Pa.

3.   Variable inlet guide vanes provide a way of reducing the capacity of a fan-duct system, in a manner analogous to that discussed in section 12.13 for centrifugal compressors in refrigeration systems. Such vanes are positioned in the inlet eye of a centrifugal fan in a way that permits them to assist or retard the airflow through the impeller. Each vane is radially mounted and is hinged along a radial centre-line, being thus able to vary its inclination to the direction of airflow. The inclination may be such that the swirl imparted to the airstream changes its angle of entry to the impeller blades. This is accomplished without appreciable loss of power over a large range of the fan capacity. The result of partly closing the vanes is not the same as a throttling action. Instead, it rotates the pressure–volume characteristic curve for the fan about the origin, in a clockwise direction. Figure 15.28 shows that a duty Q_2 at p_{tF2} can be obtained if the vanes are closed enough to shift the point of operation from P_1 to P_2. If the vanes were capable of complete closure, then, when they were so closed, the pressure–volume curve for the fan would be the ordinate through the origin. This is a practical impossibility since leakage always occurs past the vanes.

Most axial flow fans are direct-coupled to the driving motor, which offers the choice of several fixed speeds of rotation, related to the number of pairs of poles of the motor. The required duty is then obtained by selecting an appropriate blade angle. Fans driven by a given motor shift the position of their pressure–volume characteristic curve as their blade angle is changed, somewhat like the rotation of the characteristic curve of a centrifugal fan as its inlet guide vanes close (see Figure 15.28). The efficiency of axial flow fans is much better than that of centrifugal fans with forward-curved impellers and this advantage is retained over a wide range of volumetric duties.

4. Varying the effective impeller width of a centrigugal fan. The principle of running fans in parallel (see section 15.24) can be used to vary the capacity of a forward-curved centrifugal fan. If a disc is mounted on the fan shaft, within the impeller wheel of a forward-curved fan, it divides the impeller width into two parts: a section adjacent to the inlet eye that can handle airflow and a section next to the backplate of the casing that is shielded from the inlet and can handle virtually no air. See Figure 15.30(*b*). It follows that if the position of the disc is adjustable the effective width of the impeller can be varied with a corresponding change in the airflow rate. The pressure–volume characteristic curve will then move as the position of the disc is altered. For example, with the disc in the mid-position on the fan shaft, the fan curve could be regarded as that shown for a single fan in Figure 15.30(*a*), cutting the system curve at the point P_1. If the disc was then moved to make available the full width of the impeller, the fan curve would also move and cut the system curve at the point P_2, in Figure 15.30(*a*). It follows that movement of the disc, in a direction parallel to the fan shaft, will change the effective impeller width and give a modulating control over fan capacity. For the full impeller width the fan and system curves intersect at the point P_2. As the disc is moved towards the fan inlet eye, less of the impeller width is used and, at the point P_1, only the front half is effective. Figure 15.30(*a*) shows a broken line representing in general the movable fan curve. The system curve is cut at a point Q which can be moved from P_2, through P_1, towards the origin at O, as the position of the disc is altered. The method is termed disc throttling.

15.23   The effect of opening and closing branch dampers

In any system it is usual to ensure that the correct quantity of airflows along the branches off the index run by partially closing regulating dampers in the various branches. (The alternative to this would be to size down the branch ducts so that they absorbed the correct amounts of energy by virtue of their reduced diameters.) Only the index run has its damper fully open. Thus, in a simple system, as illustrated in Figure 15.29(*a*), if the index run is A-B-C, then the natural energy loss along BB′ is less than it is along BC, for the flow of the design air quantities. This situation is clearly anomalous; only one total pressure can exist at B, hence the loss along BB′ must be artificially increased by the partial closure of a regulating damper. Then

$$\Delta p_{tbc} = \Delta p_{tbb'}$$

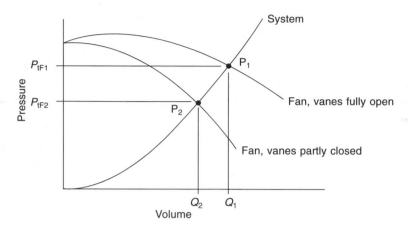

Fig. 15.28   The effect of changing the inclination of guide vanes at the inlet of a centrifugal fan.

Two interesting questions arise. What happens if the damper in BB′ is fully closed? What happens if it is fully opened? The answers are best provided by means of an example.

EXAMPLE 15.15

Suppose the simple system shown in Figure 15.29(*a*) is designed to deliver 1 m³ s⁻¹ with a total pressure loss of 625 Pa, the air quantity handled being divided equally between the two duct runs, BC and BB′. The total pressure loss is made up as follows:

> Plant   375 Pa when 1.0 m³ s⁻¹ is flowing
> Duct A′B   60 Pa when 1.0 m³ s⁻¹ is flowing
> Duct BC   190 Pa when 0.5 m³ s⁻¹ is flowing
> Duct BB′   60 Pa when 0.5 m³ s⁻¹ is flowing and its regulating damper is wide open.

Assume that there is negligible energy loss across the damper in branch BB′ when it is completely open.
Calculate:

(*a*) the pressure drop across the damper in BB′ when it is adjusted to give an airflow rate of 0.5 m³ s⁻¹ in the branch,
(*b*) the total airflow rate handled when the damper in BB′ is fully closed, and
(*c*) the total airflow rate handled when the damper in BB′ is fully open.

Answer

(*a*) Pressure to be absorbed by the partially closed damper in BB′

$$= \Delta p_{\text{tbc}} - \Delta p_{\text{tbb}'}$$
$$= 190 - 60$$
$$= 130 \text{ Pa}$$

(*b*) Any assumption can be made about the air quantity flowing along BC in order to plot a new system characteristic for the revised arrangement with the damper in BB′ closed. It is, however, most convenient to assume an airflow rate of 1 m³ s⁻¹ along BC, since the loss of energy from A to B is already known for this rate.

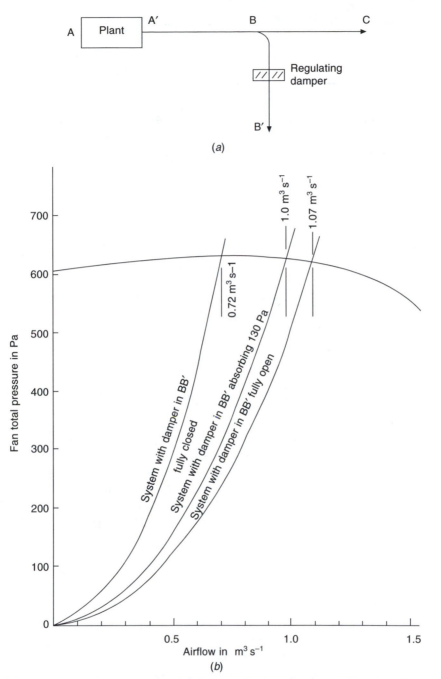

Fig. 15.29   (a) Plant and duct diagrams for example 15.15. (b) System and fan characteristics for example 15.15.

$\Delta p_{\text{tab}} = 375 + 60 = 435$ Pa when 1 m^3 s^{-1} is flowing

$\Delta p_{\text{tbc}} = 190 \times (1.0/0.5)^2 = 760$ Pa when 1 m^3 s^{-1} is flowing

Hence

$\Delta p_{\text{tac}} = 435 + 760 = 1195$ Pa when 1 m^3 s^{-1} is flowing

A new system characteristic can now be established:

m^3 s^{-1}	0.2	0.4	0.6	0.8	1.0
Pa	48	191	430	765	1195

This is plotted in Figure 15.29(*b*). The intersection of this new system curve with the fan characteristic shows that 0.72 m^3 s^{-1} is handled when the damper in BB′ is fully closed.

(*c*) If the damper in BB′ is fully open, then the loss of total pressure from B to C must equal that from B to B′. Any value for this loss of pressure may be assumed and a new system curve established. For ease of computation, assume that the pressure drop along BB′ is 190 Pa. Hence the air quantity flowing along BB′ may be calculated:

$$Q_{bb'} = 0.5 \text{ m}^3 \text{ s}^{-1} \text{ with a pressure drop of 60 Pa}$$

$$Q_{bb'} = 0.5 \times \sqrt{\frac{190}{60}}$$

$$= 0.89 \text{ m}^3 \text{ s}^{-1} \text{ with a pressure drop of 190 Pa}$$

When 0.89 m^3 s^{-1} flows through BB′ the amount passing through AB is 0.89 + 0.50 = 1.39 m^3 s^{-1}. Since the design pressure drop from A to B is 375 + 60 = 435 Pa when 1.0 m^3 s^{-1} is handled we have

$$\Delta p_{tab} = 435 \times (1.39/1.0)^2 = 840 \text{ Pa}$$

and the total system loss is 840 + 190 = 1030 Pa. This is the starting point for establishing a new system characteristic curve, according to an assumed square law, as shown in Figure 15.29(*b*). The intersection with the fan curve occurs at 1.07 m^3 s^{-1}, and this is the total air quantity handled by the system, when the damper in BB′ is left fully open. Thus, for the particular conditions of this example, an overload of 7 per cent is imposed on the system.

It is clear that a risk exists, when a system is first started, of overloading the fan motor if the branch dampers are open. This could cause the motor to burn out. Accordingly, it is sound practice to include a main damper which can be shut when the system is first started and then opened gradually until the approximate design air quantity is handled. It is also essential to add a margin of 25 to 35 per cent to the fan power when selecting the motor required to drive the fan. Incidentally, this margin provides something in hand if the fan does not at first deliver the required air quantity, and the speed of its rotation must be increased. If the power characteristic is known for a fan running at a certain speed, then the power curve for any other speed can be determined and plotted by means of Fan Law A.3, as given in section 15.16. See the discussion on proportional balancing in section 13.14 and on margins in section 15.18.

15.24   Fans in parallel and series

A pair of identical fans connected in series handles the same volume as a single fan but develops twice the fan total pressure. Thus, the characteristic data for a single fan running at 8 rev s^{-1} in example 15.14 would become, for two in series:

m^3 s^{-1}	0	0.5	1.0	2.0	2.5	3.0	3.5
Pa	720	770	820	866	848	800	686

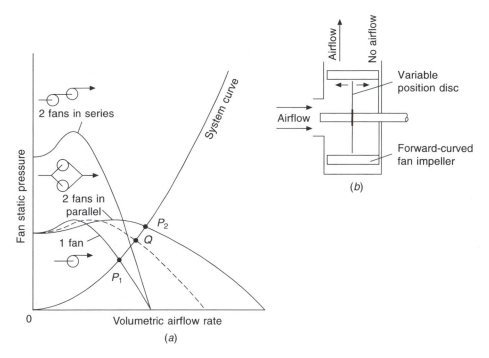

Fig. 15.30 (*a*) Characteristic curves for: a single fan, 2 fans in series and 2 fans in parallel. (*b*) Section through a forward-curved centrifugal fan showing a movable position disc. If the fan is regarded as having a characteristic like that of 2 fans in parallel then its capacity can be reduced by moving the disc towards the inlet eye. This is seen by the broken line in sub-figure (*a*).

When identical fans are connected in parallel they handled twice the volume of a single fan at the same fan total pressure. This normally causes no problems but, occasionally, with forward curved fans operating in the vicinity of their point of inflection (see Figure 15.20(*a*)), the performance can hunt between the three possible airflow rates for a given total system resistance. The solution is to increase the system resistance a little in order to rotate the system characteristic out of the area of instability, with the loss of some airflow rate.

Exercises

1. A system of duct and plant has a total pressure loss of 600 Pa when 1300 litres s^{-1} of air flows. A fan running at 11 rev s^{-1} is coupled to the system and has the following characteristic:

Litres s^{-1}	0	100	200	300	400	500	600	700	800	900	1000	1100
Pa	450	457	462	465	461	445	417	376	327	266	194	100

(*a*) Determine the actual duty achieved and the speed at which the fan must run to get the design duty.
(*b*) If two of these fans are connected in series what will be the duty for the system mentioned if they each run at 11 rev s^{-1}?

Answer

(*a*) 883 litres s^{-1} at 275 Pa, 16.2 rev s^{-1}, (*b*) 1015 litres s^{-1} at 360 Pa.

2.   The total pressure loss through an air handling plant and supply duct index run has been wrongly calculated as 250 Pa for a design flow rate of 5 m^3 s^{-1}. A fan having the following characteristic performance when running at 7 rev s^{-1} has been selected and installed:

m^3 s^{-1}	0	1	2	3	4	5
Pa	235	272	284	284	273	250
Efficiency %	—	58	63	67	69	70

During commissioning the measured duty was 3.45 m^3 s^{-1} at a fan total pressure of 280 Pa.

(*a*) Determine the correct pressure through the plant and system for the design airflow rate and the speed at which the fan should run to deliver this.

(*b*) If the maximum speed at which the fan may safely run is 9 rev s^{-1} and if the pulleys are changed to obtain this speed, what will then be the airflow rate and the absorbed fan power?

(*c*) If the original filter in the plant had a pressure loss of 200 Pa when passing 3.45 m^3 s^{-1} and if, in an attempt to increase the duty at the new fan speed, it is replaced by a different type of filter with a pressure drop of 150 Pa when passing 5 m^3 s^{-1}, what should the airflow rate and absorbed fan power then be? Assume that the pressure loss through all parts of the plant and system is proportional to the square of the volumetric flow rate.

Answer

(*a*) 588 Pa, 10 rev s^{-1}, (*b*) 4.44 m^3 s^{-1}, 3.02 kW, (*c*) 4.53 m^3 s^{-1}, 1.70 kW.

3.   (*a*) State three fan laws that describe the behaviour of a fan for: a given fan size, a given duct system and a fixed air density.

(*b*) State three additional laws for the case of: a given fan, a given duct system, a given speed and varying density.

(*c*) A fan has a pressure–volume characteristic at 10 rev s^{-1} as follows:

m^3 s^{-1}	0	0.5	1.0	1.5	2.0	2.5	3.0	3.5
Pa	400	435	460	487	483	474	450	393

Assuming a square law relating pressure drop and flow rate in the duct system to which the fan is coupled, determine the quantity of air handled and the fan total pressure if the system has a fan total pressure of 580 Pa when 3 m^3 s^{-1} is flowing. At what speed must the fan be run to give 3.0 m^3 s^{-1}? If the fan efficiency is then 75 per cent what will the power absorbed be?

Answer

(*c*) 11.5 rev s^{-1}, 2.32 kW.

Notation

Symbol	Description	Unit
A	cross-sectional area of a duct or an air jet	m^2
A'	reduced cross-sectional area at a vena-contracta	m^2
A_{fo}	area across the flanges at fan outlet	m^2
a	area, or overall dimension of a flat oval duct	m^2 or mm
b	overall dimension of a flat oval duct	mm
C	duct pressure drop constant or the curve ratio of a bend	–
C_A	coefficient of area	–
C_E	coefficient of entry or flow coefficient	–
C_V	coefficient of velocity	–
D	diameter of a fan inlet eye	m
D_e	equivalent duct diameter	m or mm
D_h	mean hydraulic diameter $= 4A/P$	m or mm
d	internal duct diameter	m or mm
f	dimensionless coefficient of duct friction (Fanning)	–
f_c	coefficient of friction for a circular duct	–
f_r	coefficient of friction for a rectangular duct	–
f'	coefficient for determining an approximate value of f	–
g	acceleration due to gravity	$m\ s^{-2}$
H	head lost in metres of fluid flowing	m
h	height of a duct or a turning vane	m or mm
K	ratio of throat to heel radius for a bend	–
k_s	absolute roughness of a duct wall material	m or mm
L	length of straight duct	m
L/D	number of equivalent duct diameters	–
l	length of straight duct	m
m	mass	kg
N_3	constant $= \pi^3/(32\rho) = 0.308\ 42\rho^{-1}$	$m^3\ kg^{-1}$
N_4	constant $= 1.255\pi\mu/4\rho = 0.985\ 67\mu\rho^{-1}$	$m^2\ s^{-1}$
n	number of turning vanes, dimensionless constant or fan speed	– or rev s^{-1}
P	perimeter of a duct or a section of a jet of air	m or mm
p	pressure	Pa
p_{at}	atmospheric pressure	Pa
p_s	static pressure	Pa
p_{sF}	fan static pressure	Pa
p_{so}	static pressure at fan outlet	Pa
p_t	total pressure	Pa
p_{tF}	fan total pressure	Pa
p_{ti}	total pressure at fan inlet	Pa
p_{to}	total pressure at fan outlet	Pa
p_v	velocity pressure	Pa
p_{vo}	velocity pressure at fan outlet	Pa
Δp	pressure loss or rate of pressure loss	Pa or Pa m^{-1}
Δp_s	change of static pressure	Pa
Δp_t	total pressure loss or rate of total pressure loss	Pa or Pa m^{-1}

Q	airflow rate	$\text{m}^3\ \text{s}^{-1}$
Q'	theoretical airflow rate without loss at a suction opening	$\text{m}^3\ \text{s}^{-1}$
(Re)	Reynolds number	–
R_c	centre-line radius of bend	m or mm
R_o	throat radius of a bend	m or mm
R_t	throat radius of a bend	m or mm
R_{n+1}	heel radius of a bend	m or mm
r	fraction of static regain	–
s	space between turning vanes	mm
t	dry-bulb temperature	°C
V	mean velocity of airflow	m s^{-1}
V'	theoretical velocity without loss at a suction opening	m s^{-1}
v_{fo}	notional mean air velocity at fan outlet	m s^{-1}
W	width of a duct	mm
w_a	air power	kW
w_f	fan power	kW
y	duct length between supply air branches	m
δ_q	small quantity of air	m^3
δ_t	small quantity of time	s
η_t	total fan efficiency	%
ζ	pressure loss coefficient, applied to a velocity pressure	–
ζ_a	pressure loss coefficient for an abrupt expander	–
ζ_g	pressure loss coefficient for a gradual expander	–
θ	included angle of a duct transition piece	°
μ	absolute viscosity	$\text{kg m}^{-1}\ \text{s}^{-1}$
v	kinematic viscosity	$\text{m}^2\ \text{s}^{-1}$
ρ	density	kg m^{-3}

References

ACGIH (1995): *Industrial ventilation: A manual of recommended practice*, 22nd edn, American Conference of Governmental Industrial Hygienists. Lansing, MI.

Altshul, A.D. and Kiselev, P.G. (1975): *Hydraulics and Aerodynamics*, Stroisdat Publishing House, Moscow, Russia.

AMCA (1973): *Fans and systems*, Publication 201, Air Movement and Control Association, Arlington Heights, Il, USA.

ASHRAE Handbook (1993): Fundamentals SI, 32.21–32.46.

ASHRAE Handbook (1997) Fundamentals, SI, 32.7.

BS 848: 1980: *Fans for general purposes*, Part 1, Methods of testing performance.

BS 5141: 1983: *Specification of air heating and cooling coils*: Part 1: 1975 (1983), Methods of testing for rating of cooling coils.

BS 6583: 1985: *Methods for volumetric testing for ratings of fan sections in central station air handling units (including guidance on rating)*.

Carrier (1960): *System design manual—Part 2, Air distribution, Air duct design*. Chapter 2, 17–63df, Carrier Corporation, Syracuse, NY.

Chun Lun, S. (1983): Simplified static regain duct design procedure, *ASHRAE Trans.* **89**(2A), 78.

CIBSE (1986a): TM8, Design Notes for Ductwork.

CIBSE (1986b): Commissioning Code, series A, Air distribution systems.

CIBSE (1986c): Guide C4, Flow of fluids in pipes and ducts.

Colebrook, C.F. and White, C.M. (1937): Experiments with Fluid Friction in Roughened Pipes, *Proc. R. Soc. (A)* **161**, 367.

Colebrook, C.F. and White, C.M. (1939): Turbulent Flow in Pipes with Particular Reference to the Transition Region between the Smooth and Rough Pipe Laws, *Proc. ICE.* **II**, 133.

Harrison, E. and Gibbard, N. (1965): Balancing airflow in ventilating duct systems, *JIHVE*, Oct., **33**, 201.

HVCA (1998): DW/144, Specification for sheet metal ductwork.

Idelchik, I.E., Malyavskaya, G.R., Martynenko, O.G. and Fried, E. (1986): *Handbook of hydraulic resistance*, 2nd edn, Hemisphere Publishing Corporation, subsidiary of Harper & Row, New York.

Keith Blackman (1980): *Centrifugal Fan Guide,* Keith Blackman Ltd.

Lewitt, E.H. (1948): *Hydraulics and the Mechanics of Fluids*, Pitman.

Prandtl, I. (1953): *Essentials of Fluid Dynamics*, Blackie.

Shataloff, N.S. (1966): Static regain method of sizing for ducts, *ASHRAE Journal,* May, **43**.

Tsal, R.J. (1989): Altshul–Tsal friction factor equation, *Heating, Piping and Air Conditioning* (August).

Bibliography

1. *Fan Application Guide,* 2nd edition, Hevac Association, 1981.
2. W.C. Obsborne, *Fans,* Pergamon Press, 1966.
3. B. Eck, *Fans,* Pergamon Press, 1973.
4. *Fan Engineering,* Buffalo Forge Company, Buffalo, NY.
5. Carrier Air Conditioning Company, *Air Conditioning System Design Manual,* McGraw-Hill Book Company, 1965.

16

Ventilation and a Decay Equation

16.1 The need for ventilation

The minimum amount of fresh air required for breathing purposes is really quite small; about 0.2 litres s^{-1} per person (see section 4.10). For comfort conditioning, however, it is insufficient to supply this small amount of fresh air; other factors enter into consideration and enough fresh air must be delivered to achieve the following:

(1) Meeting the oxygen needs of the occupants.
(2) The dilution of the odours present to a socially acceptable level.
(3) The dilution of the concentration of carbon dioxide to a satisfactorily low level.
(4) Minimising the rise in air temperature in the presence of excessive sensible heat gains.
(5) Pressurising escape routes in order to prevent the spread of smoke in the event of a fire.
(6) Dealing with condensation.

Odours may be diluted to a socially acceptable level by the introduction of odourless air from outside or by the use of activated carbon filters (see section 17.10). The original work on the use of fresh air for this purpose was done by Yaglou *et al.* in 1936. This was based on research in American schools and it was found that the quantity of outside air needed to give a satisfactory reduction of odours depended on the number of people present and their standards of personal hygiene. It was also found that odours disappeared more rapidly, with a given ventilation rate, when there was more volume of the room for each person present. (This is still inexplicable: it may be due to a chemical breakdown of the constituents of the odours or to their adsorption at the room surfaces.) It was further evident that increasing the air change rate gave diminishing returns, the efficiency of odour decay reducing with an increase in the rate because of a failure to scour the whole room. Some fresh air leaves without mixing with the odours at all and so gives no dilution effect. Yaglou was able to show that school children of average socio-economic background required from 5.5 to 13.5 litres s^{-1} each as the room volume per person decreased from 14 m^3 to 3 m^3, to produce acceptable odour control. Sedentary adults of a similar background needed rather less fresh air, namely, 3 to 12 litres s^{-1} each.

Smoking in a room plays a very significant part in determining the necessary fresh air allowance and Brundrett (1975) has produced a survey on this topic. Although the unpleasant odours and the reduced visibility are obvious consequences of atmospheric pollution from smoking, the three main short-term hazards to health seem to arise from the amount of carbon monoxide (CO) generated, the particulate matter liberated and the quantity of acrolein produced. There are longer-term risks of course and it is to be noted that non-smokers in a smoking environment are subject to all these dangers to a smaller but still significant extent. Hence the need for proper ventilation.

Acrolein is a toxic lachrymator with an instantaneous effect of eye irritation and a further, similar effect on throats. Its threshold limit value for an eight hour exposure is 0.1 parts per million (abbreviated ppm), which requires a fresh air dilution rate of at least 3 m^3 per cigarette although eye irritation can still be experienced with a rate as high as 16 m^3 per cigarette. Note that the threshold limit value (abbreviated TLV) is defined as the airborne concentration of a substance that represents acceptable conditions to most working people, without adverse effect.

The particulate matter is mostly smaller in size than 0.7 μm and is respirable, according to Hoegg (1972). Depending on the production rate from the cigarette, Bridge and Corn (1972) say that a fresh air rate of from 3.5 to 5.5 m^3 per cigarette is necessary to achieve a TLV of 10 mg m^{-3}.

The most dangerous pollutant from smoking is CO. Although the TLV for an eight hour exposure is 55 ppm, much less than this is highly desirable to cater for variations in the composition of any general population, which is likely to contain a proportion of very young persons, old people and some who are sick. This minority would be more adversely affected by the presence of CO so the American recommendation for the TLV is 9 ppm and the Russians apparently require as little as 1 ppm according to Brundrett (1975). A dilution rate of 9 m^3 per cigarette is needed to limit the concentration to 9 ppm in the view of Hoegg (1972).

The effect of cigarette smoke on visibility is another issue and Leopold (1945) considered that 1 m^3 per cigarette was required as a minimum dilution rate in a sports stadium.

The conclusion to be drawn is that although a dilution rate of 20 m^3 per cigarette is enough to give satisfactory conditions for the average person this should be increased to 40 m^3 per cigarette to cover 98 per cent of the population. A typical smoker in the UK smokes an average of 1.3 cigarettes per hour but the figure is more in the United States. There are differences in the composition of the tobacco, the weight of the cigarette and the length of the stub remaining but, nevertheless, it is possible to estimate the supply of fresh air necessary to deal with a population of mixed smokers and non-smokers.

EXAMPLE 16.1

Calculate the fresh air ventilation rate needed for an office in the UK, (*a*) if everyone smokes and (*b*) if only 50 per cent of those present smoke. Take the necessary dilution rate as 40 m^3 per cigarette to satisfy the comfort and health of 98 per cent of the population.

Answer

(*a*) $$\frac{(1.3 \text{ cigarettes per h per person}) \times (40 \text{ m}^3 \text{ per cigarette}) \times 1000}{3600}$$

 = 14.4 litres s^{-1} per person.

(*b*)   If only half the people smoke this reduces to 7.2 litres s^{-1} per person.

For private offices the assumption might be that everyone could be smoking and so the higher rate would be appropriate. For boardrooms and the like, where very heavy smoking may sometimes still occur, the rate of cigarette consumption could be double that taken as an average for the population. In such cases 29 litres s^{-1} per person would be the choice. In open plan office areas not everyone will smoke. Taking the view of Brundrett (1975) that 50 per cent of a population are likely to smoke, on an average in the UK, 7.2 litres s^{-1} per person would then be a reasonable fresh air allowance. It is worth noting that if a small number of offices in a sizeable building contain a high number of smokers, recirculating air from them for mixing with air brought in from outside will not significantly reduce the dilution ability of the mixed air, provided the heavily polluted air is only a small proportion of the total amount of the recirculated air.

Note also that smoking is a simple contaminant that is not influenced by the volume of space per person, unlike body odour which is so influenced. Equation (16.9) describes the way in which the concentration of a contaminant is dependent on the ventilation rate. We can see that the only place in the equation where the volume of the room appears is in exponent n, which equals Qt/V, V being the volume of the room, Q the fresh air supply rate and t the elapsed time. In the steady state n is very large and terms in e^{-n} become vanishingly small, the expression degenerating to the form of equation (16.10), in which the volume of the room is absent.

Although a fall in the oxygen content of the atmosphere to as little as 13 per cent escapes notice, depth and frequency of breathing being unchanged, alterations in the concentration of carbon dioxide (CO_2) are more significant in their effect. An increase in the CO_2 content is not immediately apparent but when it has risen from the normal value of about 0.032 per cent (in fresh air) to 2 per cent there is an increase of 30 per cent in the depth of respiration according to the Health and Safety Executive (1979) and at 3 per cent this has gone up to 60 per cent. The same reference concludes that up to 2 per cent CO_2 causes little discomfort but another view from the BRE (1977) is that it is tolerable up to 4 per cent. A conclusion by Brundrett (1975) is that for concentrations of CO_2 exceeding 3 per cent to 5 per cent there is a conscious and objectionable need felt for an increased respiratory effort. The TLV for an eight hour exposure is 0.5 per cent, say the ACGIH (1971), but it is generally thought that an upper acceptable level is 0.1 per cent. If we take the average rate of production of CO_2 by a human being as 4.72×10^{-3} litres s^{-1} the steady state concentration of CO_2 approaches 0.1 per cent in an office of the statutory minimum volume of 11.33 m^3 (in the UK), as example 16.3 shows.

Refer to Figure 16.1. Formerly, ANSI/ASHRAE (1989) recommended that a minimum of 7.5 litres s^{-1} of uncontaminated outside air should be supplied for each person in an occupied space. This was based on research by Berg-Munch *et al.* (1986) showing that fewer than 20 per cent of people first entering an occupied space would notice undesirable body odour if 7.5 litres s^{-1} per person of uncontaminated outside air were provided. Supporting evidence by Brundage *et al.* (1988) suggested that supplying about 7.5 litre s^{-1} per person significantly reduced respiratory infections in occupied spaces.

More fresh air than this may be required in many instances to dilute the concentration of other emissions to acceptable levels. Typical of such contaminants are the volatiles emitted from wall finishes, floor coverings, curtains, furniture, cleaning materials, plastics, office equipment etc. Such contaminants, together with the presence of housemites and their droppings, and other small particulate matter, have been associated with what is termed the Sick Building Syndrome, according to WHO (1983). This is the group of symptoms (such as irritation of the mucous membranes and eyes, headaches, lethargy, tight chests, the perception of stuffy and stale air, etc.), complained of by some of the occupants in some buildings and related to poor indoor air quality. See section 4.11.

It is to be noted further (see example 16.6) that supplying about 7.5 litre s^{-1} of fresh air

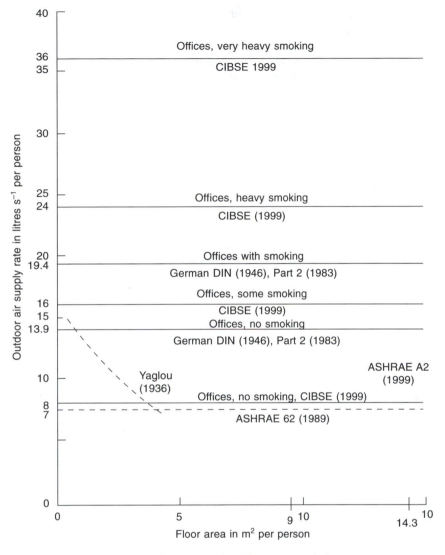

Fig. 16.1 Various outdoor air recommendations.

per person reduces the carbon dioxide content in an occupied space to approximately 0.1 per cent.

ASHRAE Standard 62 (1989) did not distinguish between places where smoking occurred and those where it did not (see Grimstrud and Teichman (1989)) and neither does ASHRAE Standard 62 (1999). The latter argues that indoor air quality depends on many factors and offers alternative procedures for choosing a rate of supply of outdoor air. The first is the Ventilation Rate Procedure which prescribes outdoor air supply rates, related to density of occupancy, for a range of commercial and other applications, in litres s^{-1} per person or in litres s^{-1} m^{-2}, as appropriate. For offices, 10 litres s^{-1} are proposed, based on a maximum population density of 14.3 m^2 per person. Among the caveats is the reasonable one that compliance will not necessarily give acceptable indoor air quality. The alternative is the indoor air quality procedure. This does not specify outdoor air supply rates. Instead it proposes restricting all known, relevant, indoor air contaminants and offers guidelines for achieving acceptable indoor air quality. Design documentation must be kept and made available for operating the system.

Fanger (1988) proposed the use of a pollution balance in an occupied space to establish an appropriate ventilation rate and suggested the following equation.

$$Q_o = 10[G/(C_{ia} - C_{oa})e_v] \qquad (16.1)$$

where

Q_o = outdoor air supply rate in litres s^{-1}
G = sensory pollution load in the space in olf
c_{ia} = perceived indoor air quality in decipol
c_{oa} = perceived outdoor air quality in decipol
e_v = ventilation effectiveness factor

Fanger (1988) introduced the concepts of the olf (to express pollution) and the decipol (to express the perception of air quality). They are defined as follows:

One olf is the pollution generated by a standard, sedentary, non-smoking person in a state of thermal neutrality (comfort).

One decipol is the perceived quality of air in a space wherein the pollution source strength is one olf and the ventilation rate with clean outdoor air is 10 litres s^{-1} (i.e. 1 decipol = 0.1 olf/litres s^{-1}). The pollution generated by a smoker is 6 olf and interpolation is allowed for mixed populations of smokers and non-smokers.

The equation applies to the steady state and if the fresh air supplied mixes completely with the air in the room then the value of the ventilation effectiveness factor, e_v, is 1.0. If some of the air supplied short circuits and does not mix with the room air then the value of e_v is defined by European Concerted Action (1992) as

$$e_v = c_e/c_r \qquad (16.2)$$

where

c_e = pollution concentration in the exhaust air
c_r = pollution concentration in the occupied zone in a room

Values suggested for e_v are given in Table 16.1.

According to Fanger (1988) and European Concerted Action (1992) an exponential relationship exists between the percentage of dissatisfied people and the ventilation rate for a pollution of one olf, produced by a standard person. This is given by

$$P = 395 \exp(-1.83q^{0.25}) \qquad (16.3)$$

where

P = percentage of people dissatisfied
q = ventilation rate in litres s^{-1} olf^{-1}

The equation applies when $q > 0.32$ litres s^{-1} olf^{-1}.

EXAMPLE 16.2

Determine the percentage of people dissatisfied by a pollution of one olf from a standard person when the ventilation rate is (a) 7.5 litres s^{-1} and (b) 0.32 litres s^{-1}.

Table 16.1   Suggested values for ventilation effectiveness factors

Supply-exhaust relationship	Temperature difference between the supply air (t_s) and the room air in the occupied zone (t_r)	Ventilation effectiveness factor, e_v
Supply and exhaust points above the occupied zone	$(t_s - t_r) < 0$ $(t_s - t_r) = 0$ to 2 $(t_s - t_r) = 2$ to 5 $(t_s - t_r) > 5$	0.9 to 1.0 0.9 0.8 0.4 to 0.7
Supply point above the occupied zone, exhaust point at low level in the occupied zone	$(t_s - t_r) < -5$ $(t_s - t_r) = -5$ to 0 $(t_s - t_r) > 0$	0.9 0.9 to 1.0 1.0
Supply point at low level in the occupied zone, exhaust point above the occupied zone	$(t_s - t_r) < 0$ $(t_s - t_r) = 0$ to 2 $(t_s - t_r) > 2$	1.2 to 1.4 0.7 to 0.9 0.2 to 0.7

Reproduced by kind permission of the CIBSE from Guide A, Environmental Design (1999).

Answer

By equation (16.3):

(a) $P = 395 \exp(-1.83 \times 7.5^{0.25}) = 19\%$
(b) $P = 395 \exp(-1.83 \times 0.32^{0.25}) = 100\%$

The relationship between the percentage of people dissatisfied and the perceived air quality in decipol is given by European Concerted Action (1992) as:

$$c_{ia} = 112[\ln(P) - 5.98]^{-4} \tag{16.4}$$

where

c_{ia} = perceived indoor air quality in decipol
P = percentage people dissatisfied

EXAMPLE 16.3

Determine the indoor air quality in decipol if the percentage of people dissatisfied is 19 per cent.

Answer

By equation (16.4):

$$c_{ia} = 112[\ln(19) - 5.98]^{-4} = 1.3 \text{ decipol}$$

If, as European Concerted Action (1992) propose, the concentration of carbon dioxide, c_{CO_2}, in an occupied space is used as an indicator of the pollution caused by bioeffluents from people, equation (16.3) can be modified to read

$$P = 395 \exp(-15.15 \, c_{CO_2}^{-0.25}) \tag{16.5}$$

This refers to sedentary, non-smoking people and excludes emissions from room furnishings etc.

The use of equations (16.3) and (16.4) makes it possible to determine values for the percentage of people dissatisfied, the perceived quality of the indoor air and the ventilation rate needed to achieve it. This procedure has yielded fresh air supply rates that are unacceptably high, because of the absence of reliable information on the emission of bioeffluents from people and volatiles from furnishings and the like. The method has not found acceptance.

EXAMPLE 16.4

Determine the percentage of people dissatisfied in an occupied space if the concentration of carbon dioxide is 0.1 per cent (1000 ppm).

Answer

By equation (16.5)

$$P = 395 \exp(-15.5 \times 1000^{-0.25}) = 25\%$$

This appears to be inconsistent with the generally held view that 0.1 per cent CO_2 is acceptable. See example 16.6.

The German standard, given in DIN 1946, Part 2 (1983), requires 13.9 litres s^{-1} per person in an open plan office without smoking (corresponding to 1.54 litres s^{-1} m^{-2}) but is increased to 19.4 litres s^{-1} per person (corresponding to 2.16 litres s^{-1} m^{-2}), when there is smoking. See Figure 16.1.

One other aspect of ventilation is that a minimum rate is needed to dilute the radioactive gas emission (radon) from some building materials used for construction according to Swedjermark (1978). The small radioactive constituent in substances such as granite may need a minimum rate of about half an air change per hour to produce an acceptably low concentration.

The three methods of expressing ventilation rate, air changes per hour, litres s^{-1} per person and litres s^{-1} m^{-2}, must be applied in the proper way. The use of litres s^{-1} per person is appropriate where the number of people can be established with some assurance, as in a theatre or restaurant. For offices or other areas where a precise occupancy is not known but a typical population density (e.g. 9 m^2 per person) can be adopted, the use of litres s^{-1} m^{-2} of floor area is suitable. Values commonly used in the UK have been 1.3 litres s^{-1} m^{-2} in the past, which has since risen to 1.4 litres s^{-1} m^{-2}. It would be uneconomical to use air changes per hour for rooms with large floor-to-ceiling heights but appropriate for lavatories where heights are conventional. Lavatories are rather a particular case, especially when there are no openable windows, and it is recommended that at least 12 air changes per hour of mechanical extract ventilation be provided, desirably with a mechanical supply at a lesser rate to ensure a slightly negative pressure. A commonly used rate of 6 air changes per hour is inadequate, in the author's opinion.

The CIBSE (1999) recommends outdoor air supply rates for sedentary occupants that are related to the production of body odour and the extent to which smoking occurs, as given in Table 16.2.

Smoking intensity is considered to be: some—when 25 per cent of the population are smokers; heavy—when 45 per cent are smokers; very heavy—when 70 per cent are smokers.

Uncooled outdoor air can also be introduced by mechanical means to mitigate the rise in room air temperature in warm weather. The air suffers a rise in temperature as it passes

Table 16.2 Recommended outdoor air supply rates for sedentary occupants

Condition	Recommended outdoor air supply rate litres s^{-1} per person
with no smoking	8
with some smoking	16
with heavy smoking	24
with very heavy smoking	36

Reproduced by kind permission of the CIBSE from Guide A, Environmental Design (1999).

through the supply fan (see section 6.5) and it follows that, ignoring any possible beneficial effects from the thermal inertia of a heavy building structure, the room air temperature can never be less than the outside air temperature. It seems likely that providing more than about eight or nine air changes per hour of uncooled mechanical ventilation gives diminishing returns and is not worth while. It is essential that openable windows are provided in buildings that are mechanically ventilated, to give relief for the occupants in warm weather.

Escape corridors and staircases are often slightly pressurised, by supplying more air than is extracted, in order to discourage the entry of smoke and so facilitate the escape of the occupants in the event of a fire. In the case of escape staircases more than thirty air changes per hour of outside air may be needed but it must be remembered that the system only operates in an emergency when a staircase free of smoke is the aim, not comfort.

The provision of sufficient air from outside can be used to deal with condensation problems. See section 6.3 and equation (6.8).

16.2 The decay equation

Although from the point of view of a purist it might be better to derive the decay equation in general terms, it is certainly easier to understand a derivation phrased with reference to a particular application. Accordingly, the equation is derived in terms of the rate at which a contaminant decays in a ventilated room under the influence of a diluting influx of fresh air. The contaminant chosen is carbon dioxide.

Consider a room, as shown in Figure 16.2, having a volume V m^3, in which the concentration of carbon dioxide is c, expressed as a fraction (e.g. parts of carbon dioxide per million parts of air). Suppose that during time Δt, a small quantity of air, Δq, entirely free of carbon dioxide (for simplicity), enters the room. A similar small quantity, Δq, of contaminated air is forced out of the room. The concentration within the room is therefore reduced by an amount $(\Delta q/V)c$. This reduction in concentration can be expressed by Δc, defined as follows:

$$\Delta c = -\frac{\Delta q}{V} c \tag{16.6}$$

The negative sign is used because the concentration is decreasing.

It follows that the rate of change of concentration is given by $\Delta c/\Delta t$, defined as follows:

$$\frac{\Delta c}{\Delta t} = -\frac{c\Delta q}{V\Delta t} \tag{16.7}$$

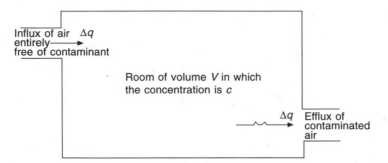

Fig. 16.2   A notional room, used as a model for the derivation of a decay equation.

But $\Delta q/\Delta t$ is the rate of influx of ventilating air and is a constant, Q, say. Thus the rate of change of the concentration with respect to time may be written as

$$\frac{dc}{dt} = -\frac{cQ}{V} \qquad (16.8)$$

The physical problem has now been phrased as a simple differential equation, and a solution to this will be of practical value in determining the answers to real problems.

By integration, the solution to equation (16.8) is

$$\int \frac{dc}{c} = -\int \frac{Q}{V}\, dt$$

$$\log_e c = -\frac{Qt}{V} + \log_e A$$

where $\log_e A$ is a constant of integration.
Hence,

$$\log_e c - \log_e A = -\frac{Qt}{V}$$

and

$$A\, e^{-(Qt/V)} = c$$

The value of the constant A is established by considering the boundary condition $c = c_0$ (the initial concentration in the room) at $t = 0$ (the instant that the ventilation began).

The solution to equation (16.8) is therefore

$$c = c_0\, e^{-(Qt/V)} \qquad (16.9)$$

If it is observed that in a ventilated room, where V is in m³, Q is in m³ s⁻¹ and t is in seconds, Qt/V is the number of air changes, then

$$c = c_0\, e^{-n} \qquad (16.10)$$

where n is the number of times the cubical content of the room is changed.

The graph of equation (16.10) is an exponential curve, as shown in Figure 16.3. It can be seen that concentration of the contaminant decays rapidly if the ventilating air is entirely free of the contaminant. After one air change it is 36.8 per cent of its initial value and after three air changes it is only 5 per cent.

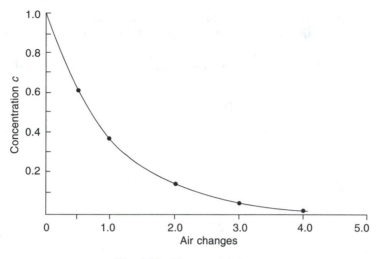

Fig. 16.3   Exponential decay.

EXAMPLE 16.5

If the air in a room has an initial concentration of 1000 ppm (parts per million) of hydrogen, how many air changes are required to reduce the value of this to 50 ppm?

Answer

From equation (16.10) we can write

$$n = \log_e(c_0/c) = \log_e 20 = 3$$

Suppose a more practical case where, with carbon dioxide as a contaminant, the fresh air used for ventilating purposes also contains some carbon dioxide. Suppose further, that there are people in the room who top up the level of carbon dioxide continually by respiratory activity.

Let c = the concentration of carbon dioxide in the room at any instant, expressed as parts per million of air,

Q' = the rate of fresh air supply in m^3 s^{-1} per person,

V' = the volume of the room in m^3 per person,

t = the time in seconds after the beginning of occupancy and ventilation,

c_a = the concentration of carbon dioxide present in the ventilating air, expressed in parts per million of air,

and V_c = the volume of carbon dioxide produced by breathing on the part of the occupants, in m^3 s^{-1} per person.

The volume of air entering the room in time Δt is $Q'\Delta t$, expressed in m^3 per person. Hence, the increase of carbon dioxide in time Δt, due to the contamination of the ventilating air, is $(Q'\Delta t)(c_a/10^6)$, in m^3 per person. (This is because even if the air in the room were initially free of the contaminant the fraction of the entering airstream which is carbon dioxide is $c_a/10^6$ and, therefore, this fraction of the entering volume is an addition to the contamination in the room in m^3.)

The volume of air forced out of the room by the entering airstream is also $Q'\Delta t$ and so, in a similar way, the amount of carbon dioxide leaving the room in m^3 per person is $(Q'\Delta t)(c/10^6)$.

Thus, a balance can be drawn up for the net change of carbon dioxide, in time Δt, expressed in m^3 per person:

$$\text{Net change of } CO_2 = V_c\Delta t + (Q'\Delta t c_a/10^6) - (Q'\Delta t c/10^6)$$

In this equation, $V_c\Delta t$ is the volume of CO_2 produced in time Δt by one person breathing, and is expressed in m^3 per person.

Since the concentration is the volume of CO_2 divided by the volume of the room, we can write the change in concentration, per person, as

$$\frac{\text{net change of } CO_2 \text{ in } m^3 \text{ per person}}{\text{volume of the room in } m^3 \text{ per person}}$$

and this can be defined by

$$\Delta c = [V_c\Delta t + (Q'\Delta t c_a/10^6) - (Q'\Delta t c/10^6)]/V'$$

expressed as a fraction.

Hence

$$\frac{\Delta c}{\Delta t} = [V_c + (Q'c_a/10^6) - (Q'c/10^6)]/V' \tag{16.11}$$

also expressed as a fractional change in concentration per unit time. Equation (16.11) may be rearranged in a form which is recognisable as amenable to solution:

$$\frac{dc}{dt} + \frac{Q'c}{V'} = \frac{10^6 V_c + Q'c_a}{V'} \tag{16.12}$$

This is the differential equation which expresses the physical problem in mathematical terms, just as was equation (16.8) for the first simple case dealt with. Equation (16.12) is obtained from equation (16.11) by multiplying throughout by 10^6, thus giving the rate of change of concentration in parts of CO_2 per 10^6 parts of air, in unit time. It may be solved by multiplying throughout by an integrating factor, $e^{(Q't/V')}$.

Then

$$\frac{dc}{dt} e^{(Q't/V')} + \frac{Q'c}{V'} e^{(Q't/V')} = \left(\frac{10^6 V_c + Q'c_a}{V'}\right) e^{(Q't/V')}$$

The left-hand side of this is the derivative of a product and so, by integrating, we get

$$A + c\, e^{(Q't/V')} = \frac{10^6 V_c + Q'c_a}{V'} \cdot \frac{V'}{Q'} e^{(Q't/V')}$$

where *A is a constant of integration.*
Write $B = AQ'$, then

$$B\, e^{(-Q't/V')} = (10^6 V_c + Q'c_a - Q'c)$$

The boundary condition is $c = c_0$ when $t = 0$, and so

$$B = 10^6 V_c + Q' c_a - Q' c_0$$

The solution to equation (16.12) is then

$$c = \left(\frac{10^6 V_c}{Q'} c_a \right) (1 - e^{-n}) + c_0 e^{-n} \qquad (16.13)$$

where n is the number of air changes after the passage of time t and is equal to $Q't/V'$. Reference to Figure 16.4 shows a graphical interpretation of this solution.

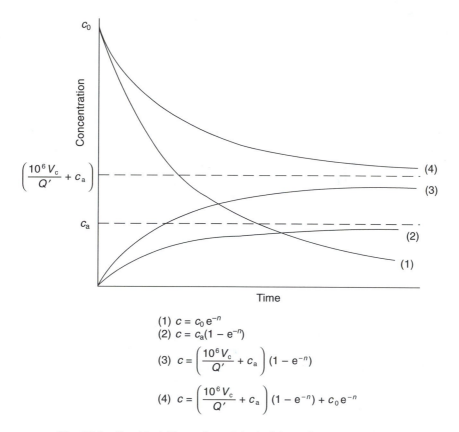

$$(1) \ c = c_0 e^{-n}$$

$$(2) \ c = c_a (1 - e^{-n})$$

$$(3) \ c = \left(\frac{10^6 V_c}{Q'} + c_a \right) (1 - e^{-n})$$

$$(4) \ c = \left(\frac{10^6 V_c}{Q'} + c_a \right) (1 - e^{-n}) + c_0 e^{-n}$$

Fig. 16.4   Graphical illustration of the build up of equation (16.13).

If the room were initially free of CO_2, the concentration in the room would change along curve (2), following the law $c = c_a(1 - e^{-n})$, and attaining an ultimate value of c_a.

If people were present but the room was initially free of CO_2, the law would be

$$c = \left(\frac{10^6 V_c}{Q'} c_a \right) (1 - e^{-n})$$

and the curve (3) would be followed. The ultimate concentration attained would be $(10^6 V_c/Q') + c_a$. When the room volume is very small the same ultimate concentration is approached more rapidly. This would be shown in Figure 16.4 by another curve, starting at the origin and lying between curve 3 and the upper broken line. When the room volume is very large the approach is slower and the curve would then lie beneath 3 in Figure 16.4.

If the initial concentration in the room were c_0, people were present, and the ventilating air had a concentration of c_a, then curve (4) would be followed according to equation (16.13), and the concentration in the room would approach a value of $(10^6 V_c/Q') + c_a$.

In the steady state, when n is very large ($\to \infty$), e^{-n} in equation (16.13) approaches zero and equation (16.13) degenerates to

$$c = (10^6 V_c/Q') + c_a$$

which becomes

$$Q' = 10^6 V_c/(c - c_a) \tag{16.14}$$

and is the same as a dilution equation given in the *CIBSE Guide*. The ventilation effectiveness factor, e_v, from Table 16.1, could be applied to the value of the air change rate, n.

EXAMPLE 16.6

If local government regulations stipulate that the minimum amount of fresh air which may be supplied to a place of public entertainment is 8 litres s^{-1} per person and that the minimum amount of space allowable in the room is 12 m^3 per person, calculate the concentration of carbon dioxide present after one hour, expressed as a percentage. Assume that fresh air contains 0.032 per cent of carbon dioxide and that human respiration produces 4.72×10^{-3} litres s^{-1} of carbon dioxide per person. Also that the supply and exhaust points are above the occupied zone and that $(t_s - t_r) < 0$. Then, from Table 16.1, $e_v = 0.9$ to 1.0. Take this as 0.95.

Answer

$\qquad Q' = 0.008$ m^3 s^{-1} per person,
$\qquad c_a = 320$ parts per million,
$\qquad V_c = 4.72 \times 10^{-6}$ m^3 s^{-1} per person,
$\qquad n = e_v Qt/V'$ (for minimum ventilation),
$\qquad V' = 12$ m^3 per person.
Hence,

$$n = 0.95 \times 0.008 \times 3600/12 = 2.28 \text{ air changes}$$

From equation (16.13)

$$c = \left(\frac{10^6 \times 4.72 \times 10^6}{0.008} + 320 \right)(1 - e^{-2.28}) + 320\, e^{-2.28}$$

$\qquad = 849.7$ ppm
$\qquad = 0.085$ per cent

EXAMPLE 16.7

A garage measures 60 m × 30 m × 3 m high and contains a number of motor cars which produce a total of 0.0024 m^3 s^{-1} of carbon monoxide.

(a) If the maximum permissible concentration is to be 0.01 per cent of carbon monoxide, what number of air changes per hour are required if the garage is in continual use?

(b) If the garage is in use for periods of 8 hours only, and if at the start of any such period the concentration of carbon monoxide is zero, what number of air changes per hour is needed if the concentration is to reach 0.01 per cent only by the end of the 8-hour period?

(c) What is the concentration after the first 20 minutes of an 8-hour period?

(d) If at the end of an 8-hour period the concentration is 0.01 per cent, for how long should the ventilation plant be run in order to reduce the concentration to 0.001 per cent?

Assure $e_v = 1.0$ for the case of supply at high level and exhaust at low level, with $(t_s - t_r) > 0$.

Answer

(a) $c_{max} = 100$ ppm

Since the garage is in continual use, c_0 also equals one hundred ppm.

The number of persons present is irrelevant but, for ease of understanding, it may be convenient to assume an occupancy of one person.

$$Q' = V'n/3600, \text{ since } t \text{ is one hour}$$
$$V' = 5400 \text{ m}^3$$
$$c_a = 0 \text{ and } V_c = 0.0024 \text{ m}^3 \text{ s}^{-1}$$

Hence, from equation (16.13),

$$100 = \left(\frac{10^6 \times 0.0024 \times 3600}{5400n} + 0 \right) (1 - e^{-n}) + 100 \, e^{-n}$$

$$1 = \frac{16}{n} (1 - e^{-n}) + e^{-n}$$

$$(1 - e^{-n}) = \frac{16}{n} (1 - e^{-n})$$

Hence $n = 16$

(b) From equation (16.13)

$$100 = \left(\frac{10^6 \times 0.0024 \times 3600}{5400 \times n} + 0 \right) (1 - e^{-n}) + c_0 e^{-n}$$

In this equation, n is the total number of air changes that has occurred after 1 hour, if the other terms of the equation are for one hour. It follows that a factor of 8 must be introduced in the appropriate places:

$$100 = \left(\frac{10^6 \times 0.0024 \times 3600 \times 8}{5400 \times n} \right) (1 - e^{-n})$$

where n is the total number of air changes that has occurred after 8 hours.

$$1 = \frac{128}{n} (1 - e^{-n})$$

Therefore,

$$\frac{n}{128} = (1 - e^{-n})$$

If n equals 128, e^{-n} is almost zero and so we can conclude that the required air change rate is 16 per hour.

(c) From equation (16.13)

$$c = \left[\left(\frac{10^6 \times 0.0008 \times 3600}{5400 \times 5.33} \right) (1 - e^{-5.33}) + 0 \right] \times e^{-5.33}$$

$$= 99 \text{ ppm}$$

(In the above solution, 0.0008 is the production of CO in m^3 in twenty minutes.)

(d) Equation (16.9) is used because there is no production of carbon monoxide after the eight-hour period and the fresh air used to ventilate contains no contaminant of this sort.

Observing that 0.001 per cent is a concentration of 10 ppm, we can write

$$10 = 100 \text{ } e^{-n}$$

because c_0 is 100 ppm.

Hence,

$$n = 2.3$$

At a rate of 16 air changes an hour, it takes 8.6 minutes to effect 2.3 air changes.

16.3 An application of the decay equation to changes of enthalpy

Throughout this book it has been usual to consider the steady-state case. Air has been supplied to the room being conditioned at a nominally constant temperature and since this temperature is lower than the design temperature in the room, the sensible heat gains to the room are offset. Under these conditions the cooling capacity of the airstream exactly matches the sensible heat gain to the room, and if the sensible gain stays unaltered, the temperature of the air in the room also remains unaltered.

The picture is different, however, when the air conditioning plant is first started up. Suppose that with the plant not running the conditions within the room are equal to those outside and that no heat gains are occurring. When the air conditioning system is started it delivers air to the room at a temperature very much below the value of the temperature prevailing there. Thus, the initial cooling capacity of the airstream is very large and the temperature of the air in the room is rapidly reduced. As this reduction is effected, the difference between supply air temperature and room air temperature decreases, and so the cooling capacity of the supply airstream diminishes.

This is a somewhat simplified picture of what is occurring. For example, heat gains are not solely due to transmission; solar radiation, electric lighting and occupants provide additional sensible gains and there are, of course, latent gains as well. There are, thus, complicated changes of load occurring.

By a process similar to that used to derive equation (16.13) it is possible to formulate, according to Jones (1963), a differential equation which represents the physical situation and to obtain a solution to it in terms of the relevant heat exchanges, enthalpies and masses.

The solution is

$$H = M\{[h_a + H(t)](1 - e^{-n})\} + H_0 e^{-n} \tag{16.15}$$

In this equation the following notation has been used:

M = mass of air contained in the room, in kg

H_0 = initial enthalpy of the air in the room, in kJ. (Hence $H_0 = Mh_0$, where h_0 is the specific enthalpy in kJ kg^{-1})

h_a = specific enthalpy of the air supplied to the room in kJ kg^{-1}

$H(t)$ = the enthalpy gain to the room, expressed in kJ s^{-1} per kg s^{-1} of air supplied, at any time t. (Hence the units of $H(t)$ are kJ kg^{-1}. In general, $H(t)$ is a function of time and the ease with which equation (16.15) may be made to yield a useful answer depends on how complicated this function is.)

G_a = rate of mass flow of the air supplied to the room, in kg s^{-1}

The use of this equation is not limited to problems relating to changes of temperature or moisture content in an air-conditioned room. It may be used to solve a variety of problems involving unsteady state operation. In using the equation it must be remembered that the influence of the thermal inertia of the building itself (as expressed by the concept of admittance in the *CIBSE Guide*) is ignored. The errors arising from this are usually quite small in the short term, particularly if the floor is carpeted and the walls are lightweight.

EXAMPLE 16.8

A room measures 3 m × 6 m × 3 m high and is air conditioned. Making use of the information given below, determine the dry-bulb temperature and relative humidity in the room three minutes after the air conditioning plant has been started, assuming that the initial state in the room is the same as the state outside. Ignore the effect of the thermal inertia of the building.

Sensible heat gain	2 kW
Latent heat gain	0.2 kW
Outside state	28°C dry-bulb, 20°C wet-bulb (screen)
Design inside state	22°C dry-bulb, 50 per cent saturation
Constant supply air state	13°C dry-bulb, 8.055 g kg^{-1}
Constant supply air quantity	0.217 kg s^{-1}
Specific heat capacity of humid air	1.025 kJ kg^{-1} K^{-1}

Answer

The humid volume at a state midway between the inside and the outside design states is about 0.8572 m^3 kg^{-1} of dry air and this establishes the fact that the mass of air contained in the room is about 63 kg.

Consider first the change of temperature that occurs when the air conditioning plant is started. Since temperature alone is the concern, the terms in equation (16.15) involving enthalpy may be conveniently modified so as to express only the sensible components of the enthalpy.

The information available may now be summarised—

$$G_a = 0.217 \text{ kg s}^{-1}$$
$$M = 63 \text{ kg}$$
$$H_0 = 63 \times 1.025 \times (28 - 0) = 1807 \text{ kJ}$$
$$H(t) = \frac{2.0 \text{ kW}}{0.217 \text{ kg s}^{-1}} = 9.22 \text{ kJ kg}^{-1}$$

484   *Ventilation and a decay equation*

$$h_a = 1 \times 1.025 \times (13 - 0) = 13.32 \text{ kJ kg}^{-1}$$
$$n = G_a t/M = 0.217t/63 = 0.00344t$$

Then, from equation (16.15),

$$H = 63(13.32 + 9.22)(1 - e^{-0.003\ 44t}) + 1807\ e^{-0.003\ 44t}$$

As t approaches a value of infinity, H approaches a value of 63×22.54 kJ, and so the specific sensible enthalpy of the air in the room tends towards a value of 22.54 kJ kg^{-1}. This means that the dry-bulb temperature of the air in the room eventually reaches a value of 22°C, as designed for. This is its 'potential value' (see section 13.10).

This may not seem very satisfactory. It must be remembered though, that when the system is started up it does not usually face its full design load. There is, thus, an opportunity for the system to pull down the room air temperature to 22°C under conditions of partial load.

After three minutes, for the case under consideration—

$$t = 3 \text{ minutes} = 180 \text{ seconds}$$
$$e^{-0.003\ 44 \times 180} = e^{-0.62} = 0.538$$

Hence,

$$H = (63 \times 22.54)(1 - 0.538) + 1807 \times 0.538$$
$$= 1627 \text{ kJ}$$

Thus

$$\text{Dry-bulb temperature} = \frac{1627}{63 \times 1.025} = 25.2°C$$

A similar approach, with appropriate modifications to the values of enthalpy so that latent heat is taken into account, but not the sensible component, yields a figure for the change in moisture content. From tables the specific enthalpy of dry air at 28°C is 28.17 kJ kg^{-1} and at the outside design state it is 55.36 kJ kg^{-1}, hence

$$G_a = 0.217 \text{ kg s}^{-1}$$
$$M = 63 \text{ kg}$$
$$H_0 = 63 \times (55.36 - 28.17) = 1712 \text{ kJ}$$
$$H(t) = 0.2/0.217 = 0.9215 \text{ kJ kg}^{-1}$$
$$h_a = 33.41 - 13.08 = 20.33 \text{ kJ kg}^{-1} \text{ where } 33.41 \text{ kJ kg}^{-1} \text{ is the enthalpy of the}$$
supply air and 13.08 kJ kg^{-1} is the enthalpy of dry air at 13°C

Then, from equation (16.15), as before,

$$H = 63(20.33 + 0.9215)(1 - e^{0.003\ 44t}) + 1712\ e^{-0.003\ 44t}$$

As t approaches infinity, H tends to 63×21.25, and so the specific enthalpy in the room tends to $21.25 + 22.13 = 43.38$ kJ kg^{-1}, the design value, where 22.13 kJ kg^{-1} is the enthalpy of dry air at 22°C, after an infinitely long period of time. After three minutes, as before, we get a solution as follows—

$$H = 63 \times 21.25(1 - e^{-0.62}) + 1712\ e^{-0.62}$$
$$= 1541 \text{ kJ or } 1541/63 = 24.5 \text{ kJ kg}^{-1}$$

At 25.2°C the sensible component of enthalpy, from tables, is 25.25 kJ kg^{-1} and so the total enthalpy is 49.75 kJ kg^{-1}.

At a dry-bulb temperature of 25.2°C and an enthalpy of 49.75 kJ kg^{-1}, the relative humidity is about 48 per cent.

Exercises

1. A classroom having a cube of 283 m^3 undergoes $1\frac{1}{2}$ air changes per hour from natural ventilation sources. The concentration of CO_2 in the outside air is 0.03 per cent and the production of CO_2 per person is 4.72×10^{-6} m^3 s^{-1}. Ignore the influence of the ventilation effectiveness coefficient, e_v.

(*a*) What is the maximum occupancy if the CO_2 concentration is to be less than 0.1 per cent at the end of the first hour, assuming that the initial concentration is 0.03 per cent?

(*b*) What is the maximum occupancy if the classroom is continuously occupied and the concentration must never exceed 0.1 per cent?

Answers

(*a*) 22, (*b*) 17.

Notation

Symbol	Description	Unit
A	constant of integration	
B	constant	m^3 s^{-1} per person
c	concentration of a contaminant, as a fraction, at any time	
c_a	concentration of CO_2 in the outside air, as a fraction	
c_{CO_2}	concentration of CO_2 in an occupied space, as a fraction	
c_e	pollution concentration in the exhaust air, as a fraction	
c_{ia}	perceived indoor air quality	decipol
c_0	initial concentration of a contaminant at time zero, as a fraction	
c_{oa}	perceived outdoor air quality	decipol
c_r	pollution concentration in the occupied zone of a room, as a fraction	
e_v	ventilation effectiveness coefficient	
G	sensory pollution load in the space	olf
G_a	mass airflow rate	kg s^{-1}
g_r	moisture content of room air	g kg^{-1}
H	enthalpy after time t	kJ
$H(t)$	enthalpy gain as a function of time	kJ kg^{-1}
H_0	initial enthalpy at time zero	kJ
h_a	specific enthalpy of air supplied to a room	kJ kg^{-1}
M	mass of air contained in a room	kg
n	number of air changes in a given time	
P	percentage of people dissatisfied	%
Q	volumetric airflow rate	m^3 s^{-1}

Q'	volumetric rate of fresh air supply per person	$m^3 s^{-1}$ per person
Q_0	volumetric rate of fresh air supply	litres s^{-1}
q	Fresh air ventilation rate for a pollution of one olf from a standard person	litres s^{-1} olf^{-1}
t	time	s
V	volume	m^3
V'	volume of a room per person	m^3 per person
V_c	rate of production of CO_2 through respiration	$m^3 s^{-1}$ per person
Δ_c	small change of concentration	
Δq	small volume of air	m^3
Δt	small interval of time	s

References

ACGIH (1971): American Conference of Governmental Industrial Hygienists, Committee on Threshold Limits.

ANSI/ASHRAE (1989): Ventilation for acceptable indoor air quality, ANSI/ASHRAE Standard 62–1989, Atlanta, Ga, USA.

ASHRAE (1999): Ventilation for acceptable indoor air quality, ASHRAE, Atlanta, Ga, USA.

ASHRAE Handbook, SI Edition (1997): Odours—Fundamentals, 13.1–13.6.

Berg-Munch, B., Clausen, G.H. and Fanger, P.O. (1986): Ventilation requirements for control of body odour in spaces occupied by women, *Environmental Int.* **12**, 195–199.

Bridge, D.P. and Corn, M. (1972): Contribution to the assessment of exposure of non-smokers to air pollution from cigarette and cigar smoke in occupied spaces, *Environmental Research* **5**, 192–209.

Brundage, J.F., Scott, R.M., Lodnar, W.M., Smith, D.W.A. and Miller, R.N. (1988): Energy efficient buildings pose higher risk of respiratory infection: study, *Journal of American Medical Association*, April 8, **259**, 2108.

Brundrett, G.W. (1975): Job No. 019, *Ventilation Requirements in Rooms Occupied by Smokers: A Review*, The Electricity Council Research Centre, Capenhurst, Cheshire, December.

BRE (1977): Ventilation requirements, *BRE Digest 206*, October.

CIBSE 1999: Guide A, Environmental design, Table 1.10, 1–19.

DIN 1946 (1983): Part 2, *Air conditioning health requirements* (VDI ventilation rules), January.

European Concerted Action, Indoor Air Quality and its Impact on Man (1992): Report No. 11, Guidelines for ventilation requirements in buildings, Commission of the European Communities, Director General for Science, Research and Development, Joint Research Centre—Environment Institute.

Fanger, P.O. (1988): Introduction of the olf and the decipol units to quantify air pollution perceived by humans indoors and outdoors, *Energy and Buildings* **12**, 1–6.

Grimstrud, D.T. and Teichman, K.Y. (1989): The scientific basis of Standard 62–1989, ASHRAE Journal, October, **9**, 51–54.

Health and Safety Executive (1979): Ventilation of Buildings: Fresh Air Requirements. Guidance Note EH22 from the Health & Safety Executive, *Environmental Hygiene*, March, **22**.

Hoegg, U.R. (1972): Cigarette smoke in confined spaces, *Environmental Health Perspective*, October, **2**, 117–128.

Jones, W.P. (1963): Theoretical aspects of air conditioning systems upon start-up, *JIHVE* **31**, 218–223.

Leopold, C.S. (1945): Tobacco smoke control—a preliminary study, *ASHVE Trans.* **51**, 255–270.

Pejtersen, J., Bluyssen, P., Kondo, H. and Fanger, P.O. (1989): *Air pollution sources in ventilation systems*, Lyngby, Denmark: Technical University of Denmark.

Swedjermark, A. (1978): Ventilation requirements in relation to the emanation of radon from building materials, World Health Organisation Symposium: Indoor Climate, Copenhagen.

WHO (1983): Indoor air pollutants, exposure and health effects, *Euro Reports and Studies*, 78, 23–26.

WHO (1987): Ventilation for acceptable indoor air quality, World Health Organisation, Copenhagen.

Yaglou, C.P., Riley, E.C. and Coggins, D.I. (1936): Ventilation Requirements, *ASHVE Trans.* **42**, 133.

Bibliography

1. I. Turiel and J.V. Rudy, Occupant-related CO_2 as an indicator of ventilation rate, *ASHRAE Trans*, 1982, **88**, Part 1, 197–210.

2. D.S. Becker, Organic contaminants in indoor air and their relation to outdoor contaminants, Phase 2—Statistical Analysis, *ASHRAE Trans.*, 1982, **88**, Part 1, 491–502.

3. J.L. Repace and A.H. Lowrey, Ventilation and indoor air quality, *ASHRAE Trans*, 1982, **88**, 895–914.

4. W.S. Cain *et al.*, *Ventilation requirements for control of occupancy odour and tobacco smoke odour*, Laboratory Studies LBL—12589, NTIS, Springfield, VA.

5. *ASHRAE Handbook* 1997 Fundamentals, chapters 12, 13 and 25.

6. *Occupational Exposure Limits*. Health and Safety Executive, HMSO (Guidance Note, Environmental Hygiene/40), 2000, updated annually.

7. *The Offices, Shops and Railway Premises Act*, 1963.

17

Filtration

17.1 Particle sizes

Atmospheric contaminants fall into four classes: solid, liquid, gaseous and organic. Distinction between some members of these classes is not clear cut and not particularly important, but recognition of their existence is relevant. For this reason the following broad statements are made; they should not be regarded as a set of definitions.

In speaking of small particles—which all atmospheric contaminants are—it is customary to use the micrometre as a unit of measurement of their size. To give a perspective, the diameter of human hair is in the range 30 to 200 μm.

(a) Dusts (<100 μm)
These are solid particles by natural or man-made processes of erosion, crushing, grinding or other abrasive wear. Dusts do not agglomerate, except under the influence of electrostatic forces, but settle on the ground by the force of gravity.

(b) Fumes (<1 μm)
These also are solid particles but formed in a different way from dusts. Fumes are produced by the sublimation, or by the condensation and subsequent fusion, of gases which are solids at normal temperature and pressure. Metals can be made to produce fumes. The term is often misused to indicate merely a gaseous substance which has a pungent smell. Fumes agglomerate into large clusters with ageing.

(c) Smokes (< 1 μm)
Smokes may be regarded as small solid particles which are the product of incomplete combustion or, more truly, as a mixture of solid, liquid and gaseous particles resulting from partial combustion. Excluding the gaseous particles, which are molecular in size, industrial and domestic furnaces produce particles which vary in size from 0.1 μm to 1.0 μm, but tobacco smokes are much smaller, existing in the range 0.01 to 1.0 μm. Hence the difficulty in their effective removal from air streams.

(d) Mists and fogs (<100 μm)
The distinction between mists and fogs is somewhat blurred. However, they are both airborne droplets which are liquid at normal temperature and pressure. Their normal range of size is 15 to 35 μm.

(e) Vapours and gases

A distinction between the two has already been drawn in section 2.2. From a filtration point of view, they are substances which are in the gaseous phase at normal temperature and pressure, but whereas a vapour may be removed by cooling to below the dew point, a gas cannot. Both gases and vapours diffuse uniformly throughout an enclosing space. Separation by inertia is not possible. The common atmospheric gases have molecule sizes from 0.0003 µm to 0.000 45 µm and mean free paths at atmospheric pressure of 0.06 to 0.2 µm, implying that only a very small percentage of the volume of a gas is occupied by the molecules themselves.

(f) Organic particles

The commonest of these are: bacteria (0.2–5.0 µm), pollen (5.0–150 µm), the spores of fungi (1.0–30 µm) and viruses (much less than 1.0 µm in size). Bacteria are generally larger in size than 1.0 µm and rely on dust particles as a mode of transport. Hence the importance of dust filtration in controlling the spread of infection by bacteria. Viruses are very small indeed (0.03 to 0.06 µm), and for this reason many have never been identified. Some are transported by airborne liquid droplets.

(g) Raindrops

These vary in size from 500 to 5000 µm.

(h) Aerosols

This is the name given to a semi-stable dispersion of small liquid droplets, or solids, in a gas, with a range of sizes from less than 0.01 to 100 µm. Although they follow a general pattern of movement in the parent gas they may coagulate or be deposited on surfaces by normal gravitational, inertial and other forces acting on them.

   Individual particles, other than viruses, of size less than 0.1 µm are not thought to be of great importance. They may, however, become permanent atmospheric impurities, particularly as smells. Other very much larger particles, such as insects and even birds, must be kept out of air-conditioning systems but, except in the case of insects and electric filters, they do not usually constitute a special filtration problem.

17.2   Particle behaviour and collection

Air resistance is a major factor in the settlement of particles under gravitational forces, the one balancing the other at a terminal velocity approximately proportional to the square of the particle size (diameter) up to about 60 µm. Because of this, gravitational separation is only of use for particles exceeding roughly 1 µm in size, for which the terminal velocity is approximately 0.3×10^{-1} m s^{-1}, and so the filters of interest in air conditioning rely on other forces.

   There are four basic ways in which dust particles are collected:

(1) *Diffusion.* The natural (Brownian) motion of the molecules of air is sufficient to impart movement to very small particles of dust by collision with them. The particles deviate from the direction of flow of the mainstream to be collected at the filter surfaces. This is the major mode of filtration in high efficiency filters where air velocities are too low for the effective inertial separation of particles less than about 0.5 µm in size.

(2) *Straining.* Particles larger than the space between the fibres of the filter material are strained out of the airstream, largely at the upstream face.

(3) *Interception*. Also termed impingement and impaction this is an aspect of inertial separation: larger, heavier dusts are removed by collision with the filter material, the background airstream of lighter and smaller particles flowing around the collection surfaces.

(4) *Electrostatic*. Apart from the high voltage fields deliberately engineered in electrostatic filters similar effects may exist naturally within filter materials, particles being collected at surfaces of opposite sign. Dust can become charged by collision with ionised molecules or by friction, and coagulate. Acoustic coagulation at ultrasonic frequencies is also possible but exceedingly large sound intensities, of the order of 1 kW m^{-2} are needed according to Brandt *et al.* (1937).

As a general principle, inter-fibre distances should be large compared with fibre diameters in a filtration material, otherwise the filter will rapidly clog but, after the dust particles collide with a fibre they must adhere to it for effective collection.

The influences on surface adhesion are:

(1) London–van der Waals forces. Dorman (1975) has explained that these are apparently electrical in origin and are more important for very small particles sticking to surfaces.

(2) Electrostatic.

(3) Surface tension. This arises if films of moisture are present on the surface. Adhesion is increased by the presence of water vapour and humidity can play a significant part for larger dusts (>1 μm), adhesion diminishing as humidity falls. Ma (1965) has shown that so-called capillary condensation can occur within filters under suitable circumstances, when humidities exceed 70 per cent. Although this assists adhesion, it can favour bacterial growth.

(4) The nature of the surface. This affects the van der Waals forces and influences the extent to which vapours and gases are adsorbed.

(5) Surface contamination.

(6) Shape and size of the dust particles.

(7) Duration of contact. (Adhesive forces increase with time.)

(8) Temperature. It is believed by Corn (1966) that, by altering the forces of surface tension, surface adhesion could be modified.

The extent to which particles penetrate a filter increases as the velocity rises and is greatest for particles of 0.1 to 0.3 μm diameter at velocities of a few cm s^{-1}, although filter material and packing density have an effect. The presence of a pin-hole in a filter greatly reduces its effectiveness in preventing the penetration of microbes or toxic materials. The penetration measured at normal velocities (see section 17.3) is usually unaffected by the presence of pin-holes and their detection must be achieved by a test at very low velocity when the proportional penetration through a pin-hole is much greater than that through the rest of the filter.

17.3 Efficiency

(*a*) *Synthetic test dusts*. The staining of walls and fabrics in rooms is not a function solely of the mass of dust present in the air, it also depends on the size of the particle, small particles of small mass stain surfaces just as much as larger and heavier particles. Straightforward gravimetric tests according to BS EN 779: 1993 measure arrestance, which expresses the weight of dust retained by a filter as a percentage of the weight fed into it and do not describe efficiency because large, massive particles are more readily collected than smaller ones which are the prime cause of staining in rooms. Furthermore, indoor air quality depends on contaminant particles of small size, as well as large. The discolouration test, in one form or another, has therefore been extensively adopted for the expression of efficiency whilst the ability of a filter to remove a mass of dust is referred to in terms of an arrestance test.

To achieve uniformity in the expression of arrestance test results, a standard dust has been developed in the United States and this has been adopted in Europe. In the past, simple synthetic dusts, such as crystals of sodium chloride, have been used with success as standards in the UK for testing purposes with high efficiency filters and ground aluminium oxide for gravimetric tests on less efficient filters. Although sodium chloride is retained as a European standard test dust for absolute filters, see Eurovent 4/4 (1980), general filter testing is now done according to BS EN 779: 1993 with a synthetic dust developed and used by ASHRAE (1992). It comprises: 72 per cent (by weight) of standardised air cleaner dust—fine, 23 per cent Molocco black (carbon black) and 5 per cent of No. 7 cotton linters ground in a Wiley mill with a 4 mm screen.

When testing high efficiency absolute filters it is usual to speak of penetration rather than efficiency, one being the complement of the other. Sodium chloride is adopted for this in Europe (see Eurovent 4/4 (1980)) but in the United States the practice is to use di-iso-octyl-phthalate (abbreviated DOP), a smoke-like homogeneous aerosol of particles of nominal 0.3 μm size, chosen because it was considered the most difficult size for filters to remove. This test, for high efficiency (small penetration) filters, must not be confused with site tests that also use DOP smoke, generated locally. DOP test results are not readily comparable with sodium flame test figures.

(*b*) *Atmospheric dust spot efficiency.* The current method (BS EN 779: 1993), based on Eurovent 4/5 (1980) and ASHRAE 52.1–92 (1992), classifies filters as coarse, numbered G1 to G4, and fine, numbered F5 to F9. Coarse filters are defined as those with an initial atmospheric dust spot efficiency less than 20 per cent and are subject only to tests of arrestance. Average values of arrestance up to 90 per cent are quoted. Corresponding values of dust spot efficiency are irrelevant and not quoted. Fine filters are those with an initial dust spot efficiency greater than or equal to 20 per cent but not exceeding 98 per cent. Values of efficiency from 40 per cent to 95 per cent are given. Values of arrestance are irrelevant and not quoted. The efficiency of a filter varies over its life as it collects more dust. The Eurovent standard deals with this by specifying a series of tests of both efficiency and arrestance, over a period of time. Measured volumes of air are taken from a ducted airstream before and after the filter under test and passed through separate filter papers, termed targets, and staining them. Equal sample airflow rates are established by using critical flow nozzles. The downstream sampling is taken intermittently with an elapsed time meter to record 'on' time. These data are used to verify the readings on the gas flow meters in each sampling tube. There is a high pressure drop through the filter papers and vacuum pumps are required to sample the upstream and downstream air. The extent to which a beam of light passes through the filter papers is interpreted to give the filter efficiency, according to a prescribed method. Before commencing the test, the opacity, or relative light transmission, of the target paper must be established using a beam of light and a photo-sensitive cell (an 'opacity meter'). The difference between the opacities of the two unused targets must not exceed 2 per cent. During the test itself the downstream (clean) sampler runs continuously for the whole of the efficiency test period but the upstream (dirty) sampler runs intermittently for a total period of time related to the anticipated efficiency and requiring some skill in its choice. The efficiency, E, is then expressed by:

$$E = 100 \left(1 - \frac{Q_1 O_2}{Q_2 O_1} \right) \tag{17.1}$$

wherein

O_1 is the opacity of the dust spot on the upstream target, given by $O_1 = (T_{u1} - T_{u2})/T_{u1}$ and O_2 is the opacity on the downstream target, given by $O_2 = (T_{d1} - T_{d2})/T_{d1}$; T_{u1} and T_{u2} are the initial and final light transmissions as a percentage through the upstream target and T_{d1} and T_{d2} are similar transmissions through the downstream target; Q_1 and Q_2 are total volumes sampled through the upstream and downstream targets.

When carrying out the dust spot efficiency test, expressed according to equation (17.1), ordinary air from the laboratory is used, *without the addition of any synthetic dust*. The intention is to get the opacities of the upstream and downstream targets, O_1 and O_2, as nearly equal as possible by the end of the test period. Then O_2/O_1 is unity and the efficiency depends on the measured volumes of air, Q_1 and Q_2, drawn through the two targets, the quantity of dirty air, Q_1, being much less than that of the clean air, Q_2. For self-renewable filters, the tests must be done either in the period when the pressure drop across the filter is rising to its upper operating limit or during a time when there is steady-state operation.

(*c*) *Arrestance.* Arrestance tests are carried out by the injection of several increments of synthetic dust, during the total period of the test.

The answers obtained for arrestance or efficiency can then be graphed against the weight of dust fed into the test rig (see Figure 17.1). The filter to be tested and an after-filter—to be fitted downstream for catchment purposes—are each weighed and then mounted in the test rig. A known weight of test dust is fed into the rig and the dust passing through the filter under test is collected by the after-filter. This is then removed and reweighed, the increase in weight being used to establish the synthetic dust weight arrestance of the filter under test, as a percentage. A minimum of four measurements must be made and arrestance is then defined as

$$A = 100(1 - W_a/W_1)\% \tag{17.2}$$

wherein W_a is the weight of synthetic dust collected by the after filter and W_1 is the weight injected.

The average atmospheric dust spot efficiency, E_m, using the notation in Figure 17.1, is defined by

$$E_m = \frac{100}{W} [W_{AB}(E_A + E_B)/2 + W_{BC}(E_B + E_C)/2 + \dots$$

$$W_{(n-1)n}(E_{(n-1)} + E_n)/2] \tag{17.3}$$

wherein $W_{(n-1)n}$ is the $(n-1)$th increment of dust fed into the test rig.

The average arrestance, A_m, is determined from:

$$A_m = \frac{100}{W} (W_1 A_1 + W_2 A_2 + \dots W_f A_f) \tag{17.4}$$

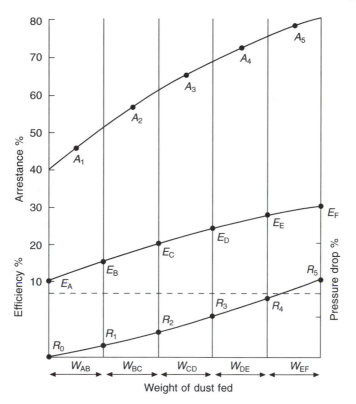

Fig. 17.1 Arrestance or efficiency rising with a test to BS EN 779: 1993.

wherein W_1, W_2, etc. are the increments of dust fed into the rig and W_f is the final increment, with $A_1, A_2, \ldots A_f$, the corresponding arrestances according to equation (17.2) and W is the total weight of dust injected.

(*d*) *Dust holding capacity.* For disposable and non self-renewable types of filter the dust holding capacity is the total of the dust increments, W, related to the pressure drop.

For self-renewable filters the dust holding capacity is expressed per m^2 of filter medium when the filter is working in a steady-state condition.

(*e*) *Sodium flame test.* Eurovent 4/4 (1980) prescribes a test where sodium chloride solution is injected into the airstream on the upstream side of the filter under test and evaporation leaves a suspension of cubical crystals having some uniformity of shape and size: the average diagonal is 0.6 µm, the largest 1.7 µm and 58 per cent are less than 0.1 µm. This test gives an instantaneous reading of the penetration of particles through the filter. A specimen of air from the downstream side of the filter is passed through a burning hydrogen flame, the intensity of the resulting yellow colour being a direct indication of the weight of sodium chloride present. The test is very stringent and therefore only used for so-called absolute filters.

Technical developments have established methods of continuous counting and sizing of particles. Most use the principles of the scattering of light, leading to the measurement of particle sizes down to 0.3 µm. With laser light sources counting and sizing is possible down to 0.1 µm. The condensation of water vapour in the atmosphere to liquid depends on

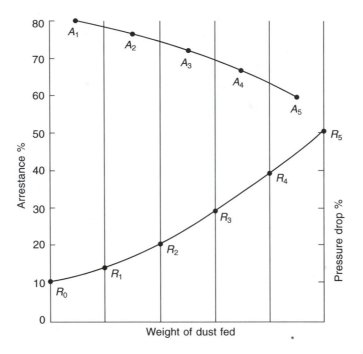

Fig. 17.2   Arrestance falling and pressure drop rising with a test to BS EN 779: 1993.
Efficiency less than 20%.

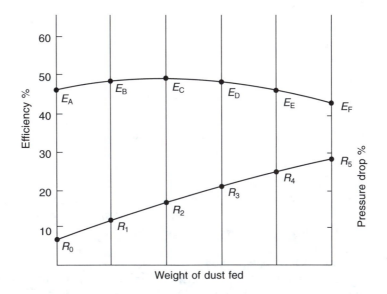

Fig. 17.3   Efficiency first rising then falling with a test to BS EN 779: 1993.

the presence of condensation nuclei and Scala (1963) has developed an instrument for their measurement. This has led to condensation nuclei counters that can deal with particle sizes of 0.01 μm. All future standards for good quality filters will be related to particle size. Ter Kuile (1997) reports developments in this direction. The effectiveness of filters can be described in terms of the ability to remove particles of a particular size, also known as their fractional efficiency. The term HEPA (high efficiency particulate air) filters has been in use for some time but with the improvement in filter standards it has become necessary to use the term ULPA (ultra low penetration air) filters. The classification in BS EN 779: 1993 then has two further classes, H10 to H14 for HEPA filters and U15 to U17 for ULPA filters. The corresponding average efficiencies are 95 per cent to 99.999 per cent at 0.3 μm for the H range and 99.9995 per cent to 99.999995 per cent at 0.12 μm for the U range. Another term used is Most Penetrating Particle Size (MPPS) for which the average efficiencies over the H and U ranges are virtually the same. Such very high efficiency filters work at lower face velocities than are usual, with capital cost implications.

17.4 Classification according to efficiency

The filters in common use have typical efficiencies and arrestances as listed in Table 17.1, according to Eurovent standards.

Table 17.1 Typical classification of arrestance and efficiency

Filter type	Face velocity m s^{-1}	Pressure drop Pa	Arrestance %	Atmospheric dust spot efficiency %
Automatic				
Viscous	1.87–2.50	100	80	—
Dry	2.54–3.58	125	70–80	—
Panel				
Cleanable, viscous	1.9–2.7	80	65–80	—
Cleanable, dry	1.72	95	70–76	—
Disposable	1.8–3.78	135	70–89	—
Bag				
Low-efficiency	3.78	220	—	30–50
Medium-efficiency	3.78	220	—	55–90
High-efficiency	2.54	220	—	90–97
Absolute				
Low-efficiency	1.40	110, 750	—	95
Medium-efficiency	1.26	250, 750	—	99.7
High-efficiency	1.26	250, 875	—	99.997
Electrostatic Agglomerator with storage section	2.54	100	—	90

Notes
1. The figures are based on averages of manufacturers' data.
2. Where two pressure drops are given they are for the clean filter and the dirty filter.
3. A single pressure drop represents an average figure.
4. Absolute (ULPA) filters with efficiencies as high as 99.999995% are also possible.

17.5   Viscous filters

Essentially, these filters have a large dust holding capacity but a low efficiency, and this defines their sphere of application; for example, they are more suitable for use in industrial areas where a high degree of atmospheric pollution prevails. Their drawback is usually the expense, particularly in automatic versions.

The principle of the viscous filter is that if the mixture of dust and air is forced to follow a tortuous path in negotiating a passage through the filtering medium, inertial separation of the more massive dust from the lighter air will occur. If the filtering medium is coated with a suitable oily fluid, the particles of dust will be trapped by the oil and retained, the air passing on. To effect this, the oil must have a surface tension low enough to permit easy entry for the dust. Once within the oil, it is desirable that the dust should disperse, not remaining in one place but being retained in depth throughout the filter (hence its large dust-holding capacity). It is thus necessary that the oil used should have a high enough capillarity to encourage dust to flow away from its point of entry to the oil. It is also essential that the oil should be non-inflammable, non-toxic, germicidal and comparatively non-evaporative. The oil should not deteriorate during its life. There are two types of viscous filter: the cell-type and the automatic type.

The cell-type consists of a cheap retaining box, open front and rear, which contains the filtering medium. The medium (often some form of industrial waste such as swarf, brass turnings, etc.) is coated with the oil, and the cells are assembled into a battery of convenient shape and placed across the airstream for filtration purposes. After use, when dirty, the cells are either thrown away and replaced with new cells or, if they have been selected with this in mind, they may be treated with a cold wash, drained, and recoated with fresh oil for further use.

The automatic viscous filter takes the form of a continuous roll of material coated with the oil and is motor-driven across the airstream. It is arranged so that the roll is drawn through a trough of oil at the bottom of the assembly. The trough serves the dual purpose of washing off the dirt and recoating the fabric of the roll with relatively clean oil. The material of the fabric takes several forms: one is an array of hinged metal plates which, dangling downwards, force the dirty airstream to change direction several times, thus achieving the inertial separation of the dust into the oil, as desired; other versions of automatic viscous filter are elaborated by the use of oil sprays and pumps. Figure 17.4 illustrates the cell and automatic types.

Pressure drops across these filters vary from 40 to 140 Pa, and recommended air velocities over the face are from 1.7 to 2.5 m s^{-1}. Viscous filters are not very effective in removing particles of size less than 0.5 μm.

17.6   Dry filters

Cotton-wool, glass-fibre fabric, pleated paper of various types, foamed polyurethane, cellular polythene and other materials are used for the construction of dry filters. As with viscous filters, there are cell-type and automatic roll-type filters available. The construction of these, in broad outline, is very similar to that of viscous filters. A certain amount of viscous, gel-like substance is sometimes applied to glass-fibre fill to help dust retention and it is common practice to increase the density of packing towards the downstream end of the filter to improve the removal of the smaller sized particles and make the filter effective through its entire depth. In general, efficiency is improved by increasing the

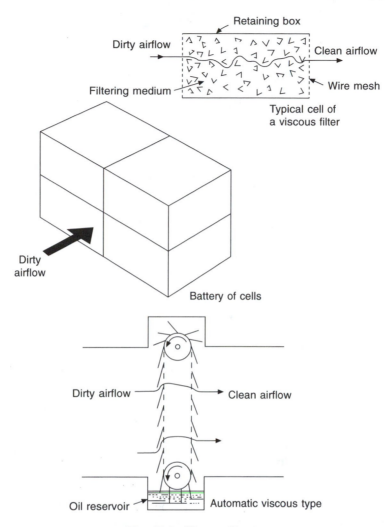

Fig. 17.4 Viscous filters.

surface area of the material offered to the airstream. In cell-types this is achieved by using pads of material placed obliquely across the airstream (Figure 17.5). An alternative to this, which can be made to yield very high efficiencies, is to use a system of pleating. This is best achieved with glass fibre paper, although other material is also used, and gives high efficiencies. Figure 17.6 shows an arrangement of pleating.

When a very large amount of material is used in a filter, the efficiency becomes very high and the filter is termed an 'absolute' filter. No filter is truly absolute but almost any desired efficiency could be achieved if sufficient filtering fabric were used; this would be associated with a high pressure drop, but one way of achieving a high efficiency without the penalty of wasted energy is to use a very large filter, that is, one with a very low face velocity.

Automatic dry-fabric filters consist of an upper roll of clean fabric wound downwards across the airstream. The dirtied material is then rewound into a roll at the bottom of the unit. Figure 17.7 shows this arrangement. The advantage of this type of filter is the low

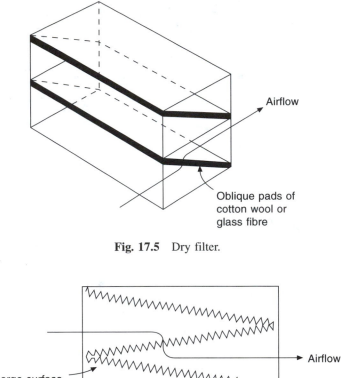

Fig. 17.5   Dry filter.

Fig. 17.6   Pleated high efficiency filter.

maintenance cost when compared with the cell-type of dry filter. Removing a dirty roll and replacing it with a clean one at relatively infrequent intervals is a cleaner task and is, therefore, easier and cheaper than changing dirty cells. The disadvantage is that the efficiency is not so high. This is overcome to some extent by using a denser material and by introducing V-form or S-shaped changes of direction in the fabric as it crosses the airstream, thus presenting more surface area. The rolls, as in automatic viscous filters, are motorised, and the geared-down electric driving motors are operated from time switches. As with all filter installations there is a risk that the filtering medium may be by-passed by dirty air. This must not be allowed.

17.7   Electric filters

An electric filter is illustrated in Figure 17.8. The principle of operation is that when air is passed between a pair of oppositely charged conductors it becomes ionised if the voltage difference between the conductors is sufficiently large. Both negative ions and positive ions are formed, the latter being in the larger quantity. By contact with the dust particles mixed with the airstream, the charge of the ions is shared with the dust. In this way, about 80 per cent of the dust particles passing through the ionising field are given a positive

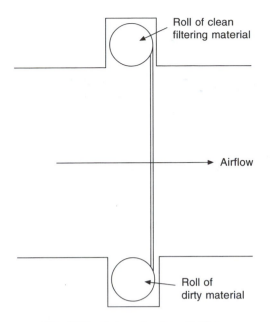

Fig. 17.7 Automatic dry roll filter.

charge and the other 20 per cent a negative charge. The ionising voltage used varies somewhat, and to achieve a given efficiency of ionisation smaller air velocities can be used with smaller voltages. However, typical ionising voltages are from 7800 to 13 000.

The electrodes which form the poles of the ionising unit mentioned above consist of alternate small diameter (of the order of 25 μm) tungsten wires and flat metal plates, spaced 25 mm or so apart. The intensity of an electric field is a function of the curvature of the charged conductor producing it, hence the small diameter of alternate poles. The tungsten wires are positively charged, the plates representing the negative electrodes being earthed.

The charged particles of dust which leave the ionising unit then pass through a collecting unit. This consists of a set of vertical metal plates, spaced about 15 mm apart. Alternate plates are positively charged and earthed and attract the negatively and positively charged dust particles, respectively. The voltage difference across the collecting plates is about 6000 or 7000, although at least one manufacturer is currently offering an electrical filter which uses about 8000 volts across the electrodes of both the ionising and collecting sections.

One of the main advantages claimed for electric filters is low maintenance cost, when compared with other filters of similar efficiency but lower capital cost. Although the low maintenance cost is debatable and if a comparison is made between filters on a basis of their owning and operating costs, the result is largely at the choice of the analyst, it is very necessary there should be automatic washing of electric filters if maintenance cost is to be minimised. To assist in the retention of the dust on the collecting plates, some manufacturers arrange for the plates to be coated with an oil. This may mean that a detergent is necessary as well as a water supply when washing is carried out. It also means that fresh oil must be put on the plates after the washing. A variation on this is to use a mixture of oil and detergent together so that when washing is done, usually with cold water, respraying with

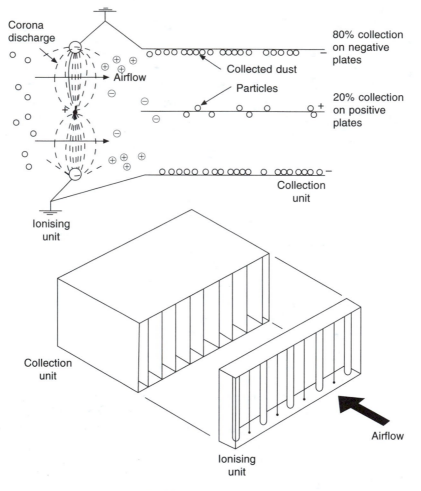

Fig. 17.8   An electric filter.

oil also provides the detergent necessary for the next wash. One manufacturer uses no detergent but asks that hot water be used for washing. Oil is not used in this case. One feature of washing common to all filters is that water at high pressure is required, gauge pressures of 2 to 3 bar being necessary. The method of washing is either from fixed stand-pipes attached to the appropriate reservoirs of water and oil or from a traversing nozzle system, attached to the reservoirs by flexible piping. The pump which feeds the water to the nozzles usually runs at 28 rev s^{-1} and is noisy, because of the requirements of a large head.

An insect screen is necessary upstream of the filter, and it is also necessary to provide either a pre-filter or an after-filter.

A pre-filter is preferred because it relieves the filtration load on the electric filter which follows. However, if it is used it must be washed automatically when the rest of the filter is cleaned, otherwise there is an increased maintenance problem. If an after-filter is used, its duty is merely to ensure that the filter fails safe. Neither the pre-filter nor the after-filter need be very elaborate; a simple screen of closely woven nylon cloth is often enough. If the filter is not washed regularly at suitably short intervals, the thickness of the dust-covering on the collection plates increases and, after a while, dust starts flaking off and is carried into the conditioned space. Hence another need for the after-filter. The ideal then is an automatically washed pre-filter and a simple after-filter.

Another type of electric filter ionises the air in the usual way and then retains the charged dust particles in a thickening layer on the collector plates. The dust is then swept off by the airstream as large agglomerated flakes which are retained in a storage section downstream consisting of an automatic, dry fabric, roll filter. Atmospheric dust-spot efficiencies of 90 per cent are achieved.

17.8   Wet filters

Washers and scrubbers of various sorts are used extensively throughout industry, largely for the absorption of soluble gases. They are not very commonly used for cleansing the air of solid dust particles. The effectiveness of a washer in removing a dust depends on the 'wettability' of the dust by water. This is a function of the surface tension of the water when it is in contact with the solid involved. Different forces of surface tension are the rule for different materials. Thus, water finds it very difficult to wet greasy materials but fairly easy to wet non-greasy materials. Examples of greasy materials are atmospheric pollutants present in urban and industrial atmospheres; it follows that washing the air is not a very effective way of cleaning it of the dust normally present. Spray-type washers are not to be regarded as filters.

As was remarked above, air washers are an effective way of dealing with undesirable atmospheric gases such as sulphur dioxide, which may be taken into solution by the water. The washer water then becomes acidic and water treatment and a continuous blow down are essential.

Air washers are a potential source of the bacteria that cause legionnaires' disease. It follows that they are a hygienic risk and are undesirable in air conditioning. See section 3.5.

17.9   Centrifugal collectors

If air is made to travel in a circular path, centrifugal force acts both on the molecules of the air and on the associated dust particles. The dust particles being the heavier, the force on them is the greater and so they are forced to the outer boundary of the curved airstream. This is the principle of the cyclone, illustrated in Figure 17.9. This method of dust removal is not in use in air conditioning systems. Its application is confined to industrial exhaust installations where the dust mixed with the airstream is relatively massive, such as wood shavings and sawdust.

17.10   Adsorption filters

The process of absorption, which is a chemical process, is to be distinguished from that of adsorption, which is a purely physical process.

A detailed explanation of adsorption is beyond the scope of this book, but Robinson and Quellet (1999) have provided a useful account of the process, which is sometimes termed gas-phase filtration and is also known as adsorption condensation. This relates to the fact that some gaseous pollutants, with suitably high boiling points, diffuse onto the suraces of

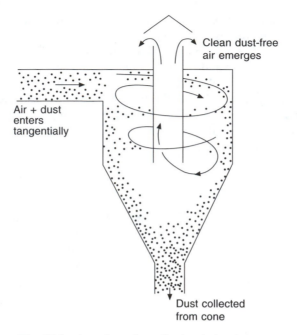

Air + dust
enters
tangentially

Clean dust-free
air emerges

Dust collected
from cone

Fig. 17.9   A cyclone for collecting industrial dust.

certain materials and condense. The temperature, molecular weight, and vapour pressure of the pollutant play a part, as also does the nature of the adsorbing surface. Pollutants with high boiling points diffuse onto the surfaces of the adsorption material and further diffusion into the surfaces within the material also occurs. The pollutants exist in the liquid phase at room temperatures. As air and pollutants flow through the filter the adsorption process continues but its effectiveness diminishes and eventually all the available receiving surface area is covered with condensed pollutant. The process is reversible and the filter can be removed and reactivated by heating.

Moisture in the air also condenses onto the surfaces, lessening their effectiveness for adsorbing the pollutant. It follows that high relative humidities are undesirable. If necessary, the air onto the adsorption filter might be warmed to reduce the humidity.

The principal adsorption filter used in air conditioning is the activated carbon filter. It is most effective in removing smells from the atmosphere, and in removing poisonous gases such as sulphur dioxide. The capacity of an activated carbon cell is expressed as an efficiency, but for sulphur dioxide, as an example, a typical cell with an efficiency of 95 per cent cannot adsorb 95 per cent of its own weight in sulphur dioxide. In fact, for the product of one particular manufacturer, a cell with a quoted efficiency of 95 per cent can adsorb 10 per cent of its own weight in sulphur dioxide. The cell contains 20 kg of activated carbon and can adsorb 2 kg of SO_2. (Activated carbon, incidentally, is prepared from the shells of coconuts and has a structure which offers an enormous surface area to any stream of gas passing over it.)

The adsorption capacity of activated carbon for ammonia, ethylene, formaldehyde, hydrogen chloride and hydrogen sulphide, which boil at between $-104°$ and $-21°C$, is inadequate for practical purposes. On the other hand, substances such as butyric acid ('body odour'), petrol, putrescine and the common mercaptans etc., which boil at the higher temperatures of 8° to 158°C, are very effectively adsorbed. The activated carbon used during the 1939–45 war for adsorbing the poisonous low boiling point substances such as arsine, hydrocyanic acid and other war gases was specially impregnated with other chemicals to increase the adsorption. This is sometimes called chemisorption. A chemical reaction occurs between the reagent added and the receiving surfaces of the adsorbent, forming a new chemical compound. Such filters cannot be regenerated but must be safely disposed of at the end of their useful lives.

On the other hand, when activated carbon has reached saturation it is removed from the filter for re-activation. This is accomplished by heating the carbon to a high temperature, of the order of 600°C or more. It is customary to return saturated cells to the supplier for re-activation, using spare replacement cells in the meantime.

17.11   Safety

With the passage of time, the filtration medium used in air filters may tend to erode under the influence of airflow and some concern has been expressed[11] about this. Glass fibre, rock wool and slag wool, used in filters, shed minute particles into the airstream and thence into the conditioned space, where they may be inhaled, with a potential risk to health.

Fibrous filter material is usually coated during manufacture to reduce significantly the risk of erosion during the life of the filter, and it does not necessarily follow that such filters are dangerous to use in air conditioning systems. However, it is noted in CIBSE (1992a) that the Health and Safety Executive gives a maximum exposure limit of 5 mg m^{-3} for inhalable micro-glass particles with a limiting fibre count of 2 fibres per minute. The electrostatic charge, naturally present in most filters, also discourages erosion and this is sometimes deliberately increased to enhance the effect. Apparently, according to CIBSE (1992b), it is possible to combine this electrostatic effect with the use of non-respirable fibres.

Fire tests on filters made of glass fibres and of synthetic, non-glass fibres have been carried out and these indicate that filters using glass fibre give a higher risk. It is to be noted that not only will the glass fibre burn but, if the igniting flame is sufficiently large, the dirt itself on the filter constitutes an increased fire hazard.

Notation

Symbol	Description	Unit
A	arrestance	%
A_1	arrestance for an initial increment of injected synthetic dust	%
A_2	arrestance for a second increment of injected synthetic dust	%
A_f	arrestance for a final increment of injected synthetic dust	%
A_m	average arrestance	%
E	atmospheric dust-spot efficiency	%
E_A	atmospheric dust-spot efficiency before the initial injection of synthetic dust	%

E_B	atmospheric dust-spot efficiency before the second injection of synthetic dust	%
E_m	average atmospheric dust-spot efficiency	%
E_n	atmospheric dust-spot efficiency before the nth injection of synthetic dust	%
O_1	opacity of a dust-spot on an upstream target	—
O_2	opacity of a dust-spot on a downstream target	—
Q_1	volume of air sampled upstream	m^3
Q_2	volume of air sampled downstream	m^3
T_{d1}	initial light transmission through a downstream target	—
T_{d2}	final light transmission through a downstream target	—
T_{u1}	initial light transmission through an upstream target	—
T_{u2}	final light transmission through an upstream target	—
W	total weight of dust injected	kg
W_a	weight of synthetic dust collected by an after-filter	kg
W_f	final incremental weight of synthetic dust injected	kg
W_n	nth incremental weight of synthetic dust injected	kg
W_{AB}	initial incremental weight of synthetic dust injected	kg
W_{BC}	second incremental weight of synthetic dust injected	kg
$W_{(n-1)n}$	$(n-1)$th incremental weight of synthetic dust injected	kg
W_1	first incremental weight of synthetic dust injected, or weight of synthetic dust injected	kg
W_2	second incremental weight of synthetic dust injected	kg

References

ASHRAE (1992): Gravimetric and dust spot procedures for testing air cleaning devices used in general ventilation for removing particulate matter, Standard 52.1–1992.

Brandt, O., Freund, H. and Hiedemann, E. (1937): *Z. Phys.* **104**, 511.

BS EN 779: 1993: Particulate air filters for general ventilation. Requirements testing, marking. British Standards Institution.

CIBSE Journal (1992a): A step forward for filtration? *Building Services The CIBSE Journal* **14**, No. 6, June, 43.

CIBSE Journal (1992b): Filters under fire, *Building Services The CIBSE Journal* **14**, No. 6, June 43.

Corn, M. (1966): *Aerosol Science* (ed. C.N. Davies), 359, Academic Press, London.

Dorman, R.G. (1975): *Dust Control and Air Cleaning*, Pergamon Press.

Eurovent 4/4 (1980): Sodium chloride aerosol test for filters using flame photometry technique, HEVAC Association.

Eurovent 4/5 (1980): Method of testing air filters used in general ventilation, 2nd edn, HEVAC Association.

Ma, W.Y.L. (1965): Air conditioning design for hospital operating rooms, *JIHVE*, 165–179, September.

Robinson, T.J. and Quellet, A.E. (1999): Filters and Filtration, *ASHRAE Journal*, April, 65–68.

Scala, G.F. (1963): A new instrument for the continuous measurement of condensation nuclei, *Analytical Chemistry* **35**(5), 702.

Ter Kuile, W. (1997): Test standards for air filtration, *Building Services Journal*, May, 51–52.

Bibliography

1. C.P. McCord and W.N. Witheridge. *Odours, Physiology and Control*, McGraw-Hill Book Company, 1949.
2. *ASHRAE Handbook* 1997 Fundamentals.
3. *ASHRAE Handbook* 1996 Systems and Equipment.

Index

CIBSE
PSYCHROMETRIC
CHART

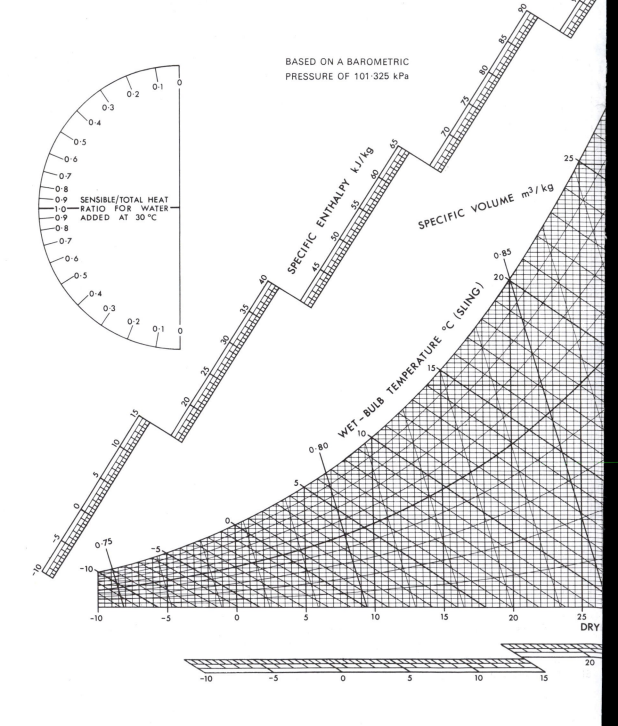

BASED ON A BAROMETRIC
PRESSURE OF 101·325 kPa

SENSIBLE/TOTAL HEAT
RATIO FOR WATER
ADDED AT 30 °C

SPECIFIC ENTHALPY kJ/kg

SPECIFIC VOLUME m³/kg

WET-BULB TEMPERATURE °C (SLING)

DRY